Lecture Notes in Physics

Edited by J. Ehlers, München, K. Hepp, Zürich, and
H. A. Weidenmüller, Heidelberg
Managing Editor: W. Beiglböck, Heidelberg

35

Proceedings of the Fourth International Conference on Numerical Methods in Fluid Dynamics

June 24–28, 1974, University of Colorado

Edited by Robert D. Richtmyer

Springer-Verlag
Berlin · Heidelberg · New York 1975

Editor:
Prof. Dr. Robert D. Richtmyer
Dept. of Physics and Astrophysics
University of Colorado
Boulder, CO. 80302, USA

Present Address:
Los Alamos Scientific Laboratories
Group T-7
P.O. Box 1663
Los Alamos, N.M. 87544, USA

Library of Congress Cataloging in Publication Data

International Conference on Numerical Methods in Fluid Dynamics, 4th, University of Colorado, 1974.
Proceedings of the Fourth International Conference on Numerical Methods in Fluid Dynamics, June 24-28, 1974, University of Colorado.

(Lecture notes in physics ; 35)
Bibliography: p.
Includes index.
1. Fluid dynamics--Congresses. 2. Numerical analysis--Congresses. I. Richtmyer, Robert D., ed. II. Series.
QA911.I54 1974 532'.05'015194 74-34043

ISBN 3-540-07139-3 Springer-Verlag Berlin · Heidelberg · New York
ISBN 0-387-07139-3 Springer-Verlag New York · Heidelberg · Berlin

Offsetdruck: Julius Beltz, Hemsbach/Bergstr.

Contents

SOME COMPUTATIONAL PROBLEMS IN BOUNDARY-LAYER FLOWS*

Herbert B. Keller

California Institute of Technology, Pasadena, California 91109

1. INTRODUCTION

Boundary-layer flows, both laminar and turbulent, have been of continuous and growing interest since their introduction by Prandtl in 1904. More recently numerical methods for the approximate solution of various boundary-layer problems have developed at an even faster rate. In this paper we discuss one such method, the Box-scheme, first developed in [9] and extended by T. Cebeci and myself [3, 4, 5, 6, 11, 12, 13], and show how it has been or could be applied to a broad variety of boundary-layer flow problems. In particular we consider laminar and turbulent two-dimensional (non-similar) flows, three-dimensional flows, internal (pipe or channel) flows, and some two-dimensional reverse flows. For turbulent flows in all the above cases we use an eddy viscosity to represent the Reynolds shear stress terms. Again this basic idea was introduced by Prandtl and we employ the formulation as modified by T. Cebeci [2]. Our numerical method has also been applied to Saffman's turbulence model equations [16, 19] for two-dimensional boundary layers. There are many additional boundary layer problems to which the method has been or could be employed. These include unsteady flows [6], chemically reacting flows [20], compressible and heat conducting flows [3], etc.

There are several distinctive features which we believe make this numerical method particularly attractive. Not the least is the ease and uniformity in setting up such a broad class of problems. The procedure does not have to be modified or augmented in going from one problem to another. Once the basic ideas are clear and an efficient subroutine for solving linear systems of a particular block-tridiagonal structure is available then the efficient and accurate solution of a huge class of boundary layer flows becomes routine.

A second basic feature of the method is that it is second order accurate on arbitrary nonuniform nets. This is achieved by formulating all problems as first order systems (of nonlinear partial differential equations). Then simple centered differences and averages over net intervals, net rectangles or net rectangular parallelopipeds are used. Thus we never difference across a net point (line or plane). Furthermore the errors in the numerical solution have asymptotic expansions in powers of the squares of the net spacing so that Richardson extrapolation is valid and yields two orders of accuracy improvement per application. We hasten to stress (again) that this permits engineering accuracy to be obtained while using unreasonably crude nets. If for some theoretical purpose high accuracy is also desired, we easily obtain that too. The second order accuracy and Richardson extrapolation on nonuniform nets can also be attained with some second order formations using(three point) Crank-Nicolsen type schemes. In particular problems these techniques, properly implemented, could be just as effective as the Box-scheme. However, the three point schemes must be modified near points of flow reversal (see for example [15]) as the properly centered forms fail near separation, special difference schemes must also be introduced to accurately compute the shear stress on boundaries and the nonuniformity in the net must be gradual. Thus these methods are not quite as universal as the Box-scheme which is just as effective with extremely rapidly varying nets and even allows jump discontinuities at netpoints.

* This work was supported by the U. S. Army Research Office (Durham) under Contract DAHC 04-68-0006. It was completed at the Applied Mathematics Summer Institute, 1974 held at U. C. Irvine and supported by O. N. R. Contract N 00014-67-A-467-0027.

The third and final basic feature of our scheme is that Newton's method is employed to solve the complete set of coupled nonlinear difference equations. Near the correct numerical solution these iterations converge quadratically so that it can be observed and the iteration error can be estimated. (This is perhaps the weakest point in most alternative implicit difference schemes. Many of them could be greatly improved by the proper application of Newton's method.) Of course to employ Newton's method we are forced to solve large order systems of linear equations. However, the linear systems in question have special block-tridiagonal forms and hence they are easily solved using a number of operations that is comparable to the number of operations required to solve scalar tridiagonal systems. We invariably check to see that quadratic convergence has occurred in the calculations, not merely that two successive iterates are within some ϵ of each other. If Newton's method does not converge within 4 or 5 iterations we suspect the results. If more than 8 or 10 iterations are required we signal "divergence" and do some thinking about the problem. (For most two and three dimensional flows we need only 2 or 3 iterations.) It is, of course, ridiculous to use a high order accurate implicit difference scheme if you cannot be assured that the iteration error is less than the truncation error. Schemes which require 10-20 iterations, not to mention those using hundreds of iterations, must be considered quite crude by present standards.

There is one crucial property common to all of the boundary layer flows to which the Box-scheme is clearly applicable, namely that all the diffusion processes occurring take place in one (coordinate) dimension. With this property in mind it is easy to determine the applicability of the scheme to an even larger class of flow problems. Indeed it can be used for special inviscid flows but we shall not discuss these applications here.

In Section 2 we formulate what we term the "standard problem," that is laminar or turbulent nonsimilar flows in two dimensions, to simply explain the basic ideas of the Box-scheme. We do this in both physical and scaled coordinates. Then in Section 3 we examine two nonstandard problems which are easily solved by iterative application of standard problems. These nonstandard problems include inverse problems in which the wall shear is specified and the pressure gradient term is unknown and internal (entrance) flows in ducts or pipes. In Section 4 we show how three dimensional boundary layer flows are computed. In Section 5 we indicate how the Box-scheme has been applied to Saffman's turbulence model equations. Finally, in Section 6 we sketch a current attempt to compute nonsimilar reverse flows.

2. STANDARD PROBLEM

Steady, incompressible, laminar or turbulent boundary layer flow over a flat plate leads to the standard (nonsimilar) boundary layer equations which we write as:

$$\begin{aligned} &\text{a)} \quad v_y = -u_x \quad &&\text{, mass cons.;} \\ &\text{b)} \quad (\nu+\epsilon)u_y = \tau \quad &&\text{, stress defn.;} \\ &\text{c)} \quad \tau_y = uu_x + v\tau/(\nu+\epsilon) + p_x \quad &&\text{, x-mom. cons.} \end{aligned} \qquad (2.1)$$

Here y is the coordinate normal to the plate which is at $y = 0$; (u, v) are the x- and y-velocity components; and

$$\text{d)} \quad \epsilon \equiv \begin{cases} 0 & \text{, in laminar flows,} \\ \epsilon[u, v, \tau], & \text{in turbulent flows.} \end{cases} \qquad (2.1)$$

The eddy diffusivity, $\epsilon[u, v, \tau]$ in (2.1d), is some nonlinear functional of the flow variables which may change from problem to problem as well as from one researcher to another. We usually employ one of the formulations due to Cebeci [2] and do not spell out the details here. (In Section 5 we do show how ϵ is computed in a specific turbulence model which does not change from one problem to another.)

The boundary conditions may be taken as:

$$\begin{aligned} &a) \quad u(x,0) = u_w(x) \ ; \\ &b) \quad v(x,0) = v_w(x) \ ; \\ &c) \quad u(x, y_\infty(x)) = u_e(x) \ . \end{aligned} \tag{2.2}$$

Here $u_w(x)$ simulates the wall motion, $v_w(x)$ simulates a mass flux (i.e. blowing) into the boundary layer and $u_e(x)$ is the free stream velocity. The "edge" of the boundary layer, at $y = y_\infty(x)$, is frequently assumed known or it may be determined by imposing the additional condition: $u_y(x, y_\infty(x)) = 0$.

At some "upstream" location, say $x = 0$, we assume known the "initial" data:

$$u(0,y) = u_0(y), \quad v(0,y) = v_0(y), \quad \tau(0,y) = \tau_0(y) \tag{2.3}$$

This data must be compatible in an appropriate sense, depending upon the problem, and is frequently determined by solving some similar (one dimensional) flow problem. If some of the initial data are obtained from experiments the remainder must then be computed. There are frequently problems of smoothing involved in this process but again we do not enter into such details here.

The standard problem is to solve (2.1) subject to (2.2) and (2.3) for (u, v, τ) in some region $0 \le x \le L$, $0 \le y \le y_\infty(x)$ being given $p_x(x)$ and the other inhomogeneous terms in (1.2) and (1.3). Of course the pressure gradient term and the edge velocity are not independent but must satisfy

$$p_x(x) = u_e(x) \frac{\partial u_e(x)}{\partial x} \ . \tag{2.4}$$

2.1 The Box-Scheme

To approximate the solution of the standard problem we place an arbitrary (nonuniform) net $\{x_n, y_j\}$ with:

$$\begin{aligned} &a) \quad x_0 = 0, \quad x_n = x_{n-1} + k_n, \quad 1 \le n \le N \\ &b) \quad y_0 = 0, \quad y_j = y_{j-1} + h_j, \quad 1 \le j < J, \quad y_J = y_\infty \end{aligned} \tag{2.5}$$

Here, for the moment we have assumed a known uniform boundary layer thickness $y_\infty(x) \equiv$ const. The elimination of this constraint is a triviality which we clarify later. The net (2.5) is indicated in Figure 1a. In Figure 1b we depict a typical mesh rectangle on which the differential equations (2.1) will be approximated by simple centered difference equations. Since (2.1a) and (2.1c) both contain x- and y-derivatives their difference approximations are "centered" at the midpoint of the mesh rectangle, say at $(x_{n-\frac{1}{2}}, y_{j-\frac{1}{2}})$. Since (2.1b) only contains a y-derivative we center it at $(x_n, y_{j-\frac{1}{2}})$, the midpoint of a vertical mesh segment. Obviously (2.1b) could just as well have been centered at $(x_{n-\frac{1}{2}}, y_{j-\frac{1}{2}})$, but we can obtain the desired higher order accuracy without doing this and hence less arithmetic is required. To complete the specification of the difference equations we need only spell out how the nonlinear terms are to be treated, that is as products of averages or averages of products. It makes no essential difference provided accurate centering is maintained.

If we denote the numerical approximations to (u, v, τ) at (x_n, y_j) by (u_j^n, v_j^n, τ_j^n) then our difference approximation to (2.1), (2.2) has the form:

$$\left.\begin{aligned}
&\text{a) } j = 0 : && u_0^n = u_w(x_n)\\
& && v_0^n = v_w(x_n)\\
&\text{b) } 1 \le j \le J: && \underset{\sim}{G}_j[(u_{j-1}^n, v_{j-1}^n, \tau_{j-1}^n), (u_j^n, v_j^n, \tau_j^n)] = \underset{\sim}{0}\\
&\text{c) } j = J : && u_J^n = u_e(x_n)
\end{aligned}\right\} \quad (2.6)$$

Here we have indicated only the variables on $x = x_n$ having assumed that all quantities for $x \le x_{n-1}$ are known. Thus for each j in (2.6b) the vector equation $\underset{\sim}{G}_j[(\ ), (\ )] = \underset{\sim}{0}$ represents the three difference equivalents of (2.1 a, b, c). In total (2.6) contains 3(J+1) nonlinear equations in the as many unknowns: (u_j^n, v_j^n, τ_j^n), $0 \le j \le J$.

To solve the difference equations we employ Newton's method. The Newton iterates $(u_j^{n,\nu}, v_j^{n,\nu}, \tau_j^{n,\nu})$ are defined recursively by using

$$\begin{pmatrix} u_j^{n,\nu+1} \\ v_j^{n,\nu+1} \\ \tau_j^{n,\nu+1} \end{pmatrix} = \begin{pmatrix} u_j^{n,\nu} \\ v_j^{n,\nu} \\ \tau_j^{n,\nu} \end{pmatrix} + \begin{pmatrix} \delta u_j^{n,\nu} \\ \delta v_j^{n,\nu} \\ \delta \tau_j^{n,\nu} \end{pmatrix}, \quad \nu = 0, 1, 2, \ldots. \qquad (2.7)$$

in (2.6) and linearizing. This gives a coupled linear system for the "correction" terms, $(\delta u_j^{n,\nu}, \delta v_j^{n,\nu}, \delta_j^{n,\nu})$, $0 \le j \le J$. If the three component correction vectors are denoted by $\underset{\sim}{\delta}_j^{n,\nu}$ at each y_j then the linearized versions of (2.6b) are easily seen to have the simple matrix two-term recursion forms:

$$L_{j-\frac{1}{2}}^{n,\nu}\ \underset{\sim}{\delta}_{j-1}^{n,\nu} + R_{j-\frac{1}{2}}^{n,\nu}\ \underset{\sim}{\delta}_j^{n,\nu} = \underset{\sim}{r}_{j-\frac{1}{2}}^{n,\nu}, \quad 1 \le j \le J \qquad (2.8)$$

Here $L^{n,\nu}_{j-\frac{1}{2}}$ and $R^{n,\nu}_{j-\frac{1}{2}}$ are the 3×3 Jacobian matrices of the $\underset{\sim}{G}_j[(\ ),(\ )]$ with respect to the first or second triad of arguments, respectively. Writing this linearized system in the order indicated in (2.6) we find that the full system on the coordinate line $x = x_n$, say,

$$\mathbb{A}^{(\nu)}_n \underset{\approx}{\Delta}^{\nu} = \underset{\approx}{r}^{\nu} \tag{2.9}$$

has the block tri-diagonal coefficient matrix:

$$\mathbb{A}^{(\nu)}_n \equiv \begin{pmatrix} M_0 & & & \\ L^{n,\nu}_{\frac{1}{2}} & R^{n,\nu}_{\frac{1}{2}} & & \\ & \ddots & \ddots & \\ & & L^{n,\nu}_{J-\frac{1}{2}} & R^{n,\nu}_{J-\frac{1}{2}} \\ & & & M_J \end{pmatrix} = [B^{n,\nu}_j, A^{n,\nu}_j, C^{n,\nu}_j]. \tag{2.10}$$

Furthermore the 3×3 blocks B_j, A_j, C_j in (2.10) have the special zero elements indicated by:

$$B^{n,\nu}_j \equiv \begin{pmatrix} * & * & * \\ 0 & 0 & 0 \\ 0 & 0 & 0 \end{pmatrix}, \quad C^{n,\nu}_j \equiv \begin{pmatrix} 0 & 0 & 0 \\ * & * & * \\ * & * & * \end{pmatrix} \tag{2.11}$$

As a result the system (2.9) can be solved by a very efficient and stable elimination procedure (which is essentially a matrix analog of the familiar scalar tri-diagonal factorization). The most complete discussion and justification of this solution method is contained in [10].

To set up and apply the Box-scheme we need only evaluate the matrix elements in (2.10) and the inhomogeneous terms in (2.9); these are most easily obtained from the form (2.8). Then an efficient code to solve the block tri-diagonal system (2.9) yields the Newton corrections. We could just as well have written the linear systems for the iterates, $u^{n,\nu+1}_j$, etc. rather than for the corrections, $\delta u^{n,\nu+1}_j$, etc. The latter is more accurate however and reduces the loss of significance while the former equations require less arithmetic to evaluate the right hand sides. The quadratic convergence of the Newton iterates, in which $\|\underset{\sim}{\delta}^{\nu+1}\| = \mathcal{O}(\|\underset{\sim}{\delta}^{\nu}\|^2)$, is easily observed in the calculations. Further, if the initial guess at the solution is such that $\|\underset{\sim}{\delta}^0\| = \mathcal{O}(\Delta x)$, say, then $\|\underset{\sim}{\delta}^2\| = \mathcal{O}(\Delta x^4)$. Thus, if fourth order accuracy is desired, it can be consistently achieved with only <u>two</u> Newton iterations. Obviously, the guess $u^{n,0}_j = u^{n-1}_j$, etc will always give an at least $\mathcal{O}(\Delta x)$ order initial estimate.

Richardson extrapolation is easily applied and yields two orders of accuracy improvement (in both Δx and Δy) with each application. These details are spelled out in [9, 10, 11] so we do not repeat them here.

2.2 Scaled Variables and Stream Function

It is well known that numerical methods are more efficient on problems with smooth or slowly varying solutions. For this reason most boundary layer flow problems should be scaled to reduce the rapid changes that occur near the walls. At the same time such scaling may resolve singularities that are present, say at the leading edge of a plate. A typical such scaling and introduction of a stream function is given by:

$$\text{a)}\quad u = \psi_y, \quad v = -\psi_x, \quad \eta \equiv \sqrt{\frac{u_e(x)}{\nu x}}\, y\,;$$

$$\text{b)}\quad f(x,\eta) \equiv [u_e(x)\nu x]^{-\frac{1}{2}}\psi(x,y)\quad ; \tag{2.12}$$

$$\text{c)}\quad P(x) \equiv \frac{x}{u_e(x)}\frac{du_e(x)}{dx}\,, \qquad b \equiv 1 + \epsilon/\nu \quad .$$

Then (2.1 a, b, c) is equivalent to the third order scalar equation (with $' \equiv \partial/\partial\eta$):

$$(bf'')' + \frac{P+1}{2} ff'' + P[1-(f')^2] = x[f'f'_x - f''f_x] \tag{2.13}$$

To apply the Box-scheme in this formulation we simply write (2.13) as the equivalent first order system:

$$\text{a)}\quad f' = u$$

$$\text{b)}\quad u' = \tau \tag{2.14}$$

$$\text{c)}\quad (b\tau)' = -\frac{P+1}{2} f\tau - P[1-u^2] + x[uu_x - \tau f_x]$$

The boundary conditions (2.2) become)

$$\text{a)}\ f(x,0) = f_w(x)$$

$$\text{b)}\ u(x,0) = u_w(x) \tag{2.15}$$

$$\text{c)}\ u(x,\eta_\infty(x)) = 1$$

Note that $u_e(x)$ does not appear in this formulation [as the u in (2.14) is a dimensionless version of the u in (2.12a)].

The standard problem (2.14), (2.15) with $f(0,\eta)$ given is differenced just as was (2.1), (2.2). Now, however, two equations (2.14 a, b) can be centered at $(x_n, y_{j-\frac{1}{2}})$ while only (2.14c) need be centered at $(x_{n-\frac{1}{2}}, y_{j-\frac{1}{2}})$; see Figure 1b.

We remark that the efficient calculation of similar boundary layer flows is obtained from the above procedure by simply setting $x = 0$ in (2.14c). Then we are essentially solving the Falkner-Skan equation. This has been done using the Box-scheme in [13].

3. NONSTANDARD PROBLEMS

We now show how our efficient numerical scheme for standard problems, say (1.14)-(1.15), is used to solve some nonstandard boundary layer flow problems. These problems are such that the pressure gradient term $P(x)$ cannot be specified in advance but rather some extra conditions or constraints are imposed. Then the stream function, $f(x, \eta)$, as well as $P(x)$ are to be determined on an appropriate (x, η)-domain. We treat two such problems:

A) Specified wall shear;

B) Entrance flows in pipes or ducts.

The former problem is, of course, of great interest for design problems in which, say, an airfoil shape is to be slightly modified to delay the onset of separation. There are many other such applications of problem A), some having to do with the reverse flow beyond separation. Entrance flows and internal flows in general do not seem to have been treated extensively, if at all, by boundary layer techniques. These methods are much simpler than the full Navier-Stokes' treatment and, as we have found, can yield excellent agreement with experiments. In short, many entrance flows are dominated by the boundary layers and even the fully developed internal flow may be adequately represented by boundary layer theory.

3.1 Specified Wall Shear

In addition to (2.14)-(2.15) and the initial data $f(0, \eta)$ we require that

$$\tau(x, 0) = S_w(x), \quad x > 0 . \qquad (3.1)$$

Now, however, $P(x)$ is not specified, but $S_w(x)$ is the required distribution of shear on the wall $y = 0$. To indicate how we solve this problem suppose $P(x)$ were known and the solution of the standard problem (2.14)-(2.15) is represented by:

$$\text{a)} \quad f(x, \eta) = \mathbb{F}(x, \eta; P(x)) . \qquad (3.2)$$

Then from (2.14 a, b) we see that our nonstandard problem is to find $P(x)$ such that

$$\text{b)} \quad \mathbb{F}_{\eta\eta}(x, \eta; P(x))\Big|_{\eta=0} = S_w(x), \quad x > 0 \qquad (3.2)$$

We seek to solve (3.2b) by Newton's method. Thus if $P^{\nu}(x)$ is the ν-th approximation we seek

$$\text{a)} \quad P^{\nu+1}(x) = P^{\nu}(x) + \delta^{\nu}(x) \qquad (3.3)$$

by using this in (3.2b) and linearizing to get:

$$\text{b)} \quad \mathbb{F}_{\eta\eta P}(x, 0; P^{\nu}(x))\delta^{\nu}(x) = S_w(x) - F_{\eta\eta}(x, 0; P^{\nu}(x)), \quad x>0 \qquad (3.3)$$

In (3.3) we have defined the "outer" iterations. To carry out one such iteration we must compute $\mathbb{F}_{\eta\eta}(x, 0; P^{\nu}(x))$ which is obtained from the solution of a standard problem. The so-called "inner" iterations are done in solving this standard problem as indicated in Section 2.1. After $\mathbb{F}_{\eta\eta}(x, \eta; P^{\nu}(x))$ is computed we can obtain $F_{\eta\eta P}(x, 0; P^{\nu}(x))$ by solving one linear variational problem. By simply differentiating (2.14)-(2.15) with respect to P we obtain this linear problem (in the laminar flow case) as:

$$\text{a)}\ F' = U$$

$$\text{b)}\ U' = T \tag{3.4}$$

$$\text{c)}\ T' = -\frac{P+1}{2}(fT+F\tau)+2PuU+x[uU_x+Uu_x-\tau F_x-Tf_x]-(1+f\tau/2)$$

subject to the boundary conditions

$$\text{a)}\ F(x,0) = 0\,, \quad \text{b)}\ U(x,0) = 0\,, \quad \text{c)}\ U(x,\eta_\infty(x)) = 0. \tag{3.5}$$

Here P, f, u, τ are known in each outer iteration and $(F, U, T) \equiv (\partial f/\partial P, \partial u/\partial P, \partial \tau/\partial P)$ are computed (with no iterations) as the solution of (3.4)-(3.5) using the Box-scheme. (The arithmetic involved is essentially that of one inner iteration.) Having solved the above with $P = P^{\nu}$ we then simply use $T(x,0) = \mathbb{F}_{\eta\eta P}(x, 0; P^{\nu}(x))$ in (3.3b) to complete the outer iteration and determine $P^{\nu+1}_{(x)}$. Some results using the above scheme or related procedures are contained in [5, 13].

3.2 Entrance Flows in Pipes or Ducts

To illustrate our application to entrance flows we consider a uniform stream incident on the channel between two parallel semi-infinite flat plates. A sketch is shown in Figure 2. The scaled variables are necessary here to eliminate the singularity at the entrance edge. Of course the pressure gradient term is not known but the incompressibility condition implies that the mass flux through each cross-section of the channel is known (constant in fact if, as we assume, there is no mass flux through the channel walls). On the center line (plane) we impose symmetry. Thus in place of (2.15) we have the boundary conditions:

$$\text{a)}\ f(x,0) = f_w(x)\,, \quad \text{b)}\ u(x,0) = 0, \quad \text{c)}\ \tau(x, \eta_{C.L.}(x)) = 0 \tag{3.6}$$

and the mass flux condition:

$$\text{d)}\ f(x, \eta_{C.L.}(x)) - f_w(x) = \eta_{C.L.}(x) \tag{3.6}$$

Here $\eta_{C.L.}(x)$ is the value of η which at a given station x is such that, in the scaling (2.12a) : $\eta_{C.L.}(x)\sqrt{\nu x/u_e} = y_{C.L.}$. Of course u_e now is the (uniform) entrance velocity.

To solve (2.14) subject to (3.6) for $f(x, \eta)$ and $P(x)$ we proceed essentially as in Section 3.1 but with (3.6d) replacing (3.1). Further the values of $\eta_{C.L.}(x)$ for x near zero are very large [see (2.12a)] so we only compute out to some $\eta = \eta_\infty$ which is well outside the boundary layer (see Figure 2). Then in $\eta_\infty \le \eta \le \eta_{C.L.}(x)$, the "core" region, the flow is uniform so we can use $f'' = 0$ there to deduce at $x = x_n$ that:

$$f(x_n, \eta_{C.L.}(x_n)) = u(x_n, \eta_\infty)[\eta_{C.L.}(x) - \eta_\infty] + f(x_n, \eta_\infty). \tag{3.7}$$

The outer iterations on (3.6d) employ (3.7) for all x in $0 \le x \le x_T$. At some transition point $x = x_T$, we switch over to physical variables and then there is no inefficiency in integrating out to $y = y_{C.L.}$ or $\eta = \eta_{C.L.}(x)$ and using (3.6d) directly.

Some results of our calculations, taken from [4] are shown in Figures 3 and 4. The first is for laminar flow in a tube and comparisons are made to the experiments of Pfenninger [18]. Next, turbulent flow in the entrance of a pipe is computed and compared to the experiments of Barbin and Jones [1].

4. THREE DIMENSIONAL FLOWS

Three dimensional steady boundary layer flows are easily treated by the Box-scheme. However there are some presently unresolved analytical questions regarding the well posedness of such problems that of course play a role in devising stable numerical methods. We shall not go into the details of these problems here but shall touch upon them as required. The basic conservation equations for an incompressible three dimensional boundary layer adjacent to a wall at $y = 0$ say, can be written as:

$$\begin{aligned}
&a)\ v_y = -u_x - w_z &&\text{, mass cons.}\\
&b)\ (\nu+\varepsilon)u_y = \tau &&\text{, x-stress}\\
&c)\ (\nu+\varepsilon)w_y = \sigma &&\text{, z-stress}\\
&d)\ \tau_y = uu_x + v\tau/(\nu+\varepsilon) + wu_z + p_x &&\text{, x-mom.}\\
&e)\ \sigma_y = uw_x + v\sigma/(\nu+\varepsilon) + ww_z + p_z &&\text{, z-mom.}
\end{aligned} \tag{4.1}$$

As in (2.1d) we employ an eddy viscosity to treat turbulent flows. The boundary conditions are, say:

$$\left.\begin{aligned}
&a)\ u(x,0,z) = u_w(x,z)\\
&b)\ v(x,0,z) = v_w(x,z)\\
&c)\ w(x,0,z) = w_w(x,z)
\end{aligned}\right\}\ \text{wall;}\quad
\left.\begin{aligned}
&d)\ u(x, y_\infty(x,z)) = u_e(x,z)\\
&e)\ w(x, y_\infty(x,z)) = w_e(x,z)
\end{aligned}\right\}\ \text{free stream} \tag{4.2}$$

Here (4.2 a, b, c) allow the simulation of wall motion and blowing or mass transfer into the boundary layer. The free stream velocity (u_e, w_e) is imposed on $y = y_\infty(x, z)$

the presumed known "edge" of the boundary layer. Again, we could determine adequate values for $y_\infty(x, z)$ during the course of the computations.

In addition we require "initial" conditions on some "upstream" surface. What constitutes an upstream surface is related to the well posedness question hinted at above. Physically it seems clear that the velocity vector and the inward normal to the initial surface (i.e. pointing into the region of computation) should make an acute angle. A simple linearized stability analysis of the system (4.1) indicates that all high frequency harmonic disturbances in the y-coordinate will not grow if the flow about which we have linearized does not turn through an angle π or greater. Combining these two reasonable heuristic conditions suggests of course that the initial surface normal and $(u_e, 0, w_e)$ on this surface have a positive inner product. How the initial surface and the flow data on it are determined will vary from problem to problem. In many cases part of the initial surface will be a plane of symmetry of the flow on which the solution can be computed using a two dimensional theory (i.e. attachment line flows) (this is the case in our two examples). Thus we assume that on some cylindrical surface: $x = x_0(s)$, $z = z_0(s)$, $0 \le y \le y_\infty(x_0, y_0)$ we are given:

$$\begin{aligned} &\text{a)}\ u(x_0, y, z_0) = u_0(s, y) \qquad &&\text{d)}\ \tau(x_0, y, z_0) = \tau_0(s, y) \\ &\text{b)}\ v(x_0, y, z_0) = v_0(s, y) &&\text{e)}\ \sigma(x_0, y, z_0) = \sigma_0(s, y) \\ &\text{c)}\ w(x_0, y, z_0) = w_0(s, y) && \end{aligned} \tag{4.3}$$

The numerical scheme proceeds by placing an arbitrary grid of points (x_n, y_j, z_m) on the volume of interest, see Figure 5a. Then on a typical net-parallelopiped, as in Figure 5b, we approximate (4.1 a, d, e) by finite differences using only values at the 8 vertices and carefully centering all terms about the mid-point of the "box". Equations (4.1 b, c) can be differenced on the segment $\overline{P_j P_{j-1}}$ since only y-derivatives occur in these equations. Here we have assumed that the numerical solution is known at the six vertices indicated by black dots in Figure 5b. Focusing attention only on the five unknown quantities at each of the points P_j and P_{j-1} the five difference equivalents of (4.1) can be written as

$$\underset{\sim}{G}_j[(u, v, w, \sigma, \tau)_{j-1}^{m,n}; (u, v, w, \sigma, \tau)_j^{m,n}] = \underset{\sim}{0}, \quad 1 \le j \le J(m, n) \tag{4.4}$$

As indicated these equations are obtained for each $j \geq 1$ on the normal to the wall through (x_n, z_m). The final point at $y = y_{J(m,n)}$ is supposed to lie just outside the boundary layer. At $j = 0$ $(y_0 = 0)$ and $j = J(m, n)$ we have the boundary conditions (4.2) at (x_n, z_m). Thus (4.4) and (4.2 yield $5[J(m, n) + 1]$ equations in as many unknowns.

We solve the nonlinear difference equations by Newton's method. The procedure and indeed most of the details are almost identical to those indicated in

(2.6)-(2.11). The major difference is that there are now five unknowns at each net point, (u, v, w, σ, τ) rather than just (u, v, τ), and so the block matrices $\mathbb{A}^{(\nu)}_{m,n} = [B_j^{m,n\nu}, A_j^{m,n,\nu}, C_j^{m,n,\nu}]$ replacing (2.10) contain 5×5 blocks. Further the boundary conditions show that the zeros are now located as in:

$$B_j^{m,n,\nu} \equiv \begin{pmatrix} * & * & * & * & * \\ * & * & * & * & * \\ 0 & 0 & 0 & 0 & 0 \\ 0 & 0 & 0 & 0 & 0 \\ 0 & 0 & 0 & 0 & 0 \end{pmatrix}, \quad C_j^{m,n,\nu} \equiv \begin{pmatrix} 0 & 0 & 0 & 0 & 0 \\ 0 & 0 & 0 & 0 & 0 \\ * & * & * & * & * \\ * & * & * & * & * \\ * & * & * & * & * \end{pmatrix} \qquad (4.5)$$

Thus on the y-coordinate line through each netpoint (x_n, z_m) all the flow quantities are computed simultaneously by evaluating several Newton iterates. Again this is done simply, efficiently and in a fairly well conditioned way by virtue of the special block structure of the $\mathbb{A}^{(\nu)}_{m,n}$.

The only remaining question for such three dimensional flows concerns the order in which the net points (x_n, z_m) in the tangential flow plane can be swept. This is not completely arbitrary since, as we have seen, we must start at some type of "corner". For example, if the "initial" surface is the positive xy- and zy-planes, then we can start at the origin and sweep along increasing coordinate lines parallel to the x- or z-axes (or both). However, stability considerations suggest that the direction of sweep should make an acute angle with the local tangential flow direction. This may be difficult or even impossible to achieve in complicated three dimensional flows and we have not yet tested it thoroughly. Indeed, in some test calculations for the flow about a circular cylinder standing on a flat plate, we have swept from the free stream toward the plane of symmetry. The results were not unstable but became progressively worse as the cylinder and symmetry axis were approached. The only other stability and sweep direction considerations similar to those we have mentioned, of which we are aware, are contained in [17].

The results of some three dimensional boundary layer calculations using the Box-scheme are shown in Figures 7a, 7b and 8b. In all of these calculations we use scaled variables and a stream function formulation. It turns out that six rather than five unknowns are required at each netpoint. Thus the "blocks" are 6×6 submatrices. The first calculations relate to the experiments of Johnston [8], whose test setup is sketched in Figures 6a and 6b, in which a jet impinges on a wall. We used from 23 to 28 netpoints across the boundary layer. The second calculations were based on experiments of East and Hoxey [7] in which a wing-like obstruction is placed in a thick two dimensional boundary layer. Both flows were turbulent and the agreement with the measured values is quite good away from separation.

5. TURBULENCE MODELS

We indicate here an application of the Box-scheme to the particular turbulence model of P. G. Saffman [19]. As in our previous discussions we consider the simplest geometry of flat plate boundary layers but now with a zero pressure gradient. The model equations proposed by Saffman can be written in this case as:

$$
\begin{aligned}
&\text{a)} \quad v_y = -u_x \quad ; && \text{, mass cons.}\\
&\text{b)} \quad (\nu+\epsilon)u_y = \tau, \quad \epsilon \equiv A\, e/\omega ; && \text{, stress defn.}\\
&\text{c)} \quad \tau_y = uu_x + v\,\tau/(\nu+\epsilon) ; && \text{, x-mom. cons.}\\
&\left.\begin{aligned}
&\text{d)} \quad (\nu+\epsilon')e_y = f, \quad \epsilon' \equiv A' e/\omega \quad ;\\
&\text{e)} \quad f_y = u e_x + vf/(\nu+e') + e\omega - \alpha'' e\,|\tau/(\nu+\epsilon)| ;
\end{aligned}\right\} && \text{turb. en.} \qquad (5.1)\\
&\left.\begin{aligned}
&\text{f)} \quad (\nu+\epsilon'')\omega^2_y = g, \quad \epsilon'' = A'' e/\omega \; ;\\
&\text{g)} \quad g_y = u\,\omega^2_x + v g/(\nu+\epsilon'') + \beta' \omega^3 - \alpha' \omega^2 |\tau/(\nu+\epsilon)|
\end{aligned}\right\} && \text{turb. vort.}
\end{aligned}
$$

Here $\epsilon = A\, e/\omega$ is the eddy-viscosity which is defined in terms of a turbulent energy, e, and a mean square turbulent vorticity, ω. Both e and ω^2 satisfy turbulent diffusion equations [(5.1 d, e) and (5.1 f, g) respectively] similar to the physical conservation laws. There are six constants in the theory, A, A', A'', α', α'' and β'', which are fixed once and for all by elementary physical and dimensional arguments, see [19]. The model equations do not hold at the wall interface but must be matched to the viscous sublayer adjacent to the wall. This is done by requiring that at some $y_0 > 50\, \nu/u_*$:

$$
\begin{aligned}
&\text{a)} \quad u(x, y_0) = \frac{u_*}{.41} \ln(y_0\, u_*/\nu) + 5u_* \;, \quad u_*(x) \equiv [\tau_w(x)/\rho]^{\frac{1}{2}} \;;\\
&\text{b)} \quad e(x, y_0) = \alpha'' u_*^2/A \quad ; \qquad \text{c)} \quad \omega(x, y_0) = \alpha'' u_*/.41\, y_0 \;; \qquad (5.2)\\
&\text{d)} \quad v(x, y_0) = -y_0 \frac{u(x, y_0)}{u_*(x)} \, \frac{du_*(x)}{dx} \;.
\end{aligned}
$$

At the edge of the boundary layer we impose

$$
\begin{aligned}
&\text{a)} \quad u(x, y_\infty) = U_e(x)\\
&\text{b)} \quad e(x, y_\infty) = 4\times 10^{-6}\, U_e^2(x) \qquad (5.3)\\
&\text{c)} \quad \omega(x, y_\infty) = \frac{2\,\omega(0, y_\infty)}{2 + \beta'\,\omega(0, y_\infty)x/U_e}
\end{aligned}
$$

Here (5.3b) implies a free stream turbulence intensity $T = (\frac{2}{3}\, e/U_\infty^2)^{\frac{1}{2}} = .00164$ and (5.3c) follows from (5.1f, g) on $y = y_\infty$ with $\omega_y^2 \equiv e_y \equiv 0$. In calculations we used $\omega(0, y_\infty) = .0374\, \omega(0, y_0)$.

The Box-scheme can be applied in a straightforward manner to the model formulated in (5.1)-(5.3) when initial data are determined at $x = 0$, say. At each netpoint there are now seven unknowns, $(u, v, \tau, e, f, \omega, g)$, and Newton's method yields block-tridiagonal systems of the form (2.9)-(2.10) but with 7×7 blocks. From (5.2) and (5.3) we see that the $B_j^{n,\nu}$ $(C_j^{n,\nu})$ have zeros in their last four (first three) rows. Initial data were taken from the experiments of Klebanoff [14]. There is some difficulty in determining compatible data for the model quantities (e, ω) and their y-derivatives. Thus some spatial oscillations occur just downstream of $x = 0$ but these are filtered out by averaging the solution at each pair of downstream stations for the first forty steps or so. It is easy to show (see [9]) that this averaging does not degrade the accuracy.

More details of these calculations will be given by D. Knight [16], who carried out the above indicated studies. In Figure 9 we show some computed downstream profiles, about 90 boundary layer thicknesses downstream, and compare them to the initial profiles obtained from Klebanoff's data.

6. REVERSE FLOWS

The Box-scheme has been employed to compute reverse-flow solutions of the Falkner-Skan equation with no difficulty in [5, 13]. Also the scheme has been used to compute up to (or extremely close to) the point of separation in Howarth's flow [11] with no special modifications (other than a decrease in the downstream step size as the point of separation is approached). These features suggest a simple adaptation of the Box-scheme to compute nonsimilar boundary layers containing a reverse flow region. Such applications are currently in progress and we sketch the basic ideas here.

We assume that the prescribed pressure gradient $p_x(x)$ and free stream velocity $u_e(x)$ in the standard problem of Section 2 with $u_w(x) \equiv v_w(x) \equiv 0$ is such that a separation-reattachment flow as indicated in Figure 10 occurs. Then there is no difficulty in computing, as in Section 2.1, the boundary layer flow upstream of the point of separation, $x \le x_0$, say, and in fact determining x_0 accurately (see [11]). Our basic idea is then to solve simultaneously all the coupled nonlinear Box-difference equations over $x_0 \le x \le x_N$, $0 \le y \le y_\infty$ where x_N is some fixed coordinate downstream of reattachment. Although the difference equations remain unaltered over the entire boundary layer the order in which we do the eliminations to solve for the (linearized) Newton iterates may change with each iteration. The change of course depends upon the location of the bounding curve of the reverse flow region. To start we may get an initial reverse flow region by solving for the local similar reverse flow solutions (i.e. Falkner-Skan flows) at each station in $x_0 < x < x_N$ for which they exist (this determines x_R initially).

Now let us assume that we have a current iterate at all net points in $x_0 \le x \le x_N$. There are 3(J+1) unknown values (u_j^n, v_j^n, τ_j^n), $0 \le j \le J$ on each netline $x = x_n$ for $1 \le n \le N$. Let us denote all these values for a given n by the vector $\underset{\sim}{U}_n$. Then the Box-scheme difference equations can be denoted by:

$$\underset{\sim}{F}(\underset{\sim}{U}_{n-1}, \underset{\sim}{U}_n) = \underset{\sim}{0} \ , \quad 1 \le n \le N . \tag{6.1}$$

We recall that $\underset{\sim}{U}_0$ is known and (6.1) for fixed n is another representation of the equations implied in (2.6). There, however, we focused attention only on the values in $\underset{\sim}{U}_n$ assuming known the values in $\underset{\sim}{U}_{n-1}$. Newton's method applied to (6.1), with $\underset{\sim}{U}_n^{\nu+1} = \underset{\sim}{U}_n^{\nu} + \delta \underset{\sim}{U}_n^{\nu}$ yields a huge linear system of the form:

$$\mathscr{L}_{n-\frac{1}{2}}^{\nu} \ \delta \underset{\sim}{U}_{n-1}^{\nu} + \mathscr{R}_{n-\frac{1}{2}}^{\nu} \ \delta \underset{\sim}{U}_n^{\nu} = - \underset{\sim}{F}(\underset{\sim}{U}_{n-1}^{\nu}, \underset{\sim}{U}_n^{\nu}), \quad 1 \le n \le N. \tag{6.2}$$

where $\delta \underset{\sim}{U}_0^{\nu} \equiv 0, \mathscr{L}_{n-\frac{1}{2}}^{\nu} \equiv \partial \underset{\sim}{F}(\underset{\sim}{U}_{n-1}^{\nu}, \underset{\sim}{U}_n^{\nu})/\partial \underset{\sim}{U}_{n-1}^{\nu} \ \& \ \mathscr{R}_{n-=}^{\nu} \equiv \partial \underset{\sim}{F}(\underset{\sim}{U}_{n-1}^{\nu}, \underset{\sim}{U}_n^{\nu})/\partial U_n^{\nu}$. Each of the matrices $\mathscr{L}_{n-\frac{1}{2}}^{\nu}$ and $\mathscr{R}_{n-\frac{1}{2}}^{\nu}$ have a block-tridiagonal structure identical to that of the $\mathbf{A}_n^{(\nu)}$ in (2.10). Now it is apparent that we could easily solve (6.2) for the $\delta \underset{\sim}{U}_n^{\nu}$ recursively starting from n = 1 and proceeding in order through n = N. This follows since (6.2) is block-lower triangular and corresponds simply to sweeping through all the net points, a line $x = x_n$ at a time, in the downstream direction. Each line involves solving a system exactly of the form (2.9). This procedure could be adequate for extremely small regions of reverse flow. However, it must at best be a very poorly conditioned procedure for solving the linear system (6.2) for reasonably large regions of reverse flow. We have at present no mathematical proof of this but it is physically or intuitively obvious from the flow interpretation of the linearized differential equations. That is, in the reverse flow region new values would be propagated opposite to the flow direction.

Thus we do not solve (6.2) in the obvious manner indicated above. Rather we rearrange the equations so that the propagation of new data is more nearly aligned with the flow direction indicated by the current iterate. This corresponds to viewing the system (6.2) as a block-tridiagonal system with block sizes varying as the width of the reverse flow region varies. In more detail those variables on $x = x_n$ for which $u_j^{n,\nu} > 0$ (i.e. outside the reverse flow region) are considered as coupled to the data on $x = x_{n-1}$ while the remaining variables are coupled to the data on $x = x_{n+1}$ (see Figure 10). In effect this means that the data on the "top" portion of $x = x_{n+1}$ are combined with the data on the "bottom" portion of $x = x_n$ to form a new vector of unknowns which are processed together. The block elimination scheme which carries out this procedure is not much more complicated than the schemes previously employed. Indeed it is a simple generalization of those codes that is presently undergoing testing.

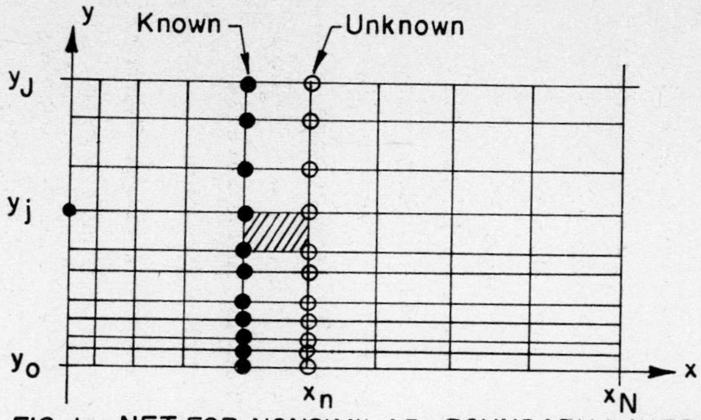

FIG. 1a NET FOR NONSIMILAR BOUNDARY LAYERS

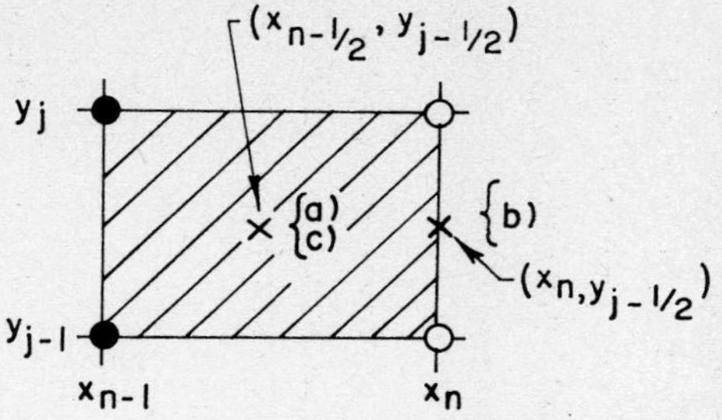

FIG. 1b TYPICAL NET-RECTANGLE FOR CENTERING OF DIFFERENCE EQUATIONS

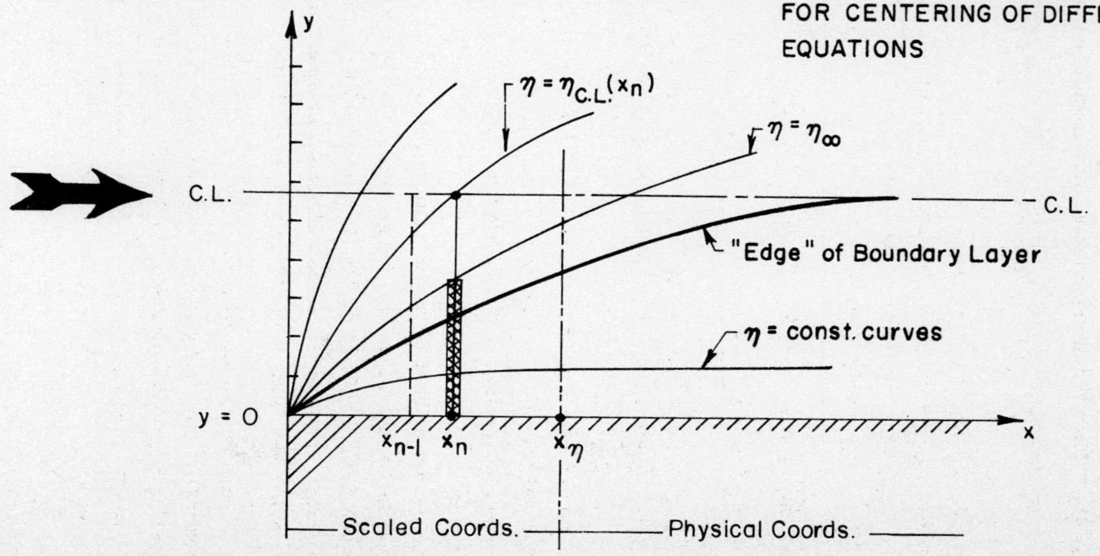

FIG. 2 ENTRANCE FLOW REGION IN A DUCT

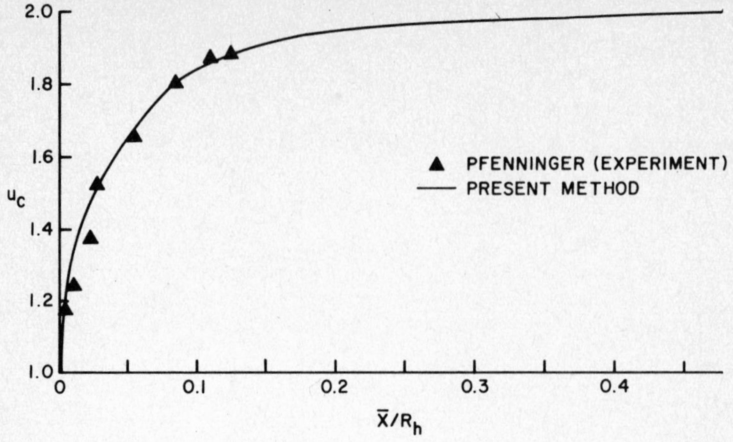

FIG. 3 LAMINAR FLOW IN A TUBE

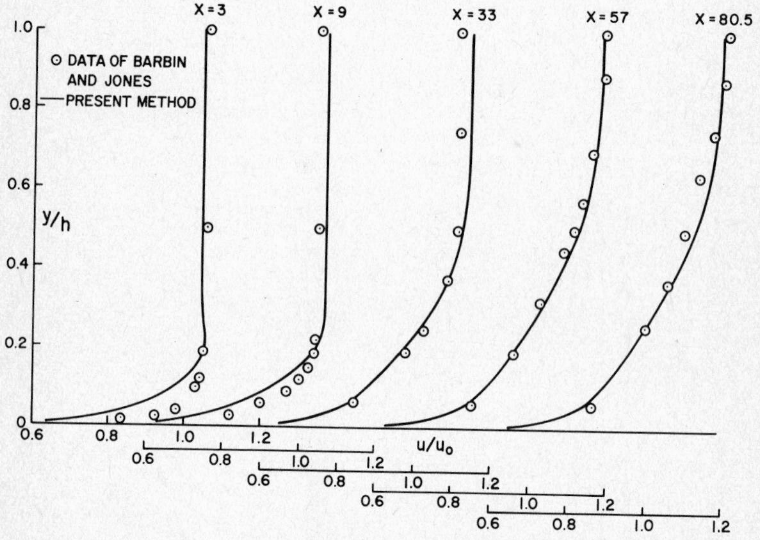

FIG. 4 TURBULENT FLOW IN A PIPE

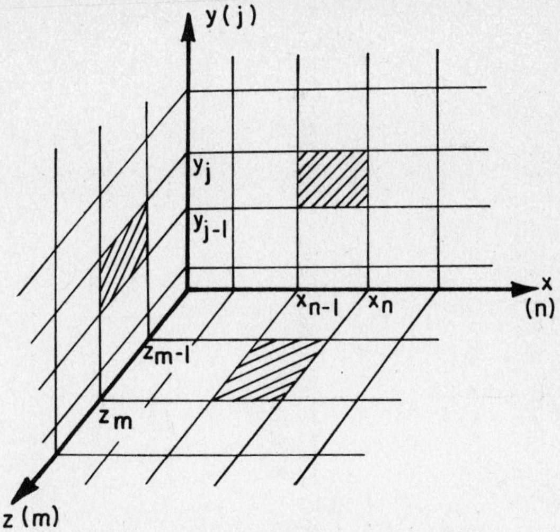

FIG. 5a NET FOR THREE DIMENSIONAL BOUNDARY LAYERS

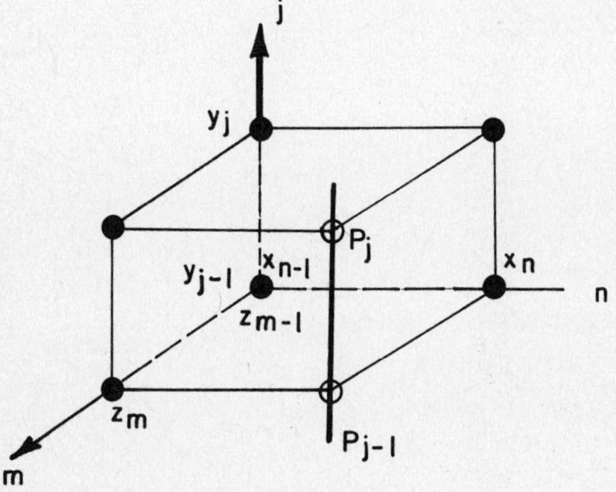

FIG. 5b TYPICAL NET-PARALLELOPIPED FOR FOR CENTERING DIFFERENCE EQUATIONS

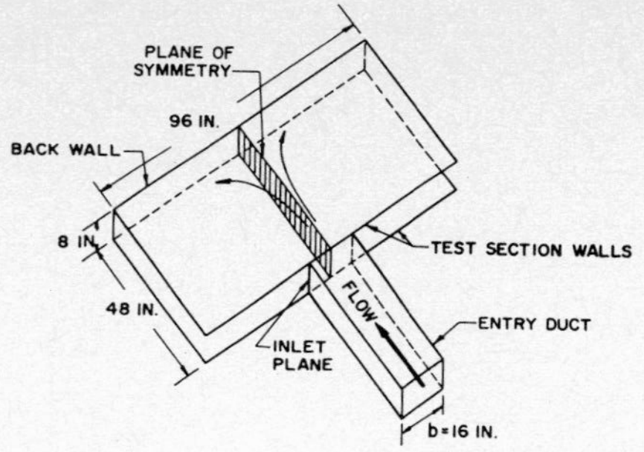

FIG. 6a SCHEMATIC DRAWING OF JOHNSTON'S TEST GEOMETRY

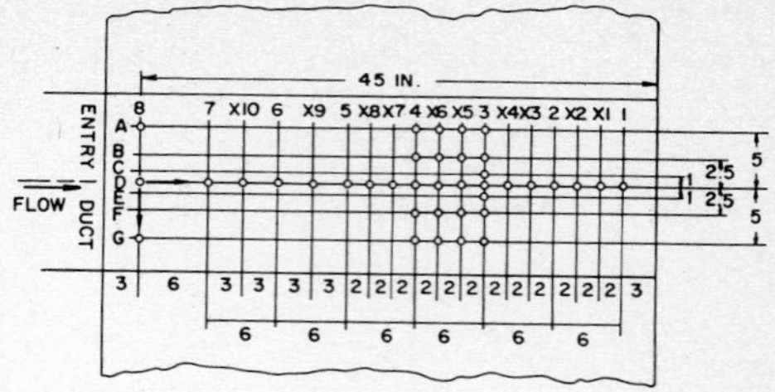

FIG. 6b SKETCH SHOWING THE MEASURED STATIONS

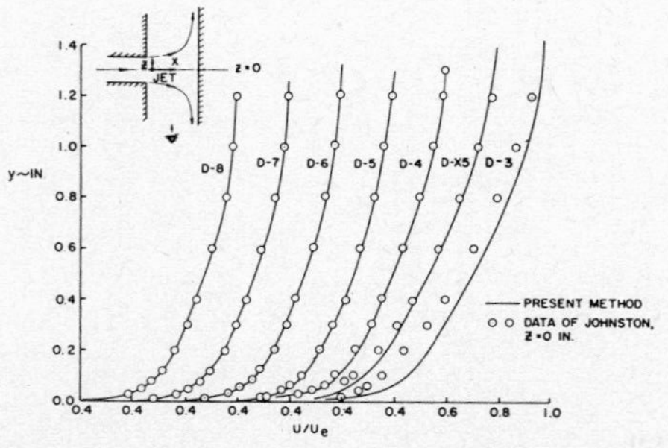

FIG. 7a RESULTS FOR THE ATTACHMENT-LINE FLOW

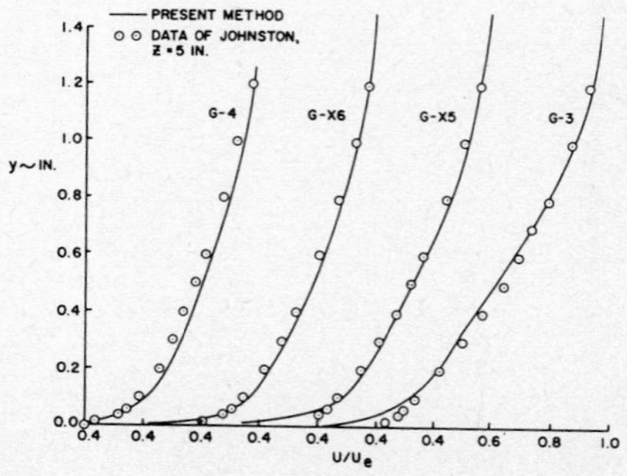

FIG. 7b RESULTS FOR THE FLOW OFF THE LINE OF SYMMETRY

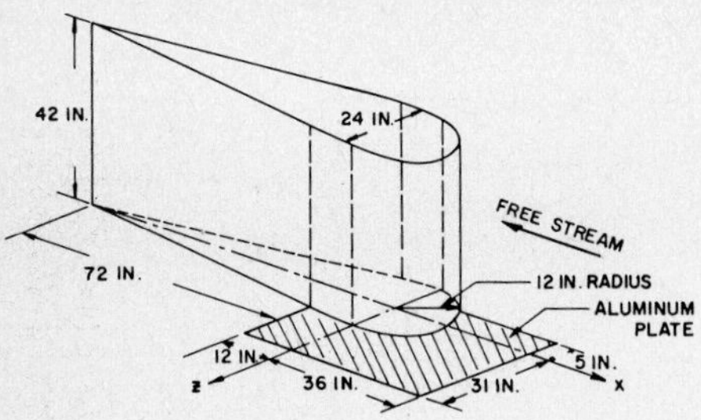

FIG. 8a SCHEMATIC DRAWING OF EAST AND HOXEY'S TEST SETUP

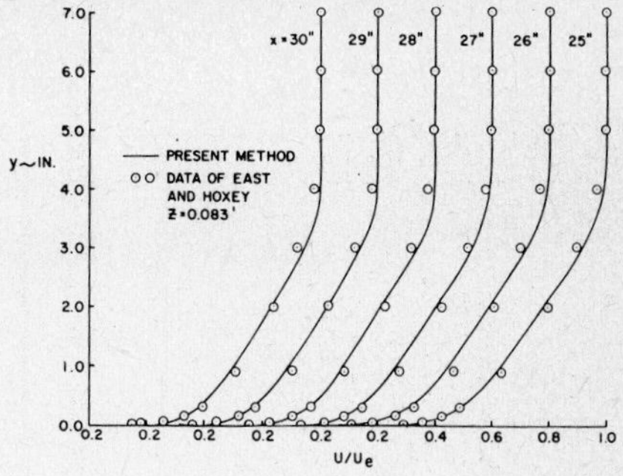

FIG. 8b RESULTS FOR THE FLOW OFF THE LINE OF SYMMETRY

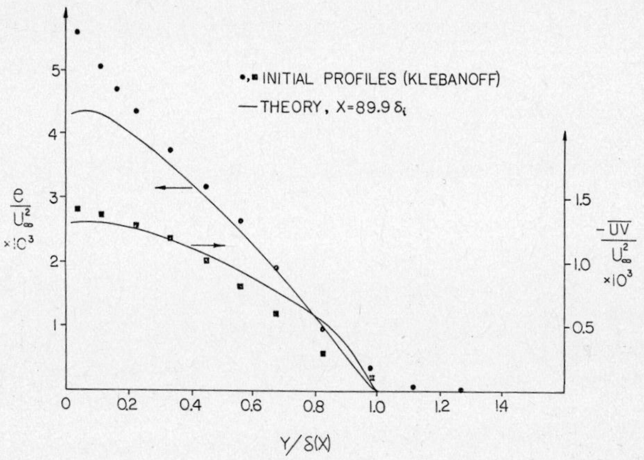

FIG. 9

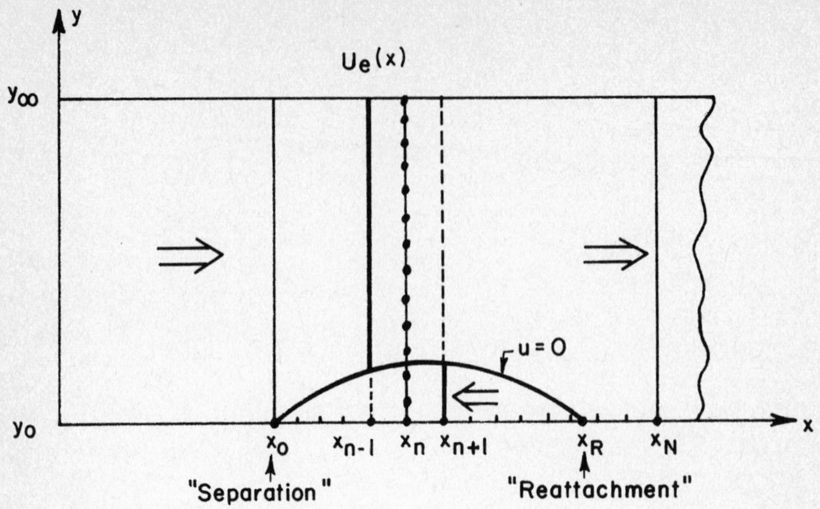

FIG. 10 REVERSE FLOW APPLICATION OF THE BOX-SCHEME

REFERENCES

[1] Barbin, A. R., and Jones, J. B., J. Basic Engr. 85 (1963), pp. 29-34.

[2] Cebeci, T. AIAA Journal 9 (1971), pp. 1091-1098.

[3] Cebeci, T., private communication (1972).

[4] Cebeci, T., Berkant, N., Silivri, I., and Keller, H. B. to appear in Computers and Fluids (1974-75).

[5] Cebeci, T. and Keller, H. B., Lecture Notes in Physics, Vol. 19, Springer-Verlag (1973), pp. 79-85.

[6] Cebeci, T. and Keller, H. B. in Recent Research in Unsteady Boundary Layers, IUTAM Symp. Vol. I, les Presses de L'Universite Laval, Quebec (1971), pp. 1072-1105.

[7] East, L. F. and Hoxey, R. P., Royal Aircraft Establishment, TR 69041 (1969).

[8] Johnston, J. P., M.I.T. Gas Turbine Lab., Rep. #39 (1957).

[9] Keller, H. B., in Numerical Solution of Partial Differential Equations, Vol. II. (ed. B. Hubbard) Academic Press, New York (1970).

[10] Keller, H. B., SIAM J. Num. Anal 11 (1974), pp. 305-321.

[11] Keller, H. B. and Cebeci, T., Lecture Notes in Physics, Vol. 8, Springer-Verlag (1971), pp. 92-100.

[12] Keller, H. B. and Cebeci, G., AIAA Journal 10 (1972), pp. 1197-1200.

[13] Keller, H. B. and Cebeci, T., J. Comp. Phys. 10 (1972), pp. 151-161.

[14] Klebanoff, P. S., NACA Report 1247 (1954).

[15] Klineberg, J. M. and Steger, J. L., paper #74-94 at AIAA meeting, Washington, D. C. (1974).

[16] Knight, D., paper in preparation, Caltech (1974).

[17] Krause, E. and Hirschel, E. H., Lecture Notes in Physics, Vol. 8, Springer-Verlag (1971), pp. 132-137.

[19] Pfenninger, W., Northrup Aircraft Report AM-133 (1951).

[19] Saffman, P. G., Studies in Appl. Math. 53 (1974), pp. 17-34.

[20] Hedman, P. O., "Particle-Gas Dispersion Effects in Confined Coaxial Jets," Brigham Young Univ., Dept. Chem. Engr. Sci. (1973).

Initial boundary value problems for hyperbolic partial differential equations

by

Heinz - Otto Kreiss

1. Systems in one space dimension

In this section we collect some well known results for problems in one space dimension. Consider a hyperbolic system

(1.1) $$\partial u/\partial t = A\,\partial u/\partial x.$$

Here $u(x,t) = (u^{(1)}(x,t),\ldots,u^{(n)}(x,t))'$ denotes a vector function and A a constant n x n matrix. Hyperbolicity implies that A can be transformed to real diagonal form, i.e., there is a nonsingular transformation S such that

(1.2) $$S\,A\,S^{-1} = \begin{pmatrix} A^{I} & 0 \\ 0 & A^{II} \end{pmatrix} = \tilde{A}$$

where

$$A^{I} = \begin{pmatrix} a_1 & 0\ldots 0 \\ 0 & a_2 \ldots 0 \\ \cdot & \cdot\,\cdot\,\cdot \\ 0\ldots 0 & a_r \end{pmatrix} < 0, \quad A^{II} = \begin{pmatrix} a_{r+1} & 0\ldots 0 \\ 0\ a_{r+2} & \ldots 0 \\ \cdot\ \cdot & \cdot\ \cdot \\ 0\ \ldots\ 0 & a_n \end{pmatrix} > 0$$

are definite diagonal matrices. We can thus introduce new variables

(1.3) $$v = Su$$

and get

(1.4) $$\partial v/\partial t = \tilde{A}\,\partial v/\partial x.$$

The last equation can also be written in partitioned form

(1.5) $$\partial v^{I}/\partial t = A^{I}\,\partial v^{I}/\partial x, \quad \partial v^{II}/\partial t = A^{II}\partial v^{II}/\partial x,$$

where $v^{I} = (v^{(1)},\ldots,v^{(r)})'$, $v^{II} = (v^{(r+1)},\ldots,v^{(n)})'$. (1.5) represents n scalar equations. Therefore we can write down its general solution:

(1.6) $$v^{(j)}(x,t) = v^{(j)}(x+a_j t)\ ,\ j = 1,2,\ldots n,$$

which are constant along the characteristic lines $x + a_j t = \text{const.}$.

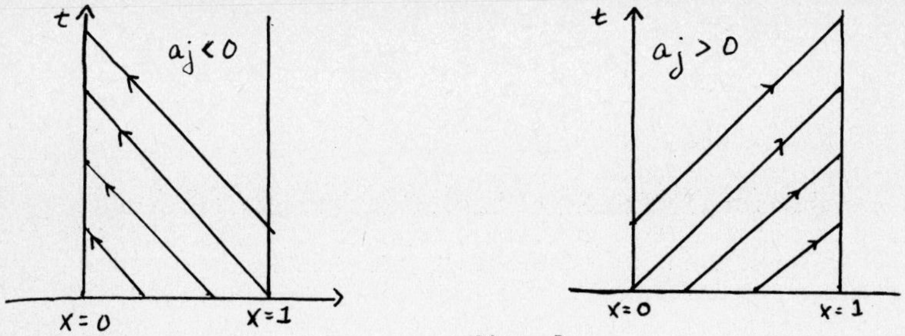

fig. 1

Now consider (1.4) in the strip $1 \geq x \geq 0$, $\infty > t \geq 0$. The solution is uniquely determined and can be computed explicitly if we specify initial conditions

$$(1.8) \qquad v(x,0) = f(x) \qquad 0 \leq x \leq 1,$$

and boundary conditions

$$(1.9) \quad v^I(0,t) = R_0 v^{II}(0,t) + g_o(t), v^{II}(1,t) = R_1 v^I(1,t) + g_1(t) .$$

Here R_0, R_1 are rectangular matrices and g_0, g_1 are given vector functions. If we consider wave propagation, then the boundary conditions describe how the waves are reflected at the boundary.

Nothing essentially is changed if $A = A(x,t)$ and $R_j = R_j(t)$ are functions of x,t. Now the characteristics are not straight lines but the solutions of the ordinary differential equations

$$dx/dt = a_j(x,t) \quad .$$

More general systems

$$(1.10) \quad \partial v/\partial t = \tilde{A}(x,t)\partial v/\partial x + B(x,t)v + F(x,t) ,$$

can be solved by the iteration

$$(1.11) \quad \partial v^{[n+1]}/\partial t = \tilde{A}(x,t)\ \partial v^{[n+1]}/\partial x + F^{[n]}$$

where

$$F^{[n]} = B(x,t)\ v^{[n]} + F$$

Furthermore, it is no restriction to assume that $\tilde{A}$ has diagonal form. If not, we can, by a change of dependent variables, achieve the form (1.10). There is also no difficulty in deriving a priori estimates. One can show

<u>Theorem 1.1</u>. There are constants K, α such that for the solutions of (1.8) - (1.10) the estimate

$$||u(x,t)|| \leq Ke^{\alpha t}||f(x)||+$$

$$(1.12) \qquad K\ \ell(\alpha,t)\{\max_{0\leq\xi\leq t}||\ F(x,\xi)||\ + \sum_{i=0}^{1}\ \max_{0\leq\xi\leq t}|g_i(\xi)|\}$$

holds. Here

$$||u|| = (\int_0^1 |u|^2\ dx)^{1/2}$$

denotes the usual L_2 - norm and $\ell(\alpha,t)$ is the function

$$\ell(\alpha,t) = \begin{cases} (1-e^{\alpha t})/\alpha & \text{if } \alpha \neq 0, \\ t & \text{if } \alpha \neq 0. \end{cases}$$

We can therefore develop a rather complete theory for initial boundary value problems by using characteristics. This has of course been known for a long time. The only trouble is, that this theory cannot be easily generalized to problems in more than one space dimension. For difference approximations it is already inadequate in one space dimension.

2. Halfplane problems in more than one space dimension for systems with constant coefficients

In this section we consider first order systems

$$(2.1)\quad \partial u/\partial t = A\, \partial u/\partial x_1 + \sum_{j=2}^{r} B_j \partial u/\partial x_j + F = P(\partial/\partial x)u + F$$

in the quarter space $t \geq 0$, $0 \leq x_1 \leq \infty$, $x_- = (x_2,\ldots,x_r) \varepsilon R_-$, where R_- denotes the $r-1$ dimensional space $-\infty < x_j < \infty$, $j=2,3,\ldots, r$.

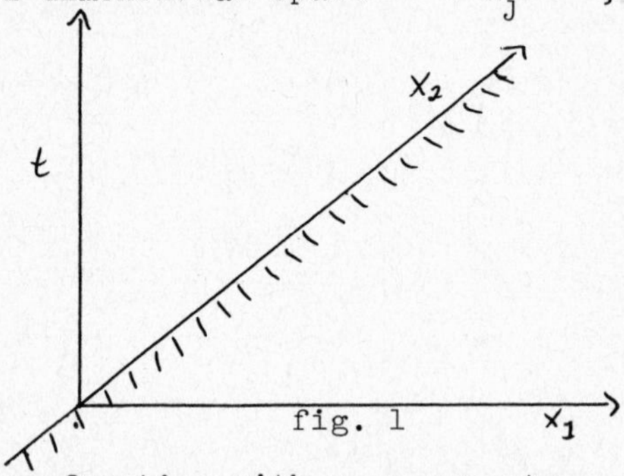

fig. 1

u is again a vector function with u components and A, B_j are constant u x n matrices. Without restriction we can assume that A is already in the diagonal form (1.2). Let the system (2.1) be strictly hyperbolic, i.e., the eigenvalues of the "symbol"

$$P(i\omega) = i(A\omega_1 + \sum_{j=2}^{r} B_j \omega_j),\quad \omega_j \text{ real},\quad |\omega|^2 = \sum_{\nu=1}^{r} |\omega_\nu|^2 = 1$$

are all purely imaginary and distinct.

To obtain a unique solution we have to prescribe for $t = 0$ initial values

$$(2.2)\qquad u(x,0) = f(x)$$

and for $x_1=0$ boundary conditions

$$(2.3)\qquad u^{I}(0,x_-,t) = R_0 u^{II}(0,x_-,t) + g.$$

Here the boundary conditions are necessarily of the form (1.9) because (2.1) reduces to a one dimensional system if the initial values do not depend on $x_2,\ldots,x_r$.

In one space dimension the initial boundary value problem is always well posed. Unfortunately, this is in general not true in more than one space dimension. However, one can describe the behaviour of our systems in terms of a few phenomena. For this purpose we consider (2.1)-(2.3) with $F \equiv g \equiv 0$.

1) The problem is not well posed if there are solutions of the form

$$w_\tau(x,t) = e^{\tau(st+iwy)}g(\tau x)$$

Here s is a complex constant with Real $s > 0$, ω_2 is real and $g(\tau x)$ is a bounded vector function which fulfills the boundary conditions (2.3). Furthermore $\tau > 0$ is any positive number, i.e., there are solutions which grow arbitrary fast with time.

2) There are waves which, when reflected at the boundary, get "rougher". Symbolically this can be expressed in the following way. If the incoming wave is given by $a(\omega)\, e^{i\omega x}$, then the reflected wave has the form $a(\omega)\omega^{\beta}e^{i\omega x}$, where $\beta>0$ is a constant. This phenomenon can be disastrous if we consider other than halfplane problems, because then we can have waves which are reflected back and forth at the boundaries and lose all their derivatives in the process.

3) The same situation as under 2). However, the outgoing wave is damped immediately. Symbolically it can be written as $a(\omega)\omega^{\beta}e^{(i\omega|\omega|^{\alpha})x}$. Thus in the interior of the domain the solution is smooth.

4) No loss of derivatives at the boundary. The reflected waves are as smooth as the incoming waves.

The last two cases represent well posed problems.

All these phenomena can be characterized algebraically. Transform (2.1) using Laplace and Fourier transform for the time and the tangential variables $x\_$ respectively. Then we get a system of ordinary differential equations

$$s\hat{u}(x_1,\omega\_,s) = A d\hat{u}(x_1,\omega\_,s)/dx_1 + i\sum_{j=2}^{r}\omega_j B_j \hat{u}(x_1,\omega\_,s) + \hat{f},$$

$$\hat{u}^{I}(0,\omega\_,s) = R_0\hat{u}^{II}(0,\omega\_,s),\omega\_ \text{ real, Real } s> 0,$$

which determines the behaviour of the initial boundary value problem. The properties of this system are determined by the associated eigenvalue problem

$$(2.4)\quad s\Psi = A\, d\Psi/dx_1 + i\sum_{j=2}^{r}\omega_j B_j\, \Psi,\ \omega_j \text{ real, } \Psi=\Psi(x_1)$$

$$\Psi^{I}(0) = R_0\Psi^{II}(0),\quad \int_0^\infty |\Psi|^2 dx_1 < \infty\ .$$

We have for example:

Theorem 2.1. The first situation arises if and only if the eigenvalue problem (2.4) has for some real vector $\omega\_ = (\omega_1,\ldots,\omega_r)$ an eigenvalue s with Real $s > 0$. The last situation is equivalent to the fact that there are no eigenvalues or generalized eigenvalues of (2.4) for Real $s \geq 0$.

As an example we consider the linearized shallow water equations

(2.5) $\partial w/\partial t = + A\partial w/\partial x + B\,\partial w/dy$, $t\geq 0$, $x\geq 0$, $-\infty<y<\infty$.

where

$$A = -\begin{pmatrix} U & 0 & 1 \\ 0 & U & 0 \\ 1 & 0 & U \end{pmatrix}, \quad B = -\begin{pmatrix} V & 0 & 0 \\ 0 & V & 1 \\ 0 & 1 & V \end{pmatrix}, \quad w = \begin{pmatrix} u \\ v \\ \phi \end{pmatrix}$$

We assume that $0<u<1$, i.e., we have "in flow" at the boundary $x=0$. Then the matrix A has two negative eigenvalues and therefore two boundary conditions have to be described at $x=0$, for example:

(2.6) $v=0$, $\beta u+\alpha\phi=0$

Choosing different values of α,β we can end up in all of the above situations. We have

Lemma 2.1. For $\alpha<-1$, $\beta=1$ the first situation occurs. For $\beta=0$, $\alpha=1$; $\beta=1$, $\alpha=0$ and $\beta=\alpha=1$ the second, third and fourth situation respectively is valid.

These results are discussed in detail in a forthcoming paper.

For symmetric hyperbolic systems there is an important class of boundary conditions for which the third or fourth situation holds. Friedrichs [2] has proved the following

Theorem 2.2. Assume that the system (2.1) is symmetric hyperbolic and that

$$A^{II} + R_0^* A^I R_0 \geq 0$$

then the problem (2.1)-(2.3) is well posed.

The reason is that an energy estimate exists. Multiply (2.1) by u. Let $F\equiv g\equiv 0$. Then partial integration and the boundary conditions give us

$$\frac{1}{2}\frac{\partial}{\partial t}\int_0^\infty\int_{R_-} |u|^2\, dx_- dx_1 = \int_0^\infty\int_{R_-} u^* Pu\, dx_- dx_1$$

$$= -\int_{R_-} u^*Au\, dx_-\Big|_{x_1=0} = -\int_{R_-} u^{I*}A^I u^I + u^{II*}A^{II}u^{II} dx_-\Big|_{x_1=0}$$

$$= -\int_{R_-} u^{II*}(A^{II}+R_o^* A^I R_o)\, u^{II}\, dx_-\Big|_{x_1=0} \leq 0.$$

3. Problems with variable coefficients in general domains

Now we consider systems (2.1)-(2.3) with variable coefficients in a general domain $\Omega \times (0\leq t\leq T)$

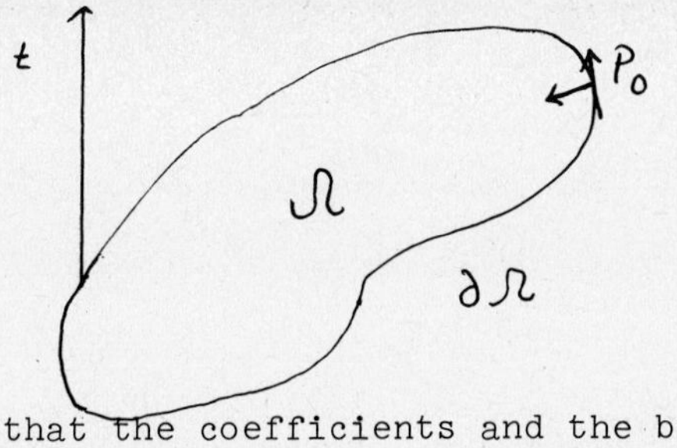

Here we assume that the coefficients and the boundary $\partial\Omega$ are sufficiently smooth. Connected with this problem are a set of halfplane problems which we get in the following way: Let $(x_o,t_o)\in \partial\Omega x(0\leq t\leq T)$ be a boundary point and let $\tilde{x} = S(x)$, $\tilde{t} = t-t_o$ with $S(x_o)=0$ be a smooth transformation which locally transforms the boundary into the halfplane $\tilde{x}_1=0$. Apply this transformation to the differential equations and the boundary conditions, and freeze the coefficients at $\tilde{x}=\tilde{t}=0$, and consider the halfplane problem for this system with constant coefficients. Under reasonable assumptions we have

Theorem 3.1. If all the halfplane problems with frozen coefficients are well posed, than the same is true for the full problem. Thus we need only to consider halfplane problems, with constant coefficients.

For the proof of Theorem 3.1 see [5].

New difficulties arise when corners are present. Not much is known except some results by Stan Osher.

4. Difference approximations in one space dimension

We start with an example which explains most of our difficulties. Consider the differential equation

$$\partial u/\partial t = \partial u/\partial x \tag{4.1}$$

in the quarter plane $x\geq 0$, $t\geq 0$ with initial values

$$u(x,0) = f(x). \tag{4.2}$$

From section 1 we know that no boundary conditions need to be specified for $x=0$, $t>0$. We want to solve the above problem using the leap-frog scheme. For that reason we introduce a time step $\Delta t>0$ and a mesh width $\Delta x>0$ and divide the x-axis into intervals of length Δx. Using the notation

$$v_\nu(t) = v(x_\nu,t)\ ,\quad x_\nu = \nu\Delta x,\ t = t_\mu = \mu\Delta t$$

we approximate (4.1), (4.2) by

$$v_\nu(t+\Delta t) = v_\nu(t-\Delta t) + 2\Delta t\ D_o v_\nu(t),\ \nu = 1,2,\ldots. \tag{4.3}$$

with initial values

$$v_\nu(0) = f(x_\nu),\ v_\nu(\Delta t) = f(x_\nu) + \Delta t\ \partial f(x_\nu)/\partial x. \tag{4.4}$$

Here

$$D_o v_\nu = (v_{\nu+1} - v_{\nu-1})/2\Delta x$$

denotes the usual centered difference operator. We assume that (4.3) is stable for the Cauchy problem, i.e., $0<\Delta t/\Delta x\leq 1$.

It is obvious that the solution of (4.3), or (4.4) is not yet uniquely determined. We must give an additional equation for v_o. For example

$$v_o = 0 \tag{4.5}$$

This relation is obviously not consistent. In general it will destroy the convergence. Let $f(x)\equiv 1$. Then $u(x,t)\equiv 0$ and

$$v_\nu(t) = 1 + (-1)^\nu y_\nu(t) ,$$

where $y_\nu(t)$ is the solution of

$$\begin{aligned} y_\nu(t+\Delta t) &= y_\nu(t-\Delta t) - 2\Delta t\, D_o y_\nu(t) , \quad \nu = 1,2,\ldots \\ y_\nu(0) &= y_\nu(\Delta t) = 0, \end{aligned} \tag{4.6}$$

with boundary conditions

$$y_0(t) = -1. \tag{4.7}$$

(4.6) and (4.7) is an approximation to the problem

$$\partial w/\partial t = -\partial w/\partial x ,$$

$$w(x,0) = 0, \quad w(0,t) = -1,$$

i.e.,

$$w(x,t) = \begin{cases} 0 \text{ for } t<x \\ -1 \text{ for } t\geq x \end{cases} .$$

Therefore

$$v_\nu(t) \sim \begin{cases} 1 \text{ for } t<x \\ 1 - (-1)^\nu \text{ for } t\geq x. \end{cases}$$

This behaviour is typical for all nondissipative centered schemes Therefore one needs to be very careful when overspecifying boundary conditions. The oscillation decays if the approximation is dissipative. However, near the boundary the error is as bad and, for systems, it can be propagated into the interior via the ingoing characteristics.

Now we replace (4.5) by an extrapolation

$$v_o(t) - 2v_1(t) + v_2(t) = 0. \tag{4.8}$$

which is the same as using for $\nu=1$ the one-sided difference formula

$$v_1(t+\Delta t) = v_1(t-\Delta t) + 2\,\frac{\Delta t}{\Delta x}\,(v_2(t) - v_1(t)) \tag{4.9}$$

The approximation is only useful if it is stable. If we choose

$$v_\nu(0) = \begin{cases} 1 \text{ for } \nu=0 \\ 0 \text{ for } \nu>0 \end{cases}, \quad v_\nu(\Delta t) \equiv 0 \text{ for all } \nu$$

as initial values then an easy calculation shows that

$$(4.10) \qquad \|v(t)\|_{\Delta x} = \text{const.}\ (t/\Delta t), \quad \|v\|_{\Delta x} = \Sigma\, |v_\nu|^2 \Delta x.$$

This growth rate is the worst possible and one might consider the approximation to be useful. However, if we consider (4.1) in a finite interval $0 \le x \le 1$ and add the boundary condition

$$(4.11) \qquad u(1,t) = v_N(t) = 0 \qquad N\Delta x = 1$$

for both the differential equation and the difference approximation, then there are solutions which grow like

$$(4.12) \qquad \|v(t)\|_{\Delta x} = \text{const.}\ (t/\Delta t)^t$$

which is not tolerable. This behaviour can be explained as follows: At the boundary $x=0$ a wave is created which grows like $t/\Delta t$. This wave is reflected at the boundary $x=1$ and is increased by another factor $t/\Delta t$ when it hits the boundary $x=0$ again, and so on.

All these difficulties can be avoided by using, instead of (4.9), the onesided approximation:

$$(4.13) \qquad v_1(t+\Delta t) = v_1(t) + \frac{\Delta t}{\Delta x}\,(v_2(t) - v_1(t))$$

or

$$(4.14) \qquad v_1(t+\Delta t) = v_1(t-\Delta t) + \frac{\Delta t}{\Delta x}\,(v_2(t) - \tfrac{1}{2}(v_1(t+\Delta t) + v_1(t-\Delta t)))$$

One can also keep (4.8) if one replaces the leap-frog scheme by the Lax-Wendroff approximation or any other dissipative approximation.

Let us discuss the general theory. Details are given in [4], [5]. We consider general difference approximations

$$v_{\nu+1}(t+\Delta t) = Q\, v_\nu(t)$$

with boundary conditions

$$B\, v_o = 0$$

such that the solution is uniquely determined by the initial values

$$v_\nu(0) = f_\nu\ .$$

One can solve this problem by using a discrete Laplace transform in time. This gives us

$$(4.15) \qquad (z-Q)\hat{v}_\nu = f_\nu\ , \quad |z|>1. \qquad Bv_o = 0.$$

The properties of this system are determined by the eigenvalue problem

$$(4.16) \qquad (z-Q)\ \phi_\nu = 0 \qquad B\phi_o = 0, \quad \|\phi\|^2_{\Delta x} = \Sigma(\phi_\nu)^2 \Delta x < \infty.$$

Corresponding to theorem 2.1 we have

Theorem 4.1. Assume that (4.16) has an eigenvalue z with $|z| \geq 1+\delta, \delta>0$. Then the approximation is not stable. If there are no eigenvalues or generalized eigenvalues z with $|z| \geq 1$, then the approximation is stable.

The above methods are for analyzing the behaviour of a given difference approximation. There are methods which systematically construct stable approximations. These techniques are discussed in [7]. The idea is the following.

Consider again the differential equation (4.1), (4.2). The problem is well posed because there is an energy equality

$$(4.17) \qquad (u, \partial u/\partial x) + (\partial u/\partial x, u) = -|u(0)|^2.$$

Therefore we want to construct approximations Q to $\partial/\partial x$ which have the corresponding property.

We define a discrete norm

$$(4.18) \qquad (u,v)_{\Delta x} = \tilde{u}^* A \tilde{v}\, \Delta x + \sum_{\nu=r}^{\infty} u_\nu^* v_\nu \Delta x$$

Here $\tilde{u} = (u_0, \ldots, u_{r-1})'$, $\quad \tilde{v} = (v_0, \ldots, v_{r-1})'$

denote the first r components of u,v and $A=A^*$ is a positive definite $r \times r$ matrix. The corresponding energy equality

$$(u,Qu)_{\Delta x} + (Qu,u)_{\Delta x} = -|u_0|^2.$$

holds if, in matrix form

$$Q = \frac{1}{\Delta x} \begin{pmatrix} A^{-1} & 0 \\ 0 & I \end{pmatrix} \begin{pmatrix} B & C \\ -C^* & D \end{pmatrix},$$

where

$$B = B_1 + \begin{pmatrix} -\frac{1}{2} & 0 \ldots . 0 \\ 0 & 0 \ldots . 0 \\ . & . \; . \; . \end{pmatrix}, \qquad B_1 = -B_1^* = \begin{pmatrix} 0 & b_{12} & \cdot\;\cdot\;\cdot & b_{1r} \\ -b_{12} & 0 & b_{23} & 0..0 \\ & & - - - & \\ -b_{1r} & \cdot\;\cdot\;\cdot & \cdot & b_{r-1r}\;0 \end{pmatrix}$$

is an $r \times r$ matrix.

$$D = \begin{pmatrix} 0 & \alpha_1 & \alpha_2 \ldots . \alpha_r & 0 \ldots . . \\ -\alpha_1 & 0 & \alpha_1 \; \alpha_2 \ldots . . & \alpha_r 0 \ldots \\ -\alpha_2 & -\alpha_1 & 0 \; \alpha_1 \; \alpha_2 \ldots . & \alpha_r 0 \ldots \\ . & . & . \quad . \quad . & . \; . \\ . & . & . \quad . \quad . & . \; . \\ . & . & . \quad . \quad . & . \; . \\ . & . & . \quad . \quad . & . \; . \end{pmatrix}$$

is an antisymmetric band matrix and

$$C = \begin{pmatrix} 0 & . & . & . & . & 0 \\ 0 & . & . & . & & 0 \\ \alpha_r & 0. & & . & . & 0 \\ \alpha_1 \ldots \alpha_r 0 \ldots & & & & & 0 \end{pmatrix}$$

This form for Q is natural. It arises from any reasonable difference approximation of the Cauchy problem when making modifications near the boundary.

Now we have only to choose the coefficients in such a way that we get the desired accuracy. An example is

$$\Delta x Q = \begin{pmatrix} -1 & 1 & 0 \ldots 0 \ldots \\ -\frac{1}{2} & 0 & \frac{1}{2} 0 \ldots 0 \ldots \\ 0 & -\frac{1}{2} & 0 \frac{1}{2} 0 \ldots 0 \ldots \\ . & . & . \quad . \quad . \\ . & . & . \quad . \quad . \quad . \end{pmatrix} = \begin{pmatrix} 2 & 0 \ldots . 0 \ldots \\ 0 & 1 0 \ldots 0 \ldots \\ 0 & 0 \; 1 0 \ldots 0 \ldots \\ . & . \quad . \quad . \\ . & . \quad . \quad . \end{pmatrix} \begin{pmatrix} -\frac{1}{2} & \frac{1}{2} & 0 \ldots . 0 \ldots \\ -\frac{1}{2} & 0 & \frac{1}{2} 0 \ldots 0 \ldots \\ 0 & -\frac{1}{2} & 0 \; \frac{1}{2} 0 \ldots 0 \ldots \\ . & . & . \quad . \quad . \quad . \\ . & . & . \quad . \quad . \quad . \end{pmatrix}$$

which is second order accurate in the interior and first order accurate on the boundary. In this case $r=1$, $A^{-1} = \frac{1}{2}$. In [ ] we have constructed methods which are accurate of order 1,2,...6 and there is no doubt that one can construct methods of any order of accuracy.

Once one has constructed these approximations for $\partial/\partial x$ there is no trouble in writing down stable approximations to systems which have the same order of accuracy.

5. Difference approximations in more than one space dimension

Nothing essentially new needs to be added to derive the theory of difference approximations in halfplanes because Fourier transforming the tangential variables x_- gives us a set of one-dimensional problems. The situation becomes much more complicated if we consider general domains with smooth boundaries. This is not the case for the differential equations because we can always introduce a local coordinate system, thus reducing the problem to a set of halfplane problems. This is not possible for difference approximations. Once we have picked the net everything is fixed. Let us consider a very simple example. We want to solve the differential equation

(5.1) $$\partial u/\partial t = -\partial u/\partial x$$

in the two dimensional domain $2y-x\leq 0$. The initial values are

(5.2) $$u(x,y,0) = f(x,y) \qquad \text{for } 2y-x\leq 0,\ t=0$$

and the boundary conditions are given by

(5.3) $$u(x,y,t) = g(x,y,t), \qquad \text{for } 2y-x=0,\ t\geq 0.$$

We introduce grid points by $x_j = j\Delta x, y_i = i\Delta y, \Delta x = \Delta y$.

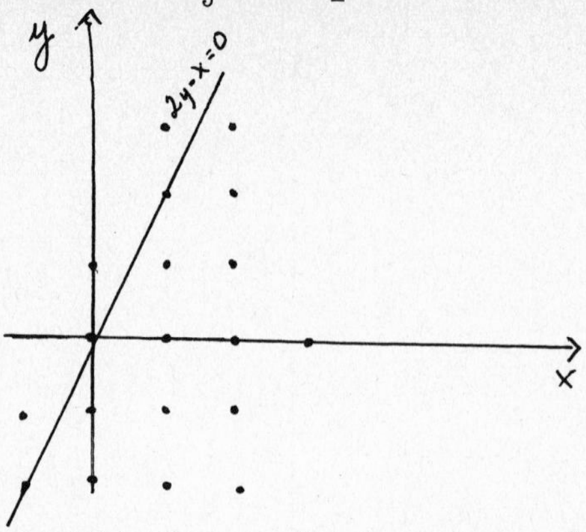

Thus, there is a gridpoint on the boundary only on every second row. Now we approximate (5.1) by the leap-frog scheme, and the boundary conditions by

$$v_{i,j} = g_{i,j}, \text{ if } 2j=i \; ; \; v_{i,j} + v_{i+1,j} = 2g_{i+\frac{1}{2},j} \; , \text{ if } 2j=i+1.$$

Here $v_{i,j} = v(i\,\Delta x, j\,\Delta y, t)$. Therefore we get two different solutions on two different meshes. As long as the solution of the differential equation is smooth the solutions of the difference equation on these different meshes fit together. However, if for example $f \equiv 0$ and $g \equiv 1$ then the solution of the differential equation is a discontinuous wave propagating into the interior. Now the solutions of the difference approximation on the different nets do not fit together.

We get oscillations in the tangential direction of the wave. There are two possible methods for remedying the situation. 1) Add dissipation to smooth out the tangential oscillations. 2) Introduce curved meshes which follow the boundary. The second procedure is much more accurate and should be preferred if possible. A lot of progress has been made in this direction. See for example [ ],...[ ].

References

[1] Amsden, A.A., and Hirt, C.W., A simple scheme for generating general curvlinear grids. Journal of Computational Physics, Vol. 11, No. 3. 1973.

[2] Friedrichs, K.O. Symmetric hyperbolic linear differential equations, Comm. Pure and Appl. Math. Vol. 7 (1954) pp. 345-392.

[3] Kreiss, H.O. Stability theory for difference approximations of mixed initial boundary value problems I, Math. Comp. Vol. 22, 1968.

[4] Gustafsson, B. Kreiss, H.O. and Sundström, A. Stability theory for difference approximations of mixed initial boundary value problems II, Math. Comp. Vol. 26, 1972.

[5] Kreiss, H.O. Initial boundary value problems for hyperbolic systems. Comm. Pure and Appl. Math. Vol. 23, 1970.

[6] Kreiss, H.O. Difference approximations for initial boundary value problems. Proc. Roy. Soc. London A 323, 1971.

[7] Kreiss, H.O. and Scherer, G. Finite elements and finite difference methods for hyperbolic partial differential equations.

NUMERICAL SOLUTION OF TURBULENT BOUNDARY LAYERS APPROACHING SEPARATION

by

D.E. Abbott
Purdue University

J.D.A. Walker
Purdue University

R.E. York
Detroit Diesel Allison

Currently available theories and predictive methods for turbulent boundary-layer development experience decreased accuracy when flows with large adverse pressure gradients must be considered, particularly when approaching separation. In the classical analysis of fully turbulent flows, the equations are time averaged in the manner first suggested by Osborn Reynolds and the composite structure of the time-mean turbulent boundary layer analyzed in terms of a two-layer (inner and outer) region. In the outer region, the mean velocity profile is observed experimentally to behave according to Coles' "law of the wake." The inner region may be further subdivided into two regions consisting of a fully turbulent layer, in which the mean profile is observed to follow the logarithmic "law of the wall," and the viscous wall layer in which the profile is adjusted to zero at the wall.

The basis for the work described here are the experimental evidence of Kline, Brodkey, and others that indicates there is an identifiable vorticity structure in both space and time in the viscous wall layer. The primary results of these experiments which are important to a mathematical modeling are the identification of longitudinal turbulent streaks which are (i) three-dimensional, but statistically determinant in the spanwise direction, (ii) time-dependent with a measurably periodic "burst" frequency, and (iii) a phenomenon which may be characterized as longitudinal vortices undergoing diffusion, convection, and stretching. Using these observations as a basic foundation, an analysis of the instantaneous velocity components in the turbulent wall layer is developed.

VORTEX-STRUCTURE MODEL

The equations analyzed are the incompressible, three-dimensional, time-dependent vorticity transport equations for instantaneous quantities. The vector form of these equations in cartesian coordinates is

$$\frac{D\overline{\omega}}{Dt} = \overline{\omega}\cdot\nabla\overline{q} + \nu\nabla^2\overline{\omega} , \quad \nabla \cdot \overline{q} = 0 \tag{1}$$

where $\overline{\omega} = \text{curl}\ \overline{q}$ is the vorticity vector. Suppose that (x,y,z) are cartesian coordinates with corresponding velocity components (u,v,w) where x and z measure distance in the streamwise and spanwise directions, respectively, and y measures distance normal to the wall. Experimental evidence clearly indicates that the turbulent boundary layer is multi-structured and we now consider the physical scalings appropriate to the turbulent wall layer.

Turbulent Wall Layer Scaling

In the innermost turbulent wall layer, the experiments suggest that the y and z scales are ν/U_τ and λ, respectively. Here ν is the kinematic viscosity, U_τ a characteristic friction velocity and λ the spanwise streak spacing. We assume that w is at most of O(v) and this leads to the following scalings:

$$\begin{aligned} x &= L\xi , \quad y = \nu y^+/U_\tau , \quad z = \lambda z^+ , \quad t = \nu t^+/U_\tau^2 \\ u &= U_\tau u^+ , \quad v = \nu v^+/L , \quad w = \nu w^+/L \end{aligned} \tag{2}$$

where L is a characteristic axial length. Substitution of (2) into (1) yields the following equation for the axial vorticity:

$$\frac{\partial\omega_z^+}{\partial t^+} + \frac{1}{R}\left(u^+\frac{\partial\omega_z^+}{\partial\xi} + v^+\frac{\partial\omega_z^+}{\partial y^+} + \frac{1}{\lambda^+}w^+\frac{\partial\omega_z^+}{\partial z^+}\right)$$

$$= \frac{1}{\lambda^+ R}\left(\frac{1}{R^2}\omega_x^+\frac{\partial w^+}{\partial\xi} + \omega_y^+\frac{\partial w^+}{\partial y^+} + \omega_z^+\frac{\partial w^+}{\partial z^+}\right) + \frac{1}{R^2}\frac{\partial^2\omega_z^+}{\partial\xi^2} + \frac{\partial^2\omega_z^+}{\partial y^{+2}} + \frac{1}{\lambda^{+2}}\frac{\partial^2\omega_z^+}{\partial z^{+2}} \tag{3}$$

$$\frac{\partial u^+}{\partial\xi} + \frac{\partial v^+}{\partial y^+} + \frac{1}{\lambda^+}\frac{\partial w^+}{\partial z^+} = 0 \tag{4}$$

where

$$\omega_x^+ = \frac{\partial w^+}{\partial y^+} - \frac{1}{\lambda^+}\frac{\partial v^+}{\partial z^+}, \quad \omega_y^+ = \frac{\partial u^+}{\partial z^+} - \frac{\lambda^+}{R^2}\frac{\partial w^+}{\partial\xi}, \quad \omega_z^+ = \frac{1}{R^2}\frac{\partial v^+}{\partial\xi} - \frac{\partial u^+}{\partial y^+}$$

with $R = U_\tau L/\nu$ and $\lambda^+ = U_\tau\lambda/\nu$. Here ω_x^+, ω_y^+, and ω_z^+ are O(1) functions of (ξ, y^+, z^+, t^+). Now the axial-length Reynolds number R is large, and the experiments of Kline & Runstadler (1959), Schraub & Kline (1965), and Kline, et al (1967) indicate that the length scale in the z-direction is much larger ($\lambda^+ \approx 100$) than the length scale in the y-direction. Consequently, to leading order only the first and next to last terms of equation (3) are retained.

Leading-Order Analysis

To leading order equation (3) applied to the wall-layer becomes:

$$\frac{\partial\omega_z^+}{\partial t^+} = \frac{\partial^2\omega_z^+}{\partial y^{+2}} \tag{5}$$

The pressure gradient appears in the boundary condition at the wall:

$$y^+ = 0: \quad \frac{\partial\omega_z^+}{\partial y^+} = -P^+ = -\frac{\nu}{\rho U_\tau^3}\frac{dp}{dx} \tag{6}$$

The initial condition for equation (5) may be written as $\omega_z^+ = f(y^+)$ for $t^+=0$. To be consistent with the present model for viscous diffusion, the wall-layer at any station x is viewed as a velocity deficient region which grows in thickness with time due to viscous diffusion until a convected disturbance disrupts the organized growth. At this point in the cycle of events an instability is presumed to occur which leads to a bursting process that ejects fluid outward and downstream. The diffusion layer then collapses and the fully turbulent outer flow is presumed to penetrate to within a distance y_1^+ of the wall (where y_1^+ is small but non-zero) whereupon the whole process repeats until it is once again interupted by a disturbance a period T later.

Thus the initial condition is taken as:

$$f(y^+) = \begin{cases} -A_o, & 0 \le y^+ \le y_1^+ \\ -\dfrac{A_o y_1^+}{y^+}, & y_1^+ \le y^+ \end{cases}$$

where A_o is a constant. Lighthill (1963) showed by dimensional arguments that the vorticity in a fully developed turbulent boundary layer varies inversely with respect to y^+. The relation for $y^+ \le y_1^+$ is required by the condition that the total vorticity of the flow remain finite for all time. Physically this initial profile corresponds to that of a vortex sheet of constant strength adjacent to the wall. The remaining boundary condition follows from Lighthill's argument and takes the form:

$$\omega_z^+ \sim -A_o y_1^+/y^+ \quad \text{for } y^+ \text{ large.} \tag{7}$$

This is simply a statement that the solution in the diffusion layer must match smoothly into the outer, fully turbulent layer for all time.

The solution of equation (5) with the initial and boundary conditions (6) and (7) is found to be

$$\omega_z^+ = -\frac{1}{2\sqrt{t^+}}\left[\{a + b\,\ell n(t^+/S^2)\}e^{-\eta^2} + 4b\,\Xi'(\eta)\right]$$
$$+ 2P^+\sqrt{t^+}\left[\pi^{-1/2}e^{-\eta^2} - \eta\,\mathrm{erfc}\,\eta\right] + O(y_1^{+2}/t^+) \qquad (8)$$

for $t^+ \gg y_1^{+2}$, where $\eta = y^+/2\sqrt{t^+}$, $S = U_\tau\sqrt{T/\nu}$, T is the period of the turbulent burst and U_τ is the local shear velocity ($U_\tau = \sqrt{\tau_w/\rho}$, with τ_w the wall shear stress and ρ the density). The function $\Xi(\eta)$ is defined in equation (10). The constants a and b are given by

$$b = \frac{1}{\kappa\sqrt{\pi}}, \qquad a = \frac{2}{\kappa\sqrt{\pi}}\{1 - \frac{\gamma_o}{2} - \ell n(y_1^+/2S)\}$$

where κ is the von Kármán's constant and γ_o is Euler's constant. Equation (8) is a simplified version of the complete solution given in Loudenback and Abbott (1973) which is valid over the entire period.

Velocity Profiles

Instantaneous velocity profiles are easily obtained from equation 8 by integration on y. The result is:

$$u^+ = \frac{\sqrt{\pi}}{2}\{a + b\,\ell n(t^+/S^2)\}\mathrm{erf}\,\eta + 4b\,\Xi(\eta)$$
$$- P^+t^+\left\{(1+2\eta^2)\mathrm{erf}\,\eta - 2\eta^2 + \frac{2\eta}{\sqrt{\pi}}e^{-\eta^2}\right\} \qquad (9)$$

where

$$\Xi(\eta) = \frac{\sqrt{\pi}}{2}\int_0^\eta e^{-\xi^2}\int_0^\xi e^{\zeta^2}\,\mathrm{erf}\,\zeta\,d\zeta\,d\xi \qquad (10)$$

The function $\Xi(\eta)$, and its derivative $\Xi'(\eta)$, has been thoroughly studied and tables of values and its asymptotic behavior for large η may be found in Loudenback & Abbott (1973) and York & Abbott (1973). The following series is found to be particularly well suited for efficient numerical evaluation:

$$\Xi(\eta) = \frac{e^{-\eta^2}}{4}\sum_{j=1}^{\infty}\frac{2^j d(j)\eta^{2j+1}}{(2j+1)!!}\ ;\ d(j) = d(j-1) + \frac{1}{j}\ ;\ d(1) = 1 \qquad (11)$$

The time average of equation (9) over the bursting period T may be found as follows:

$$U^+ = \frac{U}{U_\tau} = \frac{1}{T}\int_0^T u^+ dt$$
$$= (a-3b)\left[He^{-H^2} - \sqrt{\pi}H^2\,\mathrm{erfc}\,H + \frac{\sqrt{\pi}}{2}\mathrm{erf}\,H\right]$$
$$+ 4b\Big[\Xi(H) + H\Xi'(H) + \frac{\sqrt{\pi}}{4}\mathrm{erf}\,H + 2H^2\,[\Xi(H)$$
$$- (\frac{\sqrt{\pi}}{4}\ell n\,H + \frac{\sqrt{\pi}}{8}\gamma_o)]\Big] + 2P^+S^2\,[-\frac{1}{4}\mathrm{erf}\,H$$
$$+ H^2(1 + \frac{H^2}{3})\,\mathrm{erfc}\,H - H(\frac{H^3}{3} + \frac{5}{6})\frac{e^{-H^2}}{\sqrt{\pi}}] \qquad (12)$$

where $H = y/2\sqrt{\nu T} = y^+/2S$. It can be shown that the asymptotic form of equation (12) for large y is:

$$U^+ \sim \frac{1}{\kappa} \ln y^+ + C \tag{13}$$

where

$$C = \frac{1}{6} S^2 P^+ + \frac{\sqrt{\pi}}{2} S + \frac{1}{\kappa} (1 + \frac{\gamma_0}{2} - \ln 2S)$$

Assuming a value of C = 5.1 for zero pressure gradient ($P^+ = 0$) yields S = 10.618, which may be compared with the value of 10.25 obtained by Kline, et al (1967) by direct measurement of burst frequencies. Taking the limit of equation (9) for small time also produces equation (13), but this time the following expression is obtained for C for zero pressure gradient:

$$C = \frac{1}{\kappa} (1 - \ln y_1^+)$$

yielding the result $y_1^+ = 0.336$ with C = 5.1. This value of y_1^+ is in good agreement with the experimental values of Corino & Brodkey (1969) and Zarić (1973).

HIGHER-ORDER ANALYSIS

A higher order solution of the vorticity transport equations (3) and (4) must account for both convection and stretching of the longitudinal vortices in the wall layer. Such an analysis was carried out numerically to $O(\nu^2)$ by considering a solution of the equations for a given vortex model at a fixed x-location. A particular model which has closed streamlines in the y-z plane is given by:

$$\psi^+ = v_s^+ z_s^+ E(y^+/y_s^+) \sin(z^+/z_s^+) \tag{14}$$

where

$$E(y^+/y_s^+) = \exp(-y^+/y_s^+) - 2\exp(-2y^+/y_s^+) + \exp(-3y^+/y_s^+)$$

and v_s^+, y_s^+, and z_s^+ are the appropriate velocity and length scales for a spanwise vortex distribution of wavelength $\lambda^+ = 2\pi z_s^+$. Equation (14) was suggested by Stuart (1965) and is in substantial agreement with the experimentally deduced results of Bakewell & Lumley (1967). For this model, the velocity components become:

$$u^+ = u^+(y^+,z^+,t^+),\quad v^+ = -v_s^+ E(y^+/y_s^+) \cos(z^+/z_s^+),\quad w^+ = \frac{v_s^+ z_s^+}{y_s^+} E'(y^+/y_s^+) \sin(z^+/z_s^+)$$

Numerical solutions were obtained on planes of symmetry given by $z^+ = 2\pi/\lambda^+$ where the convective and vortex stretching terms are a maximum. To $O(\nu^2)$, equations (3) and (4) then reduce to:

$$\frac{\partial \omega_z^+}{\partial t^+} + \frac{\partial}{\partial y^+} (v^+ \omega_z^+) = \frac{\partial^2 \omega_z^+}{\partial y^{+2}} \tag{15}$$

The initial condition is taken to be given by the leading-order solution equation (8). Parametric solutions may then be obtained as a function of time for given values of v_s^+. Details may be found in Loudenback and Abbott (1973).

RESULTS

The time-mean velocity profile given by (12) has been compared with the experimental data of 23 separate experiments covering 197 x-stations of data by York and Abbott (1973). Twenty of the cases came from the 1968 Stanford Conference on Turbulent Boundary Layer Computation, see Kline, et al (1968), and four of the more challenging examples are shown here. For all non-zero pressure gradient flows, it is necessary to correlate the stability parameter S and the pressure gradient parameter P^+. A simple correlation given by

$$S = S_0[1 + 10\, \mathrm{sgn}(P^+) P^{+2/3}]^{-1}, \quad S_0 = 10.618 \tag{16}$$

has been obtained by York and Abbott (1973) and seems to fit various experimental data over a limited range of P^+.

Figure 1 shows four stations of Perry's flow in a diverging channel. This was one of the most difficult of the mandatory runs at the Stanford Conference and typical results based on the classical van Driest eddy viscosity model are shown in Fig. 1(a). The results of the present analysis, (12) and (16), are given in fig. 1(b) and show significantly improved comparisons with the data. The two stations shown in this figure are the last two x-stations just upstream of separation. In fig. 1(c), Coles' wake model has been combined with (12) in order to include the outer region of the boundary layer. (The station numbers in these figures correspond to the Stanford Conference identification system.)

In figure 2, the present model, including the wake, is compared with (a) the Herring and Norbury strongly accelerated flow, (b) the Newman Series 2 airfoil flow, and (c) Fraser's Flow B where the flow at station 12 is very close to the separation point in a conical diffuser.

Figure 3 illustrates an additional result from the present theory which has rather important implications, although it has not as yet been tested against experiment. This result is based on the fact that the present theory is time dependent, and thus there is an unsteady effect on the instantaneous velocity profiles with increasing pressure gradient as separation is approached. Figure 3(a) shows a plot of (9) at a point approaching separation, figure 3(b) at the point where the instantaneous profile first obtains zero wall shear, and figure 3(c) at a point where the time mean profile does not exhibit separation even though the instantaneous profiles show reverse flow over part of the cycle. For each case shown, the instantaneous dashed profiles are labeled for the interval 0(0.2)1.0 of t/T. The solid profile is the time average given by (12).

Figure 4 shows a plot of the vorticity distribution given by (8) over one cycle for zero pressure gradient ($P^+ = 0$). The time average vorticity $\overline{\Omega}$ is also shown. Finally, fig. 5 shows a plot of the Reynolds stress $-\rho\overline{uv}$ calculated from (9) by the relation $-\rho\overline{uv} = \tau_w - \tau(y)$, where $\tau = \mu\partial U/\partial y$ and the details are in Loudenback and Abbott (1973). For comparison, the data of Schubauer (1954) is also shown.

REFERENCES

Bakewell, H.P. & J.L. Lumley: *Physics of Fluids*, 10, 1880 (1969).

Corino, E.R. & R.S. Brodkey: *JFM*, 37, 1 (1969).

Kline, S.J. et al: *JFM*, 30, 741 (1967).

Kline, S.J. et al (ed): *Proceedings, Computation of Turbulent Boundary Layers*, Vol. II. Compiled Data, Stanford University Press (1968).

Kline, S.J. and P.W. Runstadler: *JAM*, 26, 166 (1959).

Lighthill, M.J.: *Laminar Boundary Layers*, Chap. II, L. Rosenhead (ed), Oxford University Press (1963).

Loudenback, L.D. & D.E. Abbott: "Time-Dependent Turbulent Boundary Layers in a Nominally Steady Flow," CFMTR-73-1, Purdue University (1973).

Schraub, F.A. & S.J. Kline: "A Study of the Turbulent Boundary Layer with and without Longitudinal Pressure Gradients," MD-12, Stanford University (1965).

Schubauer, G.B.: *J. Appl. Phys*, 23, 191 (1954).

Stuart, J.T.: "The Production of Intense Shear Layers by Vortex Stretching and Convection," AGARD R-514 (1965).

York, R.E. & D.E. Abbott: "The Effects of Pressure Gradient on Velocity Profiles in Turbulent Boundary Layers," CFMTR-73-2, Purdue University (1973).

Zarić, Z.: "Statistical Analysis of Wall Turbulence Phenomena," IUTAM-IUGG Symposium, University of Virginia (1973).

The authors gratefully acknowledge the support of AFOSR under Grant No. 74-2707.

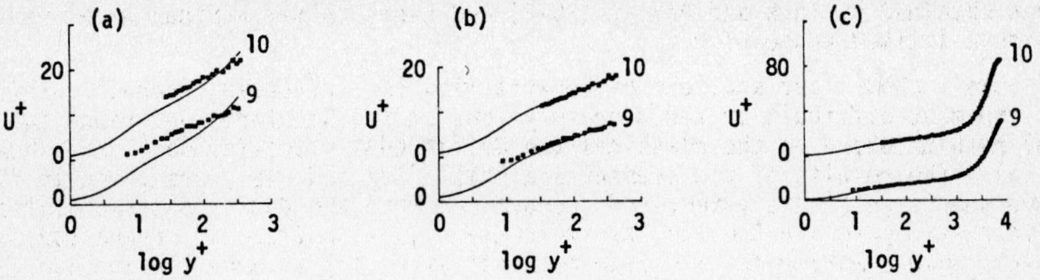

Fig. 1: Comparisons with Perry's Flow; (a) van Driest, (b) equ. (12), (c) equ. (12) with wake.

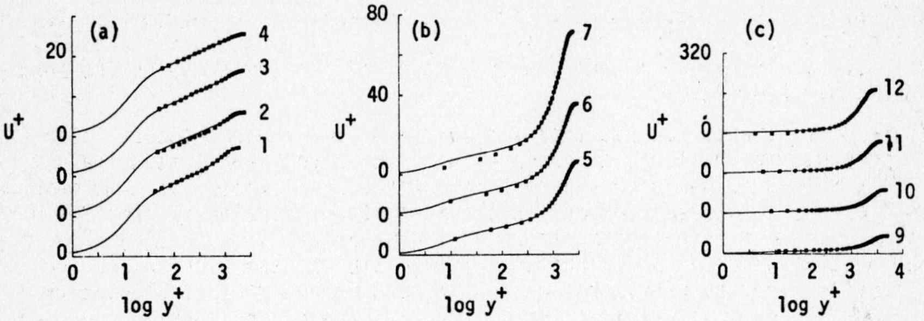

Fig. 2: Comparisons; (a) accelerated flow, (b) airfoil flow, (c) diffuser flow.

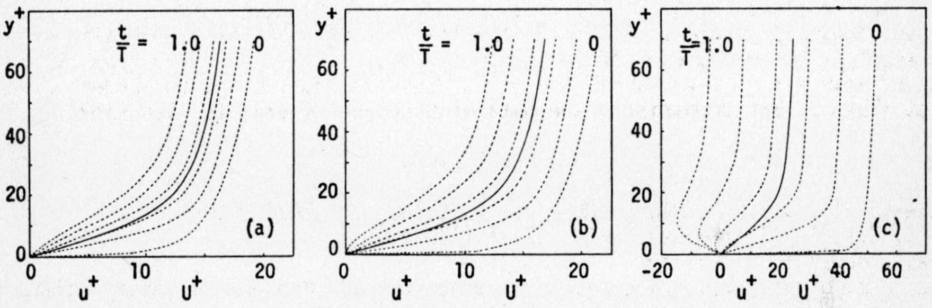

Fig. 3: Instantaneous and mean profiles; (a) $P^+=0.05$, (b) $P^+=0.079$, (c) $P^+=0.5$.

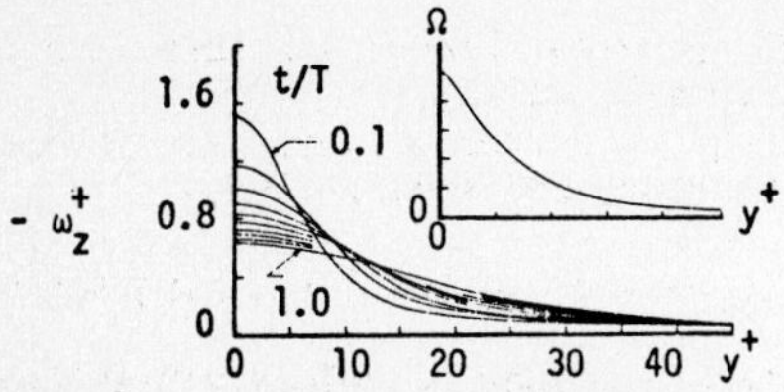

Fig. 4: Vorticity distribution.

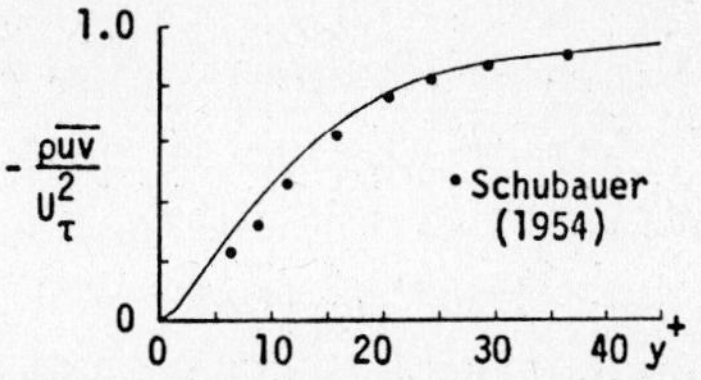

Fig. 5: Reynolds stress prediction.

NUMERICAL MODELS OF ROTATING STARS

J. F. G. Auchmuty

Department of Mathematics
Indiana University
Bloomington, Indiana 47401

1. Introduction

Rotating stars may be modelled by the equilibrium solutions of the equations for a compressible, inviscid fluid subject to self-gravitational and centrifugal forces. When the equation of state of the fluid is

$$p = f(\rho) \tag{1.1}$$

the equations are the Euler equations

$$f'(\rho) \text{ grad } \rho = \rho\left[\text{grad }(B\rho) + \frac{L(m_\rho(r))}{r^3}\hat{i}_r\right]. \tag{1.2}$$

Here we are using cylindrical polar coordinates (r,θ,z); $\hat{i}_r$ is the unit vector in the radial direction and ρ is the density.

The star is assumed to be rotating about the z-axis and L represents the square of the angular momentum per unit mass. L is a given function of some Lagrangian variable for the fluid; for our work this variable is the proportion of the mass inside a cylinder of radius r

$$m_\rho(r) = \frac{1}{M}\int_0^r\int_{-\infty}^{\infty}\int_0^{2\pi} s\rho(s,\theta,z)\,ds\,d\theta\,dz \tag{1.3}$$

with M being the total mass of the star.

Finally $B\rho$ represents the gravitational potential

$$B\rho(x) = \int_{\mathbb{R}^3} \frac{\rho(y)}{|x-y|}\,dy. \tag{1.4}$$

The usual problem is to find a density distribution ρ which satisfies (1.2), given the equation of state (1.1), the angular momentum law $L(m)$ and the mass M of the star.

In the non-rotating case $(L \equiv 0)$, equation (1.2) may be reduced to an ordinary differential equation and the solutions of those equations have been studied in great detail, see, for example, Chandrasekhar [4], chapters 3 and 4.

Even for non-rotating stars, there need not always be well-defined solutions of (1.2) of given mass. For white dwarf stars whose equation of state is given parametrically by

$$\rho = Au(2u^2 - 3)(u^2 + 1)^{\frac{1}{2}} + 3\sinh^{-1}u \qquad (1.5)$$
$$\rho = Bu^3$$

there are solutions only when the mass M is less than the Chandrasekhar mass M_3.

For rotating stars one can show that the solutions of (1.2) are given by the local extrema of the functional

$$E(\rho) = \int_{\mathbb{R}^3} A(\rho(x))dx + \frac{1}{2}\int_{\mathbb{R}^3} \rho(x)\,\frac{L(m_\rho(r))}{r^2}\,dx - \frac{1}{2}\int_{\mathbb{R}^3} \rho(x)B\rho(x)dx \qquad (1.6)$$

subject to the conditions

$$\rho(x) \geq 0 \qquad (1.7)$$

and

$$\int_{\mathbb{R}^3} \rho(x)dx = M \qquad (1.8)$$

Here

$$A(s) = s\int_0^s f(u)u^{-2}du .$$

The class of allowable functions ρ are those real-valued functions which are measurable and axially symmetric in $\mathbb{R}^3$ and obey (1.7), (1.8) and

$$\int_{\mathbb{R}^3} A(\rho(x))dx < \infty .$$

In Auchmuty and Beals [1] it was shown that there are solutions of (1.2) which minimize the functional E on the class of allowable density distributions, for any mass M, provided f and L obey the following hypotheses.

P_1: $f(s)$ is nonnegative, continuous and strictly increasing for $s > 0$.

P_2: $f(s) \geq cs^\gamma$ for $s \geq 1$ and for some constants $c > 0$ and $\gamma > \frac{6}{5}$.

P_3: $\int_0^1 f(s)s^{-2}ds < \infty$.

P_4: $L(m)$ is non-negative and absolutely continuous for $0 \leq m \leq 1$ and $L(0) = 0$.

P_5: $\lim_{s\to 0} f(s)s^{-4/3} = 0$ and $\lim_{s\to\infty} f(s)s^{-4/3} = +\infty$.

When P_5 is replaced by the condition

$$P_5': \lim_{s\to 0} f(s)s^{-4/3} = 0 \quad \text{and} \quad \lim_{s\to\infty} f(s)s^{-4/3} = K > 0$$

there is a constant M_0 depending only on K, such that the functional E has a minimum on the class of allowable functions if $M < M_0$. For white dwarf stars P_5' holds rather than P_5, and then M_0 is the same as the Chandrasekhar mass M_3.

When the functional does have a minimum on the class of allowable functions, the minimizing function is continuously differentiable with compact support and is a solution of (1.2). These results are described in [1] and [2].

The conditions on the growth of the equation of state are sharp. When the star obeys the polytropic equation of state

$$p = K\rho^{\gamma} \tag{1.9}$$

one can show that the functional is unbounded below if $\gamma < 4/3$, OR if $\gamma = 4/3$ and $M > M_0$.

These results are related to classical stability conditions for stars, as if the star is assumed to be in adiabatic equilibrium (1.6) is the energy of the star.

2. Numerical Methods

There are many difficulties in trying to solve equation (1.2) directly. The equation is a non-linear integro-differential equation. Moreover it is a free boundary problem; and some of the most interesting questions about the solutions of (1.6) are those about the boundary or "What is the shape of a rotating star?"

Problems of this type were first tackled in the time of Maclaurin and many methods have been developed to treat special cases of these problems. Recently, perturbation methods and methods of successive approximation have been especially popular. For historical references and details and comparisons of a number of recent methods see Ostriker and Mark [5] or Papaloizou and Whelan [6].

Here, a method different to all of these will be described. It is based on discretizing the functional E and the constraint (1.8) and minimizing the resulting finite dimensional function.

Suppose the solution of (1.2) is contained in a region Ω in $\mathbb{R}^3$. Let $\Omega = \Omega_1 \times [0,2\pi]$ in cylindrical polar coordinates. Then assume an approximate integration formula

$$I_\Gamma(u) = \sum_{i=0}^{K} \sum_{j=0}^{N} \omega_{ij} u(r_i, z_j) \ . \tag{2.1}$$

which is an approximation to the integral

$$I(u) = \int_\Omega u(x)dx = 2\pi \iint ru(r,z)drdz \tag{2.2}$$

$\Gamma = \{(r_i, z_j) : 0 \le i \le K, \ 0 \le j \le N\}$ is a grid defined on Ω_1.

The discrete variational problem is to minimize

$$E_\Gamma(v) = \sum_{i=0}^{K} \sum_{j=0}^{N} \left[\omega_{ij} A(v_{ij}) - \frac{1}{2} \sum_{k=0}^{K} \sum_{\ell=0}^{N} b_{ijk\ell} v_{ij} v_{k\ell}\right]$$

$$+ \frac{1}{2} \sum_{i=1}^{K} \frac{L(m_i)}{r_i^2} \left(\sum_{j=0}^{N} \omega_{ij} v_{ij}\right) \tag{2.3}$$

subject to

$$\sum_{i=0}^{K} \sum_{j=0}^{N} \omega_{ij} v_{ij} = \frac{M}{2\pi} \tag{2.4}$$

and

$$v_{ij} \ge 0 \ . \tag{2.5}$$

Here $m_0 = 0$ and $m_i = \frac{2\pi}{M} \sum_{k=0}^{i} \sum_{j=0}^{N} \omega_{kj} v_{kj}$, $1 \le i \le K$.

The array $b_{ijk\ell}$ is obtained by discretizing the integral

$$\iint_{\Omega\Omega} B_h(x,y) u(x) u(y) dxdy$$

where $$B_h(x,y) = \begin{cases} |x-y|^{-1} & \text{if } |x-y| \ge h \\ h^{-1} & \text{if } |x-y| \le h \end{cases}$$

and h is the mesh of the grid.

The function E_Γ is a continuously differentiable function defined on a closed compact subset of $\mathbb{R}^{KN}$, so it always has a minimum. To find this minimum we use a projection gradient method, that preserves the constraints (2.4) - (2.5).

This method is described in detail in [3] and there it is shown that if:

(i) the original problem has a global minimum

and (ii) one choses a sequence of grids Γ_m and associated integration formulae I_{Γ_m} such that

$$I(u) = \lim_{m\to\infty} I_{\Gamma_m}(u)$$

for all $u \in C(\Omega)$,

then the solutions of the discrete problem converge pointwise to a solution of the original problem.

Details of extensive computations using this method will appear later.

References

[1] Auchmuty, J. F. G. and Beals, R., Models of Rotating Stars, Astrophysical J., 165(1971), L79-82.

[2] Auchmuty, J. F. G. and Beals, R., Variational Solutions of some Nonlinear Free Boundary Problems, Arch. Rat. Mech. & Anal., 43 (1971), pp. 255-271.

[3] Auchmuty, J. F. G., The Numerical Solution of the Equations for Rotating Stars, to appear in Springer Verlag Lecture Notes.

[4] Chandrasekhar, S., Introduction to the Study of Stellar Structure, U. of Chicago Press, Chicago (1939).

[5] Ostriker, J. P. and Mark, J. W. K., Rapidly Rotating Stars I. The Self-Consistent Field Method, Astrophysical J. 151(1968), pp. 1075-1088.

[6] Papaloizou, J. C. B. and Whelan, J. A. J., The Structure of Rotating Stars: The J^2 method and results for Uniform Rotation, Mon. Not. R. Astr. Soc., 164(1973), pp. 1-10.

NAVIER-STOKES SOLUTIONS USING A FINITE ELEMENT ALGORITHM

A.J. Baker
Principal Scientist
Bell Aerospace Division of Textron
Buffalo, New York, USA

ABSTRACT

A finite element numerical algorithm is established for the transformed-transient, and parabolic three-dimensional Navier-Stokes equations governing turbulent flow of a variable viscosity constant density fluid. The solution algorithm is established for the characteristic, uniformly parabolic equation using the Galerkin criterion on a local basis within the Method of Weighted Residuals. The algorithm contains no requirement for computational mesh or solution domain closure regularity. General boundary condition constraints for all variables are piecewise enforceable on domain closure segments arbitrarily oriented with respect to a global reference frame. Numerical solutions verify the method for diverse problems including recirculation.

INTRODUCTION

A versatile numerical prediction capability for geometrically complex three-dimensional incompressible flows could be highly instrumental in design evaluation. The establishment of such a capability must include ready acceptance of complexly non-regular solution domain shapes to be truly useful. In addition, since laminar flows are of rather limited interest, the developed equation system must allow for quite general transport phenomena. The problem class is assumed governed by the transient, three-dimensional Navier-Stokes equations, solution of the full form of which is rarely attempted or required. Two simplifications are established for these complete equations that render numerical solution considerably more tractable while retaining the turbulent and three-dimensional description. The developed parabolic Navier-Stokes equations govern three-dimensional, generally unidirectional steady flows. Alternatively, an analytical integral transformation is applied to the general transient equation system to establish a form that can include regions of recirculation. In the derivation of both equation systems, a second order tensor turbulent transport mechanism is assumed, and mass and momentum transport coefficients may be distinct. The finite element solution algorithm is established for each equation system, including utilization of non-uniform computational meshes and direct acceptance of non-regular solution domain closures upon which a generalized boundary value description is readily applied. The results of numerical evaluation of mass dispersion, using both developed equation systems, is discussed with respect to governing turbulence and geometric factors. Further evaluations for prediction of recirculation regions and stream confluence are also discussed.

DIFFERENTIAL EQUATION SYSTEMS

Employing Cartesian tensor notation with summation on repeated indices (unless underscored), the transient Navier-Stokes equations for isoenergetic incompressible flow are:

$$0 = u_{k,k} \tag{1}$$

$$u_{j,t} = -\left[u_k u_j - \frac{p}{\rho}\delta_{jk} - \frac{1}{Re}\tau_{jk}\right]_{,k} + \frac{1}{Fr}B_j \tag{2}$$

$$Y_{,t} = -\left[u_k Y - \frac{1}{Sc}\nu^e_{jk}\, Y_{,j}\right]_{,k} + S \tag{3}$$

The non-dimensional variables have their usual interpretation with u_i the velocity vector, p the pressure, ρ the constant density, Y the mass fraction with distributed source S, B_j the body force modified by the Froude number, Re the Reynolds number, and τ_{jk} the stress tensor, assumed of the general from

$$\tau_{jk} = \nu^e_{ij}\,(u_{i,k} + u_{k,i}) \tag{4}$$

The effective kinematic diffusion coefficient, ν^e_{ij}, is assumed a linear function of a laminar (ν) and turbulent (ϵ_{ij}) contribution of the form:

$$\nu^e_{ij} \equiv \nu \delta_{ij} + \epsilon_{ij} \tag{5}$$

Note that combining Eqs. (4) - (5) for laminar flow yields Stoke's viscosity law.

Two simplifications to Equations (1) - (5) are established for solution using the finite element algorithm. For steady, unidirectional flows wherein diffusional effects parallel to the predominant flow direction are negligible, the parabolic Navier-Stokes approximation is valid. Selecting the x_1 coordinate as aligned with the general flow, and constraining the summation over repeated indices to 2 and 3, for channel flow with gravity and Coriolis body forces, the parabolic Navier-Stokes system becomes

$$u_1 u_{j,1} = \left[\frac{\nu^e_{ik}}{Re}(u_{j,i} + u_{i,j})\right]_{,k} - u_k u_{j,k} - \frac{p_{,j}}{\rho} + \frac{1}{Fr}\left[\epsilon_{jik} f_i u_k - g_j\right] \tag{6}$$

$$u_1 Y_{,1} = \left[\frac{\nu^e_{ik}}{Sc} Y_{,i}\right]_{,k} - u_k Y_{,k} + S \tag{7}$$

$$\eta(x_1, x_2) = f(u_{k,k}) \tag{8}$$

In Equation (8), $\eta(x_1, x_2)$ is the equation of the planar free surface, the location of which is established as a function of Equation (1) (Baker & Zelazny, 1974a). The second simplification, applicable to transient three-dimensional flow fields with a periodic or bounded dependence in one dimension (assumed to be x_3), is established through integral transformation of Equations (1) - (5). Integrate Equation (1) between the stream surfaces $x_3 = \xi(x_1, x_2)$ and $x_3 = \eta(x_1, x_2)$; limiting summation on repeated indices to 1 and 2, obtain for Equation (1)

$$\int_\xi^\eta u_{k,k}\, dx_3 = u_3(\xi) - u_3(\eta) \tag{9}$$

Define the integral transformation of a generalized dependent variable q, as

$$Q \equiv \int_\xi^\eta q\, dx_3 \tag{10}$$

Upon noting that $u_3(\xi)$ and $u_3(\eta)$ are obtained as the time derivatives of the surfaces ξ and η, respectively, and employing Leibniz' theorem, Equations (9) and (10) can be combined to yield

$$U_{1,1} + U_{2,2} = U_{k,k} = 0 \tag{11}$$

In Equation (11), the three-dimensional continuity equation has been transformed to an equivalent two-dimensional form written on the integral transformation of the velocity vector u_i, i.e., $q = u_i$ in Equation (10). Establishment of Equation (11) allows specification of the velocity field as the curl of a single scalar component of a vector potential function of the form (Baker, 1973)

$$U_i = \epsilon_{3ik} \Psi_{,k} \tag{12}$$

where Ψ is the integral transform equivalent of the conventional two-dimensional streamfunction, and ϵ_{3ik} is the Cartesian alternating tensor. The vanishing divergence of U_i is thus ensured; specification of its curl completes the required description. After identifying the x_3 integral transform component of vorticity, Ω, as

$$\Omega \equiv \epsilon_{3ki} U_{i,k} \tag{13}$$

the curl of Equation (12) takes the form (Baker 1973, Baker & Zelazny 1974a)

$$\Omega_{,t} = \epsilon_{3ik}\left[(\Omega\Psi_{,i})_{,k} + g\eta_{,i}\xi_{,k} - \frac{1}{Fr}\Psi_{,k} f_i\right]$$
$$+ \frac{1}{Re}\left[(\nu^e_{k\ell}\Omega_{,\ell} + \nu^e_{k\ell,j}\Psi_{,j\ell})_{,k} - \nu_{,ik}(\Omega\delta_{ik} + \Psi_{,ik})\right]$$
$$+ \frac{\epsilon_{3ik}}{Re}(\nu^e_{j3} U_{k,j})\Big|_\xi^\eta \tag{14}$$

For constant viscosity, laminar flows free from body and surface (ξ, η) forces, Equation (14) is similar to the familiar two-dimensional vorticity transport equation. Problem specification is completed by combining Equations (12) - (13) to define the streamfunction-vorticity compatibility equation

$$-\Omega = \Psi,_{kk} \tag{15}$$

Finally, Equation (3) in the integral transform computational variables becomes

$$Y,_t = \epsilon_{3ik} (Y\Psi,_i),_k + S + \left[\frac{\nu^e_{k\ell}}{Sc} Y,_k\right],_\ell + \frac{\nu^e_{i3}}{Sc} Y,_i \Big|_\xi^\eta \tag{16}$$

FINITE ELEMENT SOLUTION ALGORITHM

The derived three-dimensional equation systems are uniformly initial-boundary value problems of mathematical physics. Each equation, Equation (6)-(8) and (14)- (16) is a special case of the general second-order, nonlinear partial differential equation

$$L(q) \equiv \kappa [K_{jk} (q) q,_k],_j + f(q,q,_i,x_i) - g(q, \chi) = 0 \tag{17}$$

where q is a generalized dependent variable identifiable with each computational dependent variable. In Equation (17), f and g are specified functions of their arguments, χ is identified with x_1 or t, and x_i are the coordinates for which second order derivatives exist in the lead term. The finite element solution algorithm assumes that L(q) is uniformly parabolic within a bounded open domain Σ, i.e., the lead term in Equation (17) is uniformly elliptic within its domain R, with closure ∂R, where

$$\Sigma = R \times [\chi_o, \chi) \tag{18}$$

and $\chi_o \leqslant \chi < \infty$. Unique solutions are obtained pending specification of boundary constraints for q on ∂R and an initial condition on $R \cup \partial R$. For the former, the general form is

$$\ell(q) \equiv a^{(1)} q(\bar{x}_i, \chi) + a^{(2)} K_{jk} q (\bar{x}_i, \chi),_j n_k - a^{(3)} = 0 \tag{19}$$

In Equation (19), the $a^{(i)}$ ($\bar{x}_i$, χ) are specified coefficients, the superscript bar constrains x_i to ∂R, and n_k is the local outward-pointing unit normal vector. For an initial distribution, assume given throughout $R \cup \partial R \times \chi_o$,

$$q(x_i, \chi_o) \equiv q_o (x_i) \tag{20}$$

The finite element solution is obtained for the equation system (17) - (20) using the Method of Weighted Residuals (MWR) formulated on a local basis. Since Equation (17) applies throughout R, it is valid within disjoint interior subdomains, R_m, described by $(x_i, \chi) \epsilon R_m \times [\chi_o, \chi)$ called "finite elements," wherein $\cup R_m = R$. Form an approximate solution for q within $R_m \times [\chi_o, \chi)$, called $q^*_m (x_i, \chi)$, by expansion into a series solution of the form

$$q^*_m (x_i, \chi) \equiv \{\phi(x_i)\}^T \{Q(\chi)\}_m \tag{21}$$

wherein the functionals $\phi_k (x_i)$ are members of a function set complete in R_m. The unknown expansion coefficients, $Q_k (\chi)$, represent the χ-dependent values of $q^*_m (x_i, \chi)$ at specific locations interior to R_m and on the closure, ∂R_m, called "nodes." Establish the values of the expansion coefficients by requiring that the local error in the approximate solution to $L(q^*_m)$, and $\ell(q^*_m)$, for $\partial R_m \cap \partial R$, be rendered orthogonal to the space of the approximation functions, yielding

$$\int_{R_m} \{\phi(x_i)\} L(q^*_m) d\tau - \lambda \int_{\partial R_m \cap \partial R} \{\phi (x_i)\} \ell (q^*_m) d\sigma \equiv 0 \tag{22}$$

The lead term in Equation (17) is rearranged using the Green-Gauss Theorem. For $\partial R \cap \partial R_m$ non-vanishing, the generated segment of the closed surface integral cancels the boundary condition contribution, Equation (22), by identifying $\lambda a(2)$ with κ of Equation (17). The contributions to the closed surface integral, Equation (23), where $\partial R_m \cap \partial R = 0$ can be made to vanish (Baker, 1973). The assembled finite element solution algorithm for the representative partial differential equation becomes

$$\cup \left[-\kappa \int_{R_m} \{\phi\}_{,k} K_{jk} q^*_{m,j} \, d\tau + \int_{R_m} \{\phi\} (f^*_m - g^*_m) d\tau - \kappa \int_{\partial R_m \cap \partial R} \{\phi\} (a_m^{(1)} q^*_m - a_m^{(3)}) \, d\sigma \right] = \{0\} \tag{23}$$

The rank of the global system is identical to the total number of node points on $R \cup \partial R$ for which the dependent variable requires solution. Equation (23) is either a first-order, ordinary differential or algebraic equation system, and the matrix structure is sparse and banded. Bandwidth is a function of both selected discretization and the order of the employed approximation functional. Solution of the ordinary differential system is obtained using standard numerical integration procedures. Equation solver techniques are employed for the algebraic systems. Using linear approximation functions, Equation (21), analytical evaluation of all matrix forms, Equation (23), is directly obtained (Baker 1973, Baker & Zelazny, 1974 a, b).

NUMERICAL RESULTS

The finite element solution algorithm for both developed equation systems is operational in the COMOC computer program (Orzechowski, et al 1973, 1974), which is also capable of solution for compressible flows. A degenerate test case for the parabolic Navier-Stokes equations is two-dimensional, laminar boundary layer flow. In this case, Equation (8) is solved directly for transverse velocity, u_2. Shown in Figure 1 is computed skin friction and boundary layer thickness for M = 0.27, in comparison to the Blasius solution (Schlichting, 1960). The algorithm displays excellent long-term solution stability, solution accuracy is within 2% using coarse discretizations, and convergence with discretization refinement is numerically verified.

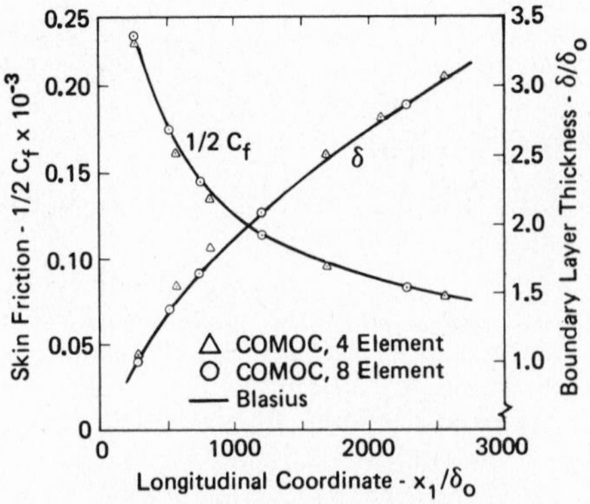

Figure 1. Boundary Layer Solution,
M = 0.272, Re = 0.63 (7)/m

Shown in Figure 2 are the results of a study on three-dimensional, turbulent waste water dispersion in a straight section of a natural waterway. Figure 2a illustrates the measured predominant-flow isovel and bottom contour distributions (Fischer, 1968). Figure 2b presents the 468 finite element discretization of the solution domain as well as the ejector location. A tensor turbulent viscosity law, Equation (5), was assumed applicable, of the form

$$\nu^e_{1j} = k_j U^* h \qquad j = 2, 3 \tag{24}$$

with principal axes parallel and perpendicular to the free surface. Values of the k_j were obtained from Fischer (1973), U^* is a friction velocity evaluated in terms of shear on the riverbed (Baker & Zelazny, 1974a), and h is local river depth. Figures 2c-d show computed mass fraction contours at 9.6 m downstream of injection as a function of k_j, for $k_2/k_3 = 3.44$ and $k_2/k_3 = 1.0$, respectively (x_3 perpendicular to free surface). Neglect of the tensor character (Figure 2d) significantly alters the predicted contours; most noticeably, the 3% contour has broadly intersected the free surface, as has the 30% contour on the riverbed. The sole difference in these evaluations was the tensor character of the turbulence model; the results amply illustrate its importance.

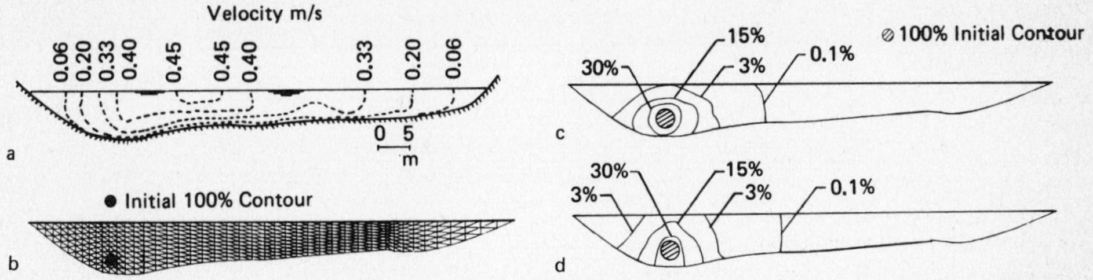

Figure 2. Predicted Mass Fraction Contours at 9.6 m Downstream of Injection, Two Diffusion Models

Not only turbulence model character, but geometrical constraints inducing large flow curvatures can influence mass transport. Figure 3a is a model of a meandering natural waterway (Fischer, 1969) assumed the outlet from a large reservoir. Solution of Equations (14)-(16) is required for this geometry due to the curvature effects; Figure 3b shows the 198 finite element discretization employed for analysis. Figure 3c shows the computed potential flow solution; the uniformly parallel equi-flow streamline distribution indicates adequate solution accuracy using this discretization. Figure 3d contains the computed steady, viscous equi-flow streamline distribution for a scalar turbulence model, Equation (5), equal in magnitude to the average of measured transverse coefficients in natural streams (Fischer, 1969). Viscous drag on the river banks have retarded the near vicinity flow; the computed streamwise velocity near the inside of a bend is about 50% larger than at the outside (Baker & Zelazny, 1974). Shown in Figure 3e is the steady-state trajectory of the maximum concentration line of a pollutant injected at the center of the reservoir outlet with an initial streamwise momentum equal to the main flow. The computed departure of the trajectory from the middle of the channel is strictly a function of the solution domain closure curvatures, as reflected in the complete Navier-Stokes solutions.

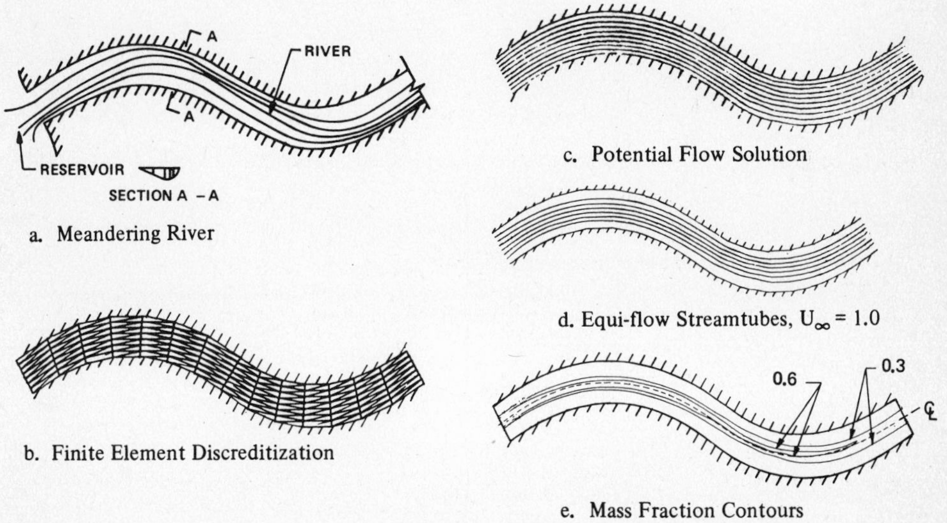

Figure 3. Flow Computations in a Meandering River

Stream confluence, and the attendant recirculation regions generated within the flow field is another solution category of interest. Shown in Figure 4 are computed streamline patterns for the classic step-wall diffuser problem for Re = 200. Figure 4a is a close-up of the corner region showing details of computed dividing streamline attachment at the wall. Note the secondary zone computed to occur at the step base. Figure 4b-c duplicate the standard case with 10% total mass flow removal and addition, respectively, at the step base. The new location of the streamlines, dividing the various recirculation zones, are indicated, and their overall appearance and wall attachment points are significantly altered. The algorithm requires no special handling to promote this behavior (Baker, 1973), and recirculation zones are computed without need for prior user interpretation.

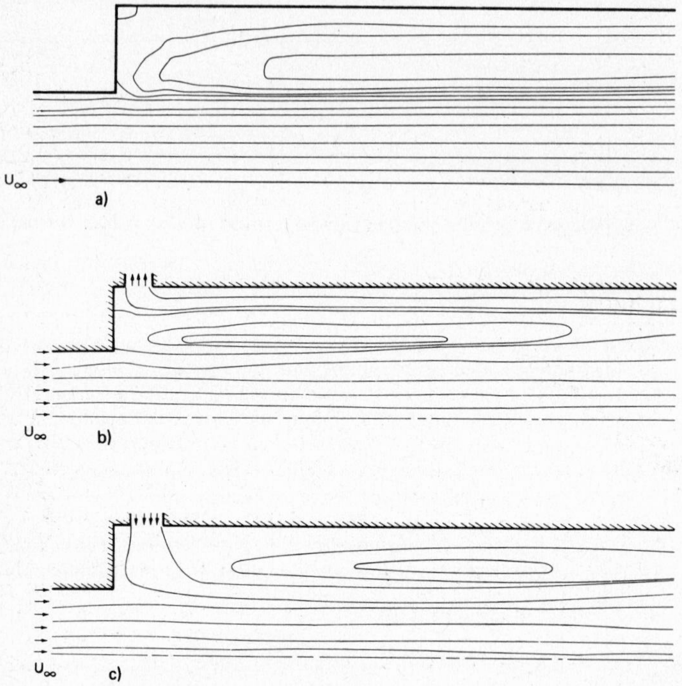

Figure 4. Flow in a Step-Wall Diffusor with Mass Addition and Removal at the Base, Re = 200

SUMMARY

A finite element solution algorithm is developed for two problem classes within the three-dimensional incompressible Navier-Stokes equations. Numerical solutions illustrate solution accuracy, versatility, and adaptiveness for a wide range of practical flows in hydrodynamics.

REFERENCES

Baker, A. J., NASA CR-2391 (1973).
Baker, A. J. and Zelazny, S. W., AIAA paper 74-7 (1974a).
Baker, A. J. and Zelazny, S. W., NASA CR- 132450 (1974b).
Fischer, H. B., Water Resources Research 5, 496-506 (1969).
Fischer, H. B., J. Flu. Mech., 53, 671-687 (1972).
Fischer, H. B., Annual Rev. Flu. Mech., 5, 59-78 (1973).
Orzechowski, J. A., Manhardt, P. D., Baker, A. J., Bell Aerospace Report D9198-954001, (1973).
Orzechowski, J. A., and Baker, A. J., NASA CR- 132449 (1974).
Schlichting, H., Boundary Layer Theory, McGraw-Hill, New York (1960).

INTERACTION OF STRONG SHOCK WAVE WITH TURBULENT BOUNDARY LAYER

By B. S. Baldwin and R. W. MacCormack

Ames Research Center, NASA, Moffett Field, Calif. 94035

INTRODUCTION

Continuing advances in numerical methods and computer capabilities now permit calculation of formerly intractable fluid flows. In the recent past, time-dependent solution of separated laminar flows based on the complete Navier-Stokes equations became feasible (MacCormack, 1971; Carter, 1973). In the present paper an extension has been made to include turbulence models in conjunction with the compressible flow equations. The turbulence transport model relations used (Saffman and Wilcox, 1974) pose a more severe numerical resolution requirement than the Navier-Stokes equations themselves. To counter this difficulty, a modification of MacCormack's (1971) numerical method is presented that produces exponential as well as second-order accuracy.

FLOW FIELD

The problem under investigation is depicted in figure 1. Air flowing from left to right forms a turbulent boundary layer on the flat plate. A shock wave incident on the boundary layer produces a separation bubble within which there is reversed flow. The experiment used for comparison is at Mach number 8.47 and Reynolds number 22.5×10^6 (Holden, 1972). The pressure rises by a factor of 83 across incident and reflected shock waves. The computations reported in this paper are confined to the rectangle BFGC, which encompasses the interaction region. The computational mesh employed is shown in figure 2. The flow field is divided into four regions with uniform spacing in each. Predictions of skin friction and heat transfer on the flat plate are made at two levels of approximation: a simple mixing-length model and a two-equation transport model of turbulence. Details of the turbulence models employed are given elsewhere (Baldwin and MacCormack, 1974). The calculations start with a uniform flow except for values imposed along the upstream and outer boundaries where they are determined from boundary-layer and oblique shock-wave theory. The turbulent boundary layer and incident shock wave grow with time and eventually a steady-state solution is achieved.

NUMERICAL METHOD

Notation

The mean flow equations and turbulence model equations can be denoted by

$$(\partial U/\partial t) + (\partial F/\partial x) + (\partial G/\partial y) = H \tag{1}$$

where U is a column vector of conserved quantities (mass, momentum, energy, turbulent energy, etc.). The fluxes F and G are column vectors containing convection and diffusion terms. The components of the source vector H associated with the mean flow equations are zero. However, nonzero source terms occur in the turbulence model equations, for example, those representing production and dissipation of turbulent energy. Complete expressions for the components of U, F, G, and H in the present investigation are given elsewhere (Baldwin and MacCormack, 1974). The fluxes and sources are functions of auxiliary variables such as the x component of velocity u and first derivatives of these variables. In the interest of brevity, all auxiliary variables derivable from the conserved quantities U will be denoted by u in this paper.

Basic Numerical Method

The present calculations use a variation of MacCormack's (1971) time-splitting method for solution of the Navier-Stokes equations. The conserved quantity U in equation (1) is advanced by a time step Δt_x as though the $\partial G/\partial y$ and H terms were absent and then by a time step Δt_y in which the $\partial F/\partial x$ term is omitted. The source term is included with $\partial G/\partial y$ because a sensitive balance develops between the source and diffusion terms in the viscous sublayer near the wall. The finite-difference operation in each case utilizes a predictor and corrector sequence. For example, the predictor step in the advancement Δt_x can be denoted by

$$\overline{U}_{ij}(t_x+\Delta t_x,t_y) = U_{ij}(t_x,t_y) - (\Delta t_x/\Delta y)[F_{ij}(t_x,t_y) - F_{i-1,j}(t_x,t_y)] \tag{2}$$

and the corrector by

$$U_{ij}(t_x+\Delta t_x,t_y) = (1/2)\{U_{ij}(t_x,t_y) + \overline{U}_{ij}(t_x+\Delta t_x,t_y) - (\Delta t_x/\Delta x)[\overline{F}_{i+1,j}(t_x+\Delta t_x,y_y) - \overline{F}_{i,j}(t_x+\Delta t_x,t_y)]\} \tag{3}$$

The bar on $\overline{F}$ indicates that predicted quantities $\overline{U}$ are to be used in the evaluation of these fluxes.

The elements of F represent fluxes and stresses that are evaluated in such a manner as to achieve second-order accuracy after the predictor-corrector sequence is completed. For example, for the cell surface lying midway between mesh points i and i+1, the flux value u is evaluated as u_i in the predictor and as u_{i+1} in the corrector, and the stress derivative of u is evaluated as $(u_{i+1} - u_i)/\Delta x$ each time.

Let $L_x(\Delta t_x/\Delta x)$ denote the pair of operations by which $U_{ij}(t_x + \Delta t_x,t_y)$ is obtained from $U_{ij}(t_x,t_y)$ and $L_y(\Delta t_y/\Delta y)$ the analogous determination of $U_{ij}(t_x,t_y + \Delta t_y)$ from $U_{ij}(t_x,t_y)$. MacCormack (1971) has shown that although the sequence $L_x(\Delta t/\Delta x)L_y(\Delta t/\Delta y)$ is accurate only to first order in Δx and Δy, symmetrical sequences such as $L_y(\Delta t/2\Delta y)L_x(\Delta t/\Delta x)L_y(\Delta t/2\Delta y)$ are accurate to second order. Computational efficiency is enhanced by use of operator sequences of the form

$$L(\Delta t) = [L_y(\Delta t/2n\Delta y)L_x(\Delta t/n\Delta x)L_y(\Delta t/2n\Delta y)]^n \tag{4}$$

where n is an integer representing the number of operations $L_yL_xL_y$ that are to be applied in one time step Δt. The advantage can be seen by noting that when mesh Reynolds numbers are greater than 2, the maximum time step for which the calculations will be stable (and time accurate) is determined by the CFL conditions

$$(\Delta t/2n\Delta y) \leq [1/(v+c)_{max}] \qquad (\Delta t/n\Delta x) \leq [1/(u+c)_{max}]$$

In regions of coarse mesh, n is set equal to 1 and larger values of n (up to 96 in this paper) are used in fine mesh regions. Most of the computing time is then spent in the finest mesh, which constitutes a small fraction of the total number of computation points. Cumulative values of the fluxes at the last cell face are stored during operation in the innermost mesh. These values are used to obtain average fluxes to be applied during operation in the next mesh outward.

Several types of nonlinear instability were encountered in the present investigation. MacCormack (1971, 1973) previously published the remedies for some of these. The severe pressure changes associated with the strong shock wave produce an erroneous drainage of conserved quantities from cells at the foot of the pressure rise. This effect is removed by inclusion of a product fourth-order damping term in the L_x operator of the form

$$\Delta U_{ij} = C(P_{ii+1,j}-2P_{ii,j}+P_{ii-1,j})(U_{i+1,j}-2U_{ij}+U_{i-1,j})/(P_{ii-1,j}+2P_{ii,j}+P_{ii-1,j}) \tag{5}$$

where P is the static pressure. In the present calculations, this term is negligible everywhere except in small regions at the beginning and end of the pressure rise where it smooths out ripples that otherwise would cause numerical instability at the foot of the pressure rise.

The turbulence model equations are unusually stiff (Lomax, 1968) in the viscous sublayer such that the calculations are unstable with a CFL number in the L_y operator greater than 0.3. Since most of the computation time is spent in the fine mesh of the viscous sublayer, it is expedient to modify the L_y operator in that region according to the (one-step) notation

$$U(t + \Delta t) = U(t)\{\exp(\Delta tH/U) + [1 - \exp(\Delta tH/U)](1/H)(\partial G/\partial y)\} \tag{6}$$

Extension of this operator to the corresponding predictor corrector sequence as in equations (2) and (3) is straightforward, but leads to expressions too lengthy for inclusion here. Use of the resulting L_y operator allows the CFL number to be increased to 0.9, thereby decreasing the computation time by nearly a factor of 3. Strictly, for a time-accurate calculation, this method should be used only in the asymptotic approach to a steady state, which consumes nine-tenths of the computation time in the present investigation. Note that the stiffness removed by equation (8) arises when H and $\partial G/\partial y$ reach large negative nearly equal values. In that event, the exponentials become negligible compared to 1.0, and the solution is driven toward a state in which H is equal to $\partial G/\partial y$. On the other hand, if the absolute value of the argument of the exponentials becomes small compared to one, the correct behavior is also recovered.

Exponential Accuracy

The turbulence quantities (e.g., turbulent energy and eddy viscosity) vary by several orders of magnitudes in the viscous sublayer near the wall, thereby posing a severe resolution requirement when a method of second-order accuracy with few mesh points is employed. In attached boundary layers this difficulty can be overcome by appealing to the universal character of the profiles near the wall. However, for separated boundary layers it is not Known a priori the extent to which the profiles will differ from their attached counterparts. To treat this difficulty, a modification of the basic method has been found that is accurate if the variations are either second-degree polynomials or exponentials.

Suppose variations with y are of the form

$$u - A = (u_j - A)\exp\{[(y-y_j)/\Delta y]\ln[(u_{j+1}-A)/(u_j-A)]\} \tag{7}$$

where A is a constant and Δy is the (uniform) mesh spacing. Evaluation of equation (7) at $y = y_{j-1}$ leads to

$$A = (u_{j+1}u_{j-1} - u_j^2)/(u_{j+1} - 2u_j + u_{j-1}) \tag{8}$$

Violation of either of the conditions

$$|u_{j+1}-2u_j+u_{j-1}| > \varepsilon(|u_{j+1}|+|u_j|) \quad , \quad (u_{j+1}-A)(u_j-A) > \varepsilon^2(u_{j+1}^2+u_j^2) \tag{9}$$

with $\varepsilon \sim 10^{-3}$ is an indication that the variation is nearly linear and therefore the basic numerical method adequate. However, if inequalities (9) are satisfied, values of u and $\partial u/\partial y$ can be computed at cell faces between mesh points from the relations

$$u(y_j + \beta\Delta y) = A + \mathrm{sgn}(u_j - A)(u_{j+1} - A)^{\beta}(u_j - A)^{1-\beta} \tag{10}$$

$$(\partial u/\partial y)_{y=y_j+\beta\Delta y} = [u(y_j + \beta\Delta y) - A]\{\ln[(u_{j+1} - A)/(u_j - A)]/\Delta y\} \tag{11}$$

If the source term H in equation (1) is zero, as it is for the Navier-Stokes equations, use of the foregoing relations in the L_y operator will lead to solutions that are accurate if the variations are either second-degree polynomials or exponentials. When the source term is not zero as in turbulence model equations, H should be multiplied by the factor

$$\overline{Q} = (1/2)[Q(z_1) + Q(z_2)] \tag{12}$$

where

$$Q(z) = (z - 1)/\ln(z) \quad , \quad z_1 = H_{j+1/2}/H_j \quad , \quad z_2 = H_{j-1/2}/H_j \tag{13}$$

Figure 3 contains results from evaluation of a simplified equation for a turbulence model quantity (square of pseudovorticity). The simplification is applicable in a part of the viscous sublayer ahead of the interaction where temperature and density variations as well as eddy viscosity can be neglected. Coordinate stretching has been used to simplify the equations listed in the figure, but the mesh spacing corresponds to that in figure 2. It can be seen in figure 3 that the method producing exponential accuracy greatly improves the finite-difference solution even though U varies as $(1 + \eta)^{-4}$ rather than exponentially. These results were computed with a time step $\Delta\tau = 0.1$. The calculations were unstable at $\Delta\tau = 0.3$. However, use of the algorithm represented by equation (6) stabilized the calculations with $\Delta\tau = 0.3$ and produced converged solutions identical to four significant figures.

Resolution of Viscous Sublayer in Compressed Region

In the region aft of reattachment, the viscous sublayer is an order of magnitude thinner than in the boundary layer ahead of the interaction. To achieve sufficient resolution, iterative solutions of the boundary layer equations are used near the wall. Periodically an inner solution is found that matches the finite-difference solution at the third row of mesh points from the wall. The inner solution then provides values of flow variables at the second row of mesh points for use in subsequent time steps. The inner solutions are repeated often enough to retain time accuracy of the calculation. Details of this procedure are given elsewhere (Baldwin and MacCormack, 1974).

RESULTS

Figures 4 and 5 contain plots of velocity profiles from the mixing-length model showing the degree of resolution achieved. The viscous sublayer is adequately resolved by the finite-difference solution in the region ahead of the interaction and in the reversed flow region. In the compressed region aft of reattachment, however, the inner solution based on the boundary-layer approximation is needed. The inner solution extended beyond the region in which it is used $(J = 3)$ continues to match the finite-difference solution farther out.

Figure 6 illustrates computed variations of eddy viscosity according to the Saffman-Wilcox (1974) turbulence transport model. Peaks in the eddy viscosity near the separation and reattachment points will probably be suppressed in later versions of the model boundary conditions. The effect of the peaks is believed to be local. Values of eddy viscosity from the mixing-length model are well below those shown in the region aft of reattachment. This results from a failure of the mixing-length model to account for nonequilibrium levels of turbulence that are convected downstream. Additional details of the calculated variations are given elsewhere (Baldwin and MacCormack, 1974). Figure 7 compares computed skin friction and heat transfer along the flat plate with measurements of Holden (1972).

REFERENCES

1. Baldwin, B. S. and MacCormack, R. W., AIAA Paper 74-558, June 1974.

2. Carter, J. E., Lecture Notes in Physics, vol. 19, Springer-Verlag, New York, 1973, P. 69.

3. Holden, M. S., AIAA paper no. 72-74.

4. Lomax, H., NASA TN D 4703, July 1968.

5. MacCormack, R. W., Lecture Notes in Physics, vol. 8, Springer-Verlag, New York, 1971, p. 151.

6. MacCormack, R. W., "Numerical Methods for Hyperbolic Systems," Lecture Notes for Short Course on Advances in Computational Fluid Dynamics, The Univ. of Tennessee Space Institute, Tullahoma, Tenn., Dec. 10-14, 1973.

7. Saffman, P. G. and Wilcox, D. C., AIAA Journal, vol. 12, no. 4, 1974.

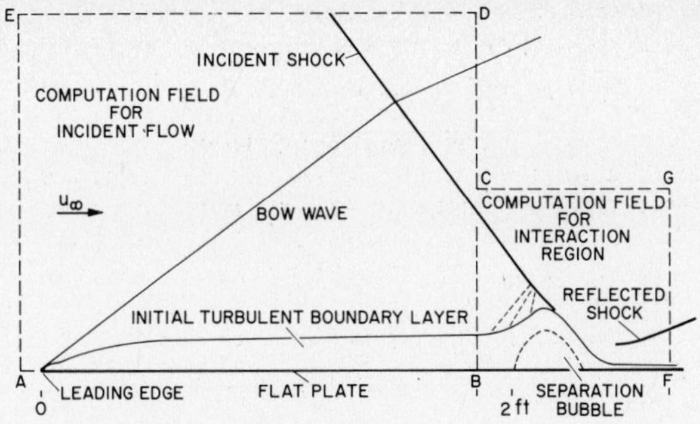

Fig. 1 Shock-induced separated turbulent flow.

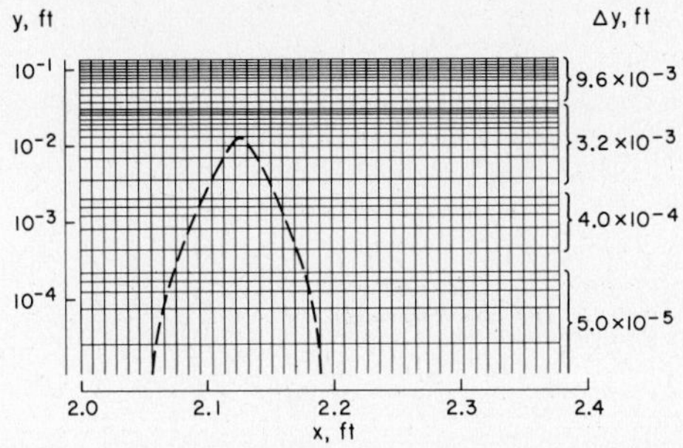

Fig. 2 Computational mesh.

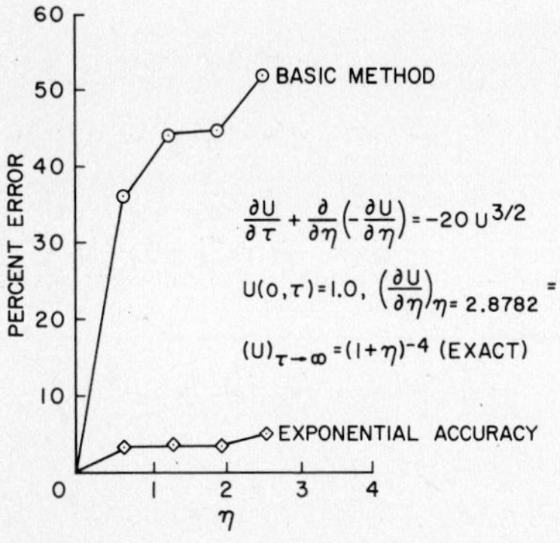

$$\frac{\partial U}{\partial \tau} + \frac{\partial}{\partial \eta}\left(-\frac{\partial U}{\partial \eta}\right) = -20\,U^{3/2}$$

$$U(0,\tau) = 1.0,\ \left(\frac{\partial U}{\partial \eta}\right)_{\eta = 2.8782} = \left(\frac{\partial U}{\partial \eta}\right)_{\tau \to \infty}$$

$$(U)_{\tau \to \infty} = (1+\eta)^{-4} \text{ (EXACT)}$$

Fig. 3 Comparison of errors in finite-difference solutions.

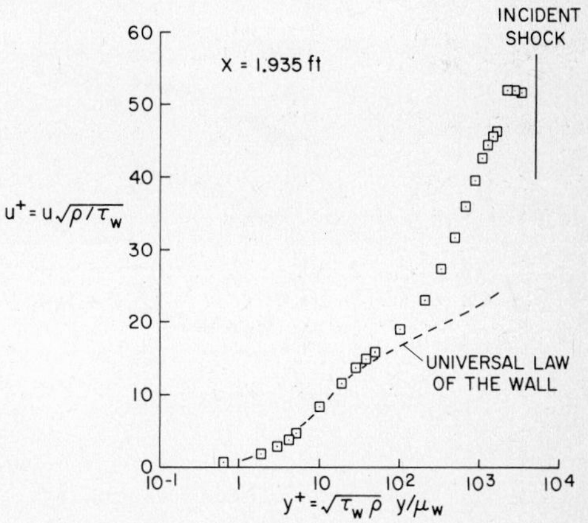

Fig. 4 Velocity profile ahead of interaction.

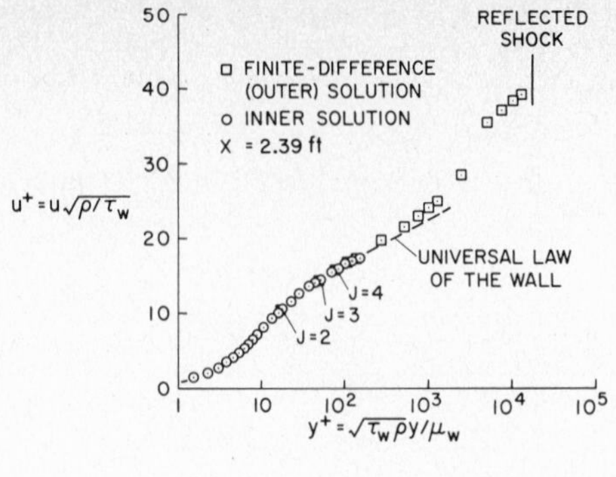

Fig. 5 Velocity profile aft of reattachment.

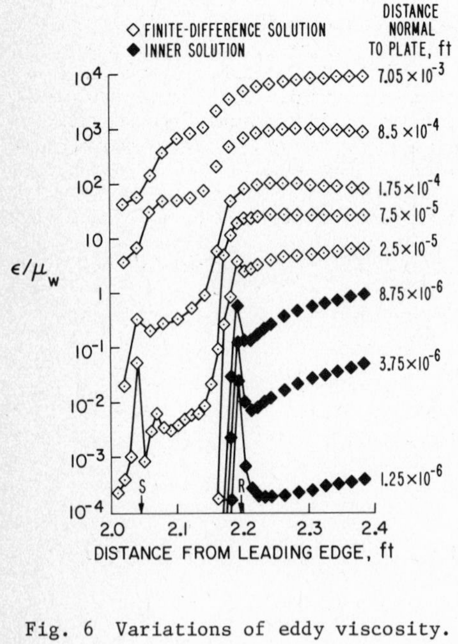

Fig. 6 Variations of eddy viscosity.

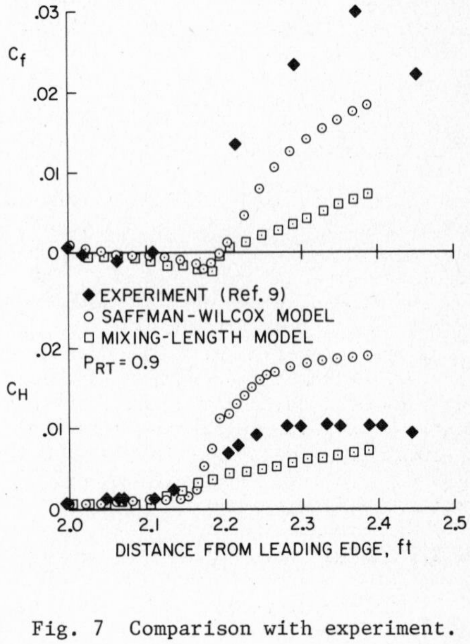

Fig. 7 Comparison with experiment.

THE NUMERICAL SIMULATION OF LOW FREQUENCY UNSTEADY TRANSONIC FLOW FIELDS

By William F. Ballhaus and Harvard Lomax

Ames Research Center, NASA
Moffett Field, Calif. 94035

INTRODUCTION

The lowest order approximation to the Euler equations for unsteady, transonic flow about a thin airfoil is

$$S_1\phi_{tt} + 2S_2\phi_{xt} = V_c\phi_{xx} + \phi_{yy} \tag{1}$$

where $V_c = S_3-(\gamma+1)M_\infty^2\phi_x-(\gamma-1)M_\infty^2K\phi_t$, ϕ = disturbance velocity potential, $S_1 = M_\infty^2(K^2/\delta^{2/3})$, $S_2 = M_\infty^2(K/\delta^{2/3})$, $S_3 = (1-M_\infty^2)/\delta^{2/3}$, M_∞ = free-stream Mach number, and δ = airfoil thickness ratio. The reduced frequency K is a measure of the degree of unsteadiness of the motion. For an airfoil of chord length c, traveling at a mean velocity U_∞ and undergoing some periodic motion of frequency ω, $K = \omega c/U_\infty$. Equation (1) is derived under the assumption $\delta^{2/3} \sim 1 - M_\infty^2 << 1$. ϕ, t, y, and x have been scaled by $c\ \delta^{2/3}U_\infty$, $1/\omega$, $c/\delta^{1/3}$, and c, respectively. Contained in (1) is a relationship for the jumps in the derivatives of the velocity potential across shock waves. This relationship is the lowest order approximation to the Rankine-Hugoniot shock relations.

The traces of the characteristics in the plane of the airfoil for the linearized form of (1) are shown in figure 1. Waves that advance in the flow direction propagate downstream with velocity $(M_\infty+1)/M_\infty K$ in the scaled coordinate system. Waves that recede (relative to particle paths) travel upstream against the flow with velocity $(M_\infty-1)/M_\infty K$, and fluid particles travel with velocity $1/K$. For low-frequency transonic flows, $1-M_\infty^2 \sim K \sim \delta^{2/3} << 1$. So disturbances traveling along the advancing waves propagate rapidly away from the airfoil, while those traveling along the receding waves remain closer to the airfoil and dominate the solution there. Hence, any numerical scheme designed to treat low-frequency, transonic flows should resolve these disturbances as accurately as possible.

A standard explicit difference scheme applied to either (1) or the Euler equations would have a stability restriction on the time step of the form $\Delta t \leq \min\ [1/(u+a),\ 1/|u-a|,\ 1/u]\Delta x$, where u and a are the streamwise velocity and speed of sound. In practice $\Delta t \leq \min\ [1/(u+a)]\Delta x$, since for transonic flows $u \sim a$. Hence, the time step depends on the propagation speed of the advancing waves. Basing Δt on the advancing rather than the receding waves typically reduces the allowable marching speed by an order of magnitude and can adversely affect the resolution of disturbances traveling along the receding waves. Two approaches are presented here that shift the dependence of Δt from the advancing waves to the receding waves. The first consists of a semi-implicit difference scheme applied to a simplified governing equation that is a suitable modification of (1) for low-frequency motion. The second retains (1) as the governing equation and uses a semi-implicit difference scheme.

DIFFERENCING OF THE LOW-FREQUENCY APPROXIMATION TO THE SMALL-DISTURBANCE EQUATION

For low-frequency, transonic flows, (1) reduces to

$$2S_2\phi_{xt} = V_c\phi_{xx} + \phi_{yy} \tag{2}$$

where $V_c = S_3 - (\gamma + 1)M_\infty^2\phi_x$. The traces of the linear characteristics of (2) in the plane of the airfoil are shown in figure 2. In this approximation, the scaled speed of sound and particle convection speed both are $O(1/K) \to \infty$. Hence, the advancing waves and particle paths in figure 2 are coincident with the x axis. The receding waves have a finite propagation speed $(M_\infty^2 - 1)/2KM_\infty^2$.

The difference scheme, shown in figure 3(a), is

$$\left[2S_2\,\frac{(D_-)_x}{\Delta t} - r(D_+D_-)_y\right]\phi^n_{j,k} = \left[2S_2\,\frac{(D_-)_x}{\Delta t} + (1-r)(D_+D_-)_y + V_c(D_+D_-)_x\right]\phi^{n-1}_{j,k} \quad (3)$$

for subsonic points, i.e., where $V_c > 0$. For supersonic points, where $V_c < 0$, the last term on the right-hand side is replaced by $V_c(D_-D_-)_x$. Here r is the fraction of the term ϕ_{yy} in (2) that is evaluated at time level n. $(D_-)_x\phi^n_{j,k} \equiv (\phi^n_{j,k} - \phi^n_{j-1,k})/\Delta x$, $(D_+D_-)_x \equiv (\phi^n_{j+1,k} - 2\phi^n_{j,k} + \phi^n_{j-1,k})/\Delta x^2$, etc. At the first mesh point downstream of a supersonic-to-subsonic shock, where the sign of V_c changes from negative to positive, Murman's shock-point operator is used to ensure that the proper shock relations are satisfied in the limit as $\Delta x, \Delta y, \Delta t \to 0$ (Murman, 1973). Equation (3) reduces to the Murman-Cole mixed-difference scheme for steady flows (Murman and Cole, 1971).

The modified partial differential equation (MPDE) for (3) is

$$2S_2\phi_{xt} - V_c\phi_{xx} - \phi_{yy} = [S_2\Delta x - \tfrac{1}{2}V_c\Delta t]\phi_{xxt} + (r - \tfrac{1}{2})\Delta t\phi_{yyt} + \text{higher order terms} \quad (4)$$

This is stable for

$$\Delta t \le \min_{j,k}\,(2S_2\Delta x/|V_{c_{j,k}}|) \qquad r \ge \tfrac{1}{2} \quad (5)$$

Taking the equality in both expressions in (5) minimizes the order of the truncation error in (4). With $r = 1/2$ the scheme is implicit in y, and (3) forms a tridiagonal matrix equation, which is solved directly for each line in the y direction, i.e., for all k, to obtain the vector ϕ^n_j. At each time level n, the scheme is marched through values of j in the stream direction x, which is consistent with the infinite downstream propagation speed of the advancing waves.

Near the leading and trailing edges of the airfoil, the small-disturbance assumption breaks down. There ϕ_x can become large, which makes the receding wave propagation speed large and, according to (5), makes Δt small. This can be overcome by using a modified version of (3) near the leading and trailing edges which evaluates a fraction m of the ϕ_{xx} term at time level n. The first truncation error term in the MPDE is now $[S_2\Delta x + (m - 1/2)V_c\Delta t]\phi_{xxt}$ and, with $r \ge 1/2$, the time step restriction at any point j,k is

$$\Delta t(1 - 2m_{j,k}) \le 2S_2\Delta x/|V_{c_{j,k}}| \quad (6)$$

For $m \ge 1/2$, the scheme is unconditionally stable. At each point we can take $m_{j,k}$ to balance $V_{c_{j,k}}$ such that the equality in (6) is satisfied. The first two truncation terms in the MPDE are thereby eliminated, and Δt is unrestricted. The scheme is implicit in both x and y and forms a matrix which is solved iteratively using successive line over-relaxation (SLOR) for y lines very close to the leading and trailing edges. The marching scheme (3) is used for the remainder of the flow field where no iteration is necessary.

DIFFERENCING OF THE COMPLETE SMALL-DISTURBANCE EQUATION

For subsonic points, $V_c > 0$, the difference scheme for equation (1), shown in figure 3(b), can be written

$$\left[\frac{S_1}{\Delta t^2} + \frac{S_2(D_-)_x}{\Delta t} + \frac{1}{2}(D_+D_-)_y\right]\phi^n_{j,k} = \left[\frac{2S_1}{\Delta t^2} + \left(V_c - \frac{\Delta x}{\Delta t}S_2\right)(D_+D_-)_x\right]\phi^{n-1}_{j,k}$$

$$+\left[-\frac{S_1}{\Delta t^2} + \frac{S_2(D_+)_x}{\Delta t} + \frac{1}{2}(D_+D_-)_y\right]\phi^{n-2}_{j,k} \quad (7a)$$

For supersonic points, $V_c < 0$,

$$\left[\frac{1}{2}\left(D_+D_-\right)_y + S_2\frac{\left(D_-\right)_x}{\Delta t}\right]\phi^n_{j,k} = -S_1\left(D_-D_-\right)_t\phi^n_{j-1,k} + \left[S_2\frac{\Delta x}{\Delta t} + V_c\right]\left(D_-D_-\right)_x\phi^{n-1}_{j,k}$$

$$+S_2\frac{\left(D_-\right)_x}{\Delta t}\phi^{n-2}_{j-1,k} + \frac{1}{2}\left(D_+D_-\right)_y\phi^{n-2}_{j,k} \qquad (7b)$$

As in (3), ϕ on each line j = constant, n = constant is solved simultaneously for all k by the solution of the tridiagonal matrix equation formed by (7). At each time level n, (7) is marched through values of j in the stream direction. The difference equations (7) reduce to the Murman-Cole mixed-difference scheme for steady flows. The time step stability constraint

$$\Delta t/\Delta x \leq \min_{j,k}\left(\left|-S_2-\sqrt{S_2{}^2+S_1V_{c_{j,k}}}\right|/\left|V_{c_{j,k}}\right|\right) \qquad (8)$$

is determined by the receding wave speed. For low-frequency motion, this time step is sufficient to adequately resolve the unsteady flow field. For high frequency motion, (7) can still be used, of course, but the time step must be reduced to a value consistent with the advancing wave speed, i.e.,

$$\Delta t/\Delta x \leq \min_{j,k}\left(-S_2+\sqrt{S_2{}^2+S_1V_{c_{j,k}}}/\left|V_{c_{j,k}}\right|\right) \qquad (9)$$

to provide adequate resolution.

COMPUTED RESULTS

Exact linear theory solutions with $K = 1$ for the unsteady, transonic, small-disturbance equation (1) and its low-frequency approximation (2) are compared with numerical results obtained using difference schemes (3) and (7). Results are given in the form of airfoil surface pressure coefficients.

$$C_p = -2\delta^{2/3}(\phi_x + \phi_t) \qquad C_p = -2\delta^{2/3}\phi_x \qquad (10)$$

for equations (1) and (2), respectively. First consider the linearized form of (2) obtained by setting $V_c = S_3$. Figure 4 compares surface pressures found using the linearized form of (3) with exact solutions to linear theory at four instances of time for an impulsively-started airfoil. A linear wave, in the form of a logarithmic singularity in the slope, propagates upstream along the receding characteristic passing through the trailing edge. This wave causes small-amplitude wiggles to appear in the finite-difference solution when $\Delta t = \Delta t_{max}$ and $r = 0.5$. Reducing Δt introduces dissipation, which eliminates the wiggles but diffuses the wave. A better approach is to take $\Delta t = \Delta t_{max}$ and $r > 0.5$ (i.e., introduce dissipation into the differencing of ϕ_{yy}), which eliminates the wiggles without diffusing the wave. For the case shown, $r = 0.6$. The same motion is considered in figure 5, which compares the exact linear and finite-difference nonlinear solutions of the low-frequency equation for a case in which the steady-state flow is entirely subsonic but the transient flow is supercritical. The solutions compare well until the flow approaches the sonic condition, denoted by $C_p{}^*$. At this point, for the nonlinear solution, a shock wave forms, propagates upstream, and decays.

In figure 6 we consider a motion of moderate frequency to accentuate the differences in the three solutions shown. At time zero, the airfoil is traveling at $M_\infty = 0.785$ and has zero thickness. It thickens according to the equation in figure 6 until it reaches a final value of $\delta = 0.10$ after traveling a distance of 2 chord lengths (for low-frequency motion, the thickening of the airfoil would take place over a distance of about 25 chord lengths). Initially the exact linear and finite-difference nonlinear solutions of the small-disturbance equation agree remarkably well considering that the large time step (8) was used, indicating that no attempt was made to resolve the advancing waves. The low-frequency approximation expands more rapidly because of the infinite propagation speed of the advancing waves. When the flow becomes supercritical, the finite-difference solutions become highly nonlinear. A shock wave forms and propagates upstream. By the time it reaches midchord, its strength has diminished considerably. The time history of the pressure at midchord shows a numerical overshoot at the shock in the small-disturbance solution and a smeared shock in the low-frequency approximation. The boundaries, at which ϕ was set equal to zero, were placed a distance of 2 chord lengths upstream and downstream and 4 chord lengths vertically from the airfoil. In the low-frequency approximation, disturbances reflected from the boundaries dissipated before they reached the airfoil. In the small-disturbance solution they did not, and they began to affect the solution after the airfoil had traveled about seven chord lengths.

CONCLUDING REMARKS

The unsteady, transonic, small-disturbance equation and its low-frequency approximation have been solved using semi-implicit difference operators. This approach significantly increased the allowable time step for stability and hence significantly reduced computer run time* without adversely affecting accuracy for low-frequency motion. For such motion, the low-frequency approximation is preferred because of its simplicity. On the other hand, the small-disturbance equation is not restricted to low-frequency motion, and, with the proposed semi-implicit difference scheme, the time step can be adjusted, based on the degree of unsteadiness of the motion, to accurately and efficiently resolve either high- or low-frequency flow fields.

REFERENCES

Murman, E. M., Proc. of AIAA Comp. Fluid Dynamics Conf. 27-40 (1973).
Murman, E. M., and Cole, J. D. AIAA J. 9, 114-121 (1971).

*Preliminary comparisons of the low-frequency approximation with MacCormack's Lax-Wendroff type explicit scheme applied to the Euler equations indicate a reduction in computer run time of between one and two orders of magnitude.

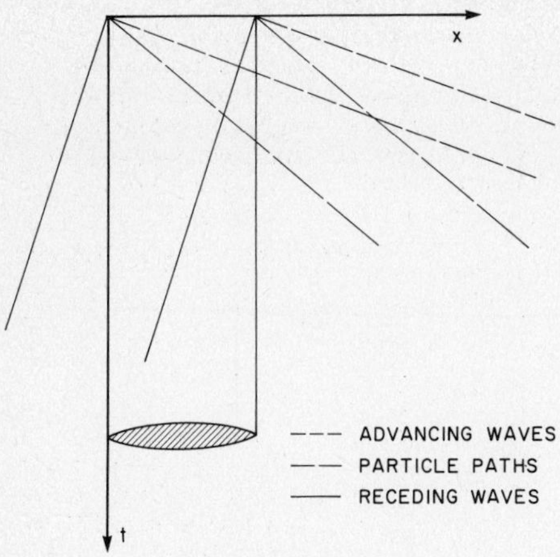

Figure 1. Characteristics for the small-disturbance equation.

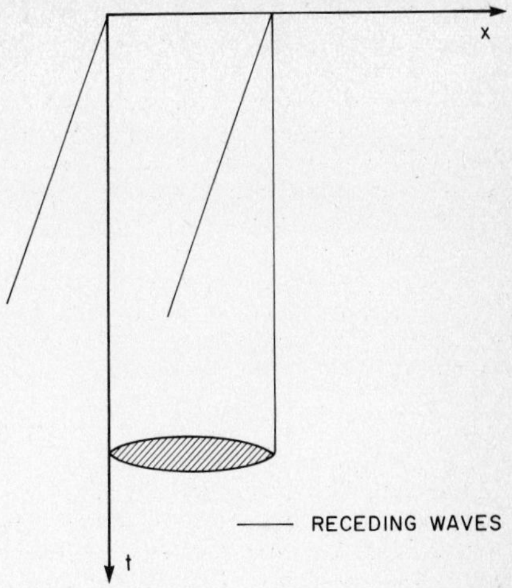

Figure 2. Characteristics for the low-frequency approximation.

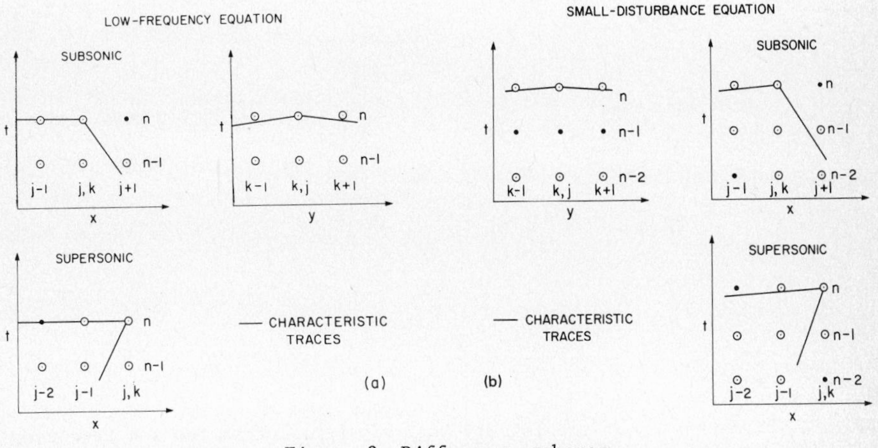

Figure 3. Difference schemes.

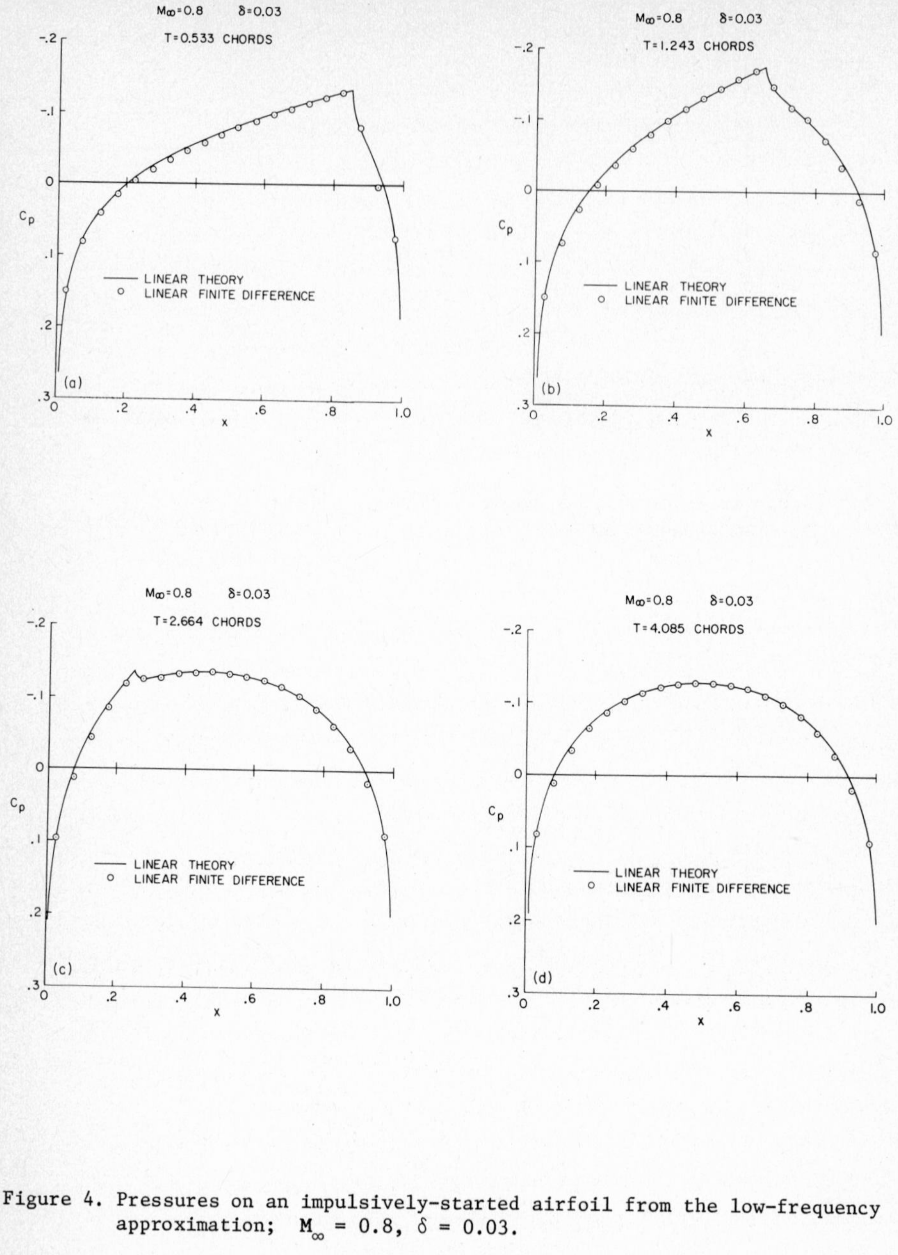

Figure 4. Pressures on an impulsively-started airfoil from the low-frequency approximation; $M_\infty = 0.8$, $\delta = 0.03$.

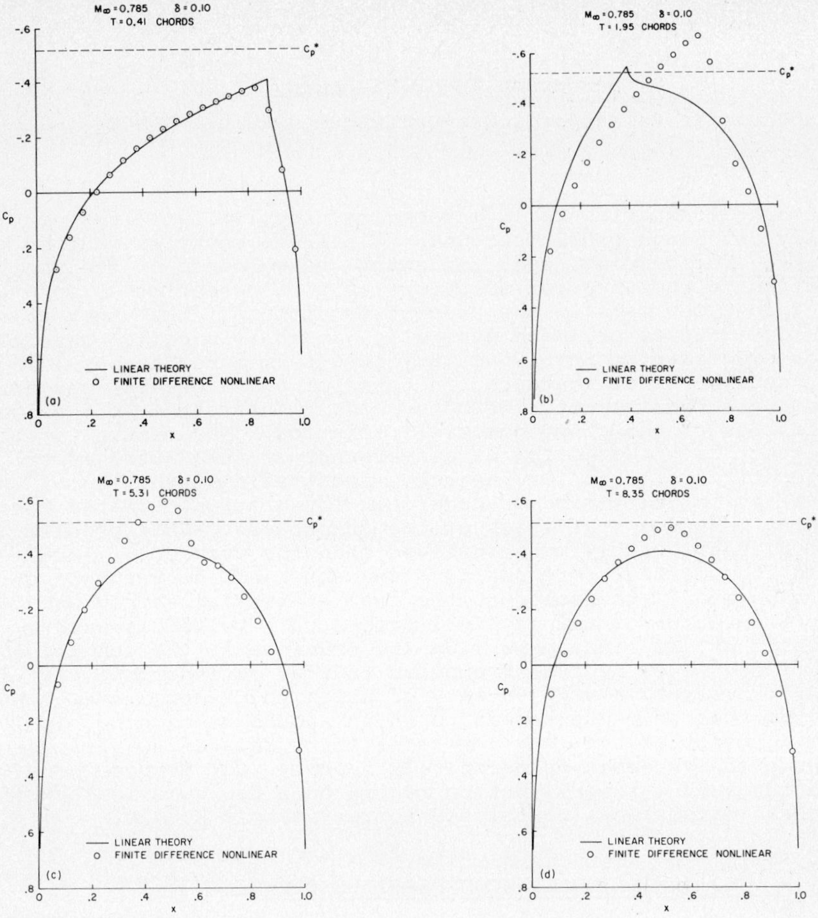

Figure 5. Pressures on an impulsively-started airfoil from the low-frequency approximation; $M_\infty = 0.785$, $\delta = 0.10$.

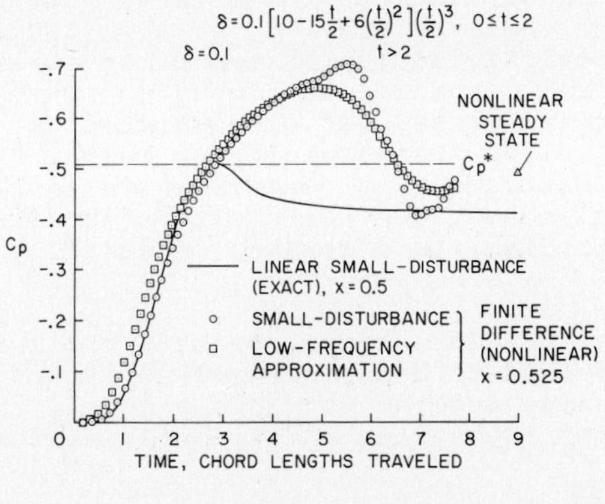

Figure 6. Time history of midchord pressures on a parabolic arc airfoil, $M_\infty = 0.785$.

QUASI-INVISCID DYNAMICS OF NUMERICAL MODELS OF TWO-DIMENSIONAL FLOWS

C. Basdevant and R. Sadourny

Laboratoire de Météorologie Dynamique, Paris FRANCE

Numerical experiments in meteorology involve two kinds of conceptually different problems. One, related to short or medium-range weather forecasting, asks for actual convergence of the approximate solution to the physical solution, up to a prescribed time - no longer than a few days in the present conditions. The other arises in climatology studies or, more generally, in what is called general circulation experiments, involving very long-term integrations : in this case, accuracy (for instance, in terms of phase errors) may not be a predominant requirement, one being mostly interested in the long-range statistics of the given numerical solution. The central problem in this respect is the question of convergence of time averages of selected quantities (like, for instance, spectral distribution of energy) over an infinite time of numerical integration, towards the corresponding measured statistics of the atmosphere. This problem appears to be considerably more involved than the problem of local convergence in numerical forecasting : one one hand, convergence of the long-range model statistics to the real statistics should ask for proper representation of all physical processes involved in such a complex system as the atmosphere. On the other hand, the non linear structure of the model by itself appears to play an important part in this matter : however, a wide variety of numerical schemes are being used, and there is no consensus as to which should be chosen. It seems that a theory of the long term statistics related to various kinds of numerical schemes is definitely needed. Our purpose here is not to provide such a theory, but to report on a few numerical experiments in these directions, restricted to very simple models.

I - TWO-DIMENSIONAL INCOMPRESSIBLE FLOWS

It is well known that two-dimensional incompressible flows are subject to two quadratic invariants, kinetic energy and squared vorticity, or enstrophy. The importance of formal conservation of both invariants in numerical calculations was pointed out by ARAKAWA (1). Detailed statistical studies of numerical models have been reported, e.g. very recently by HERRING et al. (3), among various experiments on numerical simulation of two-dimensional turbulence. Relevance of statistical mechanics to truncated systems, limited to a finite number of spectral modes, was also pointed out by KRAICHNAN (4), who gave the theoretical form of equilibrium energy spectra, in two-dimensions, based on macrocanonical ensemble averages. The question of the behaviour of time averages for a single numerical solution (pure initial value problem) is discussed in detail in BASDEVANT et al. (2).

If we consider any (truncated spectral or finite difference) model formally constrained to both energy- and enstrophy conservation, and a given initial value problem for such a model, the usual assumption that two states of the system sufficiently distant from each other in time are uncorrelated, is well verified in practice. This assumption is equivalent to assumption of ergodicity of the energy spectrum in the microcanonical ensemble. The time averaged energy spectrum for a given initial value problem is then equal to adequately chosen integrals on the intersection of the sphere $E = E_o$ and the ellipsoid $Z = Z_o$, where E_o is the initial energy, Z_o the initial enstrophy, E and Z the quadratic forms defining energy and enstrophy in the N-dimensional spectral space. Upper and lower bounds of these integrals were computed, and asymptotic forms derived when the number N of degrees of freedom in the model approaches infinity. The time averaged energy spectra for very large N, when the mesh size is small enough with respect to the scales excited in the initial fields, are the following :

$$E(\vec{k}) = \alpha^{-1}(\vec{k}) \frac{Z_o - E_o k_1^2}{k^2 - k_1^2} \qquad (k \neq k_1)$$

$$E(\vec{k}) = \alpha^{-1}(\vec{k}) \; E_o \qquad (k = k_1)$$

where k_1 is the modulus of the smallest wavevectors, and $\alpha(\vec{k})$ is the number of wavevectors having the same modulus as $\vec{k}$. These results are valid for finite difference models with formal conservation of energy and enstrophy as well as for spectral models. On the contrary, finite difference models constrained to formal conservation of energy only would lead to equipartition of energy between all modes. The answer to the question whether enstrophy conservation is enough to produce correct statistics in the case of large-scale forcing and high Reynolds numbers, is still unclear. A statistical representation of sub-grid scale motions may still be necessary.

II - TWO-DIMENSIONAL COMPRESSIBLE FLOWS

Compressible flows in two dimensions can be described by the free surface equations :

$$\frac{\partial V}{\partial t} + \vec{rot}\, \vec{V} \times \vec{V} + \vec{grad}\, (P + \frac{1}{2} V^2) = 0$$

$$\frac{\partial P}{\partial t} + \text{div}\, (P\vec{V}) = 0$$

Apart from mass conservation, there are two non-quadratic integrals of motions, the energy E and the potential enstrophy Z :

$$E = \int_S \frac{1}{2} (P + \frac{1}{2} V^2)\, P\, dS$$

$$Z = \int_S \frac{1}{2} (\frac{Z}{P})^2\, P\, dS$$

We investigated the relative importance of formal conservation of energy and potential enstrophy in finite difference models, in terms of time statistics for pure initial value problems. We used a square grid described in fig.1, and defined

$$H = P + \frac{1}{2}\left(\overline{u^2}^x + \overline{v^2}^y\right)$$

$$U = \overline{P}^x\, u$$

$$V = \overline{P}^y\, v$$

$$Z = \frac{1}{\overline{\overline{P}^x}^y}\left(\delta_x v - \delta_y u\right) \qquad (= \mathrm{rot}\, \vec{V})$$

where the bar operator means a two-point arithmetic average and δ a two-point centered derivative - both acting on immediate neighbors along the x or y directions.

We then defined two slightly different models :

$$\text{(A)} \quad \begin{cases} \dfrac{\partial u}{\partial t} - \overline{Z\, \overline{V}^x}^y + \delta_x H = 0 \\[2ex] \dfrac{\partial v}{\partial t} + \overline{Z\, \overline{U}^y}^x + \delta_y H = 0 \\[2ex] \dfrac{\partial P}{\partial t} + \delta_x U + \delta_y V = 0 \end{cases}$$

$$\text{(B)} \quad \begin{cases} \dfrac{\partial u}{\partial t} - \overline{Z}^y\, \overline{\overline{V}^x}^y + \delta_x H = 0 \\[2ex] \dfrac{\partial v}{\partial t} + \overline{Z}^x\, \overline{\overline{U}^y}^x + \delta_y H = 0 \\[2ex] \dfrac{\partial P}{\partial t} + \delta_x U + \delta_y V = 0 \end{cases}$$

Model (A) formally conserves energy, not potential enstrophy, while the reverse is true for model (B). Both models were integrated for a long time using the leapfrog approximation for time derivatives, starting from identical initial conditions.

Model (A) is shown to be consistent with the dynamics of the exact equations only up to a finite time T, inversely proportional to the square root of the initial potential enstrophy. At time T, non linear instability occurs for pure inviscid calculations, triggered by an increase in the potential enstrophy of the flow, which grows without limit. The calculation is stabilized by a small dissipative term, in which case instability is replaced by a strong dissipation of energy for $t > T$, while enstrophy saturates at a rather large value (fig.2). The saturation value of potential enstrophy approximately corresponds to an equipartition of energy among all spectral modes, which can be verified by spectral analysis. Further, T is independent of the mesh size.

We conclude that a quasi-inviscid model of two-dimensional flow which is not formally constrained to quasi-conserve potential enstrophy reaches equipartition of energy after a finite time T depending on initial conditions only. The effect of a dissipation term is just to remove the energy spuriously cascading towards the smaller scales, without preventing this cascade itself, due to the incorrect structure of non linear interactions. The fact that T does not depend on the mesh size would mean that T is an upper limit for convergence of scheme (A).

Model (B), although it does not formally conserve energy, behaves much more closely to what we can expect of the real dynamics. Energy and potential enstrophy were actually stationary in all the integrations that were performed. The time averaged energy spectra in this case correspond to an equipartition of enstrophy among the higher modes, together with an equipartition of irrotational energy.

Models (A) and (B), adapted to spherical geometry, were compared in the case of a large scale Rossby wave with wavenumber 4 in longitude. The critical time T for model (A) is approximately two months in this case: however, deterioration of the solution is already noticeable by eye after one or two weeks, and the Rossby wave appears markedly unstable. On the other hand, it looks perfectly stable when integrated by model (B), even after one hundred days.

REFERENCES

(1) ARAKAWA, A. (1966) Computational design for long term numerical integration of the equations of fluid motion : two-dimensional incompressible flow (Part I). J. Comp. Phys., 1, 119-143.

(2) BASDEVANT, C. and SADOURNY, R. (1973) Ergodic properties of inviscid truncated models of two-dimensional incompressible flows. Submitted to J. Fluid Mech.

(3) HERRING, J.R., ORSZAG, S.A., KRAICHNAN, R.H. and FOX, D.G. (1974) Decay of two-dimensional turbulence. (To be published)

(4) KRAICHNAN, R.H. (1967) Inertial ranges in two-dimensional turbulence. Phys. Fluids, 10, 1417-1423.

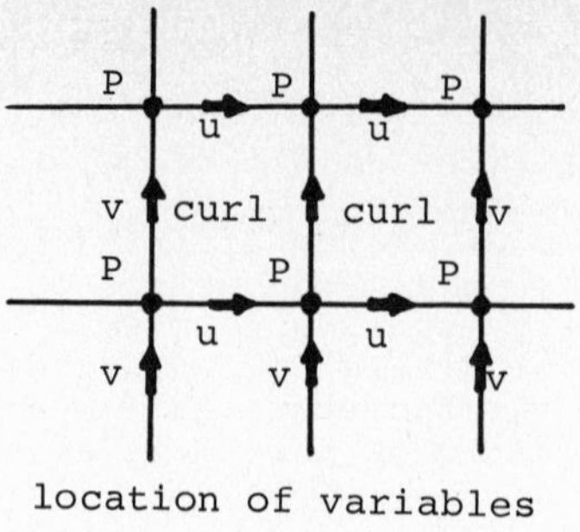

FIGURE 1

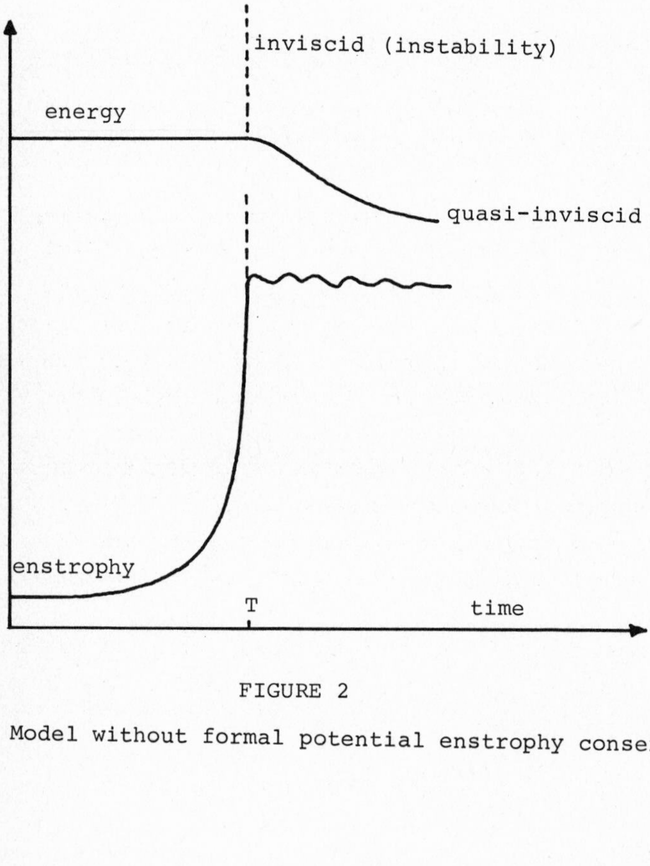

FIGURE 2

Model without formal potential enstrophy conservation.

FLAT SLENDER DELTA WINGS IN SUPERSONIC STREAM AT SMALL ANGLES OF ATTACK

A. P. Bazzhin

1. The paper deals with the problem of a flat slender delta wing placed in the supersonic stream of an inviscid gas at a small angle of attack. The most probable flow pattern over such wings at a moderate free stream Mach number will be that one in which the conical bow wave is attached to the tip of the wing only and the flows above and under the wing influence each other. This problem has been considered by many investigators in the frame of the linear theory [3]. Yet any numerical or moreover analytical solution based on the full nonlinear system of gasdynamic equations has not been received so far.

The successful application by P. Kutler et al. [4] of the shock - capturing schemes to the variety of problems and particularly to the problem of a flat delta wing with supersonic leading edges suggested the attempt to compute flow fields over slender delta wings by the use of the MacCormack scheme [5].

The additional and essential difficulty in this problem is to compute the flow field in the nearest vicinity of the leading edge. There may be at least two points of view in considering this particular question. Firstly, it is possible to use a special fine grid at the leading edge in order to reveal all the details of the flow in that region. It is also possible, on the contrary, to neglect the details of the flow near the leading edge and to use any grid that is sufficient for capturing the bow shock wave and the other main features of the flow field. In a great number of cases the loss of accuracy in the vicinity of the leading edge will not necessarily prevent from obtaining a true solution in other flow regions especially under the wing.

2. The system of gasdynamic equations governing the fictitious unsteady conical flow of an inviscid gas was used in the following conservative form:

$$\frac{\partial E}{\partial T} + \frac{\partial F}{\partial \eta} + \frac{\partial G}{\partial \zeta} + H = 0, \qquad (1)$$

where $\eta = \frac{y}{x}\tan\chi$ and $\zeta = \frac{z}{x}\tan\chi$ are two normalized independent variables (x, y, z are the cartesian coordinates, fig. 1), T - nondimensional time and the vector - columns E, F, G, H are

$$E \equiv \begin{Bmatrix} E_1 \\ E_2 \\ E_3 \\ E_4 \\ E_5 \end{Bmatrix} = \begin{Bmatrix} \rho u \\ \rho v \\ \rho w \\ \rho B - p \\ \rho \end{Bmatrix}; \qquad F = \begin{Bmatrix} \bar{v} E_1 - \eta p \\ \bar{v} E_2 + p \\ \bar{v} E_3 \\ \bar{v}(E_4 + p) \\ \bar{v} E_5 \end{Bmatrix};$$

$$G = \begin{Bmatrix} \bar{w} E_1 - \zeta p \\ \bar{w} E_2 \\ \bar{w} E_3 + p \\ \bar{w}(E_4 + p) \\ \bar{w} E_5 \end{Bmatrix}; \qquad H = \frac{2}{\tan\chi} \begin{Bmatrix} E_1^2/E_5 + p \\ E_1 E_2/E_5 \\ E_1 E_3/E_5 \\ E_1 (E_4 + p)/E_5 \\ E_1 \end{Bmatrix}. \tag{2}$$

In these expressions $\bar{v} = (E_2 - \eta E_1)/E_5$; $\bar{w} = (E_3 - \zeta E_1)/E_5$; $B = h(p, E_5) + (E_1^2 + E_2^2 + E_3^2)/2E_5^2$; h - specific enthalpy of the gas; u, v, w - velocity components in the cartesian coordinate system xyz. When considering the flow of a real gas in the state of thermodynamic equilibrium it is necessary to solve the transcendental equation to compute the pressure

$$p = h(p, E_5) \cdot E_5 - E_4 + (E_1^2 + E_2^2 + E_3^2)/2E_5 \tag{3}$$

In the case of a perfect gas there exists the simple formula for computing p:

$$p = (\gamma - 1)\left[E_4 - (E_1^2 + E_2^2 + E_3^2)/2E_5\right]. \tag{4}$$

The system of equations (1) is T-hyperbolic and the mixed Cauchy problem is well posed for it. The solution is being seeked in some rectangular domain ABCD in the plane X = 1 (fig. 2). The boundary conditions are the symmetry of the flow ($E_3 = 0$) in the plane AD, the impermeability condition ($E_2 = 0$) at the surface of the wing KK, the free stream conditions at the boundaries AB, BC, CD. The flow in the domain ABCD is given somehow at the initial moment. The problem is then to advance this solution by a time step ΔT. The desirable solution is the steady flow at time $T \to \infty$.

The finite-difference scheme of MacCormack [5] has been used for numerical integration of the system (1). The domain ABCD was divided in rectangular cells with the sides $\Delta\eta, \Delta\zeta$. The same scheme was used in the inner points of the flow field and at the surface of the wing KK. When dealing with the points at the wing the reflection principle was used for computing the lacking values of the vector F.

We do not take into consideration the possibility of arising vortices above the upper surface of the wing. It is known that there is some region of small angles of attack in which such vortices if they occur would be quite weak and their influence upon the flow would be weak as well [1]. The flow above the wing would depend on these vortices in much greater degree at the increased angles of attack, but at the same time their influence upon the flow under the wing would diminish. This conclusion seems to be true if one analyses the structure of unseparated flow of the inviscid gas above the upper surface of the wing as this structure is received by the present computations. Thus the solution for the region under the wing will probably be true at any angle of attack. As for the unseparated flow above the wing apparently it does not exist in reality. It might be considered as a background on which the gasdynamic phenomena due to viscosity of the gas arise and develop.

3. Flow fields over some delta wings have been computed to test the method. One of the variants was that of the wing with sweep angle $\chi = 70^\circ$ at $M_\infty = 6$ and $\alpha = 5^\circ$. The leading edge is supersonic in this case and there is a possibility to compare our results with those of Voskresenskiy [6], [7]. This comparison is made in fig. 3. The forms of bow shock waves under the wing, the boundaries of the disturbed region above it, the pressure distributions along the wing surface and pressure profiles in the plane of symmetry all agree quite well. (The pressure profiles are smooth curves as a result of a single smoothing of the numerical solution containing high-frequency fluctuations. Two formulas have been used for smoothing:

$$f_{i,j} =: \frac{1}{4}\left(f_{i-1,j} + 2f_{i,j} + f_{i+1,j}\right) \text{ and } f_{i,j} =: \frac{1}{4}\left(f_{i,j-1} + 2f_{i,j} + f_{i,j+1}\right)).$$

Also shown in this figure is the distribution of errors in the Bernulli integral through the grid points. It follows from this distribution that the flow field under the wing is free from such errors.

We describe now some results characterizing the inner structures of flows over flat delta wings.

Fig. 4 shows isobars in the flow field over the wing with $\chi = 80^\circ$ at $M_\infty = 6$ and $\alpha = 5^\circ$. The main features of the flow are clearly seen, first of all the bow shock wave with the intensity diminishing quickly in the region above the wing. This diminishing of the intensity of the bow shock may be considered as a result of the interference of the shock with the nearly centered fan of rarefaction waves that arised near the leading edge at the upper surface of the wing. Also seen in this figure is the embedded shock above the upper surface. This solution was obtained with the grid of 35 x (40 + 70) cells. There were twenty cells on the wing in this case.

The details of the flow field became still more distinguished (fig. 4 at right) with the grid that consisted of 70 x (80 + 20) cells. This solution was obtained from the preceeding one after interpolation it onto the finer grid and additional advancing in time until a new steady state of the flow was reached. The coordinate line $\eta = 0.25$ and the preceeding solution on it were used as the upper boundary and the corresponding steady boundary condition during this process of refining the previous solution. In the upper right corner of fig. 4 the pressure distributions on the wing are shown as obtained with the two grids. The pressure distribution according to the linear theory is shown there as well.

The lines M_s = const in the flow field over the same wing at angles of attack of 2.5 and 15° are shown in fig. 5 (M_s is the Mach number corresponding to the velocity component V_s, tangent to the sphere $\sqrt{x^2+y^2+z^2}$ = const). The transverse flow over the wing at $\alpha = 2.5^o$ resembles that around a flat plate in a transonic stream: the wing is immersed entirely in a conically subsonic flow; a local, conically supersonic zone occurs near the leading edge above the upper surface of the wing; this zone is bounded by the embedded shock wave. At the greater angle of attack, $\alpha = 15^o$, a considerable part of the upper surface is already in a conically supersonic flow. This zone is also bounded by the embedded shock wave near the wing.

The stream lines in the plane $X = 1$ are shown in fig. 6 around the wing with the sweep angle of 80° at $M_\infty = 6$ and $\alpha = 5^o$ and 15°. The unseparated flow pattern above the wing is the most interesting fact here. There is a region of very rarefied gas all above the upper surface of the wing. Every stream line approaches the plane of symmetry and then goes to the wing along it. Analysing this flow pattern one might suggest that it would not be changed essentially even if the vortices would arise in that region above the wing where the density of the gas is quite small.

The last figure 7 shows the pressure distribution along the upper and lower surfaces of the wing and on the line $\eta = 0$ beyond at angles of attack from 0° to 15°. Furthermore the pressure distribution is shown at $\alpha = 15^o$ that was received by the author in 1966 [2] by the method of integral relations. The first approximation of this method gives the pressure on the wing that differs by less than two per cent from the one obtained by the present finite-difference method.

Also shown in this fig. 7 are the aerodynamic coefficients Cy and Cx of the same wing with $\chi = 80^{\circ}$ at $M_{\infty} = 6$. The curve Cy (α) as given by the linear theory is also presented here. As in many other cases the linear theory which gives unrealistic pressure distribution along the wing is quite adequate in the case under consideration for prediction of total aerodynamic forces. The method of integral relations continues smoothly the two curves to the region of great angles of attack.

The author expresses his sincere gratitude to Miss J. Tchelysheva who composed all necessary programs for the computer and received the numerical results.

R E F E R E N C E S

1. Barsby J. E., Separated Flow past a Slender Delta Wing at Incidence "Aeron. Quart." V, v.24, N 2, 120-128 (1973)
2. Bazzhin A. P., On Computation of Flows past Flat Delta Wings at Large Angles of Attack. Izv. AN SSSR. Mechanica zidkocti i gaza, N 5, (1966)
3. Bulach B. M., Nonlinear conical flows of a gas. Moscow, "Nauka", (1970)
4. Kutler P., and Lomax H., A Systematic Development of the Supersonic Flow Fields over and behind Wings and Wing - Body Configurations Using a Shock - Capturing Finite - Difference Approach. ATAA Paper N 71-99.
5. MacCormack R. W., Numerical Solution of the Interaction of a Shock Wave with a Laminar Boundary Layer. Proc. of Sec. Intern. Conf. on Numerical Methods in Fluid Dynamics, edited by M. Holt, Springer-Verlag, Berlin, 151-163 (1971)
6. Voskresenskiy G. P., Numerical Solution for Flow past an Arbitrary Surface of a Delta Wing in the Compression Region. Izv. AN SSSR, Mechanica zhidkosti i gaza, N 4, (1968)
7. Voskresenskiy G. P., Numerical Solution for Flow past the Upper Surface of a Delta Wing in the Rarefaction Region. Prikladnaja Mechanica i Teor. Physica, N 6, (1973)

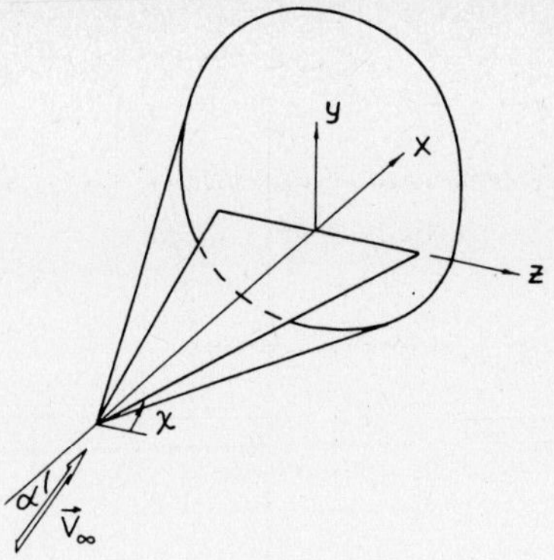

Fig. 1

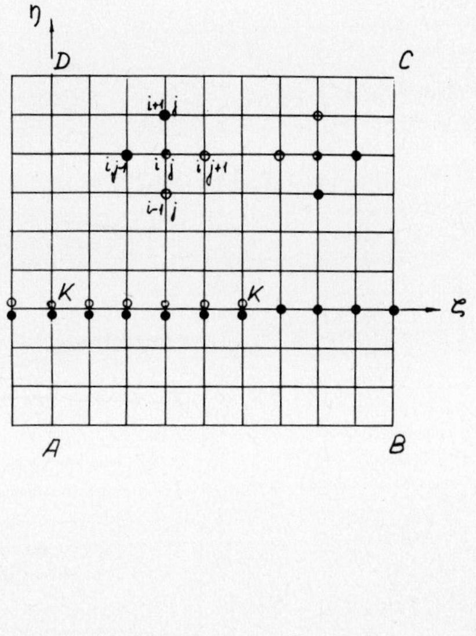

Fig. 2

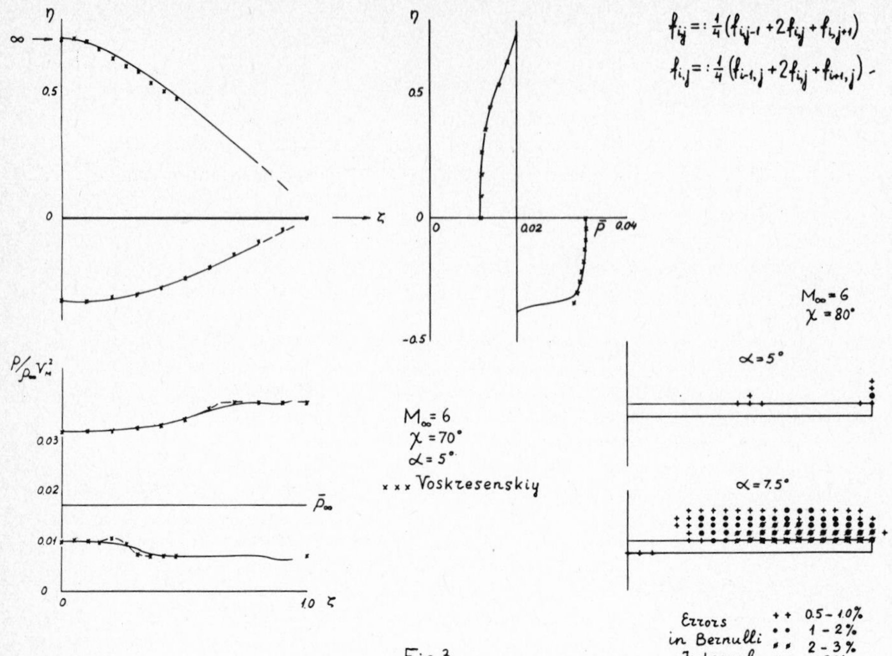

$f_{ij} =: \frac{1}{4}(f_{i,j-1} + 2f_{ij} + f_{i,j+1})$
$f_{i,j} =: \frac{1}{4}(f_{i-1,j} + 2f_{ij} + f_{i+1,j})$ -
η
∞
0.5
0
ζ
-0.5
0.02
$\bar{p}$
0.04
$p/\rho_\infty V_\infty^2$
0.03
0.02
0.01
$\bar{p}_\infty$
1.0
$M_\infty = 6$
$\chi = 70°$
$\alpha = 5°$
x x x Voskresenskiy
$M_\infty = 6$
$\chi = 80°$
$\alpha = 5°$
$\alpha = 7.5°$
Errors in Bernulli Integral
+ + 0.5 - 1.0%
• • 1 - 2%
2 - 3%
x x > 3%

Fig. 3

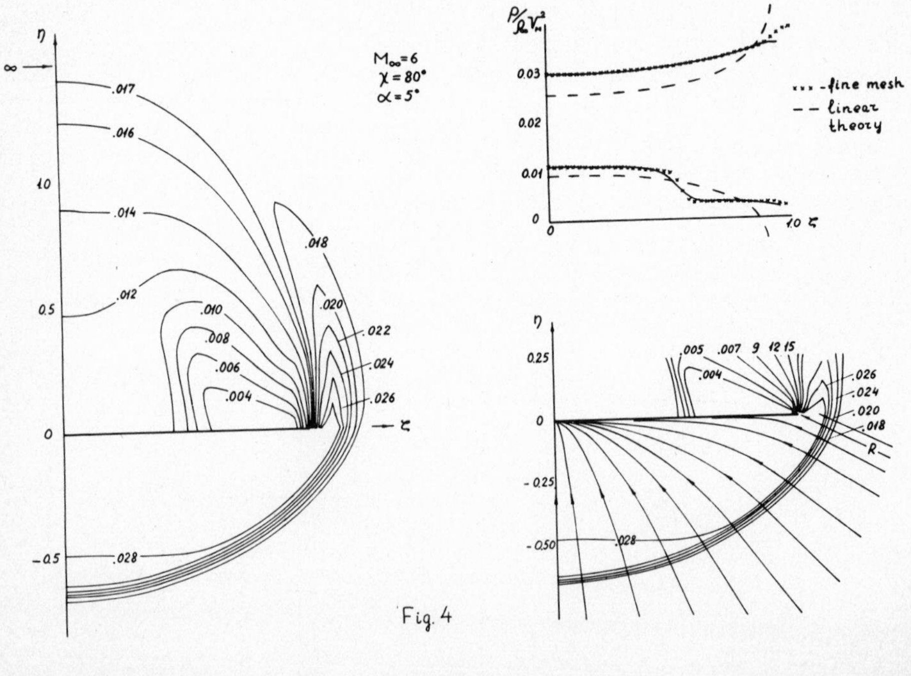

η
∞
1.0
0.5
0
-0.5
.017
.016
.014
.012
.010
.008
.006
.004
.018
.020
.022
.024
.026
.028
ζ
$M_\infty = 6$
$\chi = 80°$
$\alpha = 5°$
$p/\rho_\infty V_\infty^2$
0.03
0.02
0.01
1.0
... fine mesh
- - linear theory
0.25
-0.25
-0.50
.005
.007
9
12
15
R

Fig. 4

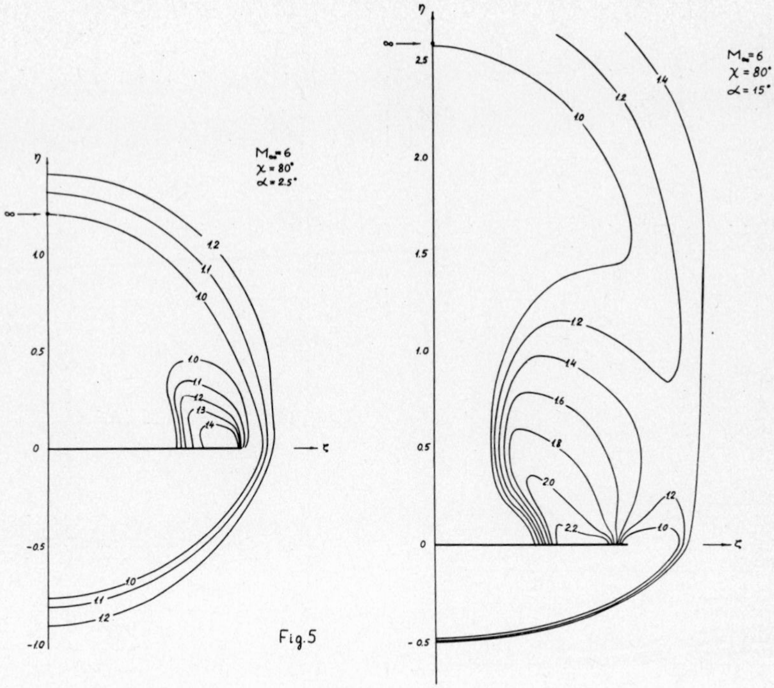

$M_\infty = 6$
$\chi = 80°$
$\alpha = 2.5°$
η
ζ
∞
1.0
0.5
0
-0.5
-1.0
1.2
1.1
1.3
1.4
$M_\infty = 6$
$\chi = 80°$
$\alpha = 15°$
2.5
2.0
1.5
1.6
1.8
2.2

Fig.5

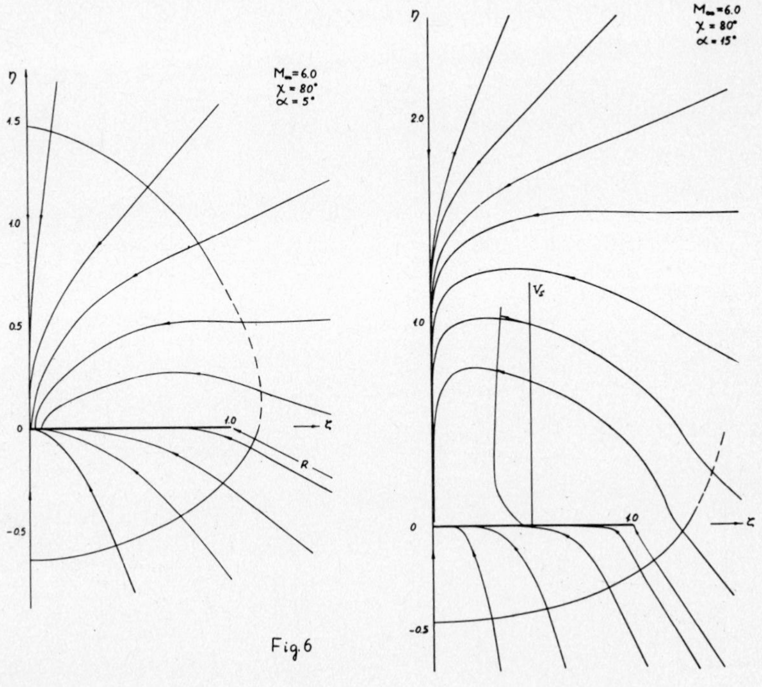

$M_\infty = 6.0$
$\chi = 80°$
$\alpha = 5°$
η
ζ
1.5
1.0
0.5
0
-0.5
R
$M_\infty = 6.0$
$\chi = 80°$
$\alpha = 15°$
2.0
V_s

Fig.6

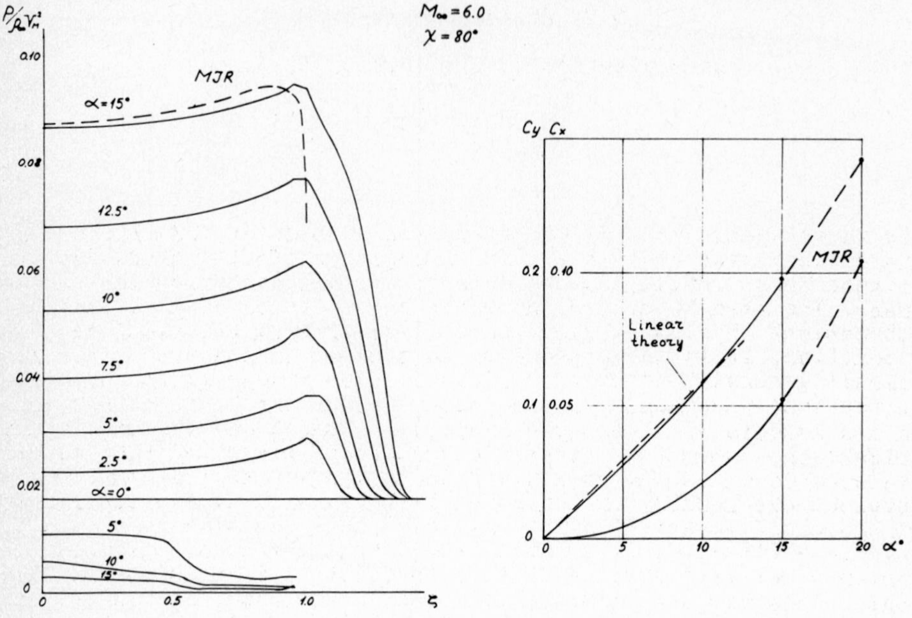
$M_\infty = 6.0$
$\chi = 80°$
$p/\rho_\infty V_\infty^2$
0.10
0.08
0.06
0.04
0.02
0
MJR
α = 15°
12.5°
10°
7.5°
5°
2.5°
α = 0°
5°
10°
15°
0
0.5
1.0
ζ
Cy Cx
0.2 0.10
0.1 0.05
0
5
10
15
20
α°
MJR
Linear theory

Fig. 7

NUMERICAL EXPERIMENT

IN GAS DYNAMICS

O.M. Belotserkovskii

Computing Center, Academy of Sciences

Moscow, U.S.S.R.

In recent years the introduction of big computers has aroused a considerably greater interest in various numerical methods and algorithms whose realization borders on carrying out numerical experiment. The need in such an approach for solution of problems of mathematical physics is prompted by ever growing practical demands; in addition, it is connected with an attempt of constructing more rational general theoretical models for the investigation of complex physical phenomena. In the past years on the basis of Harlow's and his associates' "particle-in-cell" Method [1-4] a series of numerical approaches ("large particles" method schemes) [5,6] has been developed under our direction and participation at the Computing Center of the Academy of Sciences of the U.S.S.R. These approaches are based on the splitting of physical processes by a time step and on the stabilization of a process for the solution of stationary problems. The main purpose of this research is to consider mathematical models for more complex and general gas flows in the presence of large deformations. Such an approach finds application both for the solution of Euler [5-6] and Navie-Stokes equations [7,8] .

With the help of the "large particles" method we succeeded in investigating complex gas dynamics problems - transsonic "over-critical" phenomena, "injected" flows, separation zones and so on. It is important to note that the above class of problems was regarded from a single viewpoint: sub-, trans- and supersonic flows, transition over sound velocity and the critical regime; the calculation of plane and axisymmetrical bodies was carried out by a single numerical algorithm as well.

It also seems promising to apply the main principles of the approach in question for the simulation of Boltzmann equations. The application of a statistical variant of such an approach for the solution of the Boltzmann equation is studied in [9-11] . We are not going to dwell on the description of techniques (it is given in detail in the references). There will be given only characteristic features of each approach.

The main principle of splitting a pattern into physical processes is as follows.

The medium simulated may be replaced by a system of particles (fluid particles for a continuous medium and molecules for a discrete one) which at the initial instant of time are distributed in cells of the Eulerian net in a coordinate space in accordance with the initial data.

The evolution of such a system in time Δt may be split into two stages: change of the internal state of subsystems in cells which are assumed to be "frozen" or stable ("Eulerian" stage for a continuous medium and collision relaxation for a discrete one) and subsequent displacement of all the particles proportional to their velocity and Δt without changing the internal state ("Lagrangian" stage for a continuous medium and free motion of molecules for a discrete one).

The stationary distribution of all the medium parameters is calculated after the process is stabilized in time.

1. For numerical models constructed by Yu.M. Davidov [5,6] on the basis of Eulerian equations, the mass of a whole fluid (Eulerian) cell, i.e. "a large particle" (from which the name of the method) is considered instead of the ensemble of particles in cells. Furthermore, non-stationary (and continuous) flows of these "large particles" across the Eulerian net are studied by means of finite-difference or integral representations of conservation laws. Actually, there are used here conservation laws given in the form of balance equations for a cell of finite dimensions (which is a usual procedure in deriving gas dynamics equations but without further limit transition from cell to point). As a result, we obtain divergent - conservative and dissipative-steady numerical schemes that allow us to study a wide class of complex gas dynamics problems (over-critical regimes, turbulent flows in the wake of a body, diffraction problems, transition over sound velocity etc.) [5,6,12] .

The schemes given possess internal dissipation (scheme viscosity), so that stable calculations may become possible without preliminarily singling out discontinuties (homogeneous schemes of "through" calculation). Fig.1 shows Yu.M. Davidov's pattern of a supersonic flow (M_∞ =3.5) around a finite axisymmetrical cylinder in the presence of "injection" (a sonic axial jet outflows toward the main stream from a nozzle situated on the axis of symmetry of the body). As is seen, the presence of a jet greatly complicates the flow pattern: a leading shock wave is "injected" toward the stream (its departure from the body significantly increases), a local jet supersonic region develops in front of the body, it is closed by a system of λ -shocks (lateral, oblique and front ones) and so on. A separation zone with a complicated vorticity structure occurs in the wake of the body and so on. Examples of the calculation of transsonic and "over-critical" flow regimes as well as turbulent separation zones are given in [12] .

2. The calculation of compressible viscous gas flows was performed by L.I. Severinov and A.I. Babakov with the help of the approximation of conservation laws represented in an integral form for each cell of the calculation scheme ("flows" method) [7] . Conservation laws for mass, momentum and energy of a finite volume are in the form:

$$\frac{\partial}{\partial t}\iiint_{\Omega} F\, d\Omega = -\iint_{S_\Omega} \vec{\vartheta}_F\, d\vec{s}, \quad F=\{M, X, Y, E\} \qquad (1)$$

where S_Ω - is a lateral surface Ω , where M , X , Y , E are mass, momentum and energy terms Ω , respectively, and ϑ_F is a flow density vector for each of the quantities. Eqs.(1) take account of boundary conditions and are solved numerically for each cell of the calculation region.

Consider some problems of numerical integration of eqs.(1). In approximating flow density vectors ϑ_F an essential element of the method given is that the distribution densities of additive characteristics such as densities F are calculated on the boundary S_Ω of volume Ω in a non-symmetrical way (extrapolation toward a gas flow); while the other parameters, e.g., pressure, transfer velocities of additive characteristics, derived axes u , v and T are calculated according to symmetry formulas in the viscous stress tensor and in the thermal conduction law. We believe it allows us to take account of influence regions, which is an important factor in the investigation of physical flow patterns. It is easy to see that in its essence the "flows" method is conservative both locally (for each cell of the calculation region) and integrally, i.e., for the whole region. As compared to the methods of the numerical solution of Navie-Stokes equations the transition to integral conservation laws requires approximation of lower (by one order) derivatives, which is convenient for calculations.

The characteristics of a viscous gas flow around a body of finite dimensions were systematically studied with the help of the above approach in a wide range of Reynolds numbers. In fig.2 there are given flow patterns for a sphere (separation zones of reverse-circular flows) at $M_\infty = 20$ and $550 \leq Re \leq 10^4$. More complex configurations of compressible viscous gas flows were studied as well.

3. At present, there are known rather many numerical methods for the solution of Navie-Stokes equations describing incompressible viscous flows. Most of them are developed for equations relative to flow function ψ and vortex ω. A common disadvantage of these methods is the utilization in some form of a boundary condition for a vortex on a solid surface, it is omitted in the physical formulation of the problem. The rate of the convergence of numerical algorithms is limited by the presence of an additional iteration process due to a boundary condition for a solid surface vortex. Moreover, an apparent limitation of the methods for the solution of system (ψ , ω) connected with their inapplicability for cases of space viscous flows and compressible gas flows accounts for the recent interest in the numerical solution of Navie-Stokes equations represented in natural variables.

The investigation of incompressible viscous gas flows was carried out by V.A. Gushchin and V.V. Shchennikov [8,15,16] by means of a numerical scheme of splitting analogous to the method *SMAC* given in [13] . An essential element of this method is the choice of boundary conditions on a solid surface γ : $v^n_{i,-1/2} = 0$ (non-flow); $u^n_{i+1/2,-1/2} = 0$ (attachment), from which it follows that $\tilde{u}_{i+1/2,0} = \frac{1}{2} u^n_{i+1/2,0} + \frac{1}{6} u^n_{i+1/2,1} + O(\delta y^3)$. The latter condition allows us to avoid the introduction of a layer of fictitious cells (inside a solid body), which in schemes of type *MAC* , *SMAC* and modified *MAC* [14] gives rise to an implicit calculation of the value of a solid surface vortex with the first order of accuracy. It should be noted that in terms of the approach suggested it is not obligatory to calculate the value of a solid surface vortex. It may be obtained from a calculated field of velocities. In calculating a pressure field homogeneous boundary conditions are obtained by the approach given in [14].

Thus, the difference scheme of the method involved enables us to calculate a flow field without the values of vortex and pressure on a solid surface. The calculation results manifest its effectiveness. The difference scheme of the method gives us a single algorithm for calculating both incompressible viscous flows around plane, axisymmetrical and three-dimensional bodies of complex configuration and internal flows in a wide range of Reynolds numbers [8,15,16] .

A great number of problems of external hydrodynamics is solved with the help of this method. In a wide range of Reynolds numbers $(1 \leq Re \leq 10^3)$ there are studied incompressible viscous flows around different bodies of finite dimensions: a rectangular slab and a cylinder of finite length whose axis is parallel to the velocity vector of a flow $\bar{U}_\infty$ [15] , a sphere and a cylinder with the axis perpendicular to $\bar{U}_\infty$ [8], a rectangular parallelepiped (a three-dimensional flow), as well as bodies of more complex form. Fig.3 shows flow patterns around a cylinder (a plane problem)for numbers Re =1,10,30 and 50($Re=2RU_\infty/\nu$, where R is the cylinder radius).

4. Finally, the applicability of a statistical variant of such an approach is investigated by V.E. Yanitsky[9-11]for the solution of the Boltzmann equation. The main problem in this field is the development and investigation of the model of the behaviour of a gas medium consisting of a finite number of particles. The model is based upon merging the above ideas of splitting in terms of Bird's statistical treatment [17,18] and of Katz' ideas [19] about the existence of models asymptotically equivalent to the Boltzmann equation.

As is typical of "particle-in-cell" methods, a medium simulated is replaced by a system containing a finite number N of particles of fixed mass. At a given instant of time t_α in each cell j there are $N(\alpha, j)$ particles endowed with certain velocities. The main calculation cycle comprises two stages: at the first stage particles only collide with their counterparts in a cell (collision relaxation) and at the second stage they are only displaced and interact with the boundary of a reference volume and with the surface of a body (collisionless relaxation).

The main distinction between the model suggested in [9-11] and Bird's model lies in the fact that at the first stage of calculations each group of N particles in a cell is regarded as Katz' statistical model for an ideal monoatomic gas consisting of a finite number of particles in a homogeneous coordinate space. In simulating collisions Monte-Carlo methods of the numerical solution of the main equation of Katz' model are considered in our approach.

For the realization of the second stage of calculation of the evolution of a gas simulated it is suggested in [9-11] that use should be made of the numerical algorithms for the displacement of particles utilizing incomplete information about the position of particles in a coordinate space. This reduces the need in the volume of the prossessor memory, which significantly increases the method effectiveness.

The model was tested for the solution of a problem dealing with the structure of a direct shock in a gas consisting of elastic balls in the range of Mach numbers M_∞ =1.25 ÷ 4. Fig.4 shows the graphs of density $\tilde{n}(x)$, longitudinal temperature $\tilde{T}_{\parallel}(x)$, transverse temperature $\tilde{T}_{\perp}(x)$ and total temperature $\tilde{T}(x)$ for numbers M_∞ = 2. The unit of length is the free mean path of molecules in a flow. The relation $\Delta t/\Delta x$ is chosen to satisfy sufficient reliability of stability conditions. In fig.4 for comparison there is given density $\tilde{n}(x)$ obtained by direct numerical integration of the Boltzmann equation [20] on the net Δx close to the one used in our calculations (Δx =0.2 ÷ 0.3).

In conclusion, we should say that a series of numerical algorithms developed ("numerical experiment") allows us to effectively investigate a wide class of complex gas dynamics phenomena from a single viewpoint.

References.

1. Evans, M.W., and Harlow, F.H. Los Alamos Scient.Lab., Rept. N LA - 2139 (1957).
2. Rich, M. Los Alamos Scient.Lab., Rept.N LAMS - 2826 (1963).
3. Hirt, C.W. J.Comp.Phys. 2, N 4, 339-355 (1968).
4. Gentry, R.A., Martin, R.E., and Daly, B.J. J.Comp.Phys. 1, 87-118 (1966).
5. Belotserkovskii, O.M., and Davidov, Yu.M. Preprint Comp.Center Ac. Nauk USSR, (1970).
6. Belotserkovskii, O.M., and Davidov, Yu.M. J.Vychisl. Matem. and Matem. Phys. 11, N1 182-207 (1971).
7. Belotserkovskii, O.M., and Severinov, L.I. J. Vychisl.Matem. and Matem. Phys. 13, N 2, 385-397 (1973).
8. Belotserkovskii, O.M., Gushchin, V.A., and Shchennikov, V.V. J. Vychisl. Matem. and Matem. Phys. 14, (1974) (to be printed).
9. Yanitsky, V.E. J. Vychisl.Matem. and Matem. Phys. 13, N 2, 505-510 (1973).
10. Yanitsky, V.E. J. Vychisl. Matem. and Matem. Phys. 14, N 1, 259-262 (1974).
11. Belotserkovskii, O.M., and Yanitsky, V.E. J. Vychisl. Matem. and Matem. Phys. 14, (1974) (to be printed).
12. Belotserkovskii, O.M., Davidov Yu.M. J. Vychisl. Matem. and Matem.Phys. 13, N 1, 147-171 (1973).
13. Amsden, A.A., and Harlow, F.H. Los Alamos Scient.Lab., Rept. N LA - 4370 (1970).
14. Easton, C.R. J.Comp.Phys. 9, N 2, 375-379 (1972).
15. Gushchin, V.V., and Shchennikov, V.V. Sb. Vychisl.Matem. and Matem Phys. N 2, (1974).
16. Gushchin, V.A., and Shchennikov, V.V. J.Vychisl.Matem. and Matem Phys. 14, N 2, 512-520 (1974).
17. Bird, G.A. J.Fluid. Mech. 30, P.3, 479-487 (1967).
18. Bird, G.A. Phys. Fluid. 13, N 11, 2677-2681 (1970).
19. Katz, Probability and Related Topics in Physical Sciences. Izd. Mir, (1965).
20. Cheremisin, F.G. J. Vychisl.Matem. and Matem. Phys. 10, N 3, 654-665 (1970).

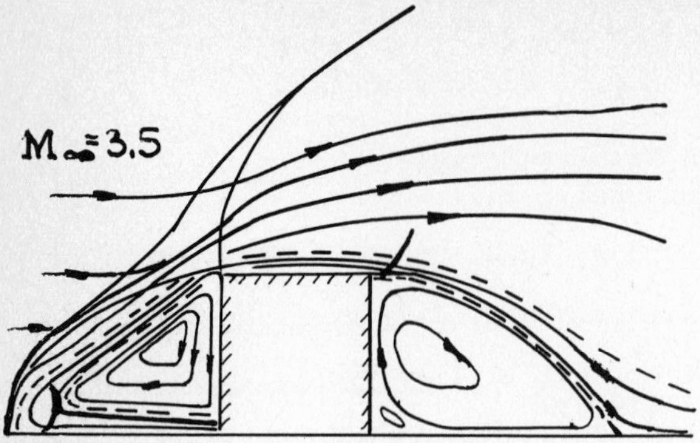

Fig 1

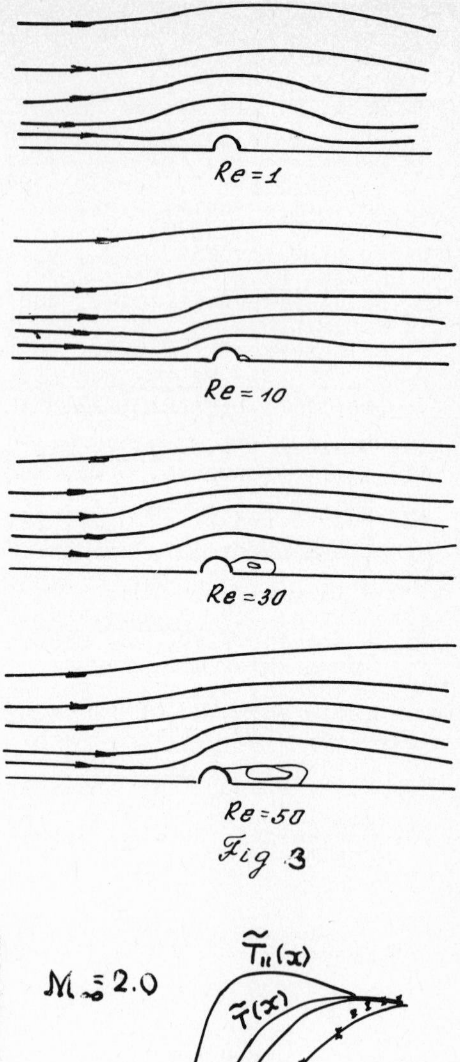

Fig 3

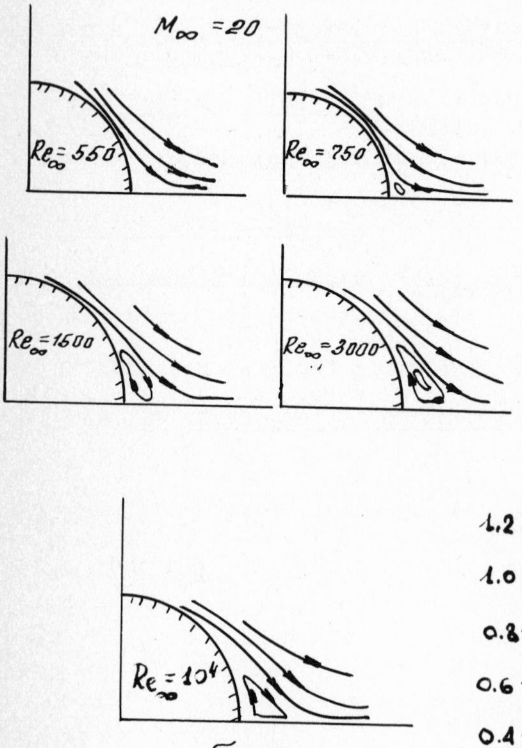

Fig 2

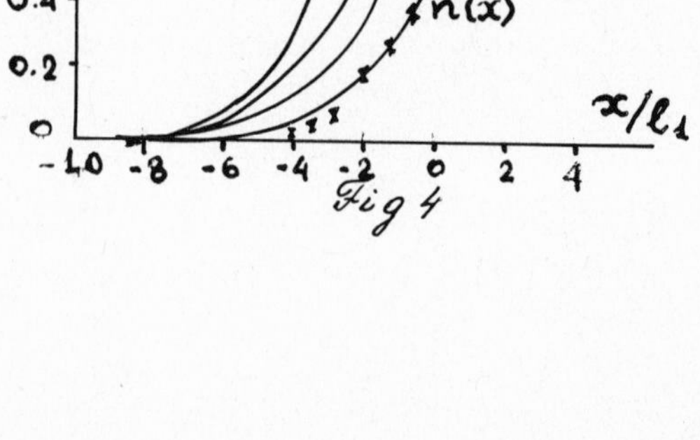

Fig 4

NUMERICAL SOLUTION
OF THE PROBLEMS OF THE MHD FLOW AROUND THE BODIES

Yu.A.Berezin, V.M.Kovenya, N.N.Yanenko

(Computer Center, Novosibirsk, USSR)

We propose absolutely stable algorithm for calculation of the two-dimensional (plane and axis-symmetric) steady supersonic flows with viscosity, thermal and electrical conductivity in a magnetic field around the finite size bodies. We use the equations of one-fluid magnetogasdynamics with viscosity, thermal and finite electrical conductivity. The nondimensional equations are represented as

$$\frac{\partial \vec{f}}{\partial t} + (\Omega + A_1 + B)\vec{f} = 0 ,$$

$$\frac{\partial \vec{H}}{\partial t} + (W + A_2)\vec{H} = 0 ,$$

$$\vec{f} = \begin{pmatrix} \rho \\ u \\ v \\ \varepsilon \end{pmatrix}, \quad \vec{H} = \begin{pmatrix} H_x \\ H_r \end{pmatrix}, \quad \Omega = \sum_{i=1}^{4} \Omega_i , \quad W = \sum_{j=1}^{2} W_j ,$$

$$\Omega_1 = \begin{pmatrix} u\frac{\partial}{\partial x} & 0 & 0 & 0 \\ 0 & u\frac{\partial}{\partial x} - \frac{4}{3Re\rho}\frac{\partial}{\partial x}\mu\frac{\partial}{\partial x} & 0 & 0 \\ 0 & 0 & u\frac{\partial}{\partial x} - \frac{1}{Re\rho}\frac{\partial}{\partial x}\mu\frac{\partial}{\partial x} & 0 \\ 0 & 0 & 0 & u\frac{\partial}{\partial x} - \frac{\gamma}{Re\rho Pr}\frac{\partial}{\partial x}\mu\frac{\partial}{\partial x} \end{pmatrix},$$

$$\Omega_2 = \begin{pmatrix} v\frac{\partial}{\partial r} & 0 & 0 & 0 \\ 0 & v\frac{\partial}{\partial r} - \frac{1}{Re\rho r^\nu}\frac{\partial}{\partial r} r^\nu \mu \frac{\partial}{\partial r} & 0 & 0 \\ 0 & 0 & v\frac{\partial}{\partial r} - \frac{4}{3Re\rho r^\nu}\frac{\partial}{\partial r} r^\nu \mu \frac{\partial}{\partial r} & 0 \\ 0 & 0 & 0 & v\frac{\partial}{\partial r} - \frac{\gamma}{Re\rho r^\nu Pr}\frac{\partial}{\partial r} r^\nu \mu \frac{\partial}{\partial r} \end{pmatrix},$$

$$\Omega_3 = \begin{pmatrix} 0 & \rho\frac{\partial}{\partial x} & 0 & 0 \\ \frac{(\gamma-1)\varepsilon}{\rho}\frac{\partial}{\partial x} & 0 & 0 & (\gamma-1)\frac{\partial}{\partial x} \\ 0 & 0 & 0 & 0 \\ 0 & (\gamma-1)\varepsilon\frac{\partial}{\partial x} & 0 & 0 \end{pmatrix},$$

$$\Omega_4 = \begin{pmatrix} 0 & 0 & \frac{\rho}{r^\nu}\frac{\partial}{\partial r} r^\nu & 0 \\ 0 & 0 & 0 & 0 \\ \frac{(\gamma-1)\varepsilon}{\rho}\frac{\partial}{\partial r} & 0 & 0 & (\gamma-1)\frac{\partial}{\partial r} \\ 0 & 0 & \frac{(\gamma-1)\varepsilon}{r^\nu}\frac{\partial}{\partial r} r^\nu & 0 \end{pmatrix},$$

$$W_1 = \begin{pmatrix} u\frac{\partial}{\partial x} & \frac{1}{Re_m r^\nu}\frac{\partial}{\partial r} r^\nu \mu_1 \frac{\partial}{\partial x} \\ 0 & u\frac{\partial}{\partial x} - \frac{1}{Re_m}\frac{\partial}{\partial x}\mu_1\frac{\partial}{\partial x} \end{pmatrix},$$

$$W_2 = \begin{pmatrix} v\frac{\partial}{\partial r} - \frac{1}{Re_m r^\nu}\frac{\partial}{\partial r} r^\nu \mu_1 \frac{\partial}{\partial r} & 0 \\ \frac{1}{Re_m}\frac{\partial}{\partial x}\mu_1\frac{\partial}{\partial r} & v\frac{\partial}{\partial r} \end{pmatrix},$$

$$A_1\vec{f} = \begin{pmatrix} 0 \\ \frac{1}{r^\nu}\left(\frac{\partial}{\partial r} r^\nu \mu \frac{\partial v}{\partial x} - \frac{2}{3}\frac{\partial}{\partial x}\mu\frac{\partial}{\partial r} r^\nu v\right) \\ \frac{\partial}{\partial x}\mu\frac{\partial u}{\partial r} - \frac{2}{3}\frac{\partial}{\partial r}\mu\frac{\partial u}{\partial x} - 2\nu v\left(\mu/r^2 + \frac{1}{3}\frac{\partial}{\partial r}\frac{r^\nu}{\mu}\right) \\ 2\mu\left[\left(\frac{\partial u}{\partial x}\right)^2 + \left(\frac{\partial v}{\partial r}\right)^2 + \nu\frac{v^2}{r^2} - \frac{1}{3}\left(\frac{\partial u}{\partial x} + \frac{1}{r^\nu}\frac{\partial}{\partial r} r^\nu v\right)^2 + \frac{1}{2}\left(\frac{\partial u}{\partial r} + \frac{\partial v}{\partial x}\right)^2\right] \end{pmatrix},$$

$$A_2\vec{H} = \begin{pmatrix} \frac{H_x}{r^\nu}\frac{\partial}{\partial r} r^\nu v & -H_r\frac{\partial u}{\partial r} \\ -H_x\frac{\partial v}{\partial x} & H_r\frac{\partial u}{\partial x} \end{pmatrix},$$

$$B\vec{f} = \frac{\varphi D}{\rho}\begin{pmatrix} 0 \\ -H_r \\ H_x \\ \frac{\mu_1 D}{Re_m} \end{pmatrix}, \quad D = \frac{\partial H_r}{\partial x} - \frac{\partial H_x}{\partial r} .$$

Velocity and magnetic field are in the same plane; $\nu = 0$ is a plane case; $\nu = 1$ is an axis-symmetric case. Gas pressure p is excluded by using equation $p = (\gamma - 1)\rho\varepsilon$.

x, r - cartesian ($\nu = 0$) or cylindrical ($\nu = 1$) coordinates (axis x is directed along the symmetry axis of a body, axis r is orthogonal to x), u, v - x, r - components of a velocity, μ - dynamical viscosity coefficient, λ - thermal conductivity coefficient, μ_1^{-1} - electrical conductivity, φ - magnetic pressure parameter. We used the following variables (lines denote dimensional functions, index 0 denotes freestream):

$$t = U_0\bar{t}/L, \quad x, r = \frac{\bar{x}, \bar{r}}{L}, \quad u, v = \frac{\bar{u}, \bar{v}}{U_0}, \quad \varepsilon = \frac{\bar{\varepsilon}}{U_0^2},$$

$$\rho = \frac{\bar\rho}{\rho_0}, \quad \mu = \frac{\bar\mu}{\mu_0}, \quad H_{x,z} = \frac{\bar H_{x,z}}{H_0}, \quad \mu_1 = \sigma_0/\bar\sigma,$$

$$Re = \rho_0 U_0 L/\mu_0, \quad Re_m = 4\pi\sigma_0 L U_0/c^2, \quad \varphi = \frac{H_0^2}{4\pi\rho_0 U_0^2} = \frac{V_A^2}{U_0^2}, \quad P_r = \frac{c_p\mu_0}{\lambda_0}$$

V_A^2 - is Alfvén velocity, U_0 - is undisturbed flow velocity, L - length of a body, σ_0 - electrical conductivity in a free flow. No slip for velocity and thermal isolation are taken on the body surface.

The stationing method is used to obtain a stationary solution, existence and uniqueness is assumed. In this case a choice of the initial conditions is arbitrary enough, and initial functions in domain were taken as freestream parameters: $u = \rho = H_x = 1$, $v = H_z = 0$, $\varepsilon = 1/\gamma(\gamma-1)M_0^2$, where $M_0 = \frac{U_0}{s_0}$, $s_0 = \sqrt{\gamma p_0/\rho_0}$. It is important to have a possibility to obtain a stationary problem solution by using a large time step τ, therefore we have constructed an implicit scheme based on the splitting according to physical processes and space variables. At first we obtain gasdynamical functions assuming a magnetic field as known, and then we obtain a new magnetic field. Finite-difference scheme for numerical solution of the system (1) is represented as

$$\begin{aligned}
&(E + \tau\Omega_1^h)\vec\xi^{\,n+1/4} = -\tau(\Omega^h + A_1^h + B^h)\vec f,\\
&(E + \tau\Omega_2^h)\vec\xi^{\,n+1/2} = \vec\xi^{\,n+1/4},\\
&(E + \tau\Omega_3^h)\vec\xi^{\,n+3/4} = \vec\xi^{\,n+1/2}, \qquad\qquad (3)\\
&(E + \tau\Omega_4^h)\vec\xi^{\,n+1} = \vec\xi^{\,n+3/4},\\
&\vec f^{\,n+1} = \vec f^{\,n} + \vec\xi^{\,n+1}.
\end{aligned}$$

Finite-difference operators Ω_i^h, $A_{1,2}^h$, B^h approximate the differential operators with first or second order of accuracy; approximation of the convective terms is made depending on a sign of velocity. System (2) is approximated by a scheme of universal algorithm:

$$(E + \tau W_1^h)\vec{\eta}^{n+1/2} = \tau (W_1^h + W_2^h + A_2^h)\vec{H}^n, \qquad (4)$$

$$(E + \tau W_2^h)\vec{\eta}^{n+1} = \vec{\eta}^{n+1/2},$$

$$\vec{H}^{n+1} = \vec{H}^n + \vec{\eta}^{n+1}.$$

The scheme (3), (4) is a total approximation scheme and realized by three-point sweepings. As shown by the linear analysis the scheme is absolutely stable. According to methodical calculations the scheme is stable at least up to the Courant number $K = \frac{\tau}{h_1 h_2}[h_1|v| + h_2|u| + s \cdot \max(h_1, h_2)] = 6$, $\quad s = \sqrt{s_0^2 + V_A^2}$.

Let us consider the results of a calculation of viscous thermal--conductive ionized flow around a L wedge with length and half angle β . Undisturbed flow is supersonic one, velocity and magnetic field upstream the wedge are directed along the body axis, $\mu = \varepsilon^{3/4}$, $\mu_1 = \varepsilon^{-3/2}$, $\gamma = 5/3$, $Pr = 0.72$. Left and upper boundaries of the domain was chosen on the distance $2 \div 4\ L$ from the body in order to have freestream conditions. Lower boundary was axis and wedge surface. Right boundary was on the distance $5 \div 6\ L$ from the wedge base and for all functions the conditions $\partial f/\partial x = 0$ were taken. The algorithm makes it possible to calculate all the flow. According to calculations there are shock zone, boundary layer on the wedge and wake with or without vortex. Shock-detachment distance x_1 as determined by the maximum gradient of density on the axis is shown in Fig. 1 $M_0 = 2$ ($\beta = 45°$, $Re = Re_m = 300$). With increasing of a magnetic field the distanse decreases at first and then increases as $\varphi \gtrsim 0.35$; growth of parameter $\varphi = V_A^2 / U_0^2$ leads to a decreasing of Mach number $M = (M_0^{-2} + \varphi)^{-1/2}$ and flow is subsonic one when $\varphi > 1 - M_0^{-2}$. Thickness of a boundary layer on the wedge is increasing with a magnetic field and coefficient $C_f = 2Re^{-1}\int_0^L \mu \frac{\partial u}{\partial n} dl$ is decreasing (l - is directed along the wedge, n - is normal to it). The same effect was detected in a case of incompressible conductive fluid too. Increasing of a magnetic field at fixed M_0, Re leads to a decreasing of the vortex zone behind the wedge. Distribution of the longitudinal velocity on the axis behind the wedge is shown in Fig.2 ($M_0 = 2$, $Re = Re_m = 500$, $\beta = 16°40'$). The flow has no vortex at $\varphi \geqslant 0.04$. In axis - symmetric case the flow is not changed in a qualitative way, but shock - detachment distance is smaller and vortex zone is wider. In Fig.3 the lines of equal values of a local Mach number $M = \{(u^2+v^2)/(\gamma(\gamma-1)\varepsilon + \varphi H^2/\rho)\}^{1/2}$ for $\varphi = 0.1$ are shown. With no magnetic field subsonic zone is bigger, but both cases are similar.

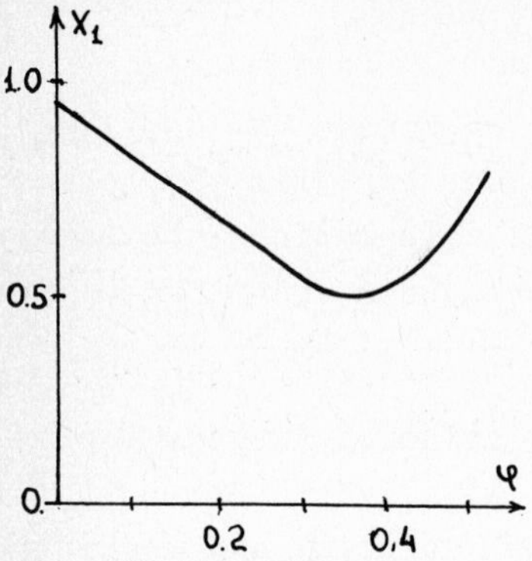

Fig. 1

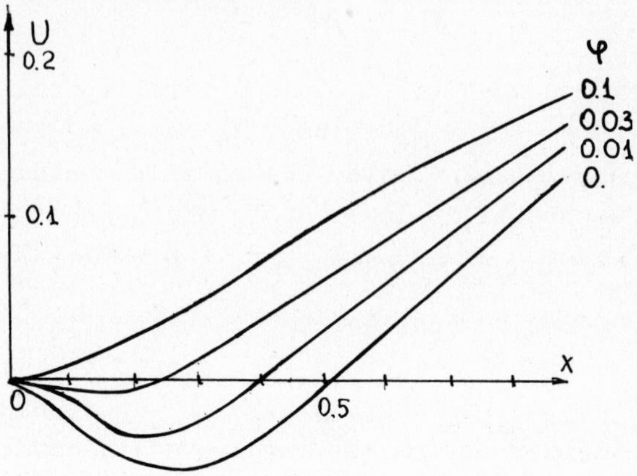

Fig. 2

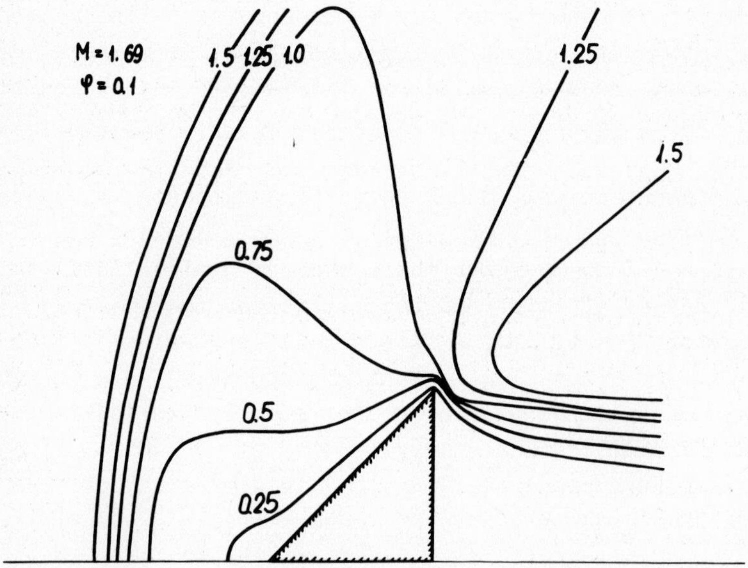

Fig. 3

NONUNIFORM GRID METHOD FOR TURBULENT BOUNDARY LAYERS*

Frederick G. Blottner
Sandia Laboratories
Albuquerque, New Mexico U.S.A. 87115

ABSTRACT

A Crank-Nicolson type finite-difference scheme with a nonuniform grid spacing has been interpreted in terms of a coordinate stretching approach to show that it is second-order accurate. The variable grid scheme is applied to a flat plate laminar to turbulent boundary layer flow with a rapidly changing grid interval across the layer. The accuracy of the solution is determined for a different number of intervals and compared to results obtained with the Keller box scheme. The influence of changing the grid spacing on the accuracy of the solutions is determined for three coordinate stretching or grid spacing relations.

INTRODUCTION

It has been recognized for sometime that a significant reduction in the number of grid points can be obtained in problems with large local gradients if a variable grid spacing or a coordinate stretching is employed in the numerical solution. The need for reducing the number of grid points is illustrated in Fig. 1 for a laminar to turbulent boundary layer flow along a flat plate. With the Levy-Lees boundary layer coordinate at the outer edge, η_e, equal to 24.25, the number of intervals required across the laminar and turbulent velocity profiles to give 1% accuracy of the wall shear stress, τ_w, has been estimated. As shown in the figure the turbulent profile requires 20 times as many grid points as the laminar profile or 80 times as many grid points if the laminar solution is stopped at $\eta_e = 6$. This example shows the necessity of finding a procedure to reduce the number of grid points required to obtain reasonably accurate solutions of turbulent boundary layer flows.

The nonuniform grid approach of Smith and Cebeci (1968) is typical of the finite-difference methods employed for solving the complete turbulent boundary layer flow. These procedures have used a slowly varying interval spacing and have required several hundred intervals across the flow. A better approach has been developed by Keller and Cebeci (1972) which employs the Keller box scheme (KBS). This method is claimed to be better than most (if not all) other numerical methods which have been employed on such problems. The KBS requires the solution of block-tridiagonal equations and becomes very time consuming when extended to the compressible boundary layer.

The purpose of this paper is to show that a Crank-Nicolson type difference scheme with an analytically prescribed grid spacing has the same accuracy as the KBS and is more efficient. The present evaluation is limited to one test case which is the laminar to turbulent boundary layer flow along a flat plate with the eddy-viscosity formulation of Keller and Cebeci utilized. To obtain an accurate solution of the problem, several solutions with a large number of intervals are calculated and then Richardson extrapolation is used to obtain an accurate result which is considered the "exact" solution. With this approach, the accuracy of the methods with various numbers of intervals is compared in a rigorous manner and the variation of accuracy with number of intervals is obtained. A complete description of the method and additional results are given in Blottner (1974).

*This work was supported by the U. S. Atomic Energy Commission.

NUMERICAL APPROACH

Consider a variable grid spacing across the boundary layer with $\eta_{j+1} = \eta_j + \Delta\eta_{j+\frac{1}{2}}$ where $\eta_1 = 0$ and $j = 1,2,3,\cdots,(J-1)$. The first and second derivatives in the η-direction occur in the boundary layer equations and require finite-difference representation. Only the first derivative will be used to illustrate the numerical approach. One finite-difference form of the first derivative with the first truncation error term included is

$$\left(\frac{\partial W}{\partial \eta}\right)_j = \frac{W_{j+1} - W_{j-1}}{\eta_{j+1} - \eta_{j-1}} - \frac{1}{2}\left(\frac{\partial^2 W}{\partial \eta^2}\right)_j (\eta_{j+1} - 2\eta_j + \eta_{j-1}) + O(\Delta\eta_j^2) \tag{1}$$

If the expression (1) is to be locally second-order accurate, the step-size variation must satisfy the relation

$$(\eta_{j+1} - 2\eta_j + \eta_{j-1}) = (\Delta\eta_{j+\frac{1}{2}} - \Delta\eta_{j-\frac{1}{2}}) = O(\Delta\eta^2_{j-\frac{1}{2}}) \tag{2}$$

If the ratio of step-sizes is expressed as

$$\Delta\eta_{j+\frac{1}{2}} = K\Delta\eta_{j-\frac{1}{2}} \quad j = 2,3,\cdots,(J-1) \tag{3}$$

which has been used by Smith and Cebeci, then Eq. (2) gives $K = 1 + O(\Delta\eta_{j-\frac{1}{2}})$. This shows that the step-size must vary slowly to make the difference relation (1) locally second-order accurate. This is the approach used in most previous non-uniform grid schemes.

The present variable grid scheme (VGS) is interpreted in terms of a coordinate stretching approach. A new coordinate N is introduced where a uniform interval ΔN is used and is related to the original coordinate η by a relation of the form

$$\eta_j = \eta(N_j) \qquad j = 1,2,3,\cdots,J \tag{4}$$

The derivative (1) is transformed into the new coordinate system and central differences are employed to obtain

$$\left(\frac{\partial W}{\partial \eta}\right)_j = \left(\frac{\partial W}{\partial N}\Big/\frac{d\eta}{dN}\right)_j = (W_{j+1} - W_{j-1})/\left[2\Delta N(d\eta/dN)_j\right] + O(\Delta N^2) \tag{5}$$

The above derivative (5) can be used to solve the governing equations in the N-coordinate system with $d\eta/dN$ obtained from Eq. (4). The present approach is to replace the coordinate derivative with finite-difference to obtain

$$(d\eta/dN)_j = (\eta_{j+1} - \eta_{j-1})/(2\Delta N) + O(\Delta N^2) \tag{6}$$

With Eq. (6) used in Eq. (5), the resulting difference equation is

$$(\partial W/\partial\eta)_j = (W_{j+1} - W_{j-1})/(\eta_{j+1} - \eta_{j-1}) + O(\Delta N^2) \tag{7a}$$

In a similar manner the second derivative can be written in finite-difference form and is

$$\left[\partial(\ell\partial W/\partial\eta)/\partial\eta\right]_j = 2\Big[\ell_{j+\frac{1}{2}}(W_{j+1} - W_j)/(\eta_{j+1} - \eta_j)$$

$$- \ell_{j-\frac{1}{2}}(W_j - W_{j-1})/(\eta_j - \eta_{j-1})\Big]/(\eta_{j+1} - \eta_{j-1}) + O(\Delta N^2) \qquad (7b)$$

The foregoing shows that the present VGS is equivalent to a coordinate stretching method if a relation of the form of Eq. (4) is used. This relation is used to specify the grid spacing in the variable grid method while it gives the relationship between the coordinates for the stretching method. In both cases the derivatives are second-order accurate in terms of ΔN.

The relation (4) for the variable grid spacing for the case given in Eq. (3) has been given by Cebeci and Smith (1970) and is written as

$$\eta_j = \eta_J\left(K^{N_j/\Delta N_o} - 1\right)\Big/\left(K^{1/\Delta N_o} - 1\right) \qquad j = 1,2,3,\cdots J \qquad (8)$$

where $N_j = (j - 1)\Delta N$ and $N_J = 1$. The values of K and ΔN_o are two parameters which are chosen to give the desired grid spacing.

APPLICATION OF SCHEME

The VGS is applied to the incompressible turbulent boundary layer equations with the eddy-viscosity formulation of Keller and Cebeci (1972) employed. The independent variables are transformed with the Levy-Lees variables $\xi = (u_e/\nu)x = Re_x$ and $\eta = \sqrt{u_e/(2x\nu)}\,y$ and new dependent variables $F = u/u_e$ and $V = 2\xi(u F\partial\eta/\partial x + \rho v/\sqrt{2\xi})/\rho u_e$ are introduced. The governing equations become

$$2\xi\ \partial F/\partial\xi + \partial V/\partial\eta + F = 0 \qquad (9a)$$

$$-2\xi F\ \partial F/\partial\xi + \partial(\ell\partial F/\partial\eta)/\partial\eta - V\partial F/\partial\eta + \beta(1 - F^2) = 0 \qquad (9b)$$

For the present investigation of a flat plate flow, the pressure gradient parameter $\beta = 0$. The coefficient $\ell = 1 + \varepsilon/\nu$ where ε is the eddy viscosity. The above system of equations is completed with the conditions at the surface ($\eta = 0$) where $F = V = 0$ and at the outer edge ($\eta = \eta_e$) where $F = 1$.

The grid point location in the ξ-direction is given as $\xi_{i+1} = \xi_i + \Delta\xi_{i+\frac{1}{2}}$ where $\xi_1 = 0$ and $i = 1,2,3\cdots I$. The derivatives and function in the continuity Eq. (9a) are evaluated using the corner values of the grid box with center at $(i + \frac{1}{2}, j - \frac{1}{2})$. Two-point central differences are used for the derivatives and the continuity equation in finite-difference form becomes

$$V_{i+1,j} = V_{i+1,j-1} - c_j(F_j + F_{j-1})_{i+1} + d_j \qquad (10)$$

where the coefficients c_j and d_j depend on known quantities. The momentum Eq. (9b) is evaluated at the point $(i + \frac{1}{2}, j)$. The nonlinear terms are determined in an iterative manner and are linearized with

$$F^2_{i,j} = (2\bar{F}F - \bar{F}^2)_{i,j} \qquad (11a)$$

$$V\ \partial F/\partial\eta = -\bar{V}\ \partial\bar{F}/\partial\eta + \bar{V}\ \partial F/\partial\eta + V\ \partial\bar{F}/\partial\eta \qquad (11b)$$

where the quantities with a bar are evaluated from a previous iteration. The momentum equation in finite-difference form with the use of Eqs. (7) becomes

$$(-A_j F_{j-1} + B_j F_j - C_j F_{j+1})_{i+1} + (aV)_{i+1,j} = D_{i+1,j} \qquad (12)$$

where the coefficients A_j, B_j, C_j, a_j and D_j at (i + 1) depend on known quantities and the dependent variables evaluated from a previous iteration.

The governing finite-difference equations (10) and (12) are coupled and are solved with a method suggested by Davis and used by Werle and Bertke (1972). This method is a slight modification of the usual tridiagonal procedure. Since the original equations are nonlinear, the linearized equations require at least one iteration at each step along the surface in order to obtain a second-order accurate scheme.

NUMERICAL RESULTS

The laminar to turbulent boundary layer flow along a flat plate is utilized as the test case and is the problem previously investigated by Keller and Cebeci with the KBS. For this problem the η-grid is determined with Eq. (8) where $\eta_\delta = \eta_J = 24.2538$, $K = 1.82$ and $\Delta N_0 = 0.1$ which gives $\Delta\eta_{1.5} = 0.05$. The ξ-grid points are: $\xi_i \times 10^{-6}$ = 0, 0.00086, 0.0043, 0.0086, 0.043, 0.17, 0.26, 0.34, 0.51, 0.68, 0.86, 1.03, 1.28, 1.50, 1.71 and 1.88.

Accuracy of the finite-difference solutions is judged by the value of the velocity gradient or the skin friction at the surface of the flat plate. The skin friction is obtained from

$$C_f = \sqrt{2/\xi}\,(\partial F/\partial\eta)_W \qquad (13)$$

The velocity gradient at the wall is obtained from

$$\ell(\partial F/\partial\eta) = a + b\eta + c\eta^2 + \cdots \qquad (14)$$

and since at the wall $F = V = 0$ and $\ell = 1$, the momentum Eq. (9b) evaluated at the wall gives $b = -\beta$. Then Eq. (14) is evaluated at $j = 1.5$ and 2.5 and the two resulting equations are used to eliminate c and obtain the wall velocity gradient.

In order to judge the accuracy of the various methods with different step-sizes, an "exact solution" of the skin friction is required. The numerical solution depends on ΔN^2 and $\Delta\xi_i^2$ as the methods being considered are second-order accurate in both coordinates. In this paper only the accuracy of the solution as ΔN is varied is of interest and the $\Delta\xi_i$ steps are the same for all problems. Therefore, the "exact solution" used to evaluate the results corresponds to the solution with $\Delta N \to 0$ but with finite $\Delta\xi_i$. This "exact solution" is obtained with Richardson extrapolation. Numerical solutions show that the percent error is decreasing by approximately $\frac{1}{4}$ as the step-size is halved which is the expected behavior of a second-order accurate scheme. Also, the percent error changes only slightly as the number of iterations is changed and only one iteration is necessary to obtain a second-order accurate scheme.

The accuracy of the VGS has been determined with the procedure described above for various number of intervals across the boundary layer. The result of this investigation is shown in Fig. 2 for the turbulent velocity profile. Also shown in this figure is the same type of investigation for the KBS. These results show that the VGS is slightly more accurate than the KBS for turbulent profiles.

The effect of the spacing of the grid points has been investigated by varying the value of K in Eq. (8) and with $\Delta N_0 = 0.1$. The influence of the value of K on the accuracy of the solution is shown in Fig. 3 for laminar velocity profiles and in Fig. 4 for turbulent velocity profiles. For laminar profiles the accuracy of the results decreases as K is increased from 1.4 while for turbulent profiles the accuracy of the results increases as K is increased. However, there is initially an increase in accuracy of the wall shear stress for the laminar profile as K goes from 1 to 1.4. With K = 2, the laminar profile results are more accurate than the turbulent profile results with the same number of intervals. These figures show that the accuracy of the turbulent results can be changed significantly when K is changed. The velocity profiles in the new coordinate N are given in Fig. 5 for the laminar case and in Fig. 6 for the turbulent case. When K = 1, the value of $N = \eta$ and the profiles appear in the original unstretched form. For both cases as K increases, a sudden change in the profile shape is introduced at the outer edge of the boundary layer which should degrade the accuracy of the numerical solution. At the same time in both cases as K increases, more grid points are introduced into the solution near the surface and better accuracy will result in this region. The combination of these two effects results in the variation of the accuracy of the skin friction as given in Figs. 3 and 4. Since the velocity gradient at the wall is used to judge the accuracy of the solutions, the results in Figs. 3 and 4 would change if another parameter is used to judge the accuracy. The results with $K \geq 1.4$ for a laminar profile indicate that by placing more of the grid points near the surface produces an adverse effect on the accuracy. It has been shown by Denny and Landis (1972) that the optimal node distribution for a laminar profile utilizes a fine grid spacing in the outer part of the boundary layer. The present result is consistent with their study. For turbulent profiles it appears that a fine grid spacing is needed mainly near the surface. When the velocity profiles are plotted as a function of N as given in Figs. 5 and 6, it is easy to see if a reasonable grid spacing has been used. The ideal profile shape would be a second-order polynomial which can be represented exactly with central differences.

Two other transformations (4) have been investigated to determine the influence on the accuracy of the turbulent results. The inverse tangent function transformation suggested for the resolution of boundary layers by Orszag and Israeli (1974)has been used. With this relation it is necessary to have approximately 80 intervals across the turbulent profile to have 1% accuracy of the skin friction at $\xi = 1.88 \times 10^6$. With Eq. (8) only 20 intervals are needed to obtain approximately the same accuracy. The following transformation has also been investigated

$$N = \alpha\eta + (1 - \alpha)\left(e^{-\eta/\epsilon} - 1\right) / \left(e^{-1/\epsilon} - 1\right) \tag{15}$$

which is a transcendental equation in terms of η. Since N is specified, an iterative procedure is used to obtain η. With $\alpha = 0.5$ and $\epsilon = 0.02$, the turbulent profile skin friction is approximately 1% accurate if 30 intervals are used across the layer. The original transformation (8) appears to be the better method of spacing the grid points for a turbulent boundary layer.

SUMMARY

The objective of this paper has been to investigate the accuracy of the VGS and the efficiency of this approach is considered now. The VGS requires modified tridiagonal equations to be solved while the KBS requires the solution of block tridiagonal equations and should require significantly more computation time. For the example problem which was solved on a CDC 6600 computer, the KBS required 2.3 times as much time as the VGS.

The main conclusions of this paper are summarized as:

1. A variable grid scheme has been developed for the incompressible turbulent boundary layer equations which has second-order accuracy and allows rapid variation of the interval spacing. This method can be extended to compressible flows with only tridiagonal equations resulting.

2. The claim of Keller and Cebeci of the superiority of the box scheme over the Crank-Nicolson type schemes is not valid. The present scheme is easier to formulate, is as accurate and requires less computer time.

3. The variable grid scheme is equivalent to a coordinate stretching approach where the coordinate transformation is used to specify the non-uniform grid.

4. The grid spacing with constant ratio of adjacent intervals is a good method for turbulent boundary layers.

REFERENCES

Blottner, F. G., Computer Methods in Applied Mech. and Engr. (to be published) (1974)

Cebeci, T., and Smith, A.M.O., J. of Basic Engr., Vol. 92, No. 3, 523-535 (1970)

Denny, V. E. and Landis, R. B., J. of Computational Physics, Vol. 9, 120-137 (1972)

Keller, H. B., and Cebeci, T., AIAA J., Vol. 10, No. 9, 1193-1199 (1972)

Orszag, S. A. and Israeli, M. in Annual Review of Fluid Mechanics, Vol. 6, Eds. M. Van Dyke and W. G. Vincenti, Annual Reviews, Inc., Palo Alto, Calif., (1974)

Smith, A.M.O., and Cebeci, T., Proceedings of the 1968 Heat Transfer and Fluid Mechanics Institute, 174-191 (1968)

Werle, M. J. and Bertke, S. D., AIAA J., Vol. 10, No. 9, 1250-1252 (1972)

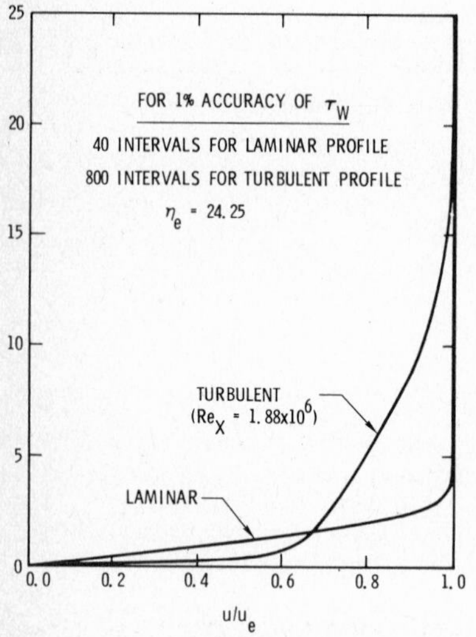

Figure 1. Laminar and Turbulent Velocity Profiles

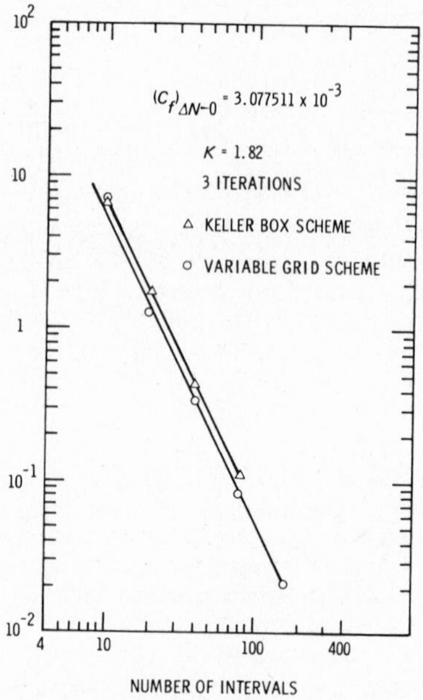

Figure 2. Accuracy of Skin Friction for Turbulent Boundary Layer at $Re_x = 1.88 \times 10^6$

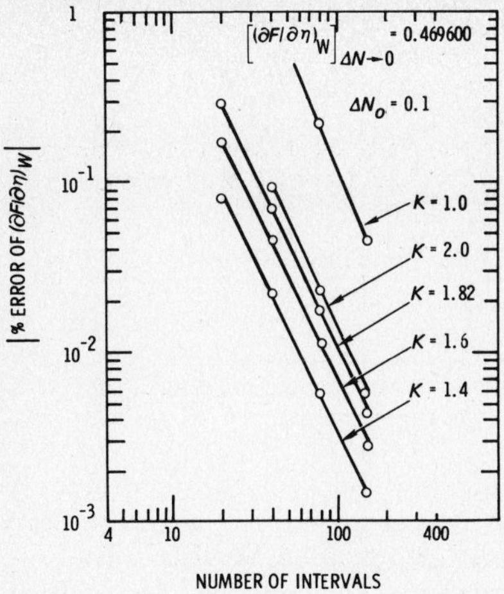

Figure 3. Accuracy of Wall Velocity with Variable Grid Scheme with Various K's for Laminar Boundary Layer at $Re_x = 0$

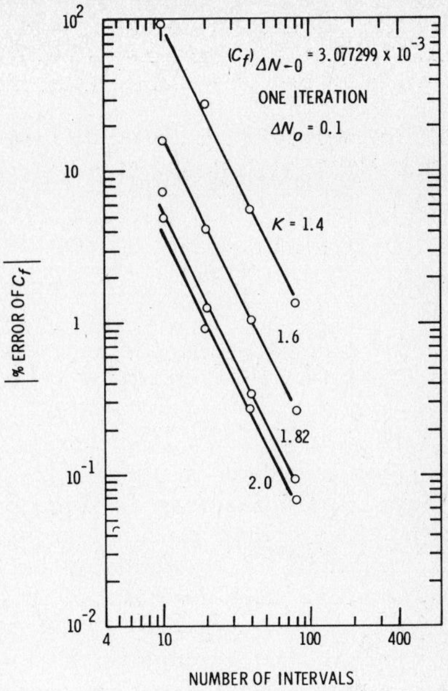

Figure 4. Accuracy of Skin Friction with Variable Grid Scheme with Various K's for Turbulent Boundary Layer at $Re_x = 1.88 \times 10^6$

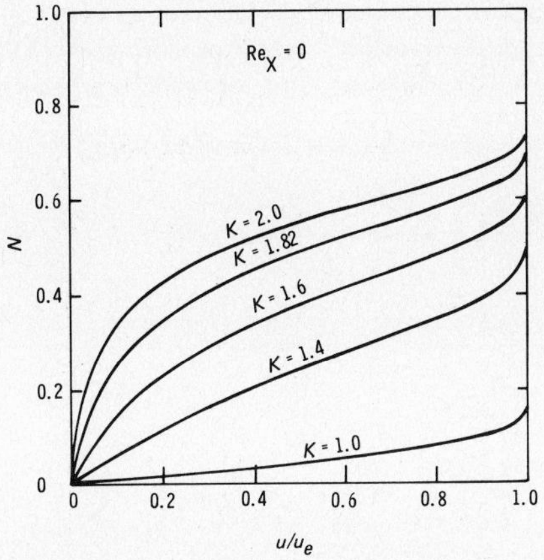

Figure 5. Velocity Profiles for Laminar Flow for Various Values of K

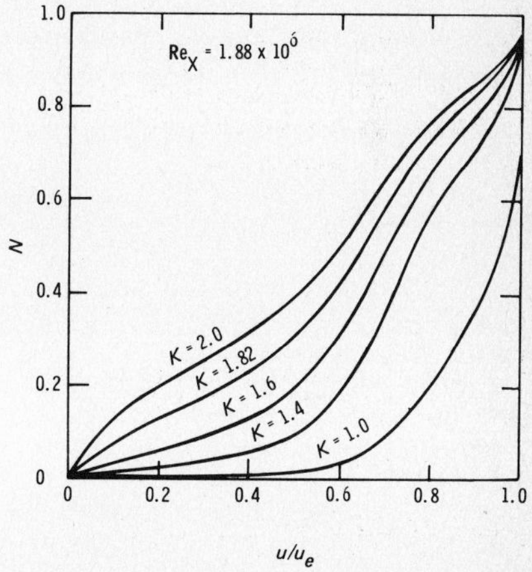

Figure 6. Velocity Profiles for Turbulent Flow for Various Values of K

NUMERICAL STUDY OF THE HEATING OF A PLASMA

JEAN-PAUL BOUJOT

Compagnie Internationale de Services en Informatique
B.P. n° 2
91190 - GIF-sur-YVETTE Saclay - France

The purpose of the T.F.R program (TOKOMAK de FONTENAY-aux-ROSES) is to study the behaviour of a plasma of toroidal shape strongly heated by a current discharge.

This program is part of the controlled fusion research program of the C.E.A. (Commissariat à l'Energie Atomique) which includes the study of TOKOMAK with non circular cross section plasma. The plasma is confined in a toroidal shell. A method for producing various non circular cross section plasmas has been developped in [2].

The M.H.D. model is the underlying basis for the numerical study, the plasma being considered as a perfect incompressible fluid. A code for simulating the evolution of two fluid is being used by plasma physicists [1] (one space dimension model).

The behaviour of a plasma subjected to the same electric discharges can also be studied by using a different cross section each time [2]. Therefore a numerical simulation code analogous to [1] must be developped without the symetry of revolution of the cylindrical model.

There are still problems in the theoretical formulation of the model in two dimensional space, particularly in the definition of the thermal diffusion coefficients. This note presents a method for solving the Maxwell equation in the M.H.D. system in two dimensional space.

I. - EQUATIONS. -

We start with the Maxwell equations

$$\text{(I)} \quad \begin{cases} \varepsilon \dfrac{\partial E}{\partial t} + \mu_0 J - \operatorname{curl} B = 0 \\ \dfrac{\partial B}{\partial t} + \operatorname{curl} E = 0 \text{ in } \Omega \subset \mathbb{R}^3 \end{cases}$$

Where :

- Ω is the domain filled by the plasma
- E the electric field, B the magnetic field and J the current density.

. $\varepsilon = \frac{1}{C^2}$, C^2 the speed of ligth.

. μ_0 the magnetic permeability.

BOUNDARY CONDITIONS.

We denote by Γ the boundary of Ω , i. e. of the plasma, and ν the unit normal vector on Γ.

Then $B \cdot \nu = 0$, $E \wedge \nu = 0$, on Γ .

INITIAL CONDITIONS.

$B(0) = B_0$ and $E(0) = E_0$ in Ω .

II. - ABOUT EXISTENCE AND UNIQUENESS OF SOLUTION. -

We use the same notations as in [3]:
Let $\mathcal{H} = (L^2(\Omega))^6$. The domain of definition $\mathcal{D}(\mathcal{A})$ of the operator is given by:

$$\mathcal{D}(\mathcal{A}) = \{\Phi,\ \Phi \in \mathcal{H},\ \text{curl}\ \varphi \in (L^2(\Omega))^3,\ \text{curl}\ \psi \in (L^2(\Omega))^3,\ \nu \wedge \varphi = 0 \mid \text{on}\ \Gamma.\}$$

for any Φ in $\mathcal{D}(\mathcal{A})$.

$$\mathcal{A}\Phi = \{-\text{curl}\ \psi, \text{curl}\ \varphi\}\ ,\quad \Phi = (\psi, \varphi).$$

for $\varepsilon \to 0$, using the results of Lions [3] , [4] it can be shown that

THEOREM

Assume that

$E_0 \in \mathcal{H}$, curl $E_0 \in \mathcal{H}$, $E_0 \wedge \nu = 0$ on Γ

$B_0 \in \mathcal{H}$ is such that curl $B_0 = \mu_0 \sigma$,
div $B_0 = 0$, $B_0 \cdot \nu = 0$ on Γ

there exist a unique pair of functions B and E such that

$E \in L^2(0, T; \mathcal{D}(\mathcal{A}))$, curl $E \in L^\infty(0, T; \mathcal{H})$,
$B \in L^\infty(0, T; \mathcal{D}(\mathcal{A}))$, $\frac{\partial B}{\partial t} \in L^\infty(0, T; \mathcal{H})$,

curl $B \in L^2(0, T; \mathcal{H})$,

Satisfying

$$(II)\quad \begin{cases} \operatorname{curl} B = \mu_0 \sigma E \\ \dfrac{\partial B}{\partial t} + \operatorname{curl} E = 0 \end{cases}$$

and, $B(0) = B_0$, $E(0) = E_0$, $\operatorname{div} B = 0$

$B.\nu = 0$, $E \wedge \nu = 0$ on Γ .

Moreover, in the particular case where Ohm's law is $J = \sigma E$ (i. e the conductivity σ is a continious, positive and bounded function of the electronic temperature), the previous system leads to

$$(III)\quad \begin{cases} \dfrac{\partial B}{\partial t} + \operatorname{curl}\left(\dfrac{1}{\mu_0 \sigma} \operatorname{curl} B\right) = 0 & \text{in } \Omega . \\ B.\nu = 0 & \text{on } \Gamma . \\ B(0) = B_0 & \text{in } \Omega . \end{cases}$$

The above theorem has been established starting from a weak formulation of the problem in suitable functional spaces. This formulation is not directly usable numerically, therefore, we are led to build a second functional formulation before attempting the numerical solutions.

III. - FUNCTIONAL FORMULATION OF THE PROBLEM.

Let $\mathbb{V} = \left[U \;:\; U \in (H^1(\Omega))^3 \text{ and } U.\nu = 0 \text{ on } \Gamma \right]$

where

$$H^1(\Omega) = \left[U \;:\; U \in L^2(\Omega), \frac{\partial U}{\partial Z} \in L^2(\Omega)^3 \right]$$

with (Z = x and y).

$$B = \begin{pmatrix} Bx \\ By \\ 0 \end{pmatrix} , \quad U = \begin{pmatrix} Ux \\ Uy \\ 0 \end{pmatrix} , \quad P = \frac{1}{\sigma \mu_0} \left(\frac{\partial By}{\partial x} - \frac{\partial Bx}{\partial y} \right)$$

then

$$\operatorname{curl}\left(\frac{1}{\sigma \mu_0} . \operatorname{curl} B\right) = {}^t\left(\frac{\partial P}{\partial y} , - \frac{\partial P}{\partial x} , 0\right)$$

Formally, for all $U \in \mathbb{V}$, the vector product of U with the first equation in III is taken and integrated over Ω , giving

$$\int_{\Omega} \frac{\partial B}{\partial t} \wedge U \quad d\omega + \int_{\Omega} \text{curl} \left(\frac{1}{\mu_0 \sigma} . \text{curl } B\right) \wedge U \, d\omega = 0$$

Then, using Stokes' formula,

$$\int_{\Omega} \left(\text{curl}\left(\frac{1}{\sigma\mu_0} \text{curl } B\right)\right) \wedge U \quad d\omega$$

$$= \int_{\Omega} \left[P . \frac{\partial Ux}{\partial x} + P \frac{\partial Uy}{\partial y} - \left(\frac{\partial}{\partial x}(P . Ux) + \frac{\partial}{\partial y}(P . Uy)\right)\right] d\omega$$

$$= \int_{\Omega} \left(P \frac{\partial Ux}{\partial x} + P \frac{\partial Uy}{\partial y}\right) d\omega + \int_{\Gamma} P \,(Ux . \nu x + Uy . \nu y) \quad d\sigma .$$

The integral on the boundary Γ is zero, because $U . \nu = 0$

The converse shows that solving II amounts to finding $B \in L^{\infty}(0, T ; (\mathfrak{D}(\mathcal{A}) \cap \mathbb{V}))$ a solution of

$$\text{(IV)} \quad \int_{\Omega} \left[\frac{\partial Bx}{\partial t} . Uy - \frac{\partial By}{\partial t} . Ux\right] d\omega + \int_{\Omega} \frac{1}{\mu_0 \sigma} \left[\frac{\partial By}{\partial x} - \frac{\partial Bx}{\partial y}\right] . \left[\frac{\partial Ux}{\partial x} + \frac{\partial Uy}{\partial y}\right] d\omega = 0 \quad \forall U \in \mathbb{V} .$$

IV. - APPROXIMATION BY FINITE ELEMENTS.

A family $(\mathcal{T}_h)$ of finite triangulations in Ω are defined such that :

$T \subset \overline{\Omega}$ for any triangle $T \in \mathcal{T}_h$

if $T, T' \in \mathcal{T}_h \implies T \cap T' = \emptyset$

either T and T' have a common side,

or T and T' have a common vertex.

h is equal to the maximum length $\mathcal{T}_h$ of sides of the triangles

Let $\Omega_h = \overset{0}{\overline{\cup \; T}}_{T \in \mathcal{T}_h}$; $\Gamma_h = \partial \, \Omega_h$.

$$\omega_h = \{P , P \in \Omega_h , \quad P \text{ vertex in } \mathcal{T}_h \text{ triangles}\}$$

$$\gamma_h = \{P , P \in \Gamma_h\} .$$

Define $\{\mathcal{T}_I\}$ to be the interior triangles that have at last two vertices in ω_h, and $\{T_\Gamma\}$ the boundary triangles that have at last two vertices in γ_h.

A regular triangulation is supposed to have been made in the domain Ω, i.e. the bound Γ is well approximated, and the triangles do not degenerate when the triangulation is refined.

1°) Functional framework.

Let $\mathbb{V}_h = \{ U_h$ such that $U_h \in C^0(\bar{\Omega}_h)$ with the U_h restriction on every $\mathcal{T}_h$ which is a polynomial, and $U_h . \nu = 0$ on $\Gamma_h \}$.

The solution of (III) can therefore be expressed as a polynomial of order less or equal to one, on interior triangles and must satisfy $B . \nu = 0$ on Γ_h.

Moreover, these polynomials must be adequate for interpolation. We give the solutions at the vertex of the triangles which are the nodes of the triangulation.

2°) Approximation of the solution on the boundary triangles.

To set the notations, consider the following figure.

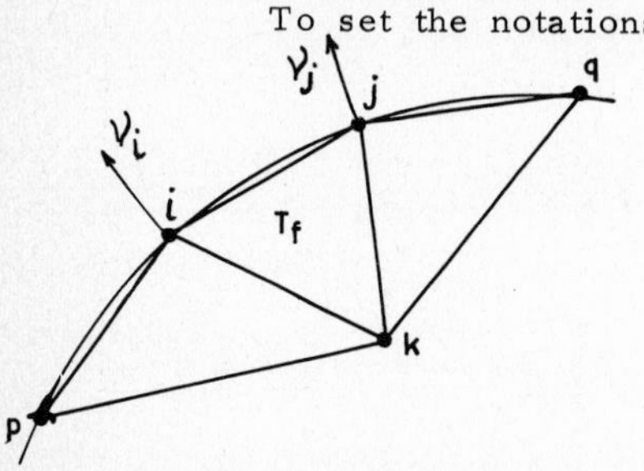

The triangle T_f has (i, j, k) for vertex.

The points p, i, j are on the parabola.

$y = \alpha_{T_f} + \beta_{T_f} x + \gamma_{T_f} x^2$ where α_{T_f}, β_{T_f}, γ_{T_f} are uniquely determined through the coordinates of p, i, j. From i, the normal ν_i to this parabola is taken with known components

$$\nu_{xi} = \left[1 + (\beta_{T_f} + 2\gamma_{T_f} x_i)^2 \right]^{-1/2}$$

$$\nu_{yi} = -\left[\beta_{T_f} + 2\gamma_{T_f} x_i \right] . \nu_{xi}$$

The same are taken on vertices i, j, q to construct ν_j.

HYPOTHESES.

The normal vectors ν_i, ν_j are supposed to be normal to the bound Γ at points i and j.

V. - CONSTRUCTION OF AN INTERPOLATION POLYNOMIAL ON T_h SATISFYING $B . \nu = 0$ -

$B = {}^t(B_x \; B_y)$ is expressed by the approximate polynomial form

$$B_P = P . \Gamma \quad \text{where}$$

$$P = \begin{pmatrix} 1 & x & y & xy & 0 & 0 & 0 & 0 \\ 0 & 0 & 0 & 0 & 1 & x & y & xy \end{pmatrix}, \quad {}^t\Gamma = (\alpha, \beta, \gamma, \delta, \bar{\alpha}, \bar{\beta}, \bar{\gamma}, \bar{\delta}).$$

B_P is given in terms of the values at nodes i, j, k in the triangle T_f under consideration and we state moreover that $B.\nu = W = 0$ at point i, j respectively.

Let

$$B_P = P \,.\, \mathbb{C}_{T_f}^{-1} \,.\, B_{T_f} \quad \text{where}$$

$${}^t B_{T_f} = (BX_i \;\; BX_j \;\; BX_k \;\; W_i \;\; W_j \;\; BY_i \;\; BY_j \;\; BY_k)$$

$$C_{T_f} = \begin{bmatrix} 1 & x_i & y_i & x_i y_i & 0 & 0 & 0 & 0 \\ 1 & x_j & y_j & x_j y_j & 0 & 0 & 0 & 0 \\ 1 & x_k & y_k & x_k y_k & 0 & 0 & 0 & 0 \\ \nu X_i & (X_i . \nu X_i) & (Y_i . \nu X_i) & (X_i Y_i . \nu X_i) & \nu Y_i & (X_i . \nu Y_i) & (Y_i . \nu Y_i) & (X_i Y_i . \nu Y_i) \\ \nu X_j & (X_j . \nu X_j) & (Y_j . \nu X_j) & (X_j Y_j . \nu Y_j) & \nu Y_j & (X_j . \nu Y_j) & (Y_j . \nu Y_j) & (X_j Y_j . \nu Y_j) \\ 0 & 0 & 0 & 0 & 1 & x_i & y_i & x_i . y_i \\ 0 & 0 & 0 & 0 & 1 & x_j & y_j & x_j . y_j \\ 0 & 0 & 0 & 0 & 1 & x_k & y_k & x_k . y_k \end{bmatrix}$$

We proceed the same way to express B on the interior triangles and on the bounded triangles for which the symetry is taken into account, i.e. $\frac{\partial B}{\partial \nu} = 0$.

VI. - APPROXIMATION. -

We return to the weak formulation of the problem given in IV, replacing it by the discretized problem :

determine B satisfying

$$\text{(V)} \quad \frac{\partial}{\partial t} \left\{ \sum_{l=1}^{\mathcal{N}} \int_{T_l} (B_x . U_y - B_y . U_x) \, d\omega \right\} + \sum_{l=1}^{\mathcal{N}} \int_{T_l} \frac{1}{\mu_0 \sigma} . \left[\frac{\partial By}{\partial x} - \frac{\partial Bx}{\partial y} \right] . \left[\frac{\partial Ux}{\partial x} + \frac{\partial Uy}{\partial y} \right] d\omega = 0 \;\; \forall \, U \in \mathbb{V}_h$$

where $\mathcal{N}$ is the total number of triangles.

For U we choose a function which has the same form as B.

The choice for U and the numbering of relevant triangles lead to the solution of a linear system of the type

$$Ax = b \qquad \text{where} \qquad A = A1 + A2$$

with

A1 =

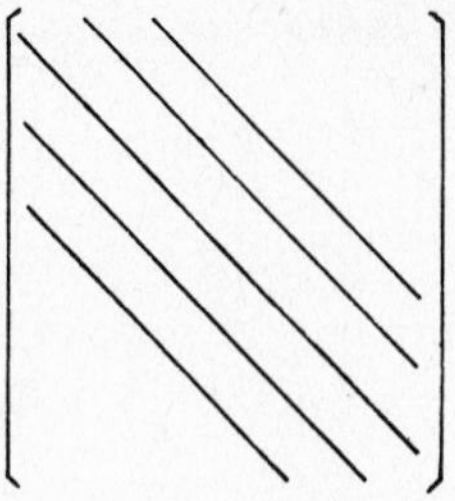

A2 =

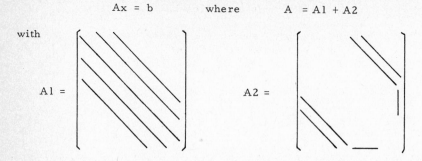

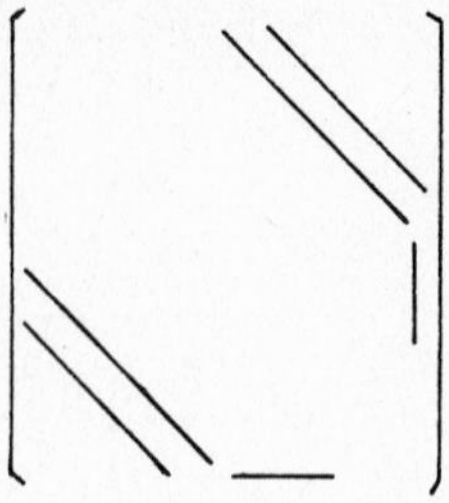

We seek a solution of this system by an iterative method of the following type :

$$A_1 \; X^{(n)} = b - A_2 . X^{(n-1)}$$

in which the solution of

$$A_1 \, X^{(n)} = G$$ uses a direct method.

This method corresponds to a fractional step [5] in the M.H.D coupled problem.

REFERENCES.

[1] J.P. BOUJOT - C. MERCIER - SOUBBARAMAYER.
Numerical simulation of the evolution of a plasma in confinement device.
First European Conference. GENEVE - 1972.

[2] J.P. BOUJOT - J.P. MORERA - R. TEMAM.
Colloque international sur les méthodes scientifiques et techniques.
I.R.I.A. - FRANCE - 1973.

[3] J.L. LIONS - G. DUVAUT.
Les Inéquations en mécanique et en physique.
DUNOD - 1971
SPRINGER - VERLAG (To appear).

[4] J.L. LIONS.
Perturbations singulières dans les problèmes aux limites et en contrôle optimal.
SPRINGER - 1973.

[5] R. TEMAM - Thèse - PARIS - 1967.

SOLUTION OF THE THREE-DIMENSIONAL COMPRESSIBLE NAVIER-STOKES EQUATIONS BY AN IMPLICIT TECHNIQUE

W. R. Briley and H. McDonald

United Aircraft Research Laboratories
East Hartford, Connecticut U.S.A.

INTRODUCTION

One of the major obstacles to the solution of the multidimensional compressible Navier-Stokes equations is the large amount of computer time generally required, and consequently, efficient computational methods are highly desirable. Most previous methods have been based on explicit difference schemes for the unsteady form of the governing equations and are subject to one or more stability restrictions on the size of the time step relative to the spatial mesh size. These stability limits usually correspond to the well-known Courant-Friedrichs-Lewy (CFL) condition and, in some schemes, to an additional stability condition arising from viscous terms. These stability restrictions can lower computational efficiency by imposing a smaller time step than would otherwise be desirable. Thus, a key disadvantage of explicit methods subject to stability limits is that the maximum time step is fixed by the spatial mesh size rather than the physical time dependence or the desired temporal accuracy. In contrast to explicit methods, implicit methods tend to be stable for large time steps and hence offer the prospect of substantial increases in computational efficiency, provided of course that the computational effort per time step is competitive with that of explicit methods. In an effort to exploit these favorable stability properties, an implicit method based on alternating-direction differencing techniques was developed and is discussed herein. The present method can be briefly outlined as follows: The governing equations are replaced by either a Crank-Nicolson or backward time difference approximation. Terms involving nonlinearities at the implicit time level are linearized by Taylor expansion about the known time level, and spatial difference approximations are introduced. The result is a system of multidimensional coupled (but linear) difference equations for the dependent variables at the unknown or implicit time level. To solve these difference equations, the Douglas-Gunn (Ref. 1) procedure for generating alternating-direction implicit (ADI) schemes as perturbations of fundamental implicit difference schemes is introduced. This technique leads to systems of one-dimensional coupled linear difference equations which can be solved efficiently by standard block-elimination methods. Complete details of the method are given in Ref. 2.

NUMERICAL METHOD

The use of implicit schemes in multidimensional fluid flow calculations is of course not without precedent, particularly for incompressible flows. Implicit methods have not always produced the desired increase in efficiency, however, since many have either been unable to take large time steps or else have required an excessive computational effort per time step. The present method derives its single-step efficiency from the use of the special linearization technique in conjunction with the ADI solution procedure, which together eliminate the need for any iteration during a time step. The linearization technique is based on an expansion of nonlinear implicit terms about the known time level t^n and leads to a one-step, two-level scheme. In one dimension, for example, the continuity equation in divergence form is

$$\partial \rho/\partial t = -\partial(\rho u)/\partial x \tag{1}$$

where ρ is density, u is velocity, x is distance, and t is time. Equation (1) is time-differenced as follows:

$$(\rho^{n+1}-\rho^n)/\Delta t = \partial/\partial x\left[\gamma(\rho u)^{n+1}+(1-\gamma)(\rho u)^n\right] \tag{2}$$

where γ is a parameter such that Eq. (2) has Crank-Nicolson form if $\gamma=\frac{1}{2}$ and backward-difference form if $\gamma=1$. The implicit term in Eq. (2) is expanded about t^n as follows:

$$(\rho u)^{n+1} = (\rho u)^n + \left(\frac{\partial \rho u}{\partial \rho}\frac{\partial \rho}{\partial t} + \frac{\partial \rho u}{\partial u}\frac{\partial u}{\partial t}\right)^n \Delta t + O(\Delta t)^2 \tag{3}$$

and time-differenced to give

$$(\rho u)^{n+1} = (\rho u)^n + \left[u^n(\rho^{n+1}-\rho^n)/\Delta t + \rho^n(u^{n+1}-u^n)/\Delta t\right]\Delta t + O(\Delta t)^2 \tag{4}$$

Finally, Eq. (4) is substituted into Eq. (2) and spatial differences are introduced using the following shorthand difference-operator notation:

$$\delta_x \rho^n \equiv (\rho^n_{j+1}-\rho^n_{j-1})/2\Delta x \tag{5}$$

where ρ^n_j denotes $\rho(x_j, t^n)$. After simplification, Eq. (2) becomes the following <u>linear</u> implicit difference approximation:

$$(\rho^{n+1}-\rho^n)/\Delta t = -\delta_x\left[\gamma(\rho^{n+1}u^n+\rho^n u^{n+1}) + (1-2\gamma)\rho^n u^n\right] \tag{6}$$

It should be emphasized that the foregoing technique for generating linear implicit difference schemes is applicable to equations of much greater generality than Eq. (1). The expansion technique of Eq. (4) is similar to the generalized Newton-Raphson method for generating linear <u>iteration</u> schemes for coupled nonlinear equations by expansion about a known <u>current guess</u> at a solution; however, there apparently has been no previous application of this idea to systems of coupled nonlinear time-dependent equations wherein the nonlinear terms are approximated to an accuracy commensurate with that of the time differencing.

The application of the foregoing procedure to the three-dimensional compressible Navier-Stokes equations is relatively straightforward but lengthy. In deriving the difference approximations for the Navier-Stokes equations, physical energy-dissipation terms and cross-derivative viscous terms are treated explicitly for convenience. Details are given in Ref. 2. Here, it is simply stated that the linear, coupled, implicit difference equations which are generated can be written in the following matrix-operator form:

$$A(\Phi^{n+1} - \Phi^n)/\Delta t = (\mathcal{D}_x + \mathcal{D}_y + \mathcal{D}_z)\Phi^{n+1} + S \tag{7}$$

where Φ^n denotes $\Phi\,(x_j, y_k, z_l, t^n)$ with x,y, and z the spatial coordinates. Here, Φ is a vector quantity representing the five dependent variables: density ρ, temperature T, and the velocity components u, v, w. A is a square matrix having one row for each of the five governing equations and one column for each dependent variable. $\mathcal{D}_x$, $\mathcal{D}_y$, and $\mathcal{D}_z$ are also 5x5 matrices (associated with the x, y, and z coordinate directions, respectively), each element of which is a three-point linear spatial difference operator. S is a vector of 5 elements containing source-like terms. Each of A, $\mathcal{D}_x$, $\mathcal{D}_y$, and $\mathcal{D}_z$, and S can be evaluated entirely from known n-level quantities. The implicit scheme (7) is solved efficiently and without iteration by an application (to coupled equations) of the Douglas-Gunn (Ref. 1) procedure for generating ADI schemes as perturbations of fundamental implicit schemes such as Eq. (7). Solution of Eq. (7) in this manner requires only the solution of one-dimensional block-tridiagonal systems of linear equations. The method requires only two levels of storage for each of the dependent variables.

DISCUSSION

The present method is intended for application to both unsteady and steady flows, as the potential for increased computational efficiency is present in both instances. For unsteady problems, the relevant question is how the stability limit compares with the time step necessary to follow the transient accurately. Greater freedom in choosing the time step is possible if a steady solution is to be computed as the asymptotic limit for large time of an unsteady solution, since in this instance there is no need to follow the transient. In view of the many factors involved, no single factor can be quoted for the relative efficiency between the present implicit method and explicit methods subject to stability conditions, however, it is argued in Ref. 2 that a gain in efficiency occurs whenever a time step greater than about twice the explicit time step is used. Thus, a potential for increased overall efficiency arises whenever the CFL or viscous stability conditions become restrictive. Since the severity of the CFL and viscous stability conditions varies from problem to problem and with the choice of grid size, the overall relative efficiency depends on the particular application. Nevertheless, certain guidelines can be established from a consideration of the effect various factors have on the stability conditions. Although in some instances the gain in efficiency would be marginal, in certain important cases the gain in efficiency would be orders of magnitude.

One of the adverse consequences of stability restrictions becomes apparent when a course-mesh solution is recomputed with a finer spatial mesh to obtain greater spatial accuracy. Assuming the maximum allowable time-step is taken, the time-step must be smaller with the finer mesh even though the physical time dependence is exactly the same. In these same circumstances, implicit methods not subject to stability restrictions can take the same time step with both spatial meshes, and consequently, the computational effort increases only linearly with the number of spatial grid points, as opposed to a quadratic increase for a CFL-limited calculation and a cubic increase for a viscous-limited calculation.

For lower Mach number flows, the CFL condition eventually becomes restrictive regardless of mesh spacing, and thus any efficient method not subject to the CFL condition (such as the present implicit procedure) is attractive for both unsteady and steady problems in the low Mach number regime. For all Mach numbers, both the CFL and viscous stability conditions become restrictive for small mesh spacing. Thus, the present method becomes increasingly attractive for both unsteady and steady flows when high resolution is necessary and especially when locally refined meshes are used, since the stability limit is usually governed by the smallest mesh spacing in the field. This latter situation is common when more than one length scale is present, as is the case for flows which are largely inviscid but have thin boundary layers requiring a locally refined mesh. Similar statements hold for (time-averaged) turbulent flows with laminar sublayers.

COMPUTED RESULTS

The present method was employed in Ref. 2 to compute solutions for three-dimensional subsonic flow in a straight rectangular duct (See Fig. 1). Two sequences of solutions were computed using a 6x6x6 grid and different time steps, as a demonstration that the stability of the method is not determined by the conventional CFL condition for explicit methods. The transient behavior of downstream centerline velocity is a suitable indicator of stability and is shown in Fig. 2 for the first sequence of solutions (Mach number M = 0.44, Reynolds number, Re = 60). The initial condition is $w_{℄} = 1$ at t = 0. It can be seen that the method gave stable solutions for CFL numbers up to 32, and that fewer time steps were required to reach steady conditions using the larger time steps (higher CFL numbers). In addition, since there is little difference between curves a and b ($N_{CFL} = 0.6$ and 1.6), it can be inferred that these solutions approximate the true transient behavior. From these two curves, the "characteristic physical time" or time required to reach steady conditions is seen to be approximately 3 for this flow case. Clearly, there is considerable truncation error in the transients for curves c, d, and e (N_{CFL} = 3.2, 16, and 32). The steady solutions, however, were the same for all five solutions in Fig. 2. Information analogous to that in Fig. 2 is given in Fig. 3 for the second sequence of solutions (M = 0.044, Re = 60), and in this case, stable solutions were obtained for CFL numbers up to 1250. Although optional artificial viscosity terms can be employed with the present method (Ref. 2) if warranted by the intended application, none were used for these stability tests. One additional solution is included here in which a relatively large secondary flow is generated by moving two parallel duct walls in a direction perpendicular to the axial flow direction. The solution has M = 0.44 and Re = 60, and the wall speed is about 25 percent of the average axial velocity. A computer-produced drawing of selected streamlines for this solution is shown in Fig. 4.

This work was supported by the Office of Naval Research under Contract N00014-72-C-0183.

REFERENCES

1. Douglas, J. and Gunn, J. E., A General Formulation of Alternating Direction Methods Numerische Math., Vol. 6, 1964, p.428.

2. Briley, W. R. and McDonald, H., An Implicit Numerical Method for the Multidimensional Compressible Navier-Stokes Equations United Aircraft Research Laboratories Report M911363-6, November, 1973.

FIG. 1 - Duct Geometry and Coordinate System

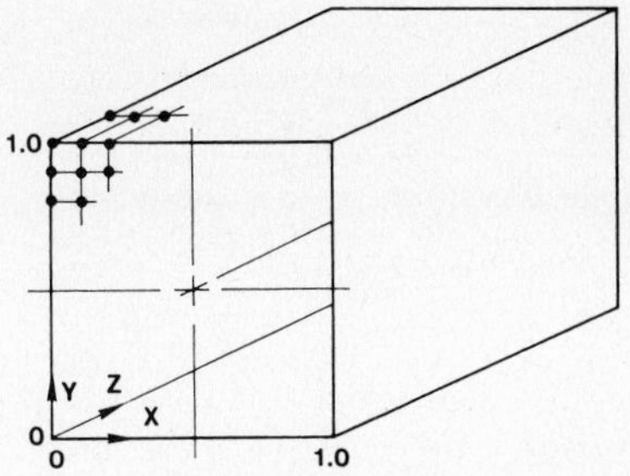

FIG. 2 - Transient Behavior of Downstream Centerline Velocity for Different Time Steps - M = 0.44, Re = 60

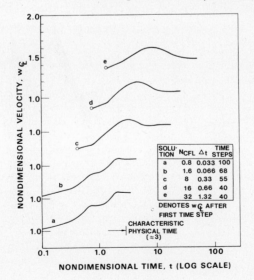

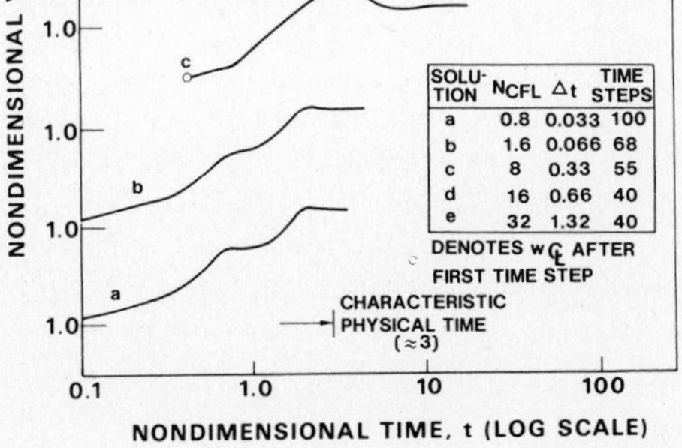

FIG. 3 - Transient Behavior of Downstream Centerline Velocity for Different Time Steps - M = 0.044, Re = 60

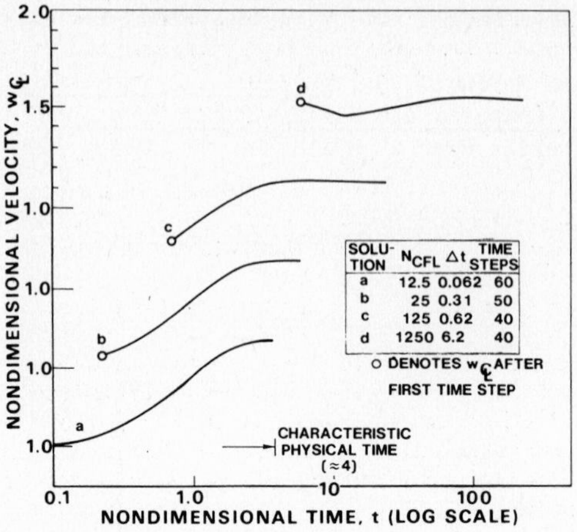

FIG. 4 - Selected Streamlines for Duct Flow with Moving Walls - M = 0.44, Re = 60

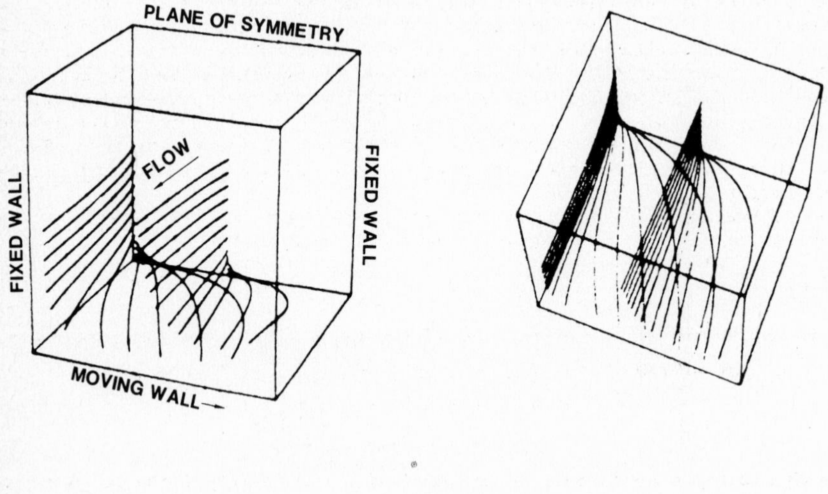

VARIATIONALLY OPTIMIZED, GRID-INSENSITIVE VORTEX TRACING

By O. Buneman
EE Department, Stanford University

Time-dependent incompressible flow with continuous vorticity has been successfully simulated on the computer as the interaction of a very large number, N, of discrete vortex filaments (Roberts and Christiansen[1]). If the N^2 interactions between these vortices are calculated individually (Abernathy and Kronauer[2]), the method soon becomes uneconomical. Interaction has to take place by way of a stream function, ψ, or the velocity field $-\partial\psi/\partial y$, $\partial\psi/\partial x$. The computer requires the information on the stream function or the field to be, likewise, supplied in discrete, finite elements. The obvious choice is a record of field values over a grid. This results in the dependence of the small-scale physics on local position: there is no translational invariance, nor is there isotropy. One gets truncation errors, enhanced noise and instabilities (Langdon[3]).

We must look for a gridless or grid-insensitive field representation. Moreover, if we express the flow-field in finite terms, it is desirable to develop some physical model for the problem which the computer actually simulates when such discretization is introduced in addition to the discretization into elementary vortices.

There is no good physical interpretation of a potential or flow-field recorded over a grid and interpolated, linearly or to higher order, between grid points. For the analogous charged particle problem, the process of depositing the sources of the field (charges/vortices) into a grid has been given the graphic description of "charge sharing" (Birdsall[4]) and the combination of linear charge sharing with linear interpolation is known as "PIC" and "CIC" codes.

H.R. Lewis[5] has put these procedures on a more solid footing by (1) introducing quadratic and higher order splines as basis functions for the continuous Eulerian field quantities employed in the theory and (2) deriving the dynamical equations for the spline-coefficients from a variational principle. We shall take over from Lewis both the idea of using splines (cubic splines in particular) and the idea of deriving the dynamical equations from an action principle. However, as will be explained, the basis functions for our scheme are not splines but Fourier harmonics: the splines are used merely as an efficient method of interpolation. So far, we have considered the 2D problem only.

In execution, our method of vortex tracing consists of recording the field components in terms of two-dimensional arrays of spline coefficients. Likewise a two-dimensional table of spline-coefficients of the vortex distribution is built up by distributing each vortex among the nearest coefficients according to spline-weighting. However, the field will be generated from the vorticity in a manner which (1) guarantees the high degree of smoothness essential for good performance from spline interpolation, and (2) permits the physical interpretation of the whole procedure as a "strict" simulation of interacting vortices with gaussian profile. This last feature, then, gets us back to a physical picture and it eliminates the non-physical aspects brought about by grids.

The following action function (to which the author has been unable to find a previous reference) describes inviscid vortex motion:

$$\iiint \tfrac{1}{2}(\nabla\psi)^2\, dx\,dy\,dt \;+\; \sum \pm \left\{ \int \psi\,(x_\mu, y_\mu)\,dt + \int \tfrac{1}{2}(x_\mu\, dy_\mu - y_\mu\, dx_\mu) \right\} \qquad (1)$$

Here μ is the vortex index. Variation of x_μ yields $dy_\mu/dt = \partial\psi/\partial x_\mu$, variation of y_μ yields $dx_\mu/dt = \partial\psi/\partial y_\mu$ and variation of ψ yields :

$$\nabla^2\psi = \sum \pm\, \delta\,(x - x_\mu)\,\delta(y - y_\mu).$$

We choose representation of ψ in terms of a finite set of Fourier harmonics, for the following reasons:

(1) The inevitable loss of information resulting from discrete, finite representations amounts to suppressing fine-scale detail but in such a way that infinite smoothness is guaranteed. All derivatives are continuous and interpolation will be efficient.

(2) There is translational invariance: locally the results do not depend on whether a vortex is on a grid line or in between - there is no grid at all.

(3) The harmonics are orthogonal and remain separable in the terms which are quadratic in the interaction - such as $(\nabla\psi)^2$ - and which lead to linear dynamical equations.

(4) Turbulence is traditionally analyzed in Fourier space.

(5) Differentiations become multiplications by k_x and k_y.

(6) Rectangular boundaries are implied in the discrete spacing δk_x, δk_y of k_x and k_y. Alternatively, large-scale periodicity can be implied ($\delta k_x = \delta k_y = 2\pi/L$) in the representation of an effectively infinite domain.

(7) If the cut-off of the spectrum is made according to the magnitude $|\vec{k}| = k_{max}$ of the $\vec{k}$-vector, the representation will be locally isotropic.

In Fourier-space the action (1) becomes:

$$-L^2 \int \tfrac{1}{2} \sum_{\vec{k}} k^2 \left|\psi_{\vec{k}}\right|^2 dt + \sum_{\mu} \pm \left\{ \sum_{\vec{k}} P(k) \int \psi_{\vec{k}}\, e^{i\vec{k}\cdot\vec{r}_\mu} dt + \int \tfrac{1}{2}(x_\mu dy_\mu - y_\mu dx_\mu) \right\} \tag{2}$$

The factor P(k) is introduced as an alternative to the abrupt cut-off at $k=k_{max}$ and will be assumed to be bell-shaped. Sharp cut-off would imply that in x,y-space our vortices acquire halos or "aliasses". The Euler-Lagrange equations resulting from varying x_μ, y_μ and ψ_k respectively are:

$$\frac{dy_\mu}{dt} = -\sum_k ik_x P(k)\, \psi_{\vec{k}} e^{i\vec{k}\cdot r_\mu} \tag{3}$$

$$\frac{dx_\mu}{dt} = \sum_k ik_y P(k)\, \psi_{\vec{k}} e^{i\vec{k}\cdot\vec{r}_\mu} \tag{4}$$

$$-L^2 k^2 \psi_{\vec{k}} = P(k) \sum_\mu e^{-i\vec{k}\cdot\vec{r}_\mu} \tag{5}$$

Equations 3 and 4 indicate that we have an effective stream function ψ^{eff} whose Fourier components are given by $\psi^{eff}_{\vec{k}} = P(k)\, \psi_{\vec{k}}$. Equation 5 indicates that the Fourier transform of each vortex which gives rise to this effective potential would be $P^2(k)$. If, therefore, P(k) is taken to be gaussian, then the source vortex will be gaussian in configuration space also.

A spread-out vortex with gaussian profile is the result of viscosity having acted, for a time proportional to the effective area of spread, on an initially ideal filamentary vortex: truncation with a gaussian P(k) can therefore be justified as a step toward implementing a "real" fluid on the computer. However, the area of spread, i.e. the "age" of the vortices is given by the inverse of the area of spread of P(k) in k-space and hence fixed. All vortices represented in this manner have the same middle-aged spread!

It is easy to simulate a larger spread in $\vec{r}$-space than that given by P(k): one introduces several satellites around a center. Thus "older" vortices can be simulated. The ingenious method of adding a random walk to each vortex displacement (A. Chorin[6]) is a simple alternative. To some extent, one can make up thinner, younger vortices by surrounding a positive (clockwise) vortex concentrically with a halo of negative (anti-clockwise) vortices. However, the limit of resolution is set by the cut-off imposed in $\vec{k}$-space, and such narrowed patterns tend to result in undesirable sidebands or aliasses. For introducing young vortices, say at some boundary layer, it is recommended that their flow field be superimposed on the continuous stream function ψ by direct summation of their contributions (logarithms of the distances), as was done by Chorin[6], and as is done by Hockney[7] in the equivalent particle problem.

In the present calculations we have not used exactly a gaussian profile, but one which is generically close to it. Since the random walk interpretation of the gaussian has already been referred to before, we explain the approximation used in terms of this concept. Gaussian <u>probability</u> distributions are the result of many random walks and J.v.Neumann recommended 12 random walks as sufficient to match the precision of most computers. This puts the effective range of the distribution just 6 times as far as the r.m.s. distance. Most of this range is weighted down very heavily and it would be a waste to carry all these Fourier harmonics. If one stops at 4 random walks, the maximum $|k|$ carried is $2\sqrt{3}$ times $\sqrt{\langle k^2\rangle}$, and the resulting profile is already very close to $\exp(-\frac{1}{2}k^2/\langle k^2\rangle)$. It is in fact the cubic spline profile, tending to zero cubically as k_{max} is approached. Since we used cubic splines for interpolation(see below) it was convenient to use this profile in $\vec{k}$-space also. It was used for the quantity $P^2(k)$, not for $P(k)$. The transform into $\vec{r}$ space is the fourth power of the sinc-function. This is positive everywhere and has very small sidebands (or aliasses): the first satellite is two thousandth of the center peak.

In vortex tracing one would not benefit from packing a very large number of vortices into an area of dimensions k_{max}^{-2} because they would effectively merge into each other. This means that the total number of ψ-harmonics will not be substantially less than the total number of vortices.

However, the labor of advancing the vortices by means of equations (3,4) would be prohibitive if each vortex required at each time step the evaluation of all the Fourier harmonics on the right-hand sides. Likewise the determination of each harmonic $\psi_{\vec{k}}$ from (5) requires the evaluation of twice as many trigonometric functions as there are vortices. All this might turn out to be as expensive as the primitive process of summing all direct interactions, ignoring stream functions.

The evaluation of the contribution to each harmonic $\psi_{\vec{k}}$ from each vortex might be done by a table look-up and some interpolation in this table, to avoid the evaluation of trigonometric functions. It is by consideration of such interpolation from a coarse table that one is led to a feasible and accurate scheme for advancing the vortices and obtaining the stream function.

Cubic spline interpolation applied to trigonometric functions is found to give remarkably good results even when the data points are spaced as far apart as a quarter period. For larger phase-shifts between data points one still reproduces qualitatively recognizable sinusoids but there are significant quantitative departures.

If $S_3(x)$ is the standard cubic spline for equidistant abscissae (unit distance apart), and if g_j are the spline coefficients for the function e^{ikx}, then the cubic spline approximant to this function is:

$$e^{ikx} \approx \sum_{j=[x]-1}^{[x]+2} g_j S_3(x-j)$$

and the g_j obey the recurrence formula $g_{j-1} + 4g_j + g_{j+1} = 6\, e^{ikj}$.

This formula is solved by $g_j = 3e^{ikj}/(2+\cos k)$, as can be checked by substitution. Hence e^{ikx} is approximated by $\sum_j \frac{3e^{ikj}}{2+\cos k} S_3(x-j)$ where j should run through a range which includes the integers $[x]-1$, $[x]$, $[x]+1$, $[x]+2$.

Let us now use the approximants in place of the genuine exponentials in the dynamical and Poisson's equations (3) - (5):

$$\frac{dy_\mu}{dt} = -\frac{\partial}{\partial x_\mu}\sum_j\sum_\ell\left(\sum_{\vec{k}}\frac{9P(k)\ \psi_{\vec{k}}\ e^{ik_xj+ik_y\ell}}{(2+\cos k_x)(2+\cos k_y)}\right)S_3(x_\mu-j)S_3(y_\mu-\ell) \qquad (6)$$

$$\frac{dx_\mu}{dt} = \frac{\partial}{\partial y_\mu}\quad \cdots\cdots \qquad (7)$$

$$-L^2k^2\psi_{\vec{k}} = \frac{9P(k)}{(2+\cos k_x)(2+\cos k_y)}\sum_j\sum_\ell e^{-ik_xj-ik_y\ell}\sum_\mu \pm S_3(x_\mu-j)S_3(y_\mu-\ell) \qquad (8)$$

In the last formula the indices j and ℓ run through the entire range of data points, but in the innermost sum (over the sources, index μ) one generates contributions $S_3(x_\mu - j)$ and $S_3(y_\mu - \ell)$ only to the nearest two data points on each side of x_μ and y_μ respectively.

The last equation (eq. 8) shows that $\psi_{\vec{k}}$ is $\dfrac{-9P(k)}{L^2k^2(2+\cos k_x)(2+\cos k_y)}$ times the discrete 2D Fourier transform of the data $\sum_\mu \pm S_3(x_\mu - j)\, S_3(y_\mu - \ell)$ recorded on the j,ℓ mesh. The equations for dy_μ/dt and dx_μ/dt show that the effective stream function is obtained by spline interpolation using a table of spline coefficients which result from Fourier back-transforming the harmonics: $9\psi_{\vec{k}}P(k)/(2+\cos k_x)(2+\cos k_y)$.

It is therefore possible to execute the vortex tracing program as if an effective stream function had been recorded over a mesh and as if a "vorticity" was being accumulated into a mesh. Each vortex is distributed over its 4x4 nearest mesh points and the local flow field is interpolated at each vortex from the nearest 4x4 mesh values. The weights in each operation are cubic spline functions, either calculated algebraically on the spot, or, better, subtabulated sufficiently finely to allow direct look-up without further interpolation. To pass from the pseudo-vorticity table to the pseudo-stream function table one:

(i) Fourier transforms in two dimensions, using an FFT,

(ii) multiplies by $-\left(\dfrac{9P(k)}{Lk(2+\cos k_x)(2+\cos k_y)}\right)^2$ and

(iii) Fourier back-transforms using an FFT.

Steps (i)-(iii) can also be implemented as a convolution with a function whose transform is $-\left(\dfrac{9P(k)}{Lk(2+\cos k_x)(2+\cos k_y)}\right)^2$, and provided only a restricted number of grid points is involved at either end (as was the case in the computation reported below), this is the fastest method. The fact that one is, in principle, handling Fourier harmonic representations of continuous functions then disappears from sight. However, when large arrays are involved, execution of (i)-(iii) as given will be cheaper, in view of the ease with which one can carry out FFT's.

The reader is reminded of the fact that the true basis functions of this scheme are Fourier harmonics and not the splines. The latter are introduced purely as an interpolation device adn for high precision one might want to use several times as many splines as harmonics.

If one employs too few splines, grid effects are liable to show up due to this interpolation: in other words the translational invariance of exact, or accurately interpolated Fourier harmonics may be lost.

An experiment has been made keeping just as many splines as Fourier harmonics. This is an extreme case: one would not wish to waste information contained in the harmonics by tabulating fewer data than the number of harmonics.

Even in our experiment there is a slight redundancy. The Fourier analysis is done in variable x and y with both k_x and k_y running through a complete range up to, and including, k_{max}. The factor $P_2(k)$, on the other hand, suppresses all harmonics outside the circle $k_x^2 + k_y^2 = k_{max}^2$ in k-space, including the periphery of the circle itself. In fact the actual number of admitted basis functions is $\pi/4$ times smaller than the number of grid points or splines.

The highest harmonics, those for which k_x or k_y is just below k_{max}, have a phase-change per grid point just short of π. For these, cubic splines will not give very good results. However, these harmonics are also very strongly attenuated by the shape factor P(k), and therefore the innacuracy of interpolation is innocuous.

The results of the experiment were displayed in a movie: a single vortex was moved through the interpolation grid and the stream lines it generated by going through steps (i)-(iii) above were calculated. Regions between neighboring streamlines were alternately made black and left blank. In the movie these streamlines are seen to remain perfectly circular and to follow the vortex center concentrically. Four stills from the movie are shown in figures 1 - 4, the first for the vortex center directly on a gridpoint, the last for the vortex center in the middle of the cell the two cases in between for typical unsymmetrical vortex positions.

The fact that the streamline pattern is insensitive to the interpolation grid encourages one to accept the $4/\pi$ redundancy as sufficient for most purposes. We are now in the process of simulating a Kelvin-Helmholtz instability with 32768 vortices and $\pi \times 32^2$ harmonics.

In the execution of dynamical vortex tracing, one is not interested in local stream function values but in local gradients. One forms these before spline interpolation by multiplying the Fourier harmonics with k_x and k_y rather than differentiating $S_3(x-j)$ and $S_3(y-j)$. One prepares one interpolation table for each component of the gradient.

An important question is whether the residual grid effects create a self-effect on each vortex, i.e. whether an unsymmetrically placed vortex will make itself rotate. (Because of the translational invariance of Fourier harmonics this cannot happen if one uses accurate interplation.) This question has been tested in one dimension. It was found that with the grid as coarse as described (spacing π/k_{max}), the interpolated gradient at the source position was so near round-off level that one wonders whether it can be theoretically shown to remain strictly zero.

REFERENCES

1. J.P. Christiansen (1973) "Numerical Solution of Hydrodynamics by the Method of Point Vortices", Journ. Comput. Phys., 13, 363.
2. F.H. Abernathy, and R.E. Kronauer (1962) J. Fluid Mech., 13, 1.
3. B. Langdon (1970) Journ. Comput. Phys., 6, 247.
4. C.K. Birdsall and D. Fuss (1969) Journ. Comput. Phys., 3, 494.
5. H.R. Lewis, Methods of Comput. Phys., vol. 9
6. A.J. Chorin and P.S. Bernard, "Discretization of a Vortex Sheet, with an Example of Roll-Up" (1973) Journ. Comput. Phys., 13, 423.
7. R.W. Hockney, S.P. Goel and J.W. Eastwood (1974) "Quiet High-Resolution Computer Models of a Plasma", Journ. Comput. Phys., 2, 148.

Work supported by United States Atomic Energy Commission.

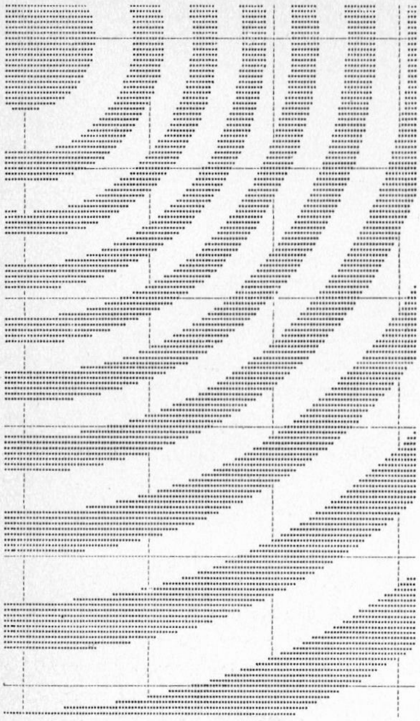

Figure 1

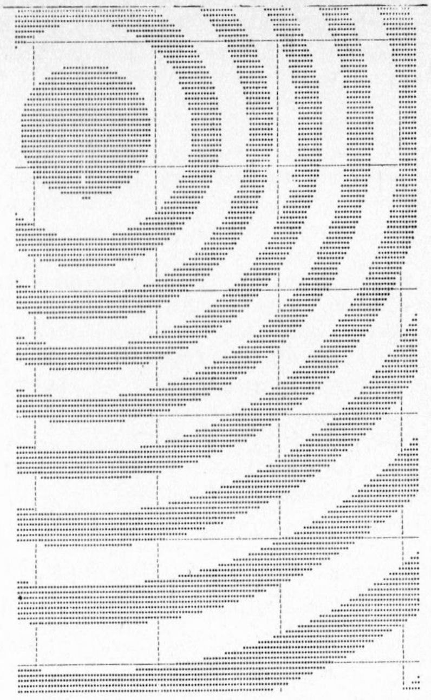

Figure 2

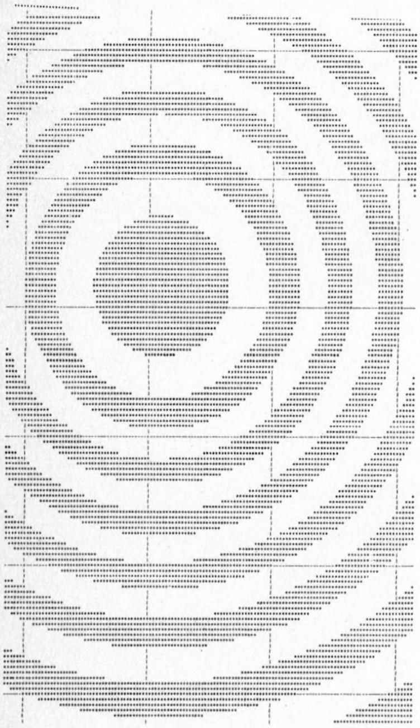

Figure 3

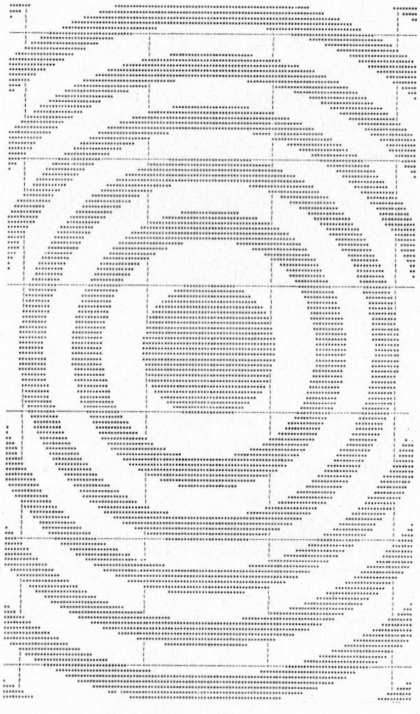

Figure 4

A NUMERICAL STUDY OF PULSATILE FLOW THROUGH CONSTRICTED ARTERIES

Bart J. Daly
Los Alamos Scientific Laboratory
University of California
Los Alamos, New Mexico

INTRODUCTION

A numerical technique has been developed for the purpose of studying the pulsatile flow of blood through a constricted artery, in order to examine the flow phenomena associated with the formation, growth and detachment of plaque material at arterial walls. The calculation procedure makes use of the Arbitrary Lagrangian-Eulerian method for the calculation of transient, multi-dimensional, viscous flow. This technique is particularly well adapted to biological flows, since it permits a virtually arbitrary and time-varying boundary configuration.

The technique development has concentrated on the effective simulation of two crucial characteristics of blood flow in large arteries: a nonisotropic and space varying elastic model of distensible arteries, and an efficient procedure for calculating pulsatile flow. The pulsatile nature of the flow requires a sophisticated boundary treatment, appropriate for the representation of forward, reversed and mixed flows. The accuracy of the procedure has been successfully tested by comparison with an analytic solution for oscillating flow through a uniform rigid tube. The method is now being used to calculate the pulsatile flow of an incompressible, Newtonian fluid through constricted, rigid walled arteries with an experimentally measured biological inflow.

While there is much to be learned from this rigid wall study, an accurate simulation of arterial flow requires the inclusion of an elastic wall treatment. For this purpose, a nonisotropic and space varying elastic model has been developed. This flexibility is necessary in order to account for the nonisotropic restraining force exerted on arteries by the surrounding tissue, and for the variation in elastic modulus between normal arterial walls and plaque-lined walls. These phenomena are thought to be important contributors in the process of plaque dislodgement and the consequent arterial occlusion.

THE NUMERICAL PROCEDURE

The Arbitrary Lagrangian-Eulerian (ALE) procedure has been described by Hirt, Amsden and Cook (1974) and by Amsden and Hirt (1973). The latter reference contains a complete description, together with flow diagrams and Fortran listing, of the YAQUI code, from which the numerical formulation employed in this study was derived. Rather than repeat the description of ALE provided in these references, the present paper will present only a brief outline of the method together with a more detailed explanation of the modifications adopted in the present investigation.

In its most general formulation the ALE procedure is applicable to flow at all speeds, but in this incompressible flow problem a considerable simplification can be achieved by considering the density to be constant and by ignoring the effects of minor fluctuations in temperature. Therefore, we restrict our attention to the continuity and momentum equations

$$\nabla \cdot \underset{\sim}{u} = 0 \quad , \tag{1}$$

$$\frac{d\underset{\sim}{u}}{dt} = \nabla \cdot \underset{\sim}{\Pi} \quad , \tag{2}$$

where $\underset{\sim}{u}$ is the fluid velocity and $\underset{\sim}{\Pi}$ is the stress tensor. In general, $\underset{\sim}{\Pi}$ includes elastic as well as pressure and viscous stresses, but we shall defer consideration of elasticity effects until a later section. Gravitational forces are negligible and therefore neglected in this study.

The philosophy of the ALE method is to subdivide a computation cycle into two phases, a Lagrangian phase and a rezone phase, in order to promote flexibility regarding the motion of the computation mesh relative to the fluid. A staggered finite difference grid is used, with coordinates and velocities specified at mesh vertices and pressure at cell centers (Fig. 1). The pressure and viscous forces acting on each vertex are computed relative to an auxiliary computation mesh, an overlapping system of momentum control volumes (the dashed lines in Fig. 1). To this end it is convenient to rewrite the conservation equations (1) and (2) in integral form,

$$D = \int \nabla \cdot \underset{\sim}{u} \, dV = 0 \quad , \tag{3}$$

$$\int \frac{d\underset{\sim}{u}}{dt} \, dV = \int \underset{\sim}{\Pi} \cdot \underset{\sim}{n} \, dS \quad , \tag{4}$$

where V is an appropriate local volume, corresponding to a computation cell for Eq. (3) and a momentum control volume for Eq. (4). S is the bounding surface and $\underset{\sim}{n}$ is the unit outward normal from this surface.

The use of an overlapping grid for the solution of Eq. (4), as shown in Fig. 1, can produce a decoupling of the solution at alternate mesh points. Therefore a fourth order node coupling procedure, based on one proposed by Chan (1973), is included in the numerical computation. This high order scheme was chosen in order to provide effective damping of the short wavelength instability, characteristic of the alternate node decoupling, without greatly effecting the longer wavelength modes. Referring to Fig. 1, we write the expression for the smoothing to be applied to the u component of velocity at a typical vertex,

$$\begin{aligned} u^j_{i_{\text{adjusted}}} &= u^j_i + \frac{\beta}{16\, m^j_i} \Big[\left(m^{j+1}_i + m^j_i\right) \left(u^j_{i+1} - u^{j+1}_{i+1} - 2u^j_i + 2u^{j+1}_i + u^j_{i-1} - u^{j+1}_{i-1}\right) \\ &\quad - \left(m^j_i + m^{j-1}_i\right) \left(u^{j-1}_{i+1} - u^j_{i+1} - 2u^{j-1}_i + 2u^j_i + u^{j-1}_{i-1} - u^j_{i-1}\right) \Big] \quad , \end{aligned} \tag{5}$$

where m^j_i is the mass associated with vertex i, j and β is a constant coefficient. The inclusion of mass weighting, which represents the only departure from Chan's formulation, makes the procedure momentum conservative.

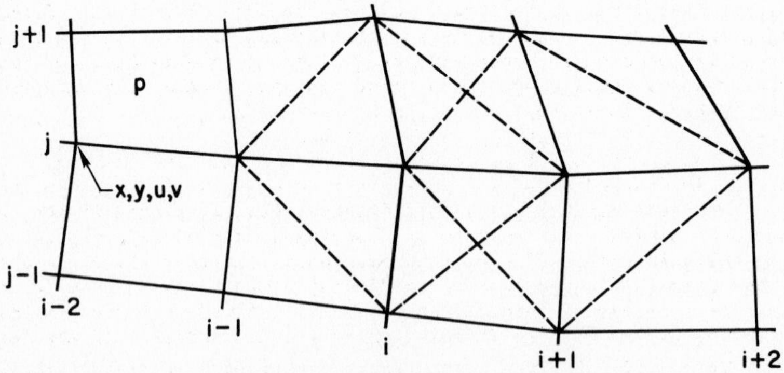

Figure 1. Details of the computation mesh, showing the placement of variables and the location of momentum control volumes (the dashed lines).

The Lagrangian phase of the calculation is subdivided into two subphases, an explicit one in which the pressure and viscous forces acting over each face of a momentum control volume are summed, and an implicit subphase in which the pressures are modified in order to ensure that the continuity equation is satisfied to within a suitable convergence criterion. While it is not the intention of this paper to describe these algorithms in detail, we do wish to mention some specific differences between the present formulation and that proposed by Amsden and Hirt (1973). One such difference has to do with the form of certain viscous terms at the axis of cylindrical symmetry. For vertices along this axis Amsden and Hirt represent the terms $\frac{2\nu}{r}\frac{\partial}{\partial r}\left(r\frac{\partial u}{\partial r}\right)$ and $\frac{\nu}{r}\frac{\partial}{\partial r}\left(r\frac{\partial v}{\partial r}\right)$, with finite difference formulations that approximate the differential forms, $2\nu\, \partial^2 u/\partial r^2$ and $\nu\, \partial^2 v/\partial r^2$, where ν is the kinematic coefficient of viscosity. In the present study the corresponding difference expressions are multiplied by a factor of 2, in which form they are consistent with the expressions used in the interior of the mesh. In a test of Poiseuille flow through a uniform rigid tube, the latter formulation produced the correct velocity profile at the axis while the former one did not.

At the completion of the explicit Lagrangian phase of the calculation the velocity field will not in general satisfy Eq. (3). A relaxation phase is required to reduce the residuals in the velocity divergence, D, to a small acceptable error. The velocity field is modified on a cell by cell basis through changes in the cell pressures, δp, computed by Newton's method,

$$\delta p = -\,\omega\, D \left(\frac{\partial D}{\partial p}\right)^{-1} \quad , \tag{6}$$

where ω is an over-relaxation parameter. The derivative, $\partial D/\partial p$, can be conveniently computed numerically, as indicated by Hirt, Amsden and Cook (1974). The finite difference expression for D in cylindrical coordinates approximates the differential form,

$$D = \left(\frac{\partial u}{\partial r} + \frac{\partial v}{\partial z}\right)\left(1 + \frac{u\,\delta t}{r}\right) + \frac{u}{r} \quad , \tag{7}$$

where δt is the time increment. The inclusion of the term, $u\,\delta t/r$, accounts for the change in volume of the cell as the cell moves with the fluid. The form of Eq. (7) was first proposed by Brackbill and Pracht (1973).

The rezoning phase of ALE affords considerable freedom in the choice of a convective transport procedure. This option has been used to advantage in the distensible wall study (see below), but in the rigid tube calculations it has been convenient to utilize an Eulerian procedure in which the grid is always moved back to its original position. An interpolated donor cell (or upwind) differencing formulation has been utilized for the convective terms, so that the momentum packet fluxed across a boundary is centered on the boundary after fluxing. The first order time error introduced by this procedure exactly balances the error caused by the lack of time centering in the calculation.

In the course of this study it was discovered that the donor cell formulation described by Amsden and Hirt (1973) was introducing asymmetries completely apart from the upwind velocity weighting. This difficulty had its source in the fact that the momentum to be fluxed across a face of a momentum control volume (Fig. 1) was obtained by a linear combination of the momenta at the two vertices located diagonally across that face. This resulted in an alternate node bias in choosing the upstream velocity. In a private communication C. W. Hirt showed that for a pure shear flow problem ($u = 0$, $\partial v/\partial x \neq 0$) this formulation introduced smearing of the form, $\partial v/\partial t \propto \partial^2 v^2/\partial x^2$. This difficulty was corrected by determining upstream velocities through interpolation along the streamlines that passed through the center of each face of a momentum control volume.

The Pulsatile Flow

The inflow boundary condition for the rigid tube calculations reported here was chosen such that the volume rate of flow through the tube corresponded at all times to that measured by McDonald (1955) in the canine femoral artery. Figure 2 shows the mean velocity of flow for one full cardiac cycle obtained from McDonald's data. Each period consists of a strong forward flow, followed by a flow reversal, a quiescent period, and a second weak flow reversal.

The use of time-varying boundary conditions in a finite difference calculation means that the flow at each calculation cycle corresponds to a new set of initial conditions. For the sake of efficiency and physical realism, these boundary conditions must be designed so that the change in the volume rate of flow is small compared to the magnitude of the flow, and is incorporated in a way that perturbs the solution as little as possible. The first condition is not difficult to satisfy, except when the flow is changing direction; then the time step must be kept small relative to normal stability considerations. Experience has shown that satisfaction of the second condition depends upon the proper timing of the flow accelerations within the computation cycle.

In the computation mesh used in this study (Fig. 3) vertices are arranged on radial lines, with the bottom row of vertices marking the boundary at which the pulsatile flow boundary condition is specified. The velocity at these boundary vertices is required to have no radial component, but it is subject to axial accelerations as a result of viscous and pressure stresses. (The stresses at these boundary vertices are not computed directly, since there is not sufficient information to do this, but are obtained from the stress calculation on the radial line next to the pulsatile flow boundary.) Thus at the end of the explicit Lagrangian phase all of the vertices in the mesh, including the boundary vertices, have been subject to viscous and (explicit) pressure accelerations. Since the pressure field generally does not vary greatly from cycle to cycle, the explicit pressure acceleration of phase one accomplishes the bulk of the pulsatile flow acceleration for that cycle. Only a minor correction to the boundary velocities is then necessary to make the volume rate of inflow correspond to the prescribed rate at that time. This is an important point, not only because it requires a relatively minor flow acceleration to be transmitted through the tube in the pressure iteration phase, but also because this minor flow correction does not greatly perturb existing flow conditions.

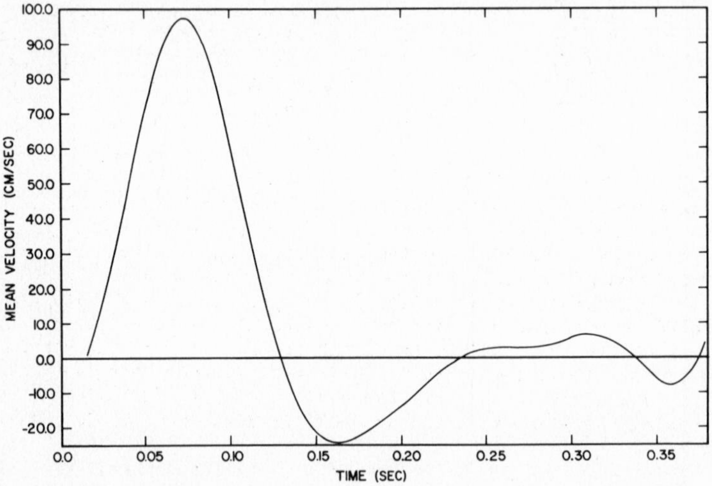

Figure 2. Pulsatile flow in the canine femoral artery, obtained from measurements by McDonald (1955).

The prescribed flow is computed each calculation cycle from data supplied by McDonald (1955). The difference between this prescribed flow and the integrated flow through the bottom boundary (after the explicit Lagrangian phase) is computed and divided by the area open to flow to determine the acceleration to be applied to these boundary vertices. The same acceleration is applied to all of the boundary vertices except the vertex lying on the rigid no slip wall. Thus all of the vorticity added to the system as a result of this acceleration is concentrated at the wall and thereafter is diffused radially. The velocities at the pulsatile flow boundary remain fixed throughout the pressure iteration and the rezone phase.

Results of the Rigid Tube Calculations

A computation mesh for the calculation of flow in a constricted rigid tube is shown in Fig. 3. The left boundary of this grid is an axis of cylindrical symmetry at which free slip conditions are maintained. The right boundary is a no slip rigid wall. Pulsatile flow conditions are maintained at the bottom boundary, while the top of the mesh is a continuative inflow-outflow boundary, where $\partial v/\partial z = \partial u/\partial z = 0$. A variable mesh is employed, with detailed resolution in the vicinity of the rigid wall and the constriction. This permits an effective resolution of the separated flow downstream from the constriction, while maintaining inflow and outflow conditions at some distance from it. In cross section this stenosis is a circular arc that reduces the arterial radius from a normal 0.16 cm to a minimum of 0.12 cm, corresponding to a 44% areal reduction. Similar calculations have been performed for 23% and 61% area restrictions, as well as for unconstricted flow.

Figure 3. Computation mesh

Figure 4 shows the results of a calculation of pulsatile flow through the constricted tube in Fig. 3 at four stages of the cardiac cycle. Referring to Fig. 2 these stages correspond to: initial flow acceleration, $\tau = 0.126$ (dimensionless time, $\tau = t/0.36$ sec.); peak flow, $\tau = 0.210$; flow deceleration, $\tau = 0.293$; and reverse flow, $\tau = 0.488$. A velocity vector plot and contour plots of the pressure and shear stress fields are shown at each time. The velocity vectors originate at mesh vertices (which lie on horizontal lines) with lengths proportional to their magnitude. The vectors are normalized by a reference velocity listed below each plot. The contour intervals are shown below the contour plots, and the H and L marks indicate the highest and lowest contours.

In the accelerating flow at $\tau = 0.126$ the velocity profile is flat over most of the tube, since viscous diffusion has not kept pace with the rapid flow acceleration. A steep boundary layer exists at the stenosis, and a stagnation region has formed in its wake. The pressure field is dominated by a strong gradient on the upstream side of the constriction and a low pressure region just downstream from the point of minimum radius. The shear stress is primarily negative in this forward directed flow and is concentrated on the upstream side of the stenosis. A local region of positive shear stress (note the H), indicating a weak reverse flow in the separated region, exists downstream. The bulge in the shear stress contour near the axis at the stenosis marks a depression in the velocity profile. This local minimum results from the strong radial variation in axial pressure gradient in the vicinity of the flow obstruction.

In the peak flow at $\tau = 0.210$ the Poiseuille-like velocity profile in the unconstricted part of the tube is an indication that the flow is no longer accelerating. Reverse flow is visible in the separated region, and the resulting positive velocity gradient is evident in the shear stress plot. The pressure gradient in the constriction is increased over that at $\tau = 0.126$ (note the larger contour interval), although the overall pressure drop along the tube is approximately the same. This maintenance of a strong pressure drop through the peak flow stage marks a departure from unconstricted flow measurements, and could provide a diagnostic tool for early detection of arterial stenoses.

The overall pressure gradient has reversed direction at $\tau = 0.293$, when the flow is in the decelerative stage. Although it is not evident in the figure, the entire separated flow above the constriction has reversed direction, so that the separation line now corresponds to a shear layer. This is indicated in the shear stress plot by the sharp change in gradient above the stenosis. Flow reversal is complete at $\tau = 0.488$, and a new separated region has formed in the wake region below the obstacle. The shear stress is mostly positive at this time, but regions of negative shear stress exist in the wake as well as in the core of the tube. The latter phenomena is an indication that the region of peak reverse flow is located at some distance from the axis. With time viscous diffusion will move this peak to the axis.

Wall Distensibility

A realistic study of arterial hemodynamics requires the inclusion of the effects of wall distensibility. Therefore, a separate concern in this investigation has been the development of an algorithm for the incorporation of a movable elastic boundary. This part of the development has proceeded separately from the pulsatile flow study, and utilizes a much simpler inflow-outflow treatment.

The fluid acceleration due to elastic forces is computed during the explicit part of the Lagrangian phase of the calculation. These forces arise from angular strains, due to changes in the radius of the artery, and from longitudinal strains resulting from a lengthwise stretching of the artery. The tethering effect of tissues surrounding the artery restricts longitudinal strains more than angular strains, so that the modulus of elasticity is nonisotropic. The radial component of the force exerted on the fluid as a result of angular strain in the arterial wall has the form

$$-E_a h s_j^n \left(r_j^n - R_j\right)/R_j \quad , \tag{8}$$

where E_a is the elastic modulus for angular strain, h is the thickness of the wall, s_j^n is the length of arc associated with the vertex on the jth row (Fig. 1) at the wall at time step n, r_j^n is the radius of that vertex, and R_j is the equilibrium position of the wall at row j. There is no axial component of angular strain.

The radial and axial components of the force on the fluid resulting from the longitudinal strains are

$$E_1 h \left[\left(\frac{s_{j+\frac{1}{2}}^n - \Delta z}{\Delta z}\right)\left(\frac{r_{j+1}^n - r_j^n}{s_{j+\frac{1}{2}}^n}\right)\left(\frac{r_{j+1}^n + r_j^n}{2}\right) - \left(\frac{s_{j-\frac{1}{2}}^n - \Delta z}{\Delta z}\right)\left(\frac{r_j^n - r_{j-1}^n}{s_{j-\frac{1}{2}}^n}\right)\left(\frac{r_j^n + r_{j-1}^n}{2}\right)\right] , \tag{9}$$

$$E_1 h \left[\left(\frac{s_{j+\frac{1}{2}}^n - \Delta z}{\Delta z}\right)\left(\frac{z_{j+1}^n - z_j^n}{s_{j+\frac{1}{2}}^n}\right)\left(\frac{r_{j+1}^n + r_j^n}{2}\right) - \left(\frac{s_{j-\frac{1}{2}}^n - \Delta z}{\Delta z}\right)\left(\frac{z_j^n - z_{j-1}^n}{s_{j-\frac{1}{2}}^n}\right)\left(\frac{r_j^n + r_{j-1}^n}{2}\right)\right] , \tag{10}$$

respectively, where E_1 is the elastic modulus for longitudinal strain, $r_{j\pm1}^n$, $z_{j\pm1}^n$ are the positions of the wall vertices above and below vertex r_j^n, z_j^n at time cycle n, and the arc lengths are given by

$$s_{j\pm\frac{1}{2}}^n = \sqrt{\left(r_{j\pm1}^n - r_j^n\right)^2 + \left(z_{j\pm1}^n - z_j^n\right)^2} \quad .$$

The arc length, s_j^n, used in Eq. (8) is the average of $s_{j+\frac{1}{2}}^n$ and $s_{j-\frac{1}{2}}^n$. In Eqs. (9) and (10) we have used the mesh spacing in the axial direction, Δz, as the equilibrium length of arc in the longitudinal direction. This usage derives from the particular method of rezoning (to be described below) used in the distensible tube calculations, in which vertices are constrained to lie on horizontal lines, and from the fact that the equilibrium radius of the artery is uniform, i.e., $R_j = \text{constant}$. The generalization of the procedure to include a non-uniform equilibrium arc length is straightforward.

This explicit calculation of elastic stresses is followed by an iterative solution for the new pressure field. The elastic boundary is treated as a free surface in this relaxation procedure, so that at the end of the Lagrangian phase of the calculation the position of the elastic wall results from a balance between pressure and elastic stresses.

In the rezone phase of these distensible tube calculations the vertices are moved radially, in order to accommodate the motion of the boundary, but they do not move axially. The trajectory of all wall vertices is computed in order to determine the new position of the wall, but the vertices are not moved to these positions. Instead, each wall vertex is moved radially to the point on the wall with axial coordinate equal to that of the vertex, and interior vertices are moved proportionately.

Results of the Distensible Tube Calculations

Figure 5 shows the mesh configuration and velocity and pressure fields obtained in a calculation of flow through a distensible tube. The left boundary in each of these plots corresponds to an axis of symmetry, while the right boundary is an elastic wall. The bottom boundary is an inflow region at which the velocity is specified to be

$$u = 0 \ , \qquad v = 0.5 \sin^2 (0.8 \pi t) \quad .$$

Continuative outflow is maintained at the upper boundary as in the rigid tube calculations.

At this stage of the calculation the boundary of the computation mesh has been distorted by the passage through the tube of two pressure waves, and a third pulse is beginning to form at the bottom boundary. The fluid at the inflow boundary is being accelerated, producing a strong pressure gradient that is causing a bulge in the elastic wall. This bulge will grow in amplitude until the wall is sufficiently distended that the elastic restoring force exceeds the pressure driving force. In this application the elastic modulus is small enough that the elastic force does not begin to restore the wall until after the incoming fluid has begun to decelerate. The overdriving of the wall past its equilibrium position results in the wave pattern visible here, with each bulge corresponding to a region of high pressure along the elastic wall.

ACKNOWLEDGMENT

The author would like to thank Professor C. K. Chu of Columbia University for very helpful discussions regarding the formulation of the elastic boundary condition.

REFERENCES

Amsden, A. A. and Hirt, C. W., University of California, Los Alamos Scientific Laboratory Report LA-5100 (1973).

Brackbill, J. U. and Pracht, W. E., J. Comp. Physics 13, 455 (1973).

Chan, R. K.-C., Science Applications, Inc. Report SAI-73-575-LJ (1973).

Hirt, C. W., Amsden, A. A. and Cook, J. L., J. Comp. Physics 14, 227 (1974).

McDonald, D. A., J. Physiol. 127, 533 (1955).

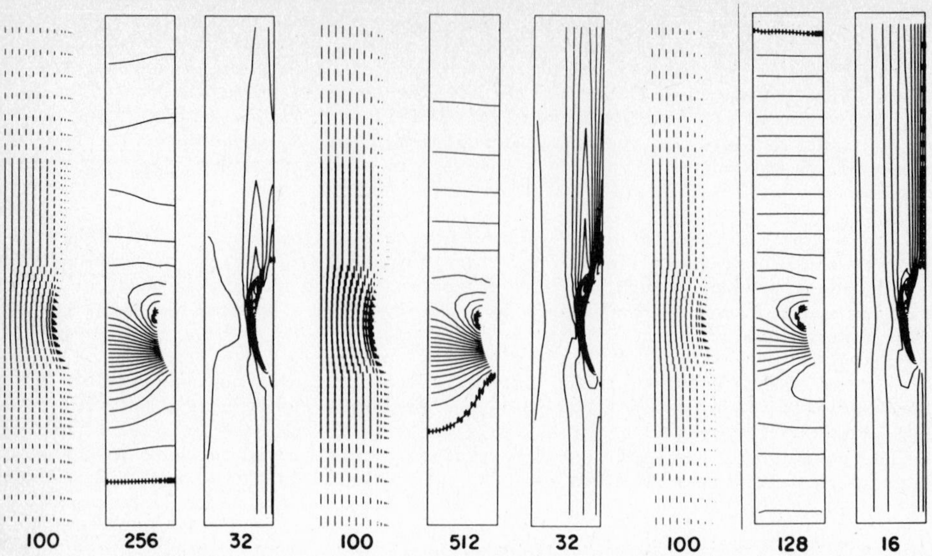

τ = 0.126   τ = 0.210   τ = 0.293

Figure 4. Velocity, pressure and shear stress plots at four stages of flow through a constricted canine femoral artery. The values listed below each plot are the normalizing velocities or contour intervals.

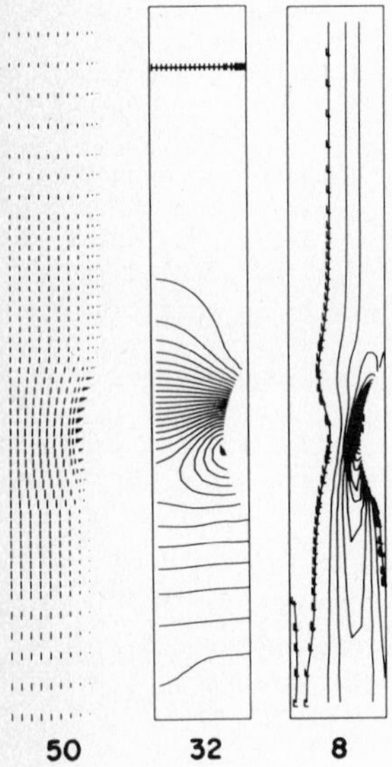

τ = 0.488

Figure 4. (continued)

Figure 5. Computation mesh, velocity vectors and pressure contours showing the motion of pressure waves through a distensible tube.

A STUDY OF A THREE DIMENSIONAL FREE JET USING THE VORTICITY/VECTOR POTENTIAL METHOD

G. de Vahl Davis*, Associate Professor and M. Wolfshtein, Senior Lecturer,
Department of Aeronautical Engineering,
Technion-Israel Institute of Technology,
Haifa, Israel.

INTRODUCTION

A finite-difference solution has been obtained of the problem of a free jet issuing from a square nozzle into a slower parallel stream. The flow is three-dimensional, steady, incompressible, and laminar and the flow domain is unbounded. The main purpose of the paper is to describe and discuss the vorticity/vector potential method, in which the Navier-Stokes and continuity equations are converted into three equations for the components of the vorticity vector. The velocity components are obtained by differentiation of the components of the vector and scalar potentials, which are found by solving three Poisson equations and a Laplace equation respectively.

THE DIFFERENTIAL EQUATIONS

A fluid discharges from a square duct of side $2L$ with a velocity W_d into surroundings in which the fluid velocity W_∞ is parallel to the axis of the duct z. The flow is governed by the equations of motion and continuity for laminar incompressible flow which take the following non-dimensional form

$$\frac{D\underset{\sim}{u}}{Dt} = - \nabla P + \frac{1}{Re} \nabla^2 \underset{\sim}{u} , \tag{1}$$

$$\nabla \cdot \underset{\sim}{u} = 0 . \tag{2}$$

where the scale factors are L and W_d, and $Re = W_d L/\nu$. Since $(\nabla \times \underset{\sim}{u}) \times \underset{\sim}{u} \equiv \underset{\sim}{u} \cdot \nabla \underset{\sim}{u} - 1/2 \nabla \underset{\sim}{u}^2$, (1) can be written

$$\frac{\partial \underset{\sim}{u}}{\partial t} = - (\nabla \times \underset{\sim}{u}) \times \underset{\sim}{u} - \nabla P + \frac{1}{Re} \nabla^2 \underset{\sim}{u} \tag{3}$$

where $P = p + 1/2 \rho \underset{\sim}{u}^2$ is the total pressure.

As proposed by Hirasaki and Hellums (1968, 1970) we introduce a vector potential $\underset{\sim}{\psi}$ and a scalar potential ϕ which satisfy

$$\underset{\sim}{u} = \nabla \times \underset{\sim}{\psi} - \nabla \phi . \tag{4}$$

It can be shown that $\underset{\sim}{\psi}$ may be solenoidal, i.e.

$$\nabla \cdot \underset{\sim}{\psi} = 0$$

in which case taking the divergence of (4) yields

$$\nabla^2 \phi = - \nabla \cdot \underset{\sim}{u} = 0 \tag{5}$$

Now, by taking the curl of Eq. (4), the vorticity $\underset{\sim}{\zeta}$ is related to the vector potential $\underset{\sim}{\psi}$ by

$$\underset{\sim}{\zeta} = - \nabla^2 \underset{\sim}{\psi} \tag{6}$$

Taking the curl of (3) yields

---

* Permanent address: School of Mechanical and Industrial Engineering, University of New South Wales, Kensington, Australia 2033.

$$\frac{\partial \underset{\sim}{\zeta}}{\partial t} = - \nabla \times (\underset{\sim}{\zeta} \times \underset{\sim}{u}) + \nabla^2 \underset{\sim}{\zeta}/Re \tag{7}$$

Equations (5),(6),(7) form a set for the determination of ϕ , $\underset{\sim}{\psi}$ and $\underset{\sim}{\zeta}$ respectively.

Taking into account the three components each of $\underset{\sim}{\zeta}$ and $\underset{\sim}{\psi}$, it can be seen that there are seven differential equations to be solved. In contrast, the primitive variable set (1), (2) contains only four equations. However, experience with the two-dimensional vorticity-stream function formulation suggests that the potential equations are very stable, and therefore the total number of iterations required to solve a given problem may be assumed to depend only on the properties of the three vorticity equations. Further, the elimination of the pressure results in the elimination of sound waves as well, and thus improves stability. Another advantage of the present system is that mass conservation is automatically satisfied.

THE BOUNDARY CONDITIONS

We are given, or can postulate plausibly, boundary conditions on $\underset{\sim}{u}$. These must be used to derive suitable conditions on $\underset{\sim}{\zeta}$, $\underset{\sim}{\psi}$ and ϕ.

If we consider ϕ first, we can choose whatever conditions we wish provided only that they are suitable for use with the Laplace equation (5) governing ϕ , and compatible with the velocity boundary conditions. Hirasaki and Hellums (1970) proposed to specify

$$\underset{\sim}{n} \cdot \nabla\phi \equiv - \frac{\partial \phi}{\partial n} = - \underset{\sim}{n} \cdot \underset{\sim}{u} \tag{8}$$

where $\underset{\sim}{n}$ is the unit vector in an outward direction normal to a boundary. The corresponding $\underset{\sim}{\psi}$ boundary conditions are that the tangential components, and the normal derivative of the normal component, are all zero on the boundary.

Finally, we consider the conditions $\underset{\sim}{\zeta}$. On z = 0, both inside and outside the duct, the flow is parallel and uniform, and therefore $\underset{\sim}{\zeta} = 0$. This is not the case, however, at the very edge of the duct where there is a discontinuity* in w and hence non-zero (in theory, infinite) values of ζ_x and ζ_y.

Consider the duct edge which is parallel to the x-axis. Here $\partial v/\partial z = 0$ and $\zeta_x = \partial w/\partial y$; $\zeta_y = \zeta_z = 0$. On the assumption that both the inside and outside boundary layers are thinner than one mesh length Δy, the vorticity flux across one mesh cell of the size $\Delta x \cdot \Delta y$ is then $\Delta x(W_\infty^2 - 1)/2$. On the further assumption that the flow remains essentially parallel to the z-axis for a distance of Δz, and neglecting transverse diffusion, this is the vorticity flux which enters the neighboring cell, which is also given by $\Delta x\Delta y((w\zeta_x)_W + (w\zeta_x)_P)/2$ where subscript "W" represents the point at the edge of the wall and "P" represents the point adjacent to it in the z-direction. Hence

$$\zeta_{x,W} = - \zeta_{x,P} w_P/w_W + (W_\infty^2 - 1)/w_W/\Delta y \tag{9}$$

where w_W is a fictitious velocity representing a local convection velocity which may be arbitrarily chosen without having any influence on the value of $\zeta_{x,W}$. The only limitation on w_W is that it should not become vanishingly small as this will imply zero convection and contrast the assumption which yielded Eq. (9). We chose $w_W \doteq 1$. Other possibilities include w_P, W_∞, and $(1 + W_\infty)/2$. Having specified w_W, $\zeta_{x,W}$ can be found from (9). Similar reasoning permits ζ_y to be computed along the duct edges parallel to the y-axis.

Upon implementation of these conditions, it was found that $\partial w/\partial z$ was not zero on the duct axis in the exit plane, a condition which we felt should exist there. Thus an adjustment was made to ζ_x and ζ_y along the duct edges. Since (as

---

* The discontinuity arises, of course, because we have not computed the boundary layers on the inside and outside surfaces of the duct.

discussed below) the entire equation set was solved iteratively, one application of this adjustment per iteration was sufficient to ensure that the condition $\partial w/\partial z = 0$ at the origin was satisfied at convergence.

On the remaining boundaries of the solution region, the conditions on ζ are straightforward. Thus on the side boundaries $x = \pm X$, $y = \pm Y$, we assume $\zeta = 0$. On the downstream boundary, the flow is assumed to be fully developed and parallel; thus

$$\zeta_x = \frac{\partial w}{\partial y}\ , \qquad \zeta_y = -\frac{\partial w}{\partial x}\ , \qquad \zeta_z = 0.$$

Finally the velocity must be specified on all boundaries. Thus, on the plane $z=0$, $u=v=0$ and $w=1$ (inside the duct), or $w=W_\infty$ (outside the duct). On the side boundaries it was assumed that the radial velocity component is inversely proportional to radial distance from the duct axis, and independent of z. Hence

$$u = \frac{K\,x}{x^2+y^2} \quad \text{and} \quad v = \frac{K\,y}{x^2+y^2} \tag{10}$$

where K was calculated to satisfy mass continuity over the entire solution region boundary.

On the downstream plane $z = Z$ it was assumed that the flow is fully developed, i.e. that $\partial w/\partial z = 0$. (Of course, this is not strictly true. However if the length Z of the solution region is sufficient, the error introduced by this assumption is small).

THE METHOD OF SOLUTION

The method of the false transient (Mallison and de Vahl Davis, 1973) was used to find the steady solution of the equations. Thus (5),(6),(7) were rewritten

$$\frac{1}{\alpha_\zeta}\frac{\partial \underset{\sim}{\zeta}}{\partial t} = -\nabla \times (\underset{\sim}{\zeta} \times \underset{\sim}{u}) + \frac{1}{Re}\nabla^2 \underset{\sim}{\zeta} \tag{11}$$

$$\frac{1}{\alpha_\psi}\frac{\partial \underset{\sim}{\psi}}{\partial t} = \underset{\sim}{\zeta} + \nabla^2 \underset{\sim}{\psi} \tag{12}$$

$$\frac{1}{\alpha_\phi}\frac{\partial \phi}{\partial t} = \nabla^2 \phi \tag{13}$$

and, after conversion to component form, were replaced by finite difference approximations (using central differences throughout) and solved using the Samarskii-Andreev (1963) ADI procedure.

An empirical stability limit

$$\frac{\Delta t}{\min(\Delta x^2, \Delta y^2, \Delta z^2)} \leq 0.8$$

has been found to apply to (11) and (12), even though, individually, each equation is unconditionally stable. Therefore Δt was chosen according to this criterion. The instability appears to arise from the coupling between the equations, and the rate of convergence can be enhanced by adjustment of the quantities $\alpha_\zeta, \alpha_\psi$ and α_ϕ, thus effectively modifying the time steps used in the various equations.

Little effort was devoted to seeking optimum values for the α's. It was found that $\alpha_\zeta = \alpha_\psi = 1$, $\alpha_\phi = 1.75$ enabled convergence (to a relative accuracy of 10^{-5}) to be obtained in about 300 "time" steps which required about 3 minutes on an IBM 370/165 computer when using 2250 mesh points. Vertical and horizontal symmetry (but not diagonal symmetry) was invoked, so that the mesh employed is comparable with a 9000 points mesh.

RESULTS

The more interesting features of the laminar square jet will be presented in this section, in terms of the decay of the center line velocity, the growth of the jet width, and the velocity profiles. The nozzle Reynolds number $Re_d = W_d \cdot 2L/\nu$ varied in the range 1-50, while the outer-stream to jet velocity ratio W_∞/W_d varied between 0 - 0.9.

The jet growth may be presented in terms of the half velocity width, $Y_{1/2}$ which is defined as the distance from the distance from the jet axis along the x or y directions where the velocity reaches the average between W_∞ and W_{max}. The jet width is shown in Fig. 1 for different Reynolds numbers, and a velocity ratio $W_\infty/W_d = 0.5$ In Fig. 2 the jet width is shown for different main stream velocities and at $Re_d = 5$. As may be expected, the jet width growth becomes smaller when the Reynolds number increases, and when W_∞ is nearer to W_d.

The influence of the nozzle Reynolds number on the maximum velocity decay is shown in Fig. 3 for a velocity ratio $W_\infty/W_d = 0.5$. It can be seen that the velocity decay is slower at higher Reynolds numbers. In Fig. 4 the velocity decay is given for $Re_d = 5$, and for different velocity ratios. The velocity has been normalized by using $(W_{max} - W_\infty)/(W_d - W_\infty)$. It is interesting to note that in this mode of presentation the influence of W_∞/W_d on the velocity decay is very small.

Another interesting feature of the square jet flow is its development into a round jet. This may be studied by comparing the jet half width in the x and y direction with that in the diagonal direction (x = y). The quantity $Y_{1/2}/Y_{1/2,\,dia}$ is plotted in Figure 5 versus z/L. It may be seen the jet approaches axial symmetry very fast. However, this approach is not monotonic and small oscilations in the value of $Y_{1/2}/Y_{1/2,\,dia}$ may be identified.

Finally the velocity profile at $z/L = 7$ is plotted for various velocity ratios, in Fig. 6. Complete similarity for various Reynolds numbers and velocity ratios has been obtained.

DISCUSSION AND CONCLUSIONS

In the present work the vorticity/vector-potential method was applied to an open integration domain. In earlier applications of this method the boundaries were always solid walls where the tangential components and the normal derivative of the normal component of the vector potential vanish. The extension of these boundary conditions to a region with free boundary required the addition of a new variable, namely the scalar potential. Hirasaki and Hellums (1970) have shown that the normal derivative of the scalar potential is equal to the velocity normal to the boundary. This boundary condition implies slow rate of convergence of the scalar potential equation when a considerable part of the boundary is opened. Further, when the velocity boundary conditions are specified in terms of velocity gradients this boundary condition becomes difficult to apply. In such cases it is recommended that an alternative set of vector and scalar potential boundary conditions be derived, which satisfies the velocity boundary conditions as well as the basic definitions of the potentials. This possibility is still under investigation and will be reported in later publications.

Utilization of the symmetry of the problem enabled a finer mesh to be used. As the jet spreads in the axial direction, z, the mesh must be spread in the x and y direction. In the present case the computer storage was the limit to computing capacity, as the method makes use of 7 primary and 3 auxiliary variables.

The main conclusion of the present research is that the vorticity/vector-potential method may be used for flows with free boundary and yield results at a reasonable price. It is envisioned that future research will reduce this price further.

ACKNOWLEDGEMENTS

The project was supported by Israel Foundations Trustees, Tel Aviv, whose help is acknowledged with thanks. Acknowledgement is also due to Mr. D. Maor and Mr. E. Ben Zabar for programming assistance.

REFERENCES

Aziz, K. and Hellums, J.D., Phys. Fl. 10, 314-324 (1967)
Hirasaki, G.J. and Hellums, J.D., Q. Appl. Math. XXVI, 331 (1968).
Hirasaki, G.J. and Hellums, J.D., Q. Appl.Math. XXVIII, 293-296 (1970)
Mallinson, G.D. and de Vahl Davis, G., J. Comp.Phys. 12, 435-461 (1973).
Samarskii, A.A. and Andreev, V.B., USSR Comp. Math. Math. Phys. 3, 1373-1382 (1963).

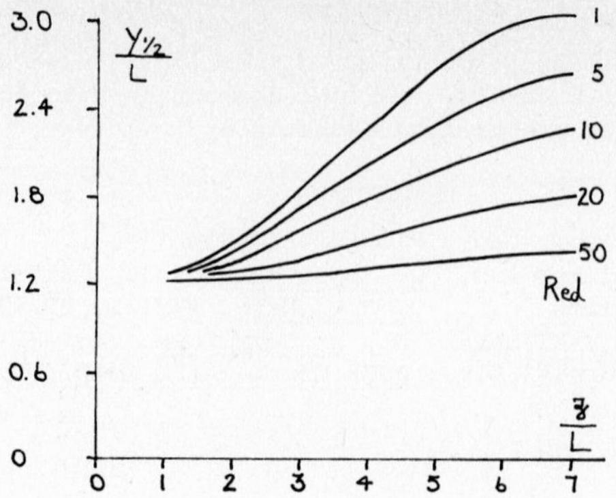

Fig. 1: Half width growth at $W_\infty/W_d = 0.5$.

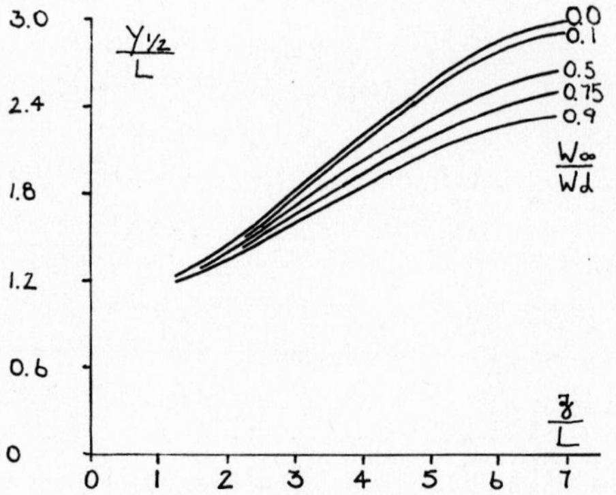

Fig. 2: Half width growth at $Re_d = 5$.

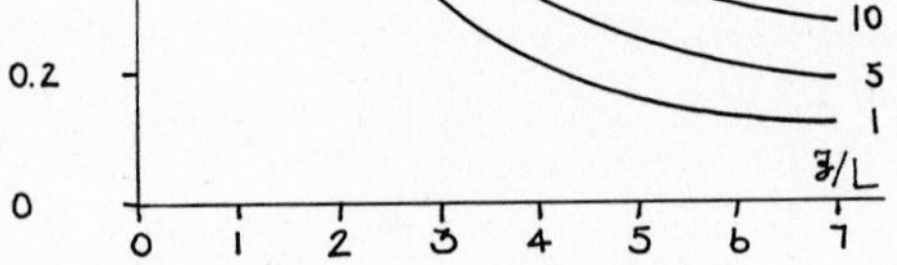

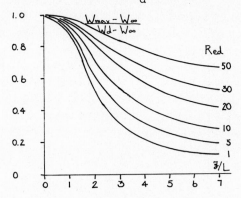

Fig. 3: Maximum velocity decay at $W_\infty/W_d = 0.5$.

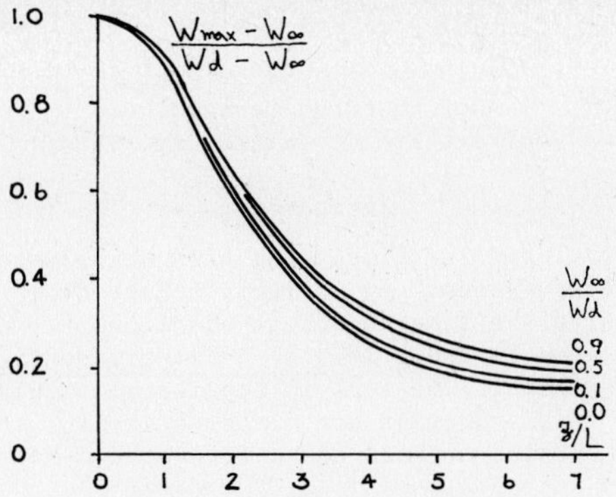

Fig. 4: Maximum velocity decay at $Re_d = 5$.

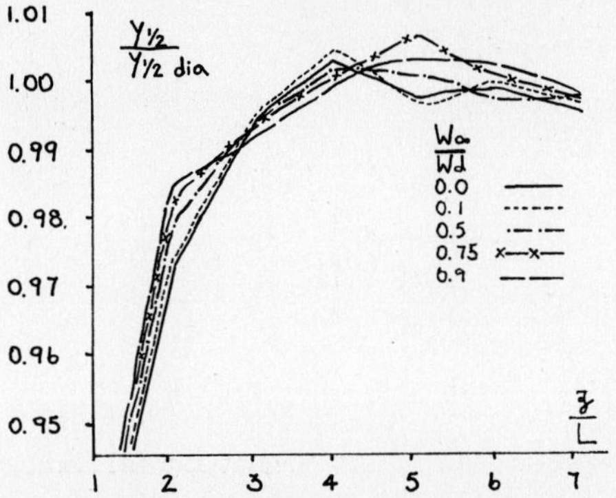

Fig. 5: Half width ratio on the diagonal and central planes at $Re_d = 5$.

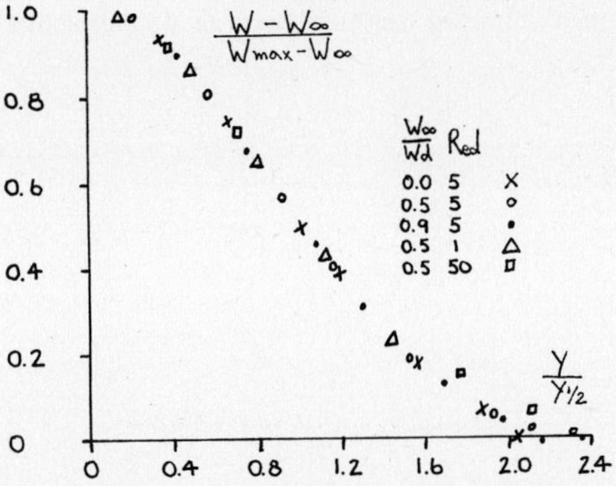

Fig. 6: Velocity profiles at $Re_d = 5$.

HIGH REYNOLDS NUMBER TRANSONIC FLOW SIMULATION

By George S. Deiwert
Ames Research Center, NASA
Moffett Field, California 94035

Introduction

A code has been developed for simulating high Reynolds number transonic flow fields of arbitrary configuration. An explicit finite-difference method with time splitting is used to solve the time-dependent equations for compressible turbulent flow. A nonorthogonal computational mesh of arbitrary configuration facilitates the description of the flow field. The code is applied to simulate the flow over a two-dimensional 18 percent thick circular-arc biconvex airfoil at zero angle of attack for several different Reynolds numbers and a free-stream Mach number of 0.775.

Analysis

Governing Equations

The flow field is described by the two-dimensional, time-dependent equations of motion written in integral form as

$$\frac{\partial}{\partial t}\int_{vol} U \, d\,vol + \int_{S} \vec{H}\cdot\vec{n}\, ds = 0 \tag{1}$$

where

$$U \equiv \begin{pmatrix} \rho \\ \rho u \\ \rho v \\ e \end{pmatrix} \qquad \vec{H} \equiv \begin{pmatrix} \rho\vec{q} \\ \rho u\vec{q} + \vec{\vec{\tau}}\cdot\vec{e}_x \\ \rho v\vec{q} + \vec{\vec{\tau}}\cdot\vec{e}_y \\ e\vec{q} + \vec{\vec{\tau}}\cdot\vec{q} - k\nabla T \end{pmatrix}$$

$$\vec{q} \equiv u\vec{e}_x + v\vec{e}_y \qquad \vec{\vec{\tau}} \equiv \sigma_x\vec{e}_x\vec{e}_x + \tau_{xy}\vec{e}_x\vec{e}_y + \tau_{yx}\vec{e}_y\vec{e}_x + \sigma_y\vec{e}_y\vec{e}_y$$

and $\vec{e}_x$, $\vec{e}_y$ are unit vectors and $\vec{n}$ is a unit normal vector.

Differencing Procedure

Equation (1) is written in the orthogonal x,y coordinate system and is satisfied for each cell of the nonorthogonal computational mesh by using the second-order-accurate, explicit finite-difference, predictor-corrector method with time splitting developed by MacCormack.[1] To evaluate the viscous derivatives for the nonorthogonal mesh, the following transformation is appropriate:

$$\frac{\partial\phi}{\partial x} = \frac{\partial\phi}{\partial\xi}\frac{\partial\xi}{\partial x} + \frac{\partial\phi}{\partial\eta}\frac{\partial\eta}{\partial x} \qquad \frac{\partial\phi}{\partial y} = \frac{\partial\phi}{\partial\xi}\frac{\partial\xi}{\partial y} + \frac{\partial\phi}{\partial\eta}\frac{\partial\eta}{\partial y}$$

where ϕ is a dummy dependent variable and (ξ,η) are the local coordinates of the nonorthogonal mesh (sketch 1).

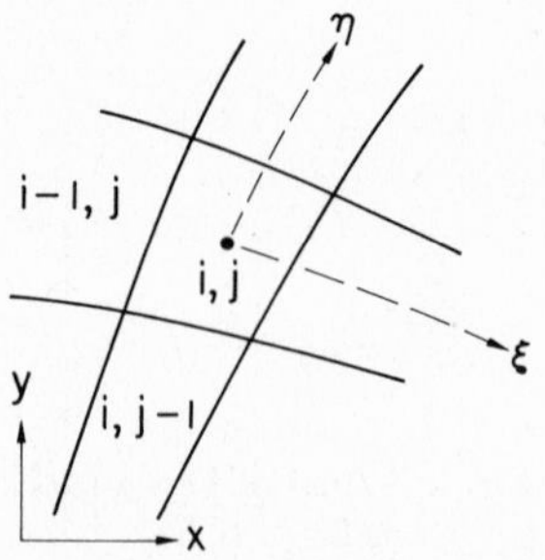

Sketch 1. Nonorthogonal Mesh Notation.

The evaluation of the viscous derivatives is as follows:

$$\frac{\partial\phi}{\partial x} = \frac{\Delta\phi_\xi\,\Delta y_\eta - \Delta\phi_\eta\,\Delta y_\xi}{\Delta x_\xi\,\Delta y_\eta - \Delta x_\eta\,\Delta y_\xi} \qquad \frac{\partial\phi}{\partial y} = \frac{\Delta\phi_\xi\,\Delta x_\eta - \Delta\phi_\eta\,\Delta x_\xi}{\Delta y_\xi\,\Delta x_\eta - \Delta y_\eta\,\Delta x_\xi}$$

where

$$\Delta\phi_\xi = \phi_{i+1,jj} - \phi_{im,jj} \qquad \Delta\phi_\eta = \phi_{ii,j+1} - \phi_{ii,jm}$$
$$\Delta y_\xi = y_{i+1,jj} - y_{im,jj} \qquad \Delta y_\eta = y_{ii,j+1} - y_{ii,jm}$$
$$\Delta x_\xi = x_{i+1,jj} - x_{im,jj} \qquad \Delta x_\eta = x_{ii,j+1} - x_{ii,jm}$$

and

$$im = \begin{cases} i & \text{for Lx} \\ i-1 & \text{for Ly} \end{cases} \qquad jm = \begin{cases} j & \text{for Ly} \\ j-1 & \text{for Lx} \end{cases}$$

$$ii = \begin{cases} i & \\ i+1 & \text{for Lx corrector} \end{cases} \qquad jj = \begin{cases} j & \\ j+1 & \text{for Ly corrector} \end{cases}$$

The Lx and Ly refer to the time-splitting operators described by MacCormack. This treatment of the viscous derivatives always results in centered differences, maintains second-order accuracy, and provides consistent treatment of discontinuous boundary conditions (as at the leading and trailing edges of the airfoil).

Control Volume, Mesh, and Boundary Conditions

An 18 percent thick circular-arc biconvex airfoil, initially at rest, is impulsively started at time zero at the desired free-stream Mach number and pressure. Figure 1 shows a typical control volume for which the flow-field development is followed in time. At a sufficient distance upstream of the leading edge (in this case 6 chord lengths), the flow is assumed uniform at the free-stream conditions ($u = U_\infty$, $v = 0$) as it is along the far transverse boundary (again, 6 chord lengths away). The downstream boundary is positioned far enough downstream of the trailing edge (9 chord lengths) so that all gradients in the flow direction may be assumed negligible ($\partial\phi/\partial x = 0$). The surface of the airfoil is impermeable, and "no slip" boundary conditions are assumed ($u = v = 0$). The airfoil is assumed adiabatic ($\nabla T \cdot \vec{n} = 0$), and the normal surface pressure gradient is zero ($\partial p/\partial n = 0$). Ahead of and behind the airfoil the flow is assumed to be symmetric. Because the airfoil is thick and the flow field is transonic, boundary-layer separation is likely. To simulate this phenomenon reliably for turbulent flow it is necessary to resolve the boundary layer to the sublayer scale. This sublayer scale is nearly proportional to $1/\sqrt{Re_c}$ so that, for the high Reynolds number flows of interest, the mesh resolution near the surface must be extremely fine. As a rule of thumb, a first mesh spacing of $\Delta y_{min} = 2/3\ c/\sqrt{Re_c}$ is adequate.

The mesh shown in figure 1 contains 50 × 38 points. In the x direction, the mesh is uniformly distributed over the surface of the airfoil (20 points) and is exponentially stretched ahead of (10 points) and behind the airfoil (20 points). In the y direction, a coarse mesh of 26 points is exponentially stretched away from the airfoil. The innermost region is further subdivided into a medium mesh of 10 exponentially stretched points and a fine mesh of 4 uniformly spaced points next to the airfoil. This results in 38 points in the y direction, the smallest spacing of which is $2c/3\sqrt{Re_c}$.

Turbulence Model

For expediency a simple mixing-length model[2] is used to describe the turbulent transport. For boundary-layer flow, the eddy viscosity is given by

$$\varepsilon = \rho\ell^2\left|\frac{\partial u}{\partial y} + \frac{\partial v}{\partial x}\right|$$

where

$$\ell = 0.4\,y\left[1 - \exp\left(-y\sqrt{\left(\frac{\rho}{\mu}\frac{\partial u}{\partial y}\right)_w}\Big/26\right)\right]$$

for the inner region, and y is measured from the point of zero tangential velocity; $\ell = 0.07\delta$ for the outer flow, and δ is determined from an arbitrary cutoff in the vorticity field. For the wake region, the turbulent viscosity is given by

$$\varepsilon = .001176\rho\delta|u_\delta - u_{\mathbb{C}}|$$

where u_δ and $u_{\mathbb{C}}$ are the velocities at the edge of the wake and its centerline, respectively.

Computational Time Step

Six different computational time steps are used in the calculation—one for the Lx operator and one for the Ly operator in each of the three mesh regions in the y direction. The time steps for each region are determined by the CFL and viscous stability requirements from the relation

$$\Delta t = \frac{h}{|V| + a + \mu+\varepsilon/\rho\ (\alpha/h)}$$

where h is the appropriate mesh spacing, V is the appropriate velocity component, a is the local speed of sound, and α is a function of the local mesh aspect ratio. In the wake behind the airfoil, the eddy viscosity ε is quite large so that the viscous stability criterion may govern the time step for the Ly operator in the fine mesh. To avoid this undesirable restriction and unnecessary resolution of the wake, the entire fine mesh region downstream of the airfoil is averaged and treated as part of the medium mesh in the Ly operator.

Results

Flow field solutions for the 18 percent thick circular-arc airfoil were computed for Reynolds numbers of $Re_c = 1\times10^6$, 2×10^6, 4×10^6, and 10×10^6 for a free-stream Mach number of 0.775. All solutions were carried out for a time corresponding to the mean flow traveling 7.5 chord lengths, and convergence to steady state was determined by monitoring the stress tensor on the body surface and in the near wake.

Figure 2 shows the pressure coefficient variation over the airfoil surface. The inviscid pressure distribution is included for comparison. All the viscous solutions lie to the left of the inviscid solution because of boundary-layer displacement effects. This effect is even more pronounced for the shock location because of flow separation predicted at the shock in the viscous solutions. At the trailing edge, the viscous pressure distributions plateau over the long separation bubble. As the Reynolds number is decreased, the shock strength decreases and the shock moves farther upstream on the airfoil. This is a result of the lower Reynolds number boundary layers being thicker, having less momentum, and thus being more susceptible to separation.

Figure 3 shows the skin friction variation over the airfoil surface. Ahead of the shock, as the Reynolds number is decreased the skin-friction coefficient is increased. Separation occurs farther upstream for the lower Reynolds number flows. Aft of the shock-induced separation region the flow tends to reattach but merges with the trailing edge separation region.

Figures 4(a) and 4(b) show the variation of displacement and momentum thicknesses, respectively. Both thickness parameters increase with decreasing Reynolds number. Just ahead of the shock-induced separation region, the thicknesses first decrease then increase dramatically over the separation bubble.

Typical boundary-layer profiles ahead of the shock are compared in figure 5 with the universal "law of the wall" for the $Re_c = 2\times10^6$ solution. All profiles are seen to have one point in the sublayer and adequately describe the "log law" and "wake flow" regions of the boundary layer. By having at least one point in the sublayer, both the skin-friction distribution and separation phenomena can be described.

Velocity profiles in the separation region are shown in figure 6 for the $Re_c = 4\times10^6$ solution. These profiles are plotted in the physical coordinate system over the aft portion of the airfoil and extend into the wake where the separation bubble is seen to close. The first separated profile is shown at 0.725 chord, indicating separation somewhat just ahead of the point. The shock location is just downstream of this profile, centered about the 0.740 chord point. The reattachment point is nearly 0.2 chord downstream of the trailing edge. Within the dividing streamline, also shown, the net mass flow is zero.

Figure 7 shows general features of the flow field: the isobars in the vicinity of the airfoil (fig. 7(a)) and the Mach lines (7(b)). The isobars are shown for $0.46 \leq p/p_\infty \leq 1.32$ in increments of 0.02 and the Mach lines for $0.40 \leq M \leq 1.40$ in increments of 0.02, both for $Re_c = 10\times10^6$. The position of the standing shock, is clearly indicated, and 7(b) shows the boundary layer and wake containing the separation bubble at the trailing edge.

Concluding Remarks

It is clear that the transonic flow field at high Reynolds number can be adequately simulated to provide detailed descriptions of (1) surface forces, (2) flow field characteristics including shock location and strength, and (3) boundary-layer and separation phenomena. The present study indicates that for an 18 percent thick circular-arc airfoil at $M_\infty = 0.775$, the separation point moves forward and the shock strength decreases with decreasing Reynolds number. While this Reynolds number effect is of second-order for this flow field, other airfoil configurations (such as supercritical airfoils) may be more sensitive.

No attempt was made to establish the validity of the turbulence model used. Since the flow is transonic, incompressible models should be reasonable ahead of the separation region. In the vicinity of the shock and in the separated flow region, however, the simple mixing length model is highly suspect and flow field details in these regions should be considered qualitative at this time.

References

1. MacCormack, R. W. and Paullay, A. J. AIAA Paper No. 72-154.
2. Launder, B. E. and Spalding, D. B. Mathematical Models of Turbulence, Academic Press, 1972.

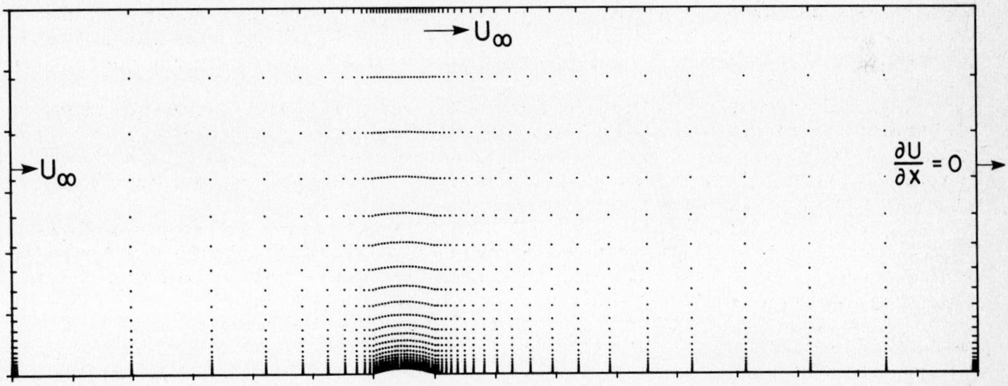

Fig. 1. Mesh configuration for 18 percent circular arc.

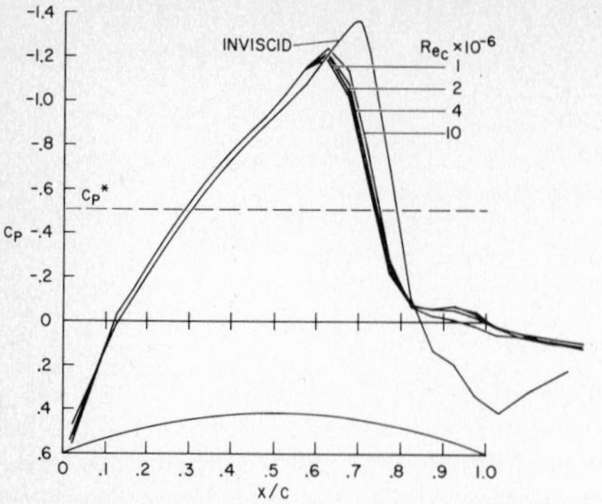

Fig.2. Pressure distribution over 18 percent circular arc, $M_\infty = 0.775$.

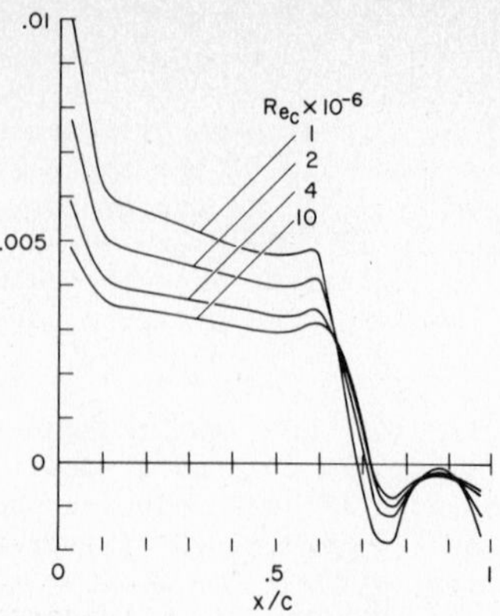

Fig. 3. Skin-friction distribution over 18 percent circular arc, $M_\infty = 0.775$.

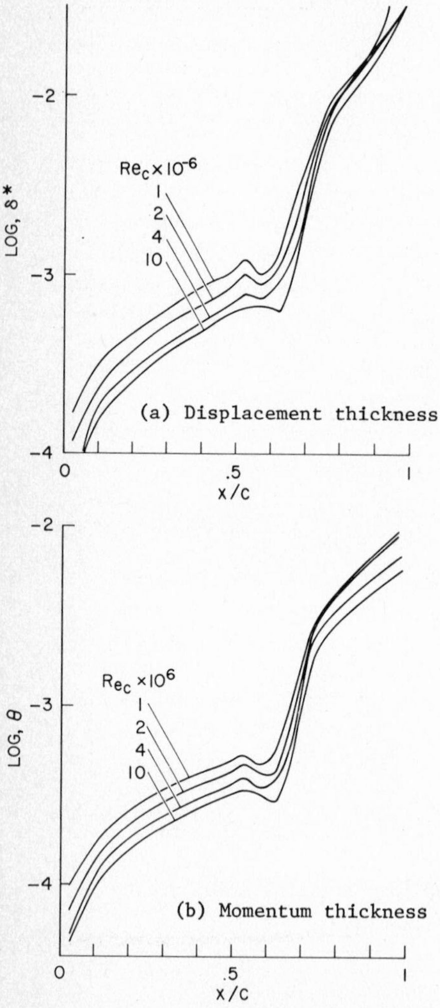

Fig.4. Thickness parameters on 18 percent circular arc, $M_\infty = 0.775$.

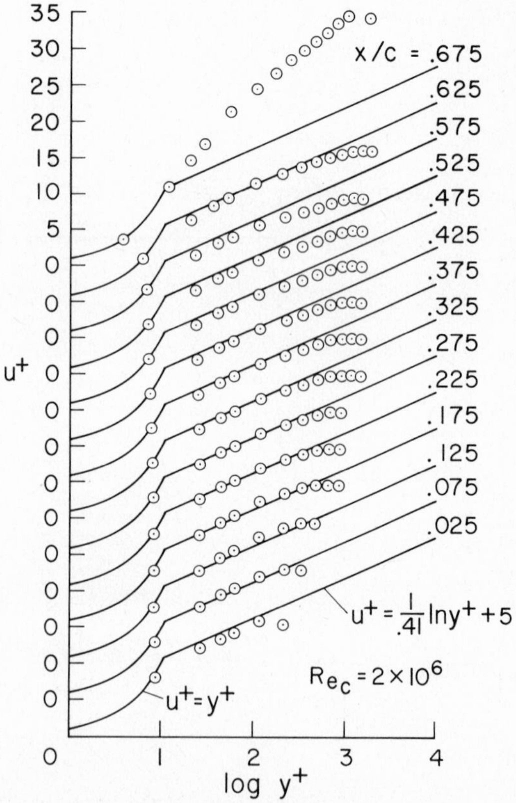

Fig. 5. Velocity profiles ahead of shock.

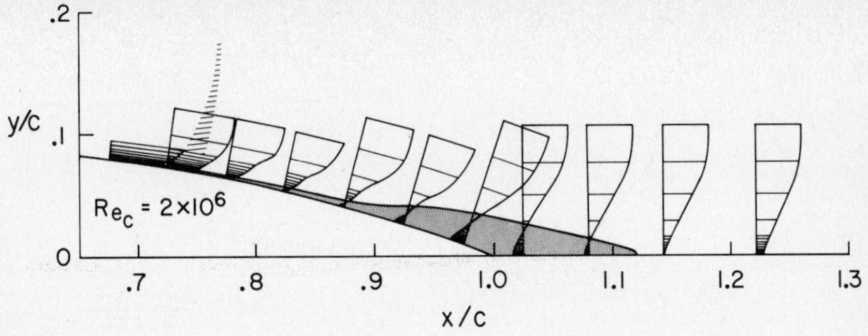

Fig. 6. Separation velocity profiles on 18 percent circular arc, $M_\infty = 0.775$.

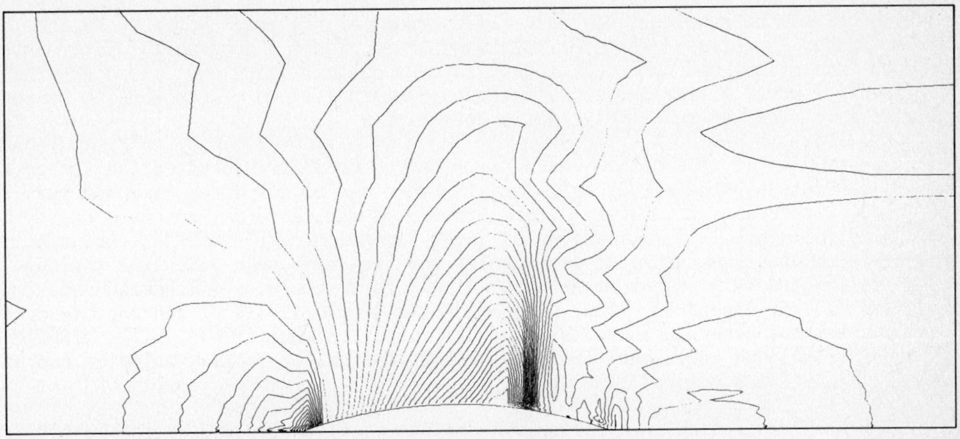

(a) Isobars

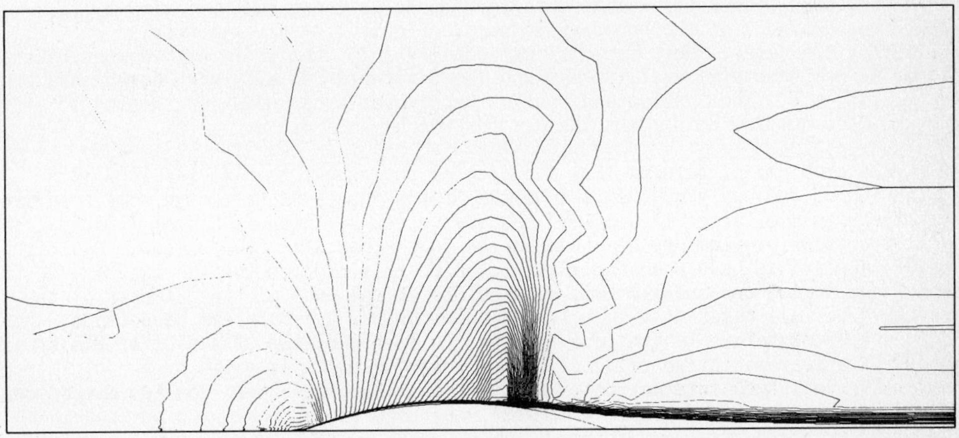

(b) Mach lines

Fig. 7. Flow field contours — $M_\infty = 0.775$, $Re_c = 10\times10^6$.

APPLICATION OF THE SERIES TRUNCATION METHOD TO TWO-DIMENSIONAL INTERNAL FLOWS

S.C.R. Dennis

Department of Applied Mathematics,
University of Western Ontario, London, Canada

INTRODUCTION

This paper is concerned with the numerical solution of the Navier-Stokes equations for the steady two-dimensional motion of a viscous incompressible fluid inside a fixed closed cylindrical body. The motion is assumed to be in a plane perpendicular to the generators of the cylinder and to be driven in some manner by the motion of the boundary enclosing the fluid or by the injection of fluid into the cylinder through the boundary. A wide number of problems of this type can be formulated and many of them are of interest since it is possible to generate various kinds of internal circulating and re-circulating flows. The study of flows of this nature contributes to the knowledge of basic fluid dynamical processes governed by the Navier-Stokes equations. Moreover, although there exists a substantial literature of both theoretical and numerical investigations of two-dimensional flows external to fixed cylinders of various cross sections, internal flows of the kind to be considered have received relatively little attention.

One problem of this nature which has received considerable attention is the motion generated inside a rectangular cavity when one of its sides is moved parallel to itself with constant velocity. This problem is indicated for a square cavity of unit side in figure 1(a), all variables being dimensionless. Numerical solutions of the Navier-Stokes equations for rectangular cavities have been given for various Reynolds numbers by a number of authors. Thus, for example, a time-dependent study has been made by Greenspan, Jain, Manohar, Noble & Sakurai (1964) and the steady flow problem has been considered by Kawaguti (1961), Simuni (1964), Mills (1965), Burggraf (1966) and Greenspan (1968). The slow motion, or Stokes, solution has been considered by Pan & Acrivos (1967).

The problem of flow in a rectangular cavity could be treated by the methods of the present paper but will not be considered further. A variation of this type of problem is shown in figure 1(b), where motion is generated inside a fixed circular boundary by moving part of the boundary with a velocity whose dimensionless radial and transverse components relative to the centre of the circle are $v_r = 0$, $v_\theta = -1$, with the remainder of the boundary at rest. Figure 1(c) gives a problem of different character. Here fluid flows radially into the circle with dimensionless velocity components $v_r = -1$, $v_\theta = 0$ and flows outward radially with components $v_r = 1$, $v_\theta = 0$. The inflow and outflow occur over arcs subtending angles 2α at the centre, the outflow being over the arc for which $-\alpha \leq \theta \leq \alpha$.

The problems of figures 1(b) and 1(c) do not appear to have received much attention on the basis of the full Navier-Stokes equations, although some treatment of a problem similar to figure 1(b) has been given by Kuwahara & Imai (1969). The inflow outflow problem has been considered on the basis of Stokes flow. Rayleigh (1893) appears to have been the first to solve this problem for the case $\alpha = 0$, in which the inflow and outflow reduce to points at the ends of the diameter of the circle. For all cases of figure 1 a Reynolds number $R = Ua/\nu$ can be defined, where a is a representative length (e.g. the radius of the circle or a side of the square) and U is a representative velocity. The Stokes solution is given by $R = 0$. It exists for all such internal flows and this distinguishes them from two-dimensional flows external to a body in an otherwise unbounded fluid, for which no Stokes solution exists.

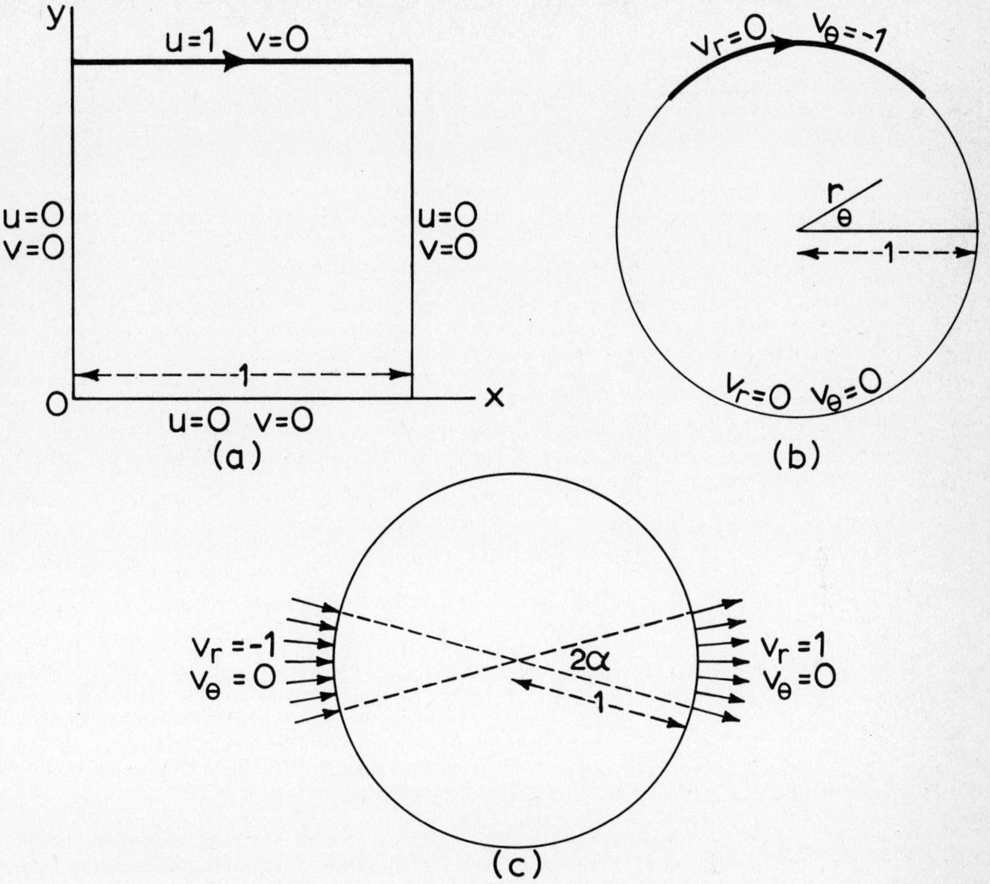

Figure 1

The existence of the Stokes solution for internal flows has the advantage of indicating an appropriate structure for the application of the series truncation method discussed in the present paper. We shall assume a series expansion for the full Navier-Stokes equations which is consistent with the form of the solution for $R = 0$ and which reduces to it when $R = 0$. We shall only treat the case of figure 1(c) in detail. This is especially simple because the flow is symmetrical about the line bisecting the arcs over which the inflow and outflow takes place, which we take as the x axis. However, the general form of the expansion is given in the next section. It can be applied to the case of figure 1(b) and also to the case of asymmetrical flows of the type given in figure 1(c) where the inflow and outflow are not symmetrically placed with respect to the diameter. It can also be applied to the case of figure 1(a) with suitable change from polar to Cartesian coordinates.

BASIC EQUATIONS AND THE SERIES TRUNCATION METHOD

For the problems depicted in figures 1(b) and 1(c) we use dimensionless polar coordinates (r,θ), where r is obtained by dividing the actual radial distance by the radius of the circle a. For two-dimensional flow we can introduce the dimensionless stream function ψ defined by

$$v_r = \frac{1}{r}\frac{\partial\psi}{\partial\theta} \quad , \quad v_\theta = -\frac{\partial\psi}{\partial r} . \tag{1}$$

Here $\psi' = Ua\psi$, $v_r' = Uv_r$, $v_\theta' = Uv_\theta$, where the primes denote dimensional quantities and U is the representative velocity. The vorticity vector is $\omega' = (0,0,\zeta')$ and we use a dimensionless scalar vorticity function ζ defined by $\zeta' = -U\zeta/a$. Thus

$$\zeta = \frac{1}{r}\frac{\partial v_r}{\partial\theta} - \frac{\partial v_\theta}{\partial r} - \frac{v_\theta}{r} . \tag{2}$$

The Navier-Stokes equations for steady incompressible flow then reduce to the equations

$$\nabla^2\psi = \zeta \tag{3}$$

$$\nabla^2\zeta = \frac{R}{2r}\left(\frac{\partial\psi}{\partial\theta}\frac{\partial\zeta}{\partial r} - \frac{\partial\psi}{\partial r}\frac{\partial\zeta}{\partial\theta}\right) , \tag{4}$$

where

$$\nabla^2 = \frac{\partial^2}{\partial r^2} + \frac{1}{r}\frac{\partial}{\partial r} + \frac{1}{r^2}\frac{\partial^2}{\partial\theta^2} .$$

The Stokes solution corresponds to putting R = 0 on the right side of (4). In both of the cases considered an exact solution of the equations (3) and (4) can be found for R = 0 in which $\psi(r,\theta)$ and $\zeta(r,\theta)$ take the form

$$\psi(r,\theta) = \frac{1}{2}F_o(r) + \sum_{n=1}^{\infty}[F_n(r)\cos n\theta + f_n(r)\sin n\theta] , \tag{5}$$

$$\zeta(r,\theta) = \frac{1}{2}G_o(r) + \sum_{n=1}^{\infty}[G_n(r)\cos n\theta + g_n(r)\sin n\theta] . \tag{6}$$

In this case the functions $F_n(r)$, $f_n(r)$, $G_n(r)$, $g_n(r)$ satisfy ordinary differential equations which are easily obtained by substituting (5) and (6) in the equations (3) and (4), with R = 0, and equating coefficients of the periodic terms on both sides of each of the resulting equations, respectively. The functions of r can then be determined to satisfy these equations with the appropriate boundary conditions, which depend upon the particular problem under consideration.

For non-zero R the same expansions (5) and (6) may be assumed and substituted in (3) and (4). The convective terms on the right side of (4) now make a non-linear contribution to the ordinary differential equations which can be evaluated explicitly, for each value of n, in terms of the functions $F_n(r)$, $G_n(r)$ (n = 0,1,2, ...) and $f_n(r)$, $g_n(r)$ (n = 1,2,3, ...). The solutions of these equations must now be carried out numerically in conjunction with the boundary conditions. We shall not proceed with the general case in the present paper, but merely illustrate it in the case of figure 1(c). This is a simplified case in that $F_n(r) \equiv 0$, $G_n(r) \equiv 0$ by virtue of the symmetry of the flow, but the general case is easily formulated for the flows described in figures 1(b) and 1(c) and, moreover, the formulation can be extended to the case of figure 1(a) by replacing the polar coordinates (r,θ) by Cartesian coordinates (x,y) and using the Cartesian form of the Navier-Stokes equations.

THE INFLOW-OUTFLOW PROBLEM

In the case of the flow given in figure 1(c), we have

$$\psi(r,\theta) = -\psi(r,-\theta), \ \zeta(r,\theta) = -\zeta(r,-\theta) ,$$

where θ is measured from the line bisecting the angle 2α. Thus

$$\psi = \zeta = 0 \quad \text{when} \quad \theta = 0, \ \theta = \pi , \ \text{for all } r \tag{7}$$

and the domain of the problem may be restricted to $0 \le r \le 1$, $0 \le \theta \le \pi$. The expansions (5) and (6) then reduce to

$$\psi(r,\theta) = \sum_{n=1}^{\infty} f_n(r) \sin n\theta , \tag{8}$$

$$\zeta(r,\theta) = \sum_{n=1}^{\infty} g_n(r) \sin n\theta . \tag{9}$$

The appropriate equations for $f_n(r)$, $g_n(r)$ $(n = 1,2,3, \ldots.)$ are

$$f_n'' + \frac{1}{r} f_n' - \frac{n^2}{r^2} f_n = g_n , \tag{10}$$

$$g_n'' + \frac{1}{r} g_n' - \frac{n^2}{r^2} g_n = t_n , \tag{11}$$

where primes denote differentiation with respect to r. The quantity t_n is the summation

$$t_n = \frac{R}{2r} \sum_{p=1}^{\infty} [\{mf_m - (n+p)f_{n+p}\}g_p' - p\{f_{n+p}' + \mathrm{sgn}(n-p)f_m'\}g_p] , \tag{12}$$

where $m = |n-p|$ and $\mathrm{sgn}(n-p)$ is the sign of n-p with $\mathrm{sgn}(0) = 0$.

The boundary conditions on r = 1 are

$$\begin{aligned} \psi &= \theta/\alpha \quad , \text{ for } 0 \le \theta \le \alpha ; \\ \psi &= 1 \quad , \text{ for } \alpha \le \theta \le \pi - \alpha ; \\ \psi &= (\pi-\theta)/\alpha , \text{ for } \pi - \alpha \le \theta \le \pi ; \\ \partial\psi/\partial r &= 0 \quad , \text{ for } 0 \le \theta \le \pi . \end{aligned} \tag{13}$$

These are sufficient, with (7), to solve the problem. Expressed in terms of the functions $f_n(r)$ and $g_n(r)$, they yield

$$\begin{aligned} f_n(0) &= g_n(0) = 0 ; \\ f_n(1) &= \frac{2}{\pi} \frac{1 - (-1)^n}{\alpha n^2} \sin n\alpha ; \\ f_n'(1) &= 0 . \end{aligned} \tag{14}$$

The conditions (14) hold for all positive integer values of n and give the four necessary conditions for each value of n to solve the sets of equations (10) and (11). In obtaining numerical solutions, the sets of equations (10) and (11) must be truncated by putting all functions $f_n(r)$, $g_n(r)$ identically zero for all $n > n_o$, which is a parameter of the numerical solutions. Thus $2n_o$ second-order differential equations must be solved for a given approximation to the flow.

The Stokes solution obtained by putting R = 0 gives $t_n = 0$ in (11) for all n. The solutions for the functions $f_n(r)$ and $g_n(r)$ are then found to be

$$\begin{aligned} f_n(r) &= \frac{[1 - (-1)^n]\sin n\alpha}{\pi \alpha n} \left(\frac{n+2}{n} r^n - r^{n+2} \right) , \\ g_n(r) &= - \frac{4[1 - (-1)^n](n+1)\sin n\alpha}{\pi \alpha n} r^n . \end{aligned} \tag{15}$$

These expressions may be used to give some check on the accuracy of the numerical methods by calculating the case R = 0 numerically. The series (15), which yield a solution which is symmetrical about $\theta = \pi/2$, can be summed to give an integral representation but we shall not consider this here.

NUMERICAL SOLUTIONS

The sets of equations (10) and (11) are solved by a standard finite-difference procedure using a grid size h. Each of the equations is approximated at each of the grid points $r = h, 2h, \ldots. , 1-h$ by expressing the typical derivatives $f_n'(r)$, $f_n''(r)$ by the central-difference approximations

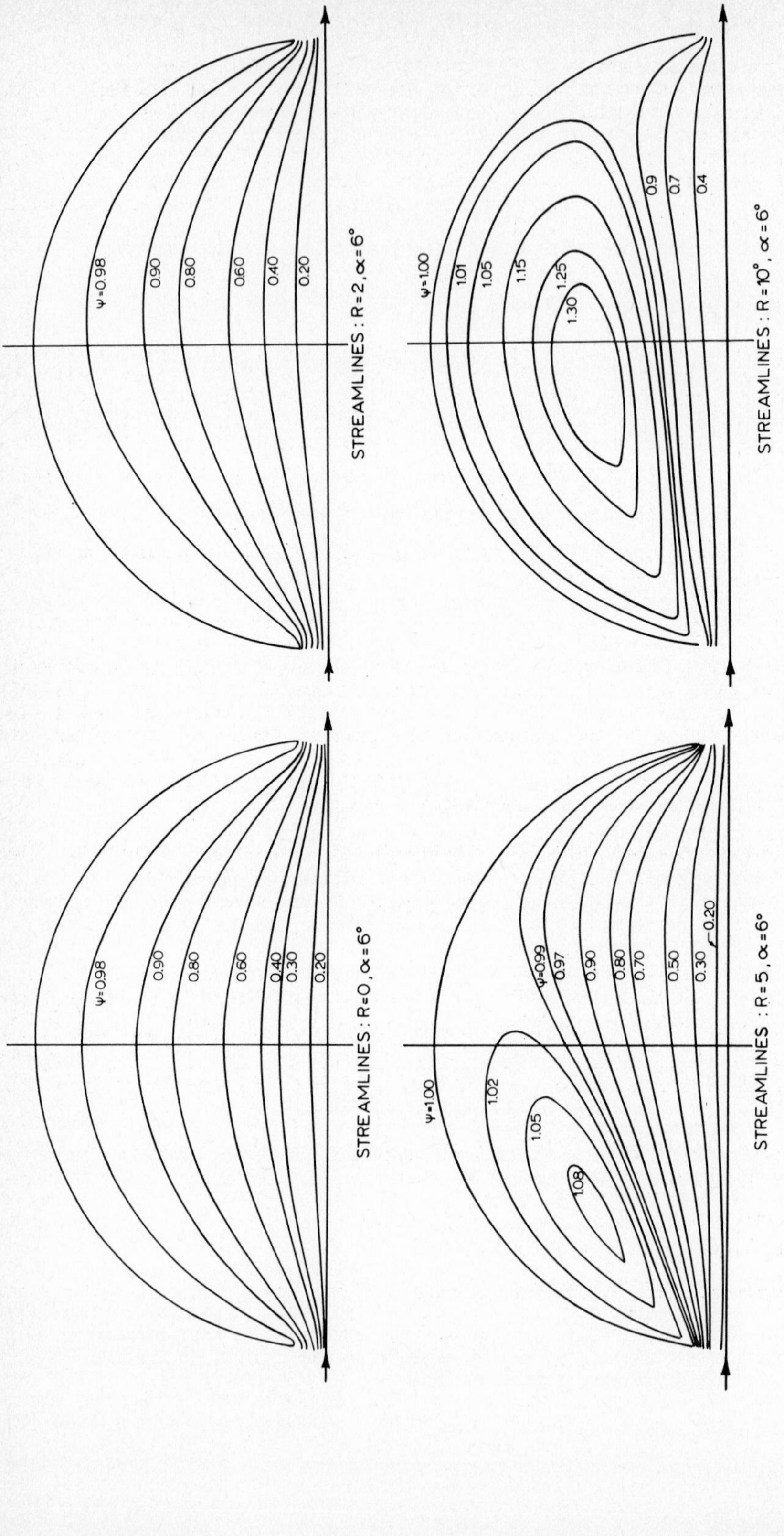

STREAMLINES : R=0, α=6°

STREAMLINES : R=2, α=6°

STREAMLINES : R=5, α=6°

STREAMLINES : R=10, α=6°

Figure 2

$$2hf'_n(r) = f_n(r+h) - f_n(r-h) \;,\; h^2 f''_n(r) = f_n(r+h) - 2f_n(r) + f_n(r-h), \qquad (16)$$

with similar approximations to derivatives of $g_n(r)$. This gives rise to sets of non-linear simultaneous equations which are solved by the Gauss-Seidel procedure. Explicit boundary conditions for (10) at $r = 0$ and $r = 1$ are given in (14) as well as a boundary condition for (11) at $r = 0$. As is quite customary in Navier-Stokes problems, a boundary condition for $g_n(1)$ to use with (11) is obtained by utilizing the last condition of (14). When this condition is expressed in terms of central differences at $r = 1$ and used in conjuction with the central-difference approximation to (11) at $r = 1$ we find that

$$h^2 g_n(1) = 2f_n(1-h) - (2+n^2h^2)f_n(1) \;, \qquad (17)$$

which allows $g_n(1)$ to be calculated.

The iterative procedure will not be described except to note that the difference equations approximating (10) and (11) are solved from $n = 1$ to n_o in a systematic cyclic procedure in which $t_n(r)$ is repeatedly re-calculated from the most recently available estimates of the various functions. The process of smoothing is used in the calculation of $t_n(r)$, i.e. if $t^*_n(r)$ is the most recently calculated value and $t_n^{(j)}(r)$ was the value previously used in (11), the new value next introduced into (11) is

$$t_n^{(j+1)}(r) = \lambda t^*_n(r) + (1-\lambda)t_n^{(j)}(r) \quad , \qquad (18)$$

where $0 < \lambda \leq 1$. If the iterative procedure diverges for a given λ, we reduce λ until convergence is obtained.

Calculations were carried out for $R = 0,2,5$ and 10 in each of the cases $\alpha = 1°$ and $6°$. Only the streamlines for $\alpha = 6°$ are given here. For each value of R, two solutions were obtained using grid sizes $h = 0.1$ and 0.05 respectively. Various numbers of terms n_o were used and the results compared for accuracy. The solutions for $h = 0.05$, $n_o = 40$ were judged to be acceptable to good accuracy and the streamlines corresponding to these solutions are shown in figure 2. The Stokes solution ($R = 0$) is symmetrical about $\theta = \pi/2$ and agrees well with the exact solution. The solution at $R = 2$ is only slightly asymmetrical about $\theta = \pi/2$, but thereafter recirculation starts as shown in the cases $R = 5$ and 10.

This work was supported by the National Research Council of Canada. The author is grateful to Mr. Eric Willis who assisted with the calculations.

REFERENCES

Burggraf, O.R. J. Fluid Mech. 24, 113 (1966).

Greenspan, D., Jain, P.C., Manohar, R., Noble, B. and Sakurai, A. Mathematics Research Center, University of Wisconsin, Technical Summary Report No. 482 (1964).

Greenspan, D. in Lectures on the Numerical Solution of Linear, Singular and Nonlinear Differential Equations, Prentice-Hall, Englewood Cliffs, New Jersey, 1968.

Kawaguti, M. J. Phys. Soc. Japan 16, 2307 (1961).

Kuwahara, K. and Imai, I. Phys. Fluids Suppl. II 12, II-94 (1969).

Mills, R.D. J. Roy. Aero. Soc. 69, 714 (1965).

Pan, F. and Acrivos, A. J. Fluid Mech. 28, 643 (1967).

Rayleigh, Lord Phil. Mag. 5, 354 (1893).

Simuni, L.M. Inzhenernii Zhournal (USSR) 4, 446 (1964).

A PHYSICALLY OPTIMUM DIFFERENCE SCHEME FOR THREE-DIMENSIONAL BOUNDARY LAYERS

H. A. Dwyer, University of California, Davis, California;
B. R. Sanders, Sandia Laboratories, Livermore, California

ABSTRACT

A new finite difference scheme has been formulated for the three-dimensional boundary layer equations based on the physics of the convective and diffusive momentum transport in the boundary layer. It is shown that the scheme is a physically optimum one and that it is consistent with the usual specification of initial conditions. A stability analysis of the linearized equations shows that a relative restriction is necessary on the step sizes along the convective coordinates even though the difference scheme is "implicit". The method is then applied to a problem that taxes the method and the laminar boundary layer equations to their limit; that problem being the supersonic flow over a spinning sharp cone at angle of attack. The results of the spinning cone calculation also yields some very useful insight into the "Magnus" problem and to the contributions to the "Magnus" force by the boundary layer flow.

INTRODUCTION

Over the past ten years there have been quite a few methods suggested for numerically solving the laminar three-dimensional boundary layer equations, References [1] through [7]. All of the methods have been successful in solving certain types of problems and have had, in some cases, good stability and convergence characteristics. However, the methods have not properly modeled the physical processes that occur in a boundary layer flow and some have had difficulty with certain types of initial conditions. In the present paper a new finite difference method is developed and applied to a problem that is on the limit of boundary layer theory (the spinning cone at angle of attack). It will be shown that the convection of momentum and heat have not been properly modeled in the past and that the new scheme does not suffer from this handicap. To further illustrate the method, a linearized stability analysis is carried out and the significant problem of flow over a supersonic spinning cone is solved.

BOUNDARY LAYER EQUATIONS

The three-dimensional boundary layer equations in their compressible form for laminar flow now will be given and discussed briefly. Shown in Figure (1) is the geometry of the spinning cone at angle of attack. With spin, the boundary layer flow that develops on the cone surface, is completely three-dimensional and no similarity exists, References [8], [9], and [10]. In terms of the coordinates along the surface x, y and the angular coordinate, s, the boundary layer approximation yields the following equations:

$$\frac{\partial}{\partial x}(\rho r u) + \frac{\partial}{\partial y}(\ r v) + \frac{1}{r}\frac{\partial}{\partial s}(\rho r w) = 0 \qquad \text{(Continuity)}$$

$$\rho u \frac{\partial u}{\partial x} + \rho v \frac{\partial u}{\partial y} + \frac{\rho w}{r}\frac{\partial u}{\partial s} - \frac{\rho r' w^2}{r} = - \frac{\partial p}{\partial x} + \frac{\partial}{\partial y}\left(\mu \frac{\partial u}{\partial y}\right) \qquad \text{(x-Momentum)}$$

$$\rho u \frac{\partial w}{\partial x} + \rho v \frac{\partial w}{\partial u} + \frac{\rho w}{r}\frac{\partial w}{\partial s} + \frac{\rho r' u w}{r} = - \frac{1}{r}\frac{\partial p}{\partial s} + \frac{\partial}{\partial y}\left(\mu \frac{\partial w}{\partial y}\right) \qquad \text{(s-Momentum)}$$

$$\rho C_p u \frac{\partial T}{\partial x} + \rho C_p V \frac{\partial T}{\partial y} + \rho C_p \frac{w}{r}\frac{\partial T}{\partial s} = u \frac{\partial p}{\partial x} + \frac{w}{r}\frac{\partial p}{\partial s} + \mu \left[\left(\frac{\partial u}{\partial y}\right)^2 + \left(\frac{\partial w}{\partial y}\right)^2\right]$$

$$+ \frac{\partial}{\partial y}\left(k \frac{\partial T}{\partial y}\right) \qquad \text{(Energy)}$$

where u, v, and w are the velocities in the x, y, and s directions, respectively, ρ - density, T - temperature, μ - viscosity, k - thermal conductivity, r' - dr/dx, p - pressure, and C_p - specific heat at constant pressure. It is convenient, especially for laminar flows, to transform the independent variables in these equations so that the numerical calculations are carried out more efficiently. For this purpose the following transformations have been employed.

$$d\eta = \left(\frac{3}{2}\frac{\rho_\infty u_\infty}{\mu_\infty x}\right)^{3/2} \left(\frac{p}{p_\infty}\right)^{1/2} \frac{T}{T_\infty} dy \qquad d\overline{x} = C^2 x^2 dx$$

where all variables with subscript infinity are arbitrary reference variables and C is equal to the sine of the cone half angle.

In terms of the physical variables the convective coordinates are x and s, and along these directions no diffusion of heat or momentum occurs. The fluid particles are only carried downstream by their convective velocities. The diffusive coordinate is y and is perpendicular to the body surface. Along this coordinate heat and momentum are diffused and then carried downstream. A finite difference scheme will now be formulated based on the nature of the coordinates just described.

FINITE DIFFERENCE METHOD AND ANALYSIS

The finite difference scheme consists of a variation of the implicit schemes that use the tridiagonal algorithm, and which have been used successfully for two-dimensional flows. It will be assumed that initial conditions are given in the y-s plane, as in the spinning cone problem. With an implicit scheme an unknown row of grid points along the y-coordinate are calculated simultaneously; and this models the diffusive nature of that coordinate. In order to illustrate the scheme, Figure (2) will be employed and the variable s will be replaced by z where r s = z.

The basic method consists of applying the equations at the unknown grid station and performing the following finite difference derivative approximation: (1) A central difference in y for the first and second derivatives; (2) a backward difference for the x-derivative; and (3) an explicit backward difference in z which takes account of the sign of the crossflow velocity. The x and z approximations are shown schematically in Figure (2). The basic novel feature of the scheme is that the crossflow derivative operator $w\,\partial/\partial z$ is always kept positive by testing and determining the direction of the crossflow velocity. In this way the convection process is properly modeled and convection always occurs in the positive direction (this process resembles windward differencing, but it is not the same thing). The z-derivative is taken to be in the positive or negative direction, depending on whether w is positive or negative, respectively.

In order to understand some of the characteristics of this scheme, a Von Neumann stability analysis is performed on a simplified version of the boundary layer equations. This version is

$$u \frac{\partial u}{\partial x} + w \frac{\partial u}{\partial z} = \alpha \frac{\partial^2 u}{\partial y^2}$$

It is convenient to define two time-like coordinates t_1 and t_2 where $\partial t_1 = \partial x/u$ and $\partial t_2 = \partial z/w$, then the above equation becomes in finite difference form

$$\frac{u_i^{n+1,m} - u_i^{n,m}}{\Delta t_i} + \frac{u_i^{n,m} - u_i^{n,m-1}}{\Delta t_2} = \frac{\alpha\, u_{i+1}^{n+1,m} - u_i^{n+1,m} + u_{i-1}^{n+1,m}}{\Delta y^2}$$

where i, n and m are the y, x and z indices, respectively. To apply the Von Neumann stability analysis it is assumed the u_i^{nm} has the form

$$u_i^{n,m} = V^n e^{ji\theta} e^{jm\phi}$$

where $j = \sqrt{-1}$. Substituting the above expression into the finite difference equation, we obtain the following equation for the amplification factor $G = V^{n+1}/V^n$

$$G = \frac{1 - \sigma_1 (1 - e^{-i\phi})}{(1 + \sigma [1 - \cos\theta])} \tag{1}$$

where $\sigma_1 = \dfrac{\Delta t_1}{\Delta t_2}$ and $\sigma = \alpha \dfrac{\Delta t_1}{\Delta y^2}$

This leads to the condition that

$$G^2 \leq 1 - 2\sigma_1 (1 - \cos\phi) + 2\sigma_1^2 (1 - \cos\phi) \quad . \tag{2}$$

Only for values of σ_1 between 0 and 1 does this expression assure that G will be less than 1. For σ_1 outside of this range, it is always possible to have G greater than one and instability occurring.

The condition that σ_1 be between 0 and 1 actually does not put a restriction on the size of the steps Δt_1 and Δt_2, but only on their ratio. Also, it should be noticed that $\Delta^2 y$ does not enter into the restriction. For σ_1 negative, which corresponds to integrating against the crossflow, the method seems to be unstable and this has been observed by Dwyer, Reference [3], in actual calculations. The condition σ_1 equals one corresponds exactly to following the zones of influence defined by the convective velocities, and would be the most desirable condition for following the convective history of the flow. However, since every y-coordinate has different velocities, it is impractical to satisfy this condition at all stations and a condition of σ_1 less than one should be employed. For σ_1 greater than one G grows significantly slower than when σ_1 is less than minus one, and many calculations have shown that useful results can be obtained when σ_1 is greater than one even though the stability analysis indicates otherwise. Although the scheme as proposed is first order, it should be possible by use of iteration to adapt to a second order method.

RESULTS FOR THE SPINNING CONE

In order to illustrate the usefullness of the proposed method, the problem of the laminar boundary layer flow over a spinning cone at angle of attack has been calculated. The geometry of the flow is shown in Figure (1), and solutions based on combined perturbation expansions and numberical methods have been obtained previously by Sedney [9] and Watkins [10]. The initial conditions for the flow consist of the boundary layer for a non-spinning cone applied at the cone tip (these solutions were obtained by the methods developed by Dwyer [2], and it is very essential that transformed coordinates be applied). The boundary conditions which are applied at the edge of the boundary layer were obtained from the cone tables developed by Jones [11]. It should be mentioned that the scheme currently proposed is completely consistent with these initial and boundary conditions as opposed to most other methods which require additional grid points to perform a numerical solution.

The spinning cone problem is particularly interesting since the surface streamline is a circle and returns back on itself. However, this streamline is the only one with this behavior, and all streamlines at a finite distance away from the wall move downstream. Therefore, when doing a numerical calculation, the first grid point away from the wall determines the maximum step size in the x-direction which can be taken. The flow conditions chosen for the numerical calculation where similar to those of Sedney, but with much higher spin rates. The flow Mach number was chosen to be 2, Pr (Prandtl Number) = 1 and an adiabatic wall condition. Also, the angle of attack (α) and spin rate (Ω) were chosen to be 2° and 30,000 rpm, respectively.

The results of the numerical calculations are shown in Figures (3) through (6). Figure (3) presents the results of the calculations for primary flow skin friction at the cone tip, or initial condition plane and it should be noticed that even without spin the numerical solution and the perturbation solution of Sedney differ considerably. This difference is due to the limited number of terms taken in the perturbation expansion for the angular dependence around the cone. The fact that there is no surface velocity is denoted by K, the spin parameter, being equal to zero, where K is defined by

$$K = \frac{\Omega r}{\bar{u}^1} \qquad \bar{u}^1 = \text{windward inviscid primary flow velocity.}$$

It should be mentioned that on the windward ray the numerical and perturbation results gave essentially identical results since C_{f_o} (numerical) equals C_{f_o} (Sedney).

Some calculations for a finite distance downstream of the tip are shown in Figure (4) with the primary flow shear again being exhibited. These results corresponded to a free stream Reynolds number (Re_∞) of $1.72 \cdot 10^6$ and a distance downstream of .8 feet, which gives a spin parameter K = .263. The value of σ_1 used was 0.95 and required a grid with 60 y, 36 s and 100 x stations and involved 20 minutes of processor time on a Burrough's 6700 machine. For a value of σ_1 = 5.0, calculations were also carried out, and gave results within 4 percent of the σ_1 = .95, and this is very surprising since Equation (2) indicates possible instability. However, many or most of the results presented in the literature on numerical solutions of 3-D boundary layers have had σ_1 greater than one. The physical results to be obtained from Figures (4) and (5) indicate that spin increases wall shear for positive ϕ (the side where the spin velocity and inviscid crossflow velocity are in the same direction), and decreases primary flow shear from negative ϕ. Also Figure (4) indicates that the perturbation solution gives the greatest error at the leeward ray when predicting the influence of spin. It should finally be mentioned that the boundary layer displacement thickness is distributed inversely proportional to the value of C_f and plays a major role in determining the "Magnus" or side force on the cone.

Other forces created in the boundary layer which influence the "Magnus" force are the centrifugal pressure distribution Δ_p, the crossflow wall shear τ_ϕ and the primary flow wall shear τ_x, where

$$\Delta_p = \int_0^{\delta} \rho \frac{w^2}{r}\, dy \qquad \tau_\phi = \mu \frac{\partial w}{\partial y} \qquad \tau_x = \mu \frac{\partial u}{\partial y} \quad .$$

The angular distribution of these quantities at $x = .8$ is shown in Figure (6) where all quantities have been normalized with respect to their windward values. From these results, it is seen that Δ_p and τ_x add while the crossflow wall shear acts in the opposite direction. In terms of actual contribution to the "Magnus" force, τ_x contributes roughly 10% of Δ_p and τ_ϕ is 60% of Δ_p. A more detailed report on the physical applications is currently being prepared.

SUMMARY

A numerical method of solving the three-dimensional boundary layer equations has been developed, and it has been shown that the method can model the convective and diffusive process in the boundary layer. The finite difference scheme itself can be classified as being implicit, but it does require an explicit evaluation of the crossflow convective derivative. Also, the direction of the crossflow derivative depends on the sign of the crossflow velocity. A Von Neumann stability analysis reveals that only a relative restriction on the convective step sizes is necessary for stability and the size of the diffusive grid size is unimportant to stability. However, numerical solutions of the spinning cone problem shows that the restriction can be violated and useful results obtained. The spinning cone solutions also show how spin influences the displacement thickness and indicates the importance of centrifugal pressure gradient, crossflow shear and primary flow shear.

REFERENCES

1. Dwyer, H. A., AIAA J., vol. 6, no. 7, pp. 1336-1342, July, 1968.

2. Dwyer, H. A., AIAA J., vol. 9, no. 2, pp. 227-287, Feb., 1971.

3. Dwyer, H. A., AIAA Paper 71-57, New York, 1971.

4. Kraus, E., AIAA J., vol. 7, no. 3, p. 575, 1969.

5. Wang, K. C., J. Fluid Mech., 43, 1, 187-209.

6. Der, J., AIAA J., vol. 9, no. 7, pp. 1294-1302, July, 1971.

7. Blottner, F. G., Sandia Laboratroies Report SLA-37-0366, April, 1973.

8. Moore, F. K., NACA TN 2279, 1959.

9. Sedney, R., J. Aero. Sci., vol. 24, pp. 430-436, June, 1957.

10. Watkins, C. B., Jr., Symposium on Fluid Dynamics, Polytechnic Institute of Brooklyn, Farmingdale, New York, Jan., 1973.

11. Jones, D. J., AGARDO GRAPH 137, Nov., 1969.

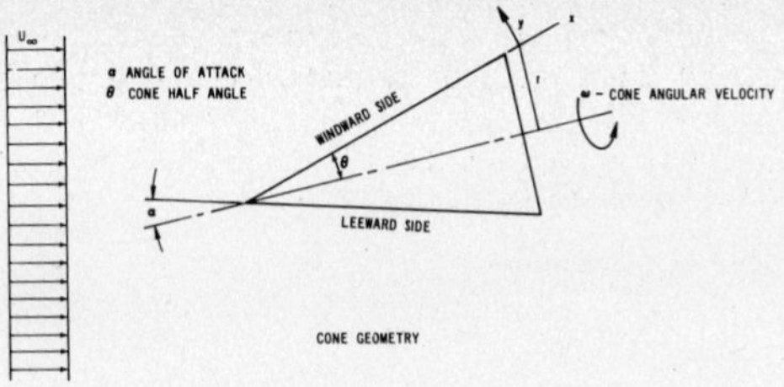

Figure 1. Problem Geometry

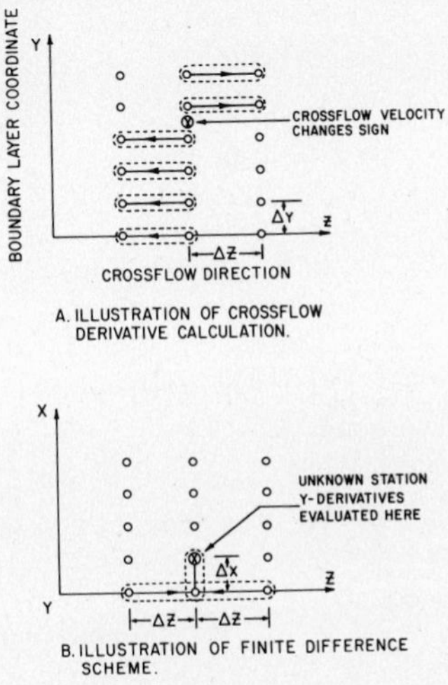

Figure 2. Finite Difference Grid

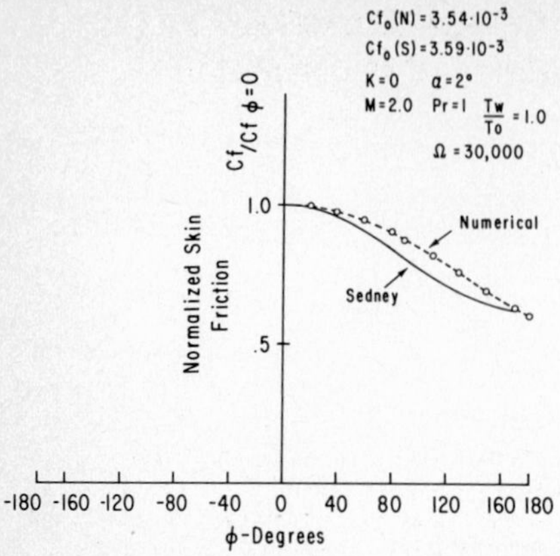

Figure 3. Primary Flow Skin Friction at Cone Tip

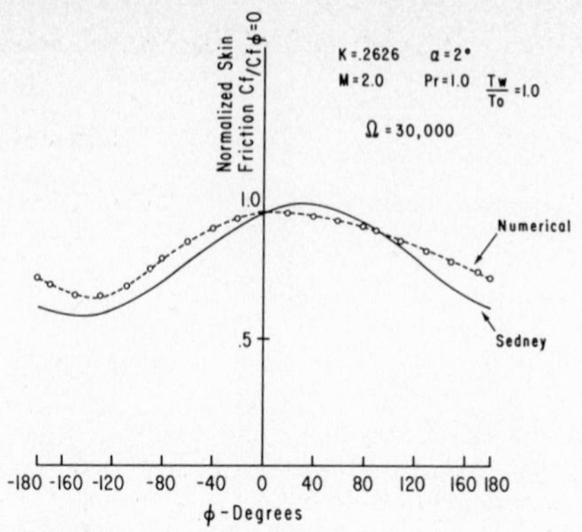

Figure 4. Primary Flow Skin Friction Away from Cone Tip

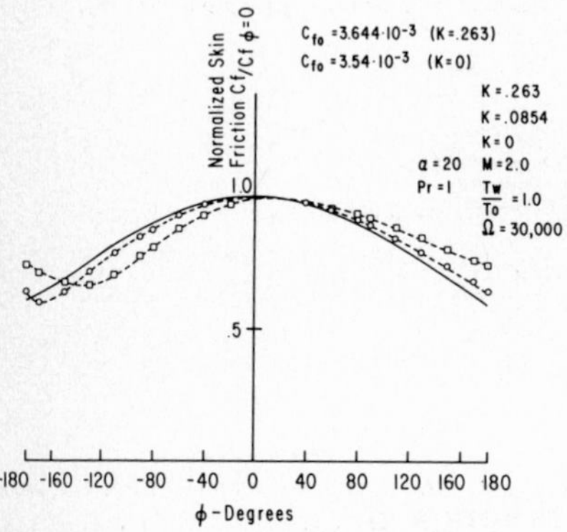

Figure 5. Variation of Wall Shear Down Cone Surface

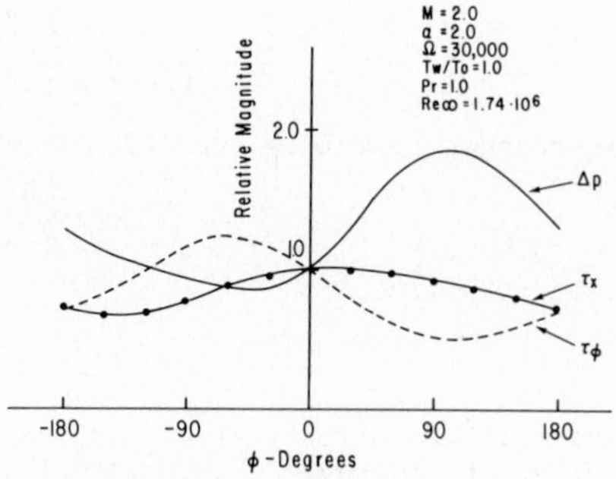

Figure 6. Boundary Layer "Magnus" Force Component Variation

NUMERICAL SOLUTION OF THE UNSTEADY NAVIER-STOKES EQUATIONS FOR THE INVESTIGATION OF LAMINAR BOUNDARY LAYER STABILITY

Hermann F. Fasel

University of Stuttgart
West Germany

INTRODUCTION

Stability and transition of laminar boundary layer flows is still a rather unresolved problem. Linear stability theory has been widely successful in predicting local conditions for stability and the initial stage preceeding transition, namely the amplification of unstable two-dimensional perturbations, can be accurately calculated. The applicability of the linear stability theory was verified by the experiments of Schubauer and Skramstad (1947), which were repeated in a somewhat more refined manner by Ross (1970). However, linear stability analysis is restricted to sinusoidal disturbances of small amplitudes and attempts to include any nonlinear effects involve enormous mathematical efforts.

Here, stability and transition phenomena of laminar, two-dimensional, incompressible boundary layer flows are investigated by introducing forced time-dependent perturbations into the steady flow along a semi-infinite flate plate. The reaction of the flow is then directly determined by numerical solution of the Navier-Stokes equations in a specified rectangular domain. This approach contains no restrictions with respect to form or intensity of the disturbances since no linearization is necessary anywhere. In this paper, however, only periodic disturbances of small amplitudes are considered to allow comparison of the calculations with both linear stability theory and experiments and thus to provide a thorough check of the numerical method.

GOVERNING EQUATIONS

The Navier-Stokes equations are used in vorticity transport form

$$\frac{\partial\omega}{\partial t} + u\frac{\partial\omega}{\partial x} + v\frac{\partial\omega}{\partial y} = \frac{1}{Re}\frac{\partial^2\omega}{\partial x^2} + \frac{\partial^2\omega}{\partial y^2} \tag{1}$$

with vorticity defined as

$$\omega = \frac{\partial u}{\partial y} - \frac{1}{Re}\frac{\partial v}{\partial x} \quad . \tag{2}$$

For the calculation of the velocity components in x- and y-direction, i.e., u and v, respectively, the following equations of Poisson-type are used

$$\frac{\partial^2 u}{\partial y^2} + \frac{1}{Re}\frac{\partial^2 u}{\partial x^2} = \frac{\partial\omega}{\partial y} \quad , \tag{3}$$

$$\frac{\partial^2 v}{\partial y^2} + \frac{1}{Re}\frac{\partial^2 v}{\partial x^2} = -\frac{\partial\omega}{\partial x} \quad . \tag{4}$$

They are derived by differentiating Eq. (2) with respect to y and x, respectively, and by subsequent use of the continuity condition

$$\frac{\partial u}{\partial x} + \frac{\partial v}{\partial y} = 0 \quad . \tag{5}$$

This research was supported by the Deutsche Forschungsgemeinschaft, Bonn - Bad Godesberg, West Germany, Research Contract Ep 5/3.

All variables in Eqs. (1) - (5) are dimensionless and relate to the corresponding dimensional quantities denoted by bars in this manner

$$x = \frac{\bar{x}}{L}\ , \qquad y = \frac{\bar{y}}{L}\sqrt{Re}\ , \qquad u = \frac{\bar{u}}{U_\infty}\ ,$$
$$v = \frac{\bar{v}}{U_\infty}\sqrt{Re}\ , \qquad \omega = \frac{\bar{\omega}L}{U_\infty\sqrt{Re}}\ , \qquad t = \frac{\bar{t}U_\infty}{L}\ , \qquad (6)$$

where L is a reference length, U_∞ a reference velocity, in this case the free stream velocity, and $Re = U_\infty L/\nu$ a Reynolds number. The y-coordinate and the y-component v of the velocity is stretched by a factor $\sqrt{Re}$ to account for the boundary layer type flow. The governing equations for our investigations are Eqs. (1), (3) and (4).

INTEGRATION DOMAIN

The governing equations are solved numerically within the rectangular integration domain A-B-C-D (Fig. 1) in the x,y-plane. The left boundary A-B is assumed to be downstream of the leading edge. The investigations described here can be considered a numerical simulation of the Schubauer-Skramstad experiments providing the flow conditions along A-B are known as resulting from the vibrating ribbon. These flow conditions can be approximated by taking, for example, perturbation profiles of the linear stability theory and periodically superimposing them on the undisturbed flow profiles, assuming that the location of the left boundary would be so far downstream of the ribbon that except for the fundamental waves all other components are negligibly small. Further we assume that for periodic disturbances the length X of the rectangular domain extends at least about four disturbance wave-lengths downstream and that the width Y is at least about three times the boundary layer thickness. Test calculations have indicated that this size of the integration domain is necessary to obtain reasonable results with the boundary conditions specified below.

INITIAL AND BOUNDARY CONDITIONS

Initial Conditions. As initial conditions for t=o we assume an undisturbed flow field in the entire integration domain

$$u(x,y,0) = u_{st}(x,y)\ , \quad v(x,y,0) = v_{st}(x,y)\ , \quad \omega(x,y,0) = \omega_{st}(x,y)\ , \quad (7a,b,c)$$

where index st refers to the undisturbed flow. The undisturbed flow field is obtained by solving the Navier-Stokes equations for the steady flow, i.e., solving Eqs. (1), (3) and (4) with $\partial\omega/\partial t = 0$ in Eq. (1).

Boundary Conditions for the Steady (Undisturbed) Flow. For the calculation of the steady, undisturbed flow field the following conditions are used on the boundaries A-B-C-D (Fig. 1)

A-B: $u_{st}(0,y) = u_{BL}(0,y)\ , \quad v_{st}(0,y) = v_{BL}(0,y)\ , \quad \omega_{st}(0,y) = \omega_{BL}(0,y)\ ;$ (8a,b,c)

B-C: $u_{st} = 0\ , \quad v_{st} = 0\ , \quad \frac{\partial\omega_{st}}{\partial x} = -\frac{\partial^2 v_{st}}{\partial y^2}\ ;$ (9a,b,c)

C-D: $\frac{\partial^2 u_{st}}{\partial x^2} = 0\ , \quad \frac{\partial^2 v_{st}}{\partial x^2} = 0\ , \quad \frac{\partial^2\omega_{st}}{\partial x^2} = 0\ ;$ (10a,b,c)

A-D: $u_{st} = 1\ , \quad \omega_{st} = 0\ , \quad \frac{\partial v_{st}}{\partial y} = 0\ .$ (11a,b,c)

Index BL in Eqs. (8) refer to the Blasius solution of the Prandtl boundary layer equations. Eq. (9c) for the calculation of ω at the wall is obtained from Eq. (5) using Eq. (9b).

Boundary Conditions for the Unsteady (Disturbed) Flow. At the left boundary (A-B) the perturbations are produced by periodically disturbing the flow profiles of the undisturbed flow (for which the Blasius solution is used)

A-B:

$$u(0,y,t) = u_{BL}(0,y) + Au'_A(y)\cos(\beta t) \quad , \tag{12a}$$

$$v(0,y,t) = v_{BL}(0,y) + Av'_A(y)\cos(\beta t + \frac{\pi}{2}) \quad , \tag{12b}$$

$$\omega(0,y,t) = \omega_{BL}(0,y) + A\omega'_A(y)\cos(\beta t) \quad . \tag{12c}$$

The amplitude distributions or perturbation profiles u'_A , v'_A , ω'_A at the left boundary are functions of y only. They can be taken from linear stability theory calculations with spatial amplification (Jordinson (1970), Kümmerer(1973)). The phase difference of $\pi/2$ between the v- and u-perturbation and the v- and ω-perturbation is consistent with linear stability theory results. However, in this formulation any variation of the phase relationships in y-direction is neglected. The common amplitude factor A allows experimentation with various perturbation amplitudes.

At the wall (B-C) the same conditions as those for the steady flow calculations are used

B-C:

$$u = 0 \quad , \quad v = 0 \quad , \quad \frac{\partial\omega}{\partial x} = -\frac{\partial^2 v}{\partial y^2} \quad . \tag{13a,b,c}$$

For the downstream boundary (C-D) and the outer boundary (A-C) conditions are employed which can be derived by assuming periodic behavior in x-direction of the perturbation flow in the neighborhood of these boundaries. Denoting the perturbation variables by a prime

$$u' = u - u_{st} \quad , \quad v' = v - v_{st} \quad , \quad \omega' = \omega - \omega_{st} \quad ,$$

the downstream conditions can be written in the form

C-D:

$$\frac{\partial^2 u'}{\partial x^2} = -\alpha^2 u' \quad , \quad \frac{\partial^2 v'}{\partial x^2} = -\alpha^2 v' \quad , \quad \frac{\partial^2 \omega'}{\partial x^2} = -\alpha^2 \omega' \quad , \tag{14a,b,c}$$

where α is the local wave number of the resulting perturbation flow. Linear stability theory and experiments (Ross (1970)) suggest that α varies in downstream direction.

For the outer boundary we have the conditions

$$\frac{\partial u'}{\partial y} = -\frac{\alpha}{\sqrt{Re}} u' \quad , \quad \frac{\partial v'}{\partial y} = -\frac{\alpha}{\sqrt{Re}} v' \quad , \quad \omega' = 0 \quad . \tag{15a,b,c}$$

The relationship for u' and v' of Eqs. (15a,b) imply asymtotic decay in y-direction. These conditions allow a relatively small integration domain in y-direction, since we do not postulate that u' and v' vanish on the outer boundary. (Linear stability theory results as well as experiments indicate that u' and v' approach zero very slowly, contrary to ω' which practically vanishes about two boundary layer thicknesses from the wall.)

The conditions for the downstream and outer boundary Eqs. (14) and (15) do not force a periodic behavior (with no damping or amplification) upon the perturbation flow near these boundaries. Test calculations have shown rather that damping or amplification is possible even on the boundaries themselves. The values for the local wave number α in the boundary conditions can also be taken from linear stability theory calculations when available. Alternatively, α can be determined in an additional iteration loop within the actual calculation of the unsteady, disturbed flow.

NUMERICAL METHOD

For reasons of numerical stability an implicit finite difference method was chosen. Stability investigations of laminar boundary layer flows require experimentation with relatively large Reynolds numbers, i.e., Reynolds numbers larger than the critical Reynolds number to allow amplification of unstable perturbations. It is essential that a numerical scheme for such investigations be numerically very stable to avoid any interaction of numerical instabilities with physically meaningful disturbances.

We are using a three-level implicit method with the following difference approximation for the time derivative $\partial\omega/\partial t$ in Eq. (1)

$$\left.\frac{\partial\omega}{\partial t}\right|^{l}_{n,m} \approx \frac{1}{2\Delta t}\left(3\omega^{l}_{n,m} - 4\omega^{l-1}_{n,m} + \omega^{l-2}_{n,m}\right) \tag{16}$$

which has a truncation error of second order. Subscripts n, m and superscript l refer to the discretization in space (Fig. 1) and time, respectively, such that for example $\omega^{l}_{n,m} \equiv \omega(n\Delta x,\ m\Delta y,\ l\Delta t)$, $0 \leq n \leq N$, $0 \leq m \leq M$, $l = 1,2,\ldots$.
For all space derivatives central differences with a truncation error of second order are applied. These difference approximations as well as u and v in the convective terms of Eq. (1) are taken at the most recent time level l.

For the solution of the non-linear difference equations a line iteration method is employed iterating simultaneously on all three equation systems resulting from the discretization of Eqs. (1), (3) and (4). The line iteration is organized such that the nodal values on lines parallel to the y-axis are determined by a direct method while proceeding iteratively in x-direction. With the suggested iteration loop, where the iteration level is denoted by index i $(i = 1,2,\ldots)$, the difference equations can be written in the form

$$\begin{aligned}&\frac{\Delta y^2}{2\Delta t}\left(t_1\omega^{l,i}_{n,m} - t_2\omega^{l-1,I_{l-1}}_{n,m} + t_3\omega^{l-2,I_{l-2}}_{n,m}\right) + \frac{\Delta y^2}{2\Delta x}u^{l,i-1}_{n,m}\left(\omega^{l,i-1}_{n+1,m} + \underline{2\omega^{l,i}_{n,m} - 2\omega^{l,i-1}_{n,m}} - \omega^{l,i}_{n-1,m}\right)\\ &+ \frac{\Delta y}{2}v^{l,i-1}_{n,m}\left(\omega^{l,i}_{n,m+1} - \omega^{l,i}_{n,m-1}\right) - \frac{1}{Re}\left(\frac{\Delta y}{\Delta x}\right)^2\left(\omega^{l,i-1}_{n+1,m} - 2\omega^{l,i}_{n,m} + \omega^{l,i}_{n-1,m}\right) - \omega^{l,i}_{n,m+1} + 2\omega^{l,i}_{n,m}\\ &- \omega^{l,i}_{n,m-1} = 0\ ,\end{aligned} \tag{17a}$$

$$\frac{1}{Re}\left(\frac{\Delta y}{\Delta x}\right)^2\left(u^{l,i-1}_{n+1,m} - 2u^{l,i}_{n,m} + u^{l,i}_{n-1,m}\right) + u^{l,i}_{n,m+1} - 2u^{l,i}_{n,m} + u^{l,i}_{n,m-1} - \frac{\Delta y}{2}\left(\omega^{l,i}_{n,m+1} - \omega^{l,i}_{n,m-1}\right) = 0\ , \tag{17b}$$

$$\begin{aligned}&\frac{1}{Re}\left(\frac{\Delta y}{\Delta x}\right)^2\left(v^{l,i-1}_{n+1,m} - 2v^{l,i}_{n,m} + v^{l,i}_{n-1,m}\right) + v^{l,i}_{n,m+1} - 2v^{l,i}_{n,m} + v^{l,i}_{n,m-1} + \frac{\Delta y^2}{2\Delta x}\left(\omega^{l,i-1}_{n+1,m} + \underline{2\omega^{l,i}_{n,m}}\right.\\ &\left.\underline{- 2\omega^{l,i-1}_{n,m}} - \omega^{l,i}_{n-1,m}\right) = 0\ ,\end{aligned} \tag{17c}$$

with $t_1=t_2=t_3=0$ for $l=0$ (steady flow calculation); $t_1=t_2=2$, $t_3=0$ for $l=1$ and $t_1=3$, $t_2=4$, $t_3=1$ for $l>1$. Integer I_l is the number of iterations necessary for convergence at time level l. The underlined terms in Eqs. (17a) and (17c) are added to assure convergence of the iteration process when central differences are used for $\partial\omega/\partial t$. The iteration is terminated when the largest relative change of all nodal values from one iteration to the next is less than a specified deviation ε. Test calculations have shown, however, that it is sufficient to apply this criterion for one variable on an arbitrary grid line parallel to the x-axis only. At a new time level, iteration is initiated with values from the previous time step. For the calculation of the steady flow starting values can be taken from the Blasius solution.

For computing the nodal values on grid lines n=constant parallel to the y-axis at a time level l and iteration level i the difference equations can be written in the form

$$-A_m\omega_{m-1} + B_m\omega_m - C_m\omega_{m+1} = D_m\ , \tag{18a}$$

$$-u_{m-1} + b_m u_m - u_{m+1} = d_m\ , \tag{18b}$$

$$-v_{m-1} + b_m v_m - v_{m+1} = f_m\ , \tag{18c}$$

thus obtaining tridiagonal coefficient matrices. The coefficients and right hand sides contain only given parameters or nodal values which were calculated at previous time or iteration levels. The solution of these equation systems could be easily achieved employing standard algorithms for tridiagonal systems. However, difficulties arise from the fact that the calculation of the vorticity at the wall according to Eq. (13c) involves grid values of v within equation system (18c). For example the second order difference relationship

$$\omega_{n,0}^{l,i} = \frac{1}{3}\left[\frac{\Delta x}{\Delta y^2}(-8v_{n,1}^{l,i} + v_{n,2}^{l,i}) + 4\omega_{n-1,0}^{l,i} - \omega_{n-2,0}^{l,i}\right] + O(\Delta x^2, \Delta y^2) \tag{19}$$

is used. It is obtained by approximating $\partial\omega/\partial x$ in Eq. (13c) with a three-point backward difference and $\partial^2 v/\partial y^2$ with a three-point difference considering that v and $\partial v/\partial y$ are both zero at the wall. Obviously, Eq. (19) requires nodal values $v_{n,1}^{l,i}$, $v_{n,2}^{l,i}$ and therefore the calculation of ω at the wall must be performed within the iteration loop. Attempts to avoid difficulties in this respect by taking the v-values in Eq. (19) at the iteration level i-1 were not successful because then the iteration procedure converges much more slowly or even fails entirely. Therefore an algorithm for solving the equation systems (18) with simultaneous calculation of ω at the wall was developed.

NUMERICAL RESULTS

In this paper some results of calculations are presented for small perturbation amplitudes (with maximum of u' being 0.01% of the free stream velocity U_∞). First, the numerical method was thoroughly checked with two test cases which are shown in the stability diagram (Fig. 2) of the linear stability theory (Jordinson (1968)). In this diagram the dimensional frequency $\bar{\beta}$ or a dimensionless frequency parameter $F = 10^4\bar{\beta}\nu/U_\infty{}^2$ is constant on rays through the origin. The conditions of the perturbation flow for various locations downstream of a constant frequency disturbance correspond to points on such a ray in such a sense that proceeding in downstream direction corresponds to moving away from the origin. The perturbations should become amplified inside the neutral stability curve and damped outside.

Both test cases are on the same ray of Fig. 2 having equal perturbation frequency F = 1.316. For the first test case the downstream location of the left boundary A-B corresponds to a point in the stability diagram where the ray crosses the lower branch of the neutral stability curve. Thus with increasing x we are proceeding into the unstable region. For the second case the left boundary corresponds to a point on the ray inside the unstable region a short distance from the intersection with the upper branch of the neutral stability curve. Then with increasing x we are proceeding into the stable region after the neutral stability curve is crossed.

Typical results for the two test cases are shown in Figs. 3 and 4 where the perturbation variables u', v', ω' are drawn for a distance $3\Delta y$ from the wall versus the downstream coordinate x. For each case this is done for four different time levels after all transient effects have disappeared and after a quasi steady state of the perturbation flow is reached with periodic character in time. The curves for the first test case in Fig. 3 clearly exhibit amplification of the perturbations in downstream direction as predicted by linear stability theory. For the second case in Fig. 4 a slight amplification can be observed first followed by decreasing amplitudes after the neutral curve of the linear stability theory is crossed. Thus the results for the two test cases show good qualitative agreement with linear stability theory, i.e., the Navier-Stokes solution gives amplification of the disturbances in the unstable region and damping in the stable region.

For quantitative comparison calculations were made traversing the entire region of instability on rays of $\bar{\beta}$=constant. The curves of u', v', ω' versus x in Figs. 5 and 6 for example are obtained from calculations where on the same ray as for the two test cases the region of instability is traversed in two steps. Step one starts in the stable region at Re^*=500 ($Re^* = U_\infty\delta_1/\nu$) and ends in the unstable region at about Re^*=1100. The curves for step 1 in Fig. 5 first exhibit damping of the perturbations and then amplification after crossing the neutral curve. For step 2 in Fig. 6 the perturbations continue to be amplified until the upper branch of the neutral curve is reached and are damped thereafter. Perturbation profiles for $Re^{*}=700$ are compared in Fig. 7 with linear stability theory calculations (Kümmerer (1973)). A comparison of the amplification curve for the maximum value of the u-perturbation profile with linear stability calculations (Jordinson (1968), Kümmerer (1973)) and measurements for $F = 1.32$ (Ross (1970)) is given in Fig. 8. The Navier-Stokes solution yields stronger amplification than linear stability theory and is in good agreement with the experimental measurements, especially for the calculations of step 2 for which perturbation profiles at the left boundary were taken from the calculations of step 1 and not from linear stability theory.

CONCLUSION

The results for perturbations of small amplitudes have shown that the numerical method for solving the Navier-Stokes equations is applicable for stability and transition studies. It is intended to apply this method for the investigation of large amplitude disturbances for which the linear stability theory is no longer valid.

REFERENCES

Jordinson, R., The transition from laminar to turbulent flow over a flat plate. Ph.D. Thesis, University of Edinburgh (1968).

Jordinson, R., The flat plate boundary layer. Part 1: Numerical integration of the Orr-Sommerfeld equation. J. Fluid Mech. 43, 801-811 (1970).

Kümmerer, H., Numerische Untersuchungen zur Stabilität ebener laminarer Grenzschichtströmungen. Diss. Universität Stuttgart (1973).

Ross, J.A., Barnes, F.H., Burns, J.G., Ross, M.A.S., The flat plate boundary layer. Part 3: Comparison of theory with experiment. J. Fluid Mech. 43, 819-832 (1970).

Schubauer, G.B., Skramstad, H.K., Laminar boundary layer oscillations and stability of laminar flow. J. Aeron. Sci. 14, 69-78 (1947).

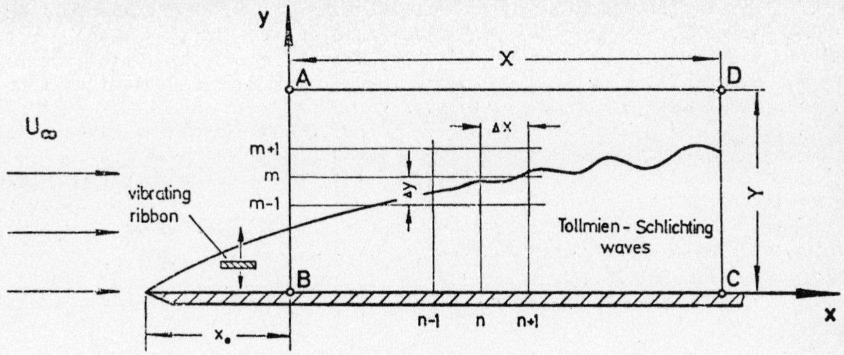

Fig. 1. Integration domain

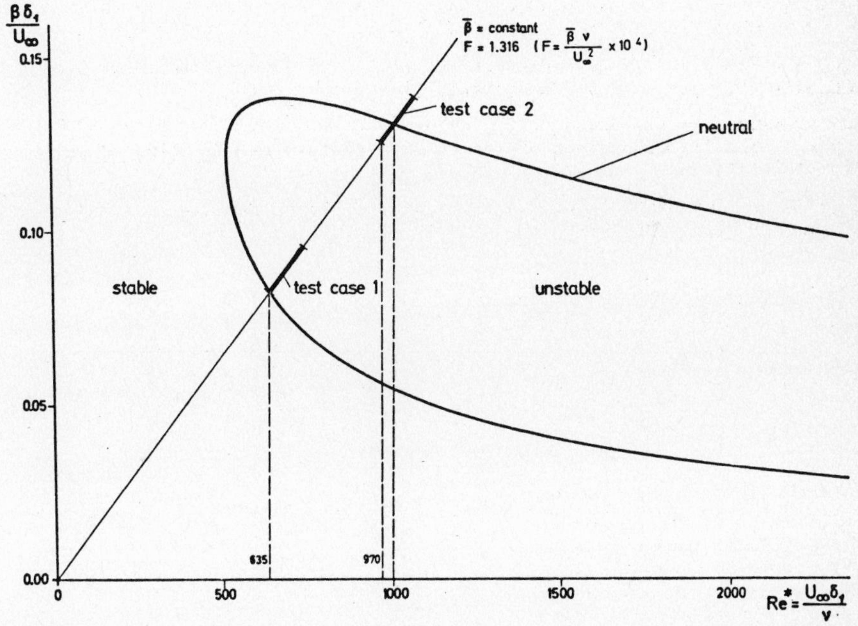

Fig. 2. Stability diagram
(linear stability theory)

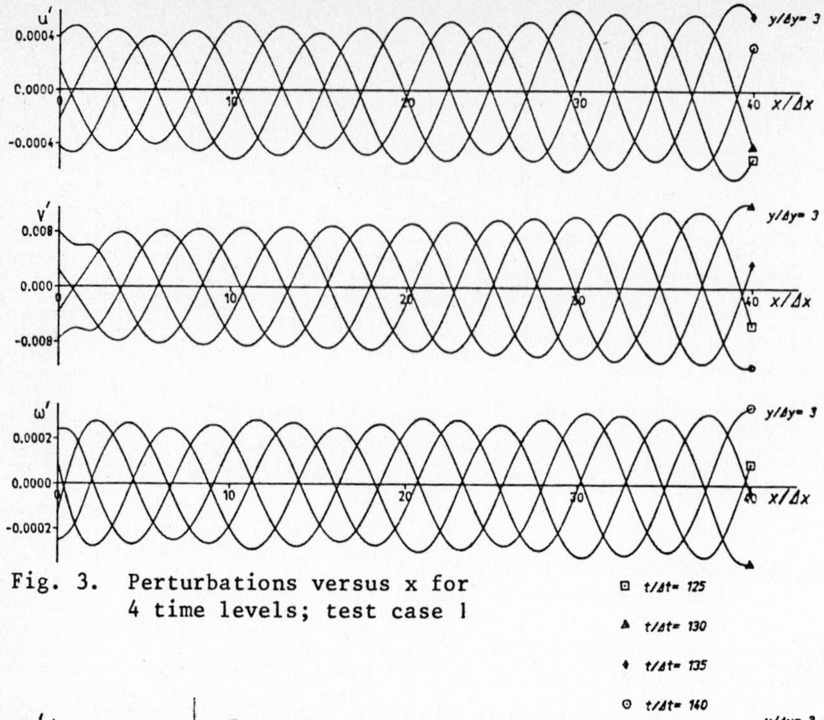

Fig. 3. Perturbations versus x for 4 time levels; test case 1

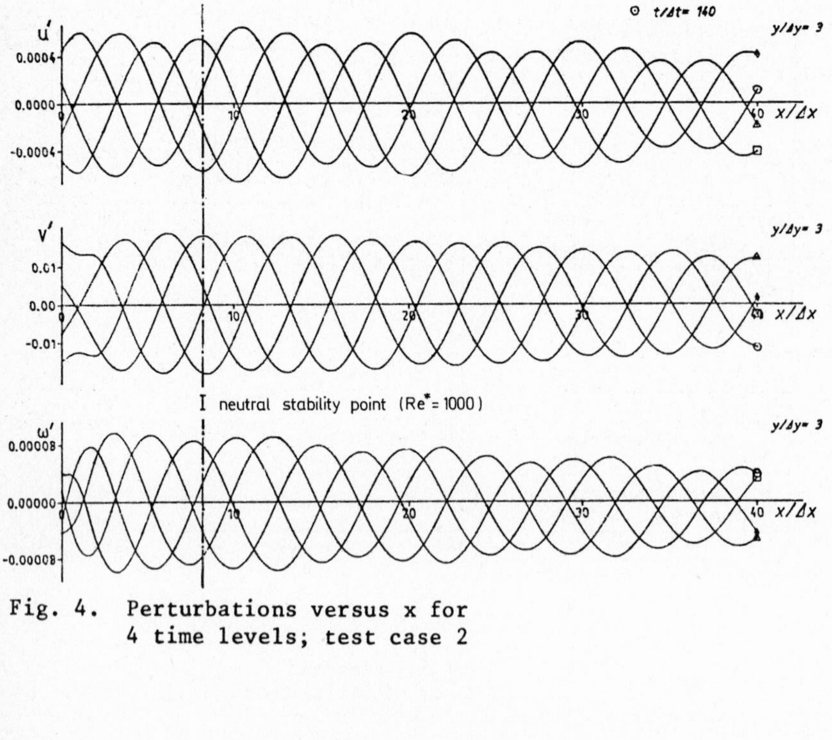

Fig. 4. Perturbations versus x for 4 time levels; test case 2

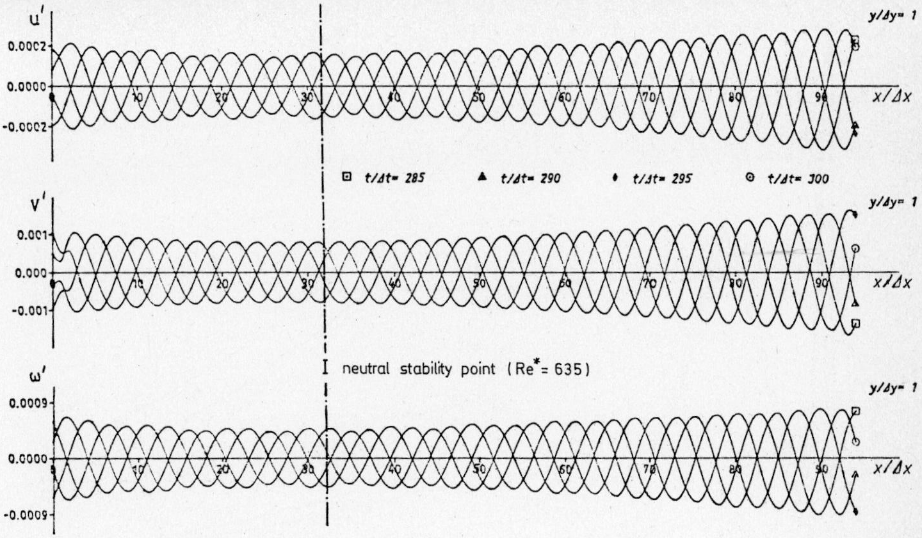

Fig. 5. Perturbations versus x for 4 time levels; step 1

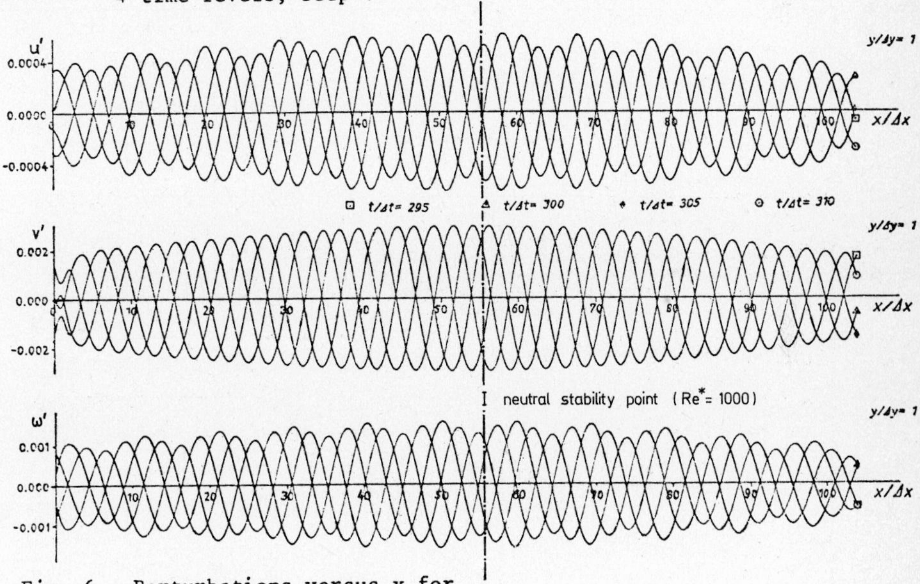

Fig. 6. Perturbations versus x for 4 time levels; step 2

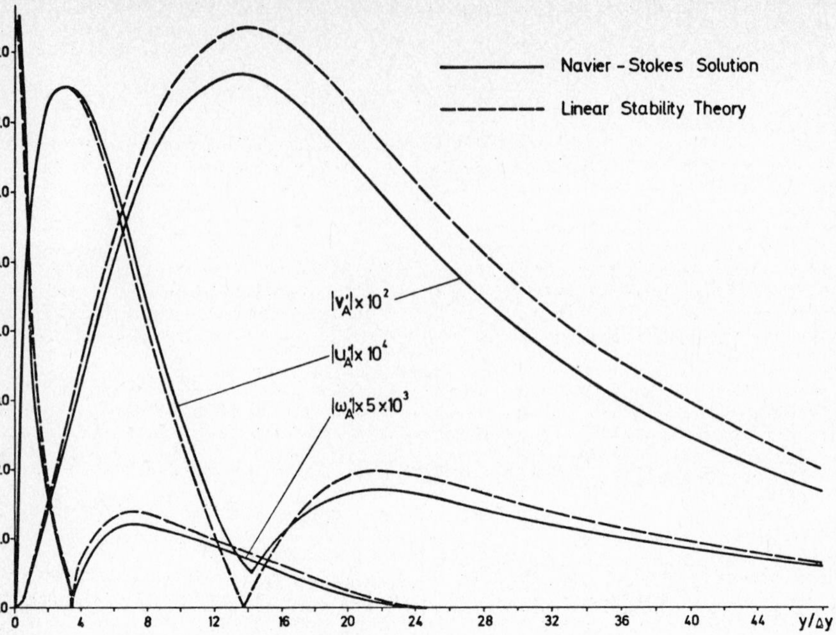

Fig. 7. Perturbation profiles for Re =700, F=1.316

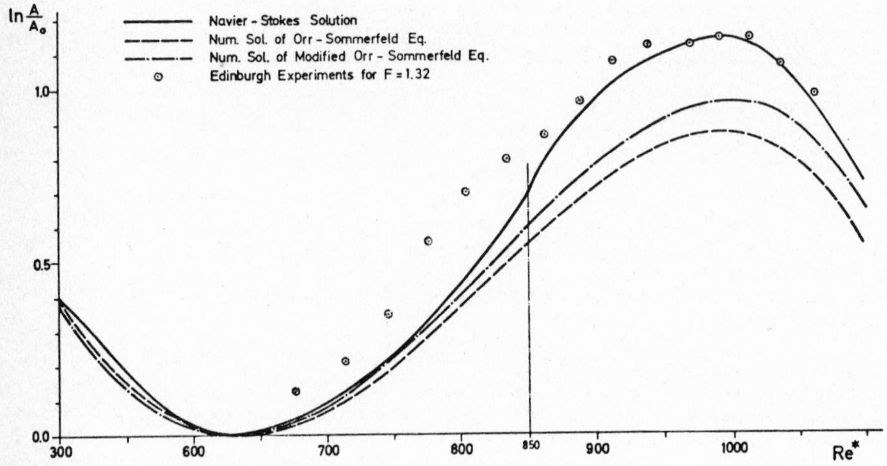

Fig. 8. Amplification curve (for maximum of u'_A)

GTT METHOD APPLIED TO CONES AT LARGE ANGLES OF ATTACK

BY

Clive A. J. Fletcher*

Weapons Research Establishment
Salisbury, South Australia

1. INTRODUCTION

Since 1948 numerical solutions, of varying degrees of sophistication, have been advanced for the inclined cone problem in supersonic flow. At the present time the most efficient method, for purely inviscid flow, is the GTT (after Gilinskii, Telenin and Tinyakov) method as used by Holt and Ndefo and by Bazzhin and Chelysheva.

The method described herein is applicable to the inviscid flow about cones at sufficiently large incidence that a substantial region of supersonic crossflow exists $(v^2 + w^2 > a^2)$. v and w are the normal and circumferential velocity components for a spherical coordinate system $\left\{ \begin{matrix} r & \theta & \phi \\ u & v & w \end{matrix} \right\}$, centered at the cone apex, with θ measured from the cone axis and ϕ measured from the windward symmetry plane.

Experimental studies indicate that, for a significant region of supersonic crossflow, an internal shock occurs. The region between the internal shock and the leeward line of symmetry has a subsonic crossflow. Prior to the present studies only the shock-capturing method of Kutler and Lomax has adequately allowed for the occurrence of an internal shock.

A feature peculiar to the inviscid flow about inclined cones is the occurrence of a vortical singularity (Ferri) at which the entropy, density and radial velocity components are discontinuous. Physically the vortical singularity is the point in the crossflow plane at which all the streamlines meet. For circular cones the vortical singularity always lies in the leeward symmetry plane and for the large angles of attack considered here it occurs away from the cone surface. Experimental studies by Feldhuhn et al. have demonstrated the occurrence of the vortical singularity even when a large leeward separated flow region occurs. Prior to the present studies no numerical method has allowed for the flow discontinuities associated with the vortical singularity.

2. FORMULATION OF THE PROBLEM

A solution is sought for the five non-dimensionalized dependent variables u, v, w, ρ and p on the surface of a unit sphere. Specification of the external parameters, freestream Mach number, M_∞, the incidence, α and the cone nose angle, θ_b completely determines the problem. The equations of motion used are continuity, three conservation of momentum equations, conservation of entropy equation and the Bernoulli equation.

At the outer shock the Rankine-Hugoniot relations give the local flow properties as a function of the freestream conditions and the local shock slope. At the cone surface the normal velocity is zero. The boundary conditions are completed by the requirement of symmetry at $\phi = 0°$ and $180°$.

The region between the outer shock and the body is transformed into a rectangle by replacing the independent variables θ and ϕ with ξ and η, where

*And University of California, Berkeley, California 94720.

$$\xi = (\theta_s(\phi) - \theta)/(\theta_s(\phi) - \theta_b), \qquad \eta = \phi/\pi, \tag{1}$$

$\theta_s(\phi)$ is the outer shock location and θ_b is the body location.

For the windward region the equations of motion are rearranged to give explicit expressions for u_ξ, v_ξ, etc. These equations are integrated simultaneously in the ξ direction along N rays which are equally spaced in the η direction. The η derivatives are represented in the following manner

$$F_\eta = \sum_k a_k \cdot F_k \quad , \tag{2}$$

where F_k is the value of F on the kth ray. In the original GTT formulation the coefficients a_k are obtained by first postulating an analytic representation for F in the η direction, e.g.,

$$F(\xi,\eta) = \sum_{j=0}^{m-1} F_j^{\,o}(\xi) \cdot \cos j \pi \eta \quad . \tag{3}$$

In the present study three ways of fixing the coefficients a_k have been examined.

a) An exact matching of the function values using a Fourier series, e.g., Eq. (3).

b) A least squares matching of the function values using a Fourier series of order m-2.

c) A fourth order finite difference representation.

For a test case of 20° cone inclined at 30° to a freestream of Mach number 7.0, all three forms gave results that agreed within the accuracy of the method.

One problem with the above formulation is that it is inherently unstable. For a model problem, the two-dimensional Laplace equation in a rectangle treated as a Cauchy problem, Gilinskii et al., found that any error grew like exp (Nx), where N is the total number of rays and x is the direction of integration. The same type of error growth may be expected in the present problem. Thus it is desirable that N should be as small as possible. However, examination of Eq. (2) indicates a truncation error proportional to $[1/(N-1)]^{N-1}$ at best. Thus the choice of N is a compromise between a large initial truncation error and a large subsequent growth of the truncation error. In the present study N = 5 has been used.

The ordinary differential equations, u_ξ, v_ξ, etc., are integrated simultaneously along five rays from the outer shock to the body (Fig. 1) using a fourth order Runge-Kutta scheme. Initially the outer shock location must be chosen somewhat arbitrarily. Integration to the cone surface normally leads to failure to satisfy the body boundary condition of zero normal velocity. An iteration scheme is required to systematically modify the outer shock location until the body condition is satisfied. A minimization scheme due to Powell has been used for this.

In the windward region a step-size of $\Delta\xi = 0.1$ has been used. Initial results indicated that reducing the step-size to 0.01 caused no significant change in the solution. Increasing the step-size to 0.2 did cause a change in the solution.

The converged solution on ray 5 (Fig. 1) forms the starting data for a characteristics solution marching in the $\phi(\eta)$ direction. The upstream interpolation version of the method of characteristics as described by Belotserkovskii and Chushkin is used in this study. Due to the hyperbolic nature of the governing equations the supersonic crossflow has no warning of the leeward symmetry requirement. Consequently, an internal shock is introduced; the region behind the internal shock has a subsonic crossflow.

The solution of the leeward region (Fig. 2) uses a modified version of the GTT method. The direction of integration is initially from the internal shock to the leeward line of symmetry. Powell's method is used to adjust the location of the internal shock until the symmetry condition of $w = 0$ at $\phi = \pi$ is satisfied. The leeward region (CDEFGC in Fig. 2) is transformed into a rectangle by introducing

$$\xi = \frac{\theta_s(\phi) - \theta}{\theta_s(\phi) - \theta_b} \qquad \text{and} \qquad \tau = \frac{\phi - \phi_{is}(\theta)}{\pi - \phi_{is}(\theta)} \quad , \tag{4}$$

where $\phi_{is}(\theta)$ is the location of the internal shock. The equations of motion are manipulated to give explicit expressions for u_τ, v_τ etc. These expressions are integrated simultaneously along ten rays of constant ξ, using a fourth order Runge-Kutta scheme and a step-size, $\Delta\tau = 0.2$. The derivatives in the ξ direction have been computed using a fourth order finite difference formula (centered where possible).

The Rankine-Hugoniot relations are used to obtain boundary conditions downstream of the internal shock. Since the internal shock does not extend to the outer shock, the solution within the region KLFGK, in Fig. 2, is obtained by interpolating the characteristics solution.

For rays of constant ξ, which do not pass through the internal shock, a quadratic distribution for w is initially assumed. Integration from the internal shock to the leeward line of symmetry is repeated, iterating on the internal shock location until $w = 0$ on $\phi = \pi$. After convergence of this first set of integrations no account has been taken of the vortical singularity.

A second set of integrations is made from the leeward line of symmetry to the internal shock location found from the first set of integrations. Powell's method is used to adjust the w_τ distribution at $\phi = \pi$ until, after integration, flow conditions match at the internal shock. Once a new w_τ distribution on $\phi = \pi$ is chosen, the equations of motion can be integrated along the line $\phi = \pi$ from the outer shock to the body. At the point on $\phi = \pi$ where $v = 0$ the correct jump conditions associated with the vortical singularity are inserted. Thus the correct boundary conditions on $\phi = \pi$ are obtained.

After convergence of the second set of integrations a third set of integrations is made from the internal shock to the leeward line of symmetry. For rays of constant ξ that do not pass through the internal shock the w distribution obtained from the second converged solution has been assumed. In order to preserve the discontinuity at the vortical singularity the u distribution obtained from the second converged solution has been assumed on all rays. It may be noted that u is tangential to the internal shock and hence it is not sensitive to the precise location of the internal shock. Converged solutions to the third set of integrations indicate that the change in the internal shock location from that found after the first set of integrations is typically less than half a degree.

3. RESULTS AND DISCUSSION

Results have been obtained in the ranges: freestream Mach number, 3 to 16, cone nose angles, 5° to 30° and angles of incidence up to 50°.

Accuracy of the present scheme is difficult to assess because of the unstable nature of the scheme. Assuming that the choice of 5 rays has made the exponential error growth negligible, the major source of error is the truncation error in the representation of the η and ξ derivatives. In the windward region the truncation error in representing the η derivatives is of order $\Delta\eta^4$ where $\Delta\eta = 0.125$. The integration of the ξ derivatives was effected by a fourth order Runge-Kutta scheme with a step-size of 0.1. In the leeward region the truncation error in representing the ξ derivatives is of order $\Delta\xi^4$ with $\Delta\xi = 0.1$. The integration of the τ

derivatives used a fourth order Runge-Kutta scheme and a step-size, $\Delta\tau$, $= 0.2$. This is approximately equivalent to $\Delta\eta = 1/90$. In the 'hyperbolic' region a second order characteristics method was used with step-sizes $\Delta\xi = 0.1$ and $\Delta\eta$ varying within the flow field between 0.001 and 0.02.

The method in all three regions is iterative; thus the execution time can vary widely depending on the external conditions: M_∞, α and θ_b. In the windward region the execution time is sensitive to how well the initial location of the external shock is determined. Thus starting from a known solution at $M_\infty = 7$, $\theta_b = 30°$ and $\alpha = 30°$, solutions to the windward region were generated for $\alpha = 31°$ to $43°$ in steps of 1° in 6.2 seconds on a CDC 7600. For the case $M_\infty = 7$, $\alpha = 30°$ and $\theta_b = 20°$ the overall execution time was approximately 7 seconds on a CDC 7600. This is of the order of a tenth the execution time for the shock capturing method applied to the same problem.

A comparison of the shock and sonic line locations, for the present method and the shock capturing method, is shown in Fig. 3. The location of the outer shock and the windward sonic line agree within 0.6%. The location and extent of the internal shock do not agree. The shock capturing method predicts an earlier and more localized internal shock than does the present method. The pressure distribution (Fig. 4) on the body is in substantial agreement even in the leeward region. For the flow variables, significant differences do occur in the leeward region, mainly due to the treatment of the vortical singularity. It can be seen from Fig. 5 that, whereas the present method retains the correct jump conditions at the vortical singularity, the shock capturing method smooths out any discontinuity.

Comparisons have also been made with the experimental results of Tracy. It is apparent from Fig. 6 that the location of the outer shock is predicted well except in the leeward region, where viscous effects are significant. The circumferential location of the internal shock is not predicted very accurately. This is not surprising in view of the viscous/inviscid interaction in the leeward region. In spite of this the agreement in pressure distribution (not shown) is good even in the leeward region.

In conclusion, it is noted that this is the first numerical method that satisfactorily accounts for both the internal shock and the vortical singularity. Figure 7, which shows the crossflow streamline pattern for a typical case, clearly shows both the role of the internal shock, namely to deflect the streamlines, and the sink-like role of the vortical singularity.

REFERENCES

A. P. Bazzhin and I. F. Chelysheva, Izv. AN SSSR Mekhanika Zhidkosti i Gaza, 2, 3, 119-123 (1967).

O. M. Belotserkovskii and P. I. Chushkin, in Basic Developments in Fluid Dynamics (ed. M. Holt), pp. 89-126, Academic Press (1965).

R. H. Feldhuhn, A. E. Winkelmann and L. Pasiuk, AIAA Journal, 9, 6, 1074-1081 (1971).

A. Ferri, NACA TR-1045 (1951).

S. M. Gilinskii, G. F. Telenin and G. P. Tinyakov, Izv. AN SSSR Mekhan. i mashinostr., 4, 9-28 (1965).

M. Holt and D. E. Ndefo, J. of Comp. Physics, 5, 3, 463-486 (1970).

P. Kutler and H. Lomax, Proc. of 2nd Int. Conf. on Num. Meth. in Fluid Dynamics, 24-29 (1970).

M. J. D. Powell, Computer J., 7, 4, 303-307 (1964).

R. R. Tracy, GALCIT Memo. No. 69 (1963).

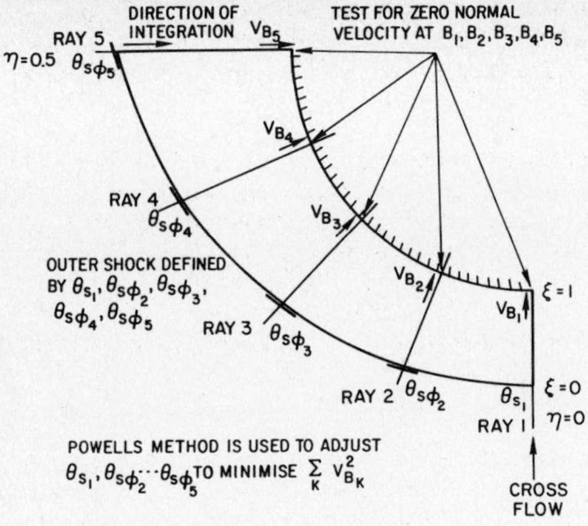

FIG. 1 SCHEMATIC REPRESENTATION OF THE NUMERICAL METHOD USED IN THE WINDWARD REGION

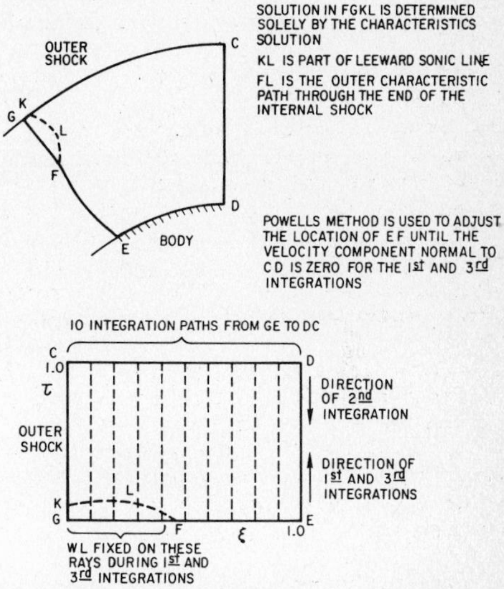

FIG. 2 SCHEMATIC REPRESENTATION OF THE NUMERICAL METHOD USED IN THE LEEWARD REGION

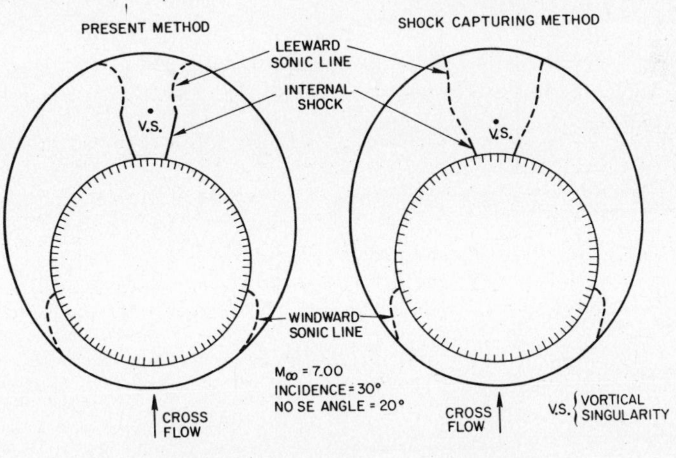

FIG. 3 COMPARISON WITH SHOCK CAPTURING METHOD- SHOCK WAVE AND SONIC LINE LOCATIONS

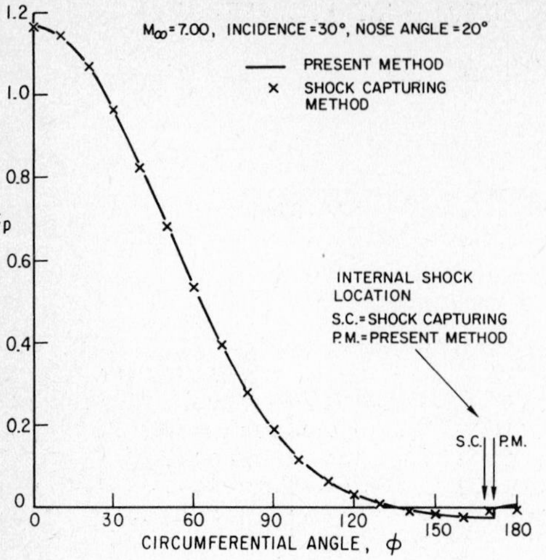

FIG. 4 COMPARISON WITH SHOCK CAPTURING METHOD – PRESSURE DISTRIBUTION

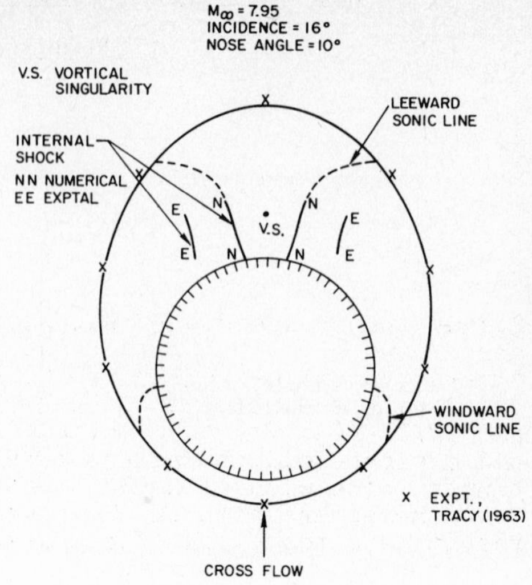

FIG. 6 COMPARISON WITH EXPERIMENT– SHOCK WAVE AND SONIC LINE LOCATIONS

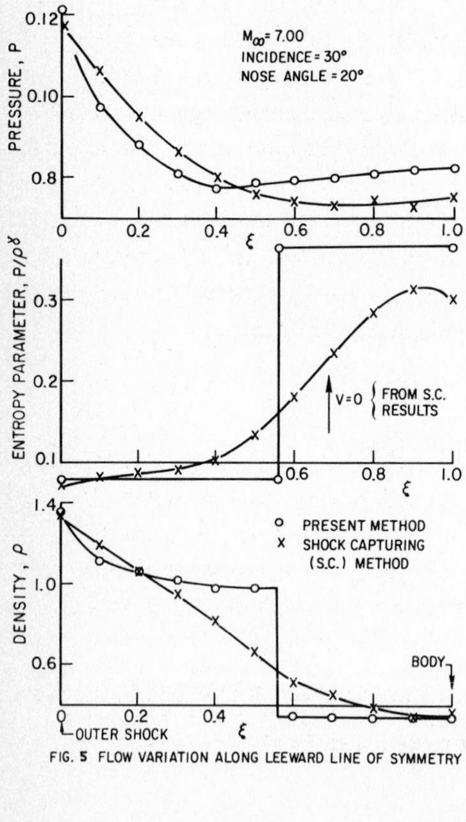

FIG. 5 FLOW VARIATION ALONG LEEWARD LINE OF SYMMETRY

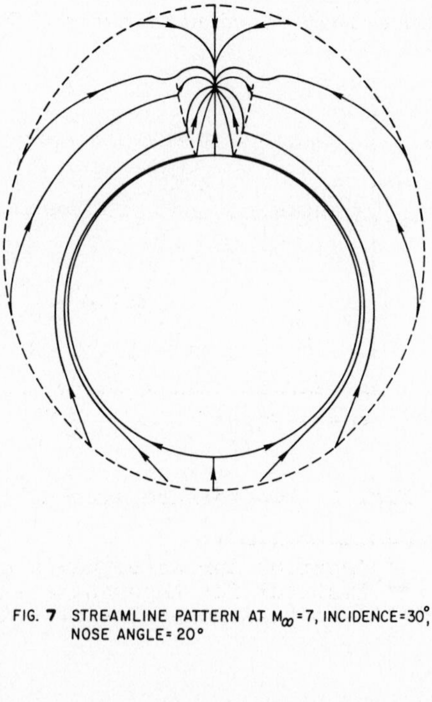

FIG. 7 STREAMLINE PATTERN AT $M_\infty=7$, INCIDENCE=30°, NOSE ANGLE=20°

THE NUMERICAL SOLUTION OF BLUNT BODY FLOWFIELDS USING RUSANOV'S METHOD

K. Förster*), K. Roesner**), C. Weiland***)

Steady three-dimensional supersonic flowfields of ideal inviscid gases around blunt bodies are calculated based on a modification of the time-dependent method due to RUSANOV [1]. Numerical results are compared with experiments of SEDNEY and KAHL [2], GOODERUM and WOOD [3], and STILP [4]. The calculations are done for the flow around a sphere and a paraboloid at Mach-numbers slightly larger than 1, and for a blunted cone with an angle of attack of 20° at the Mach-number 14.9. Some characteristics of the numerical procedure are described which result in a reduction of computing time.

I. COORDINATE SYSTEM

Starting from the time-dependent Eulerian equations - formulated in an orthogonal coordinate system - a transformation of the physical space between the bow shock and the body is defined such that $\xi = 0$ represents the body surface and $\xi = 1$ is defining the shock wave (Fig. 1.). This system of coordinates may be called 'generalized' spherical coordinate system. The transformation between the cylindrical coordinate system (r, ϑ, z) and the locally spherical coordinate system (R, ϑ, ω) is given for the body points by the formulas:

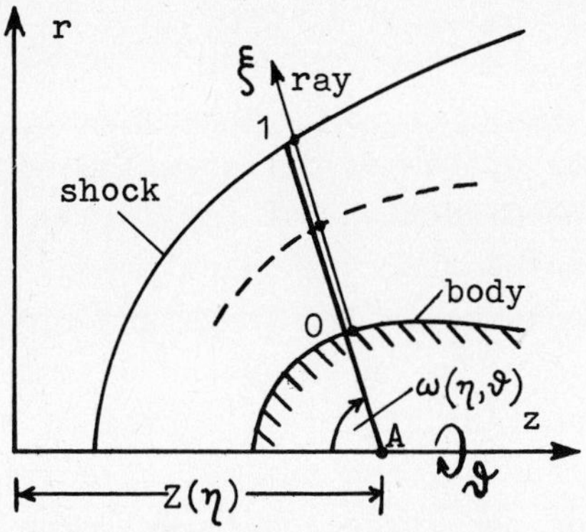

Fig. 1. Coordinate system

$$r = G \sin\omega \ ,$$
$$z = Z(\eta) - G \cos\omega \ .$$

$G(\eta, \vartheta, t)$ is the radial distance of a body point measured from the point A along the ξ-direction. The computing coordinates are chosen in such a way that

*) Institut für Aerodynamik und Gasdynamik der Universität Stuttgart,
**) Institut für Angewandte Mathematik der Universität Freiburg,
***) Aerodynamisches Institut der Technischen Hochschule Aachen

$$\xi := \frac{R - G}{F - G} ,$$

R and F are the radial distances of a field point and a shock point measured from the point A in ξ-direction.

Using this coordinate system we introduce a singularity along the axis of symmetry of the body. To avoid this in the neighbourhood of the z-axis the calculation goes back to cartesian coordinates. The advantage of the defined coordinate system is its flexibility which allows to describe bodies of various shape.

II. SYSTEM OF EQUATIONS

Transforming the Eulerian equations into the (ξ,η,ϑ) coordinate system introduces the unknown function $F(\eta,\vartheta,t)$ of the shock shape which together with its derivatives influences the coefficients of the basic system of equations. For the moving shock front we get an additional differential relation. Together with the boundary condition at the body and the Rankine-Hugoniot-relation along the shock we have to choose an initial value distribution for the flowfield quantities as a starting solution. If the solution is sought for a low Mach-number larger than 1 we may start with the solution for the Mach-number 5 e.g.

III. DIFFERENCE SCHEME

The discretization of the quasilinear system of differential equations is performed explicitly in time-direction but implicitly in the ξ-direction which remains constant during the calculation. Due to this fact RUSANOV's procedure should be called 'constant direction method'. This feature of the c.d.m. is an important advantage of the difference scheme. As the resultant algebraic system of equations is nonlinear the sweep method for solving the set of equations must be iterated along constant lines the so called 'rays'. The admissable time step is given in [1] by a sufficient stability criterion derived from the frozen-in coefficients of the nonlinear system. For our calculations this criterion was neglected for the first timesteps to jump over the very complicated history of the flowfield. As a consequence of this some thousands of iterations were saved and the computing time was drastically reduced. Another modification was used in the case of the flow around a sphere at Mach-numbers slightly larger than 1. To shorten the time for the calculation of the bow shock position we used after the first hundred iterations a nonlinear sequence transformation based on the Δ^2-Aitken-procedure.

IV. RESULTS

The modified RUSANOV-algorithm was applied to calculate the flow-fields around bodies of simple geometry and of more complicated ones with nonzero angle of attack. For axisymmetric flows the following cases were treated:

BODY	FREE STREAM MACH-NUMBER	EXPERIMENTS
Sphere	1.079 and 1.1	STILP, A. [4]
	1.3 and 1.62	GOODERUM, P.B. and WOOD, G.P. [3]
	1.5	
	5.017	SEDNEY, R. and KAHL, G.D. [2]
Paraboloid	8. and 10.	

Tab. 1. Axisymmetric flowfields

For the flow around a sphere the number of the grid points was: 17 in ξ-direction, and 34 in ω-direction. For the Mach-numbers 1.5 and 1.079 the number of grid points was reduced to 13×27.
In Fig. 2. the variation of the local shock angle σ for Mach-numbers slightly larger than 1 is compared with the experiments of STILP [4]. Starting the iteration with the solution for $M_\infty = 1.5$ the calculation stopped when the difference between two iteration steps was smaller than 10^{-5}.

Fig. 3. shows for $M_\infty = 3$ the final position of the bow shock. The dashed line gives the experiments of GOODERUM and WOOD [3]. The agreement is rather good even farther away from the body in the vicinity of the sonic line. In this case the supersonic part of the flowfield is very large. As a consequence of this the number of iterations is 3261. The stability criterion of RUSANOV was used during the whole calculations and no nonlinear sequence transformation was built in. In practice one will restrict the calculations to a small region behind the sonic line and do the rest of the calculation by another method e.g. by a marching procedure in the supersonic part.

Fig. 4. gives the density distribution along the sphere for the Mach-numbers: 1.1 , 1.3 , 1.5 , and 1.62. In the vicinity of the stagnation point the experimental data differ from the theoretical ones. The reason might be that it is hard to measure especially in that part of the flowfield.

The best agreement to measurements is shown in Fig. 5. where the pressure coefficient is plotted along the surface of a sphere.

V. CONCLUSION

Using RUSANOV's method one gets extremely good results near the body surface. Especially for the transonic region the calculations give reliable values of the flowfield quantities. Also in the three-dimensional case of a blunted cone with an angle of attack of 20° the convergence rate of the process is very fast if one confines the calculation to a small supersonic part behind the sonic line. The whole results of the calculations will be published in a report of the Gesellschaft für Weltraumforschung which supported the present work.

REFERENCES

[1] LYUBIMOV, A.N., and RUSANOV, V.V. — Gas flows past blunt bodies Part I: Calculation Method and Flow Analysis "Nauka" Press, Moscow, 1970 (NASA TT F-714, Washington, D.C. Febr.73)

[2] SEDNEY, R., and KAHL, G.D. — Interferometric study of the blunt body problem

Ball. Res. Lab., Rep. No. 1100, (1960)

[3] GOODERUM, P.B., and WOOD, G.P. — Density fields around a sphere at Mach Numbers 1.30 and 1.62

NASA TN 2173, (1950)

[4] STILP, A. — Strömungsuntersuchungen an Kugeln mit transsonischen und supersonischen Geschwindigkeiten in Luft und Frigen-Luft-Gemischen.

Ernst-Mach-Institut, Freiburg i. Br., WB Nr. 10165

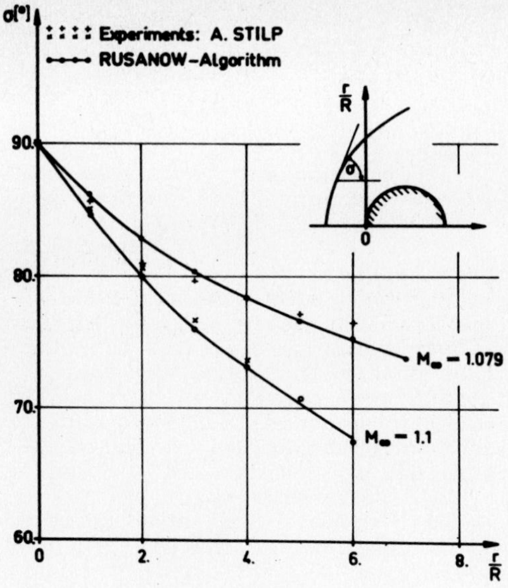

Fig. 2. Variation of local shock angle

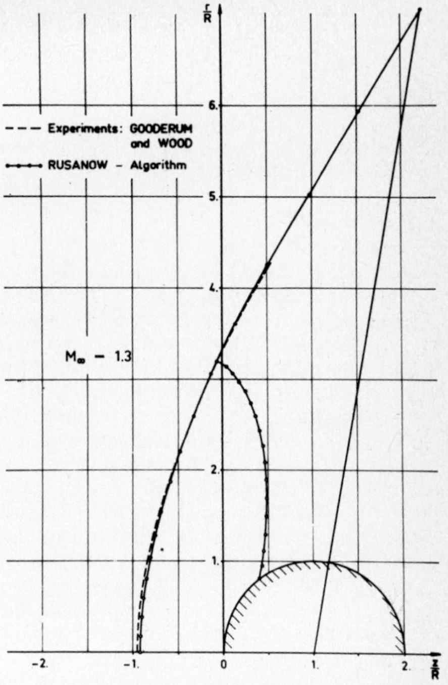

Fig. 3. Bow shock location in front of a sphere

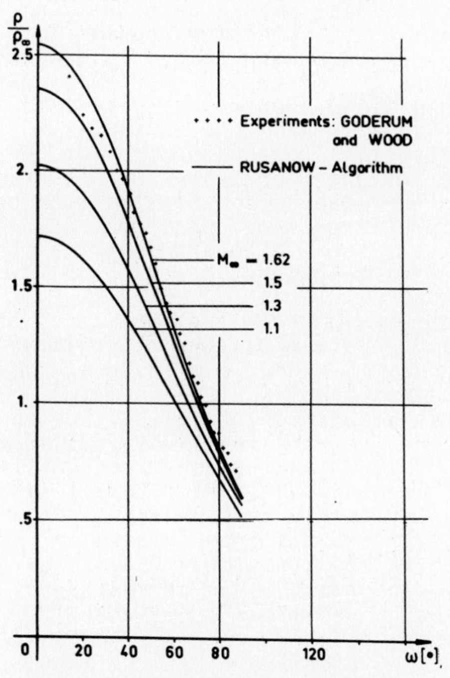

Fig. 4. Density distribution along the sphere

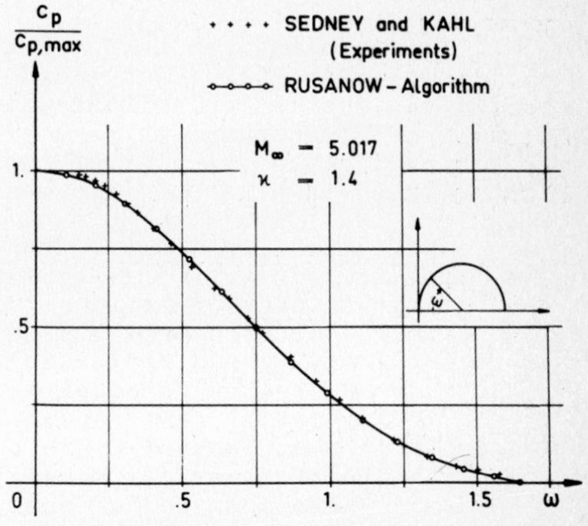

Fig. 5. Pressure coefficient along the surface

ONSET OF DISSIPATION AT ZERO VISCOSITY IN TURBULENT FLOWS

Collective work of Groupe de Recherche sur la Turbulence et les Phénomènes Aléatoires (France). Presented by Uriel Frisch (Observatoire de Nice France).

In 1949 Onsager (Suppl. Nuovo Cimento 6, 279) conjectured that the Euler equation for a perfect incompressible fluid may admit nonsmooth "turbulent" solutions for which the energy conservation law does not hold.

This question, which was not given sufficient attention at that time, can now be investigated using the method of stochastic models (Kraichnan). Such models are obtained from the Navier-Stokes equations by modifying the quadratically non-linear terms through the introduction of certain random coupling factors chosen in such a way that a) the modified equations have as much structural contact as possible with the original equations (same dimensionality and nonlinearity, conservation of energy and helicity, invariance under translations and rotations, etc) b) the modified equations lead to a closed set of "master equations" for second order moments (e.g. the energy spectrum) which can be studied by numerical and analytical tools. Such stochastic models can also be interpreted in terms of closures of the hierarchy of moment equations (see Section A below).

Investigation of the 3D Navier Stokes equations using stochastic models has led to the following results:

(i) In the presence of positive viscosity (however small) initially smooth datas remain smooth forever.

(ii) In the limit of zero viscosity initially smooth datas remain smooth up to a finite time t_* the mean square vorticity becomes infinite (as in the quasi-normal theory of Proudman and Reid); in addition a Kolmogorov type inertial range appears in the energy spectrum; energy is conserved up to t_* after which it is dissipated at a finite rate.

It is conjectured that essentially similar results hold for the original Navier Stokes equations (see Section B below). The nature of the "turbulent" singularities which occur after a finite time remains unclear although there is some experimental evidence that they affect only a negligible fraction of the total volume; it may be that they are concentrated in a set of zero measure and fractional dimension.

A) Closure Equations. The closures are usually at the level of two-point correlations (one or two times) with sometimes other ingredients like response functions (see Refs. 18 and 22 for review). Modern closures such as Orszag's Eddy-Damped-Quasi-Normal-Markovian Approximation (23) and Kraichnan's Test-Field-Model (9,11,14) are representable by stochastic model equations insuring realizability; they give $k^{-5/3}$ inertial range spectra in 3D and lead to equations easily tractable on the computer (more than the previous DIA and LHDI); agreement with direct numerical simulations is fairly good. Some applications which have been worked out recently are the error growth or predictability problem (7,14,15,20), helical MHD turbulence (5,16) and relaxation of anisotropies (8). Such closures can now be considered as operational for homogeneous problems but usually too complicated for inhomogeneous problems, although attempts have been made to reduce two-point closures to more tractable one-point closures (18). Existing two point closures assume implicitly linear relaxation of departures from gaussian state (6); for strong departures (e.g., highly intermittent flows) they cannot be trusted; neither can the $k^{-5/3}$ spectrum (12)

B) <u>Direct Approach Based on the Navier Stokes Equations</u>. The mathematical work on the Navier-Stokes equations is reviewed in Ref. 19. For the 2D and 3D Euler equation (zero viscosity) Swann (26) and Kato (10) have shown that smoothness (three square integrable derivatives) of initial datas persists at least for a finite time up to which uniqueness holds and the flow, even if random, cannot be considered as truly turbulent. In 2D the closure equations indicate that smoothness persists forever (24); for the true Euler equations uniqueness of solutions with one derivative has been proved (1); it is not known whether higher order smoothness persists forever. This question is related to possible intermittency corrections to the k^{-3} enstrophy inertial range (13). In 3D the closure equations suggest that a singularity (catastrophe) occurs at a finite time after which the Navier Stokes equations remain dissipative in the limit of zero viscosity (2,3,17). The occurrence of singularities after a finite time for the true 3D Euler equations has not been proved although it is known that vortex stretching takes place (4). Intermittency measurements suggest that (in the limit of zero viscosity) after the catastrophe the dissipation is concentrated in a set of zero measure, possibly of fractional dimension (21). Uniqueness of solutions of the 3D Navier Stokes equation for large times has not been established for small viscosities (large Reynolds numbers); however, uniqueness is known to hold (i) if dissipation is slightly increased by changing $\nu\nabla^2$ into $-\nu(-\nabla^2)^{\alpha}$ with $\alpha \geq 5/4$ (Ref.19) (ii) for the analogous but simpler problem arising from certain closure equations (2). It is likely that dissipativity, as measured by α can be decreased to some critical value, less than one, related to the exponent of the inertial range power spectrum. Bounds on this exponent n can be obtained by estimating the transfer of energy or enstrophy due to the nonlinear terms (25). It is found that $n \leq 8/3$ in 3D and $n \leq 4$ in 2D (as compared to the values 5/3 and 3 resulting from dimensional analysis and closure equations).

REFERENCES

1. C. Bardos: J. Math. Anal. and Applic. 40, 769 (1972)

2. C. Brauner, P. Penel, R. Temam: Sur une équation d'evolution non-linéaire liée à la theorie de la turbulence (I and II), to appear in Comptes Rendus Ac. Sc., (1974)

3. A. Brissaud et al.: Ann. Geophys. (Paris) 29, 539, (1974), avail. at this conf.

4. W. J. Cocke: Phys. Fluids 12, 2488 (1969)

5. U. Frisch, J. Léorat, A. Mazure and A. Pouquet: MHD Helical Turbulence. Preprint Observatoire de Nice, submitted to J. Fluid Mech. (1973)

6. U. Frisch, M. Lesieur and P. L. Sulem: le Modèle du Champ d'Epreuve (Test Field Model) de Kraichnan, presented at Round Table on Turbulence, Villard de Lans May 20-22 (1974)

7. J. R. Herring: in Proceedings of Symposium on Turbulence in Fluids and Plasmas, Culham, July (1973), to appear.

8. J. R. Herring: Approach of Axisymmetric Turbulence to Isotropy, to appear in Phys. Fluids, (1974)

9. J. R. Herring and R. H. Kraichnan: In "Statistical Models and Turbulence" p. 148, Proceedings of the La Jolla Meeting, Springer (1972)

10. T. Kato, J. Funct. Anal. 9, 296 (1972)

11. R. H. Kraichnan: J. Fluid Mech. 47, 513 and 525 (1971)

12. R. H. Kraichnan: J. Fluid Mech. 69, 305 (1974)

13. R. H. Kraichnan: Statistical dynamics of two-dimensional flows, preprint, submitted to J. Fluid Mech. (1974)

14. C. E. Leith and R. H. Kraichnan: J. Atmo. Sc. 29, 1041 (1972)

15. C. E. Leith: J. Atm. Sc. 28, 145 (1971)

16. J. Léorat, U. Frisch and A. Pouquet: Helical MHD Turbulence and the Nonlinear Dynamo Problem, presented at the Nordita Meeting, Copenhagen, June 5-7 (1974)

17. M. Lesieur: Thesis, Nice (1973)

18. D. C. Leslie: Developments in the Theory of Turbulence, Clarendon (1973)

19. J. L. Lions: Quelques Méthodes de Résolution des Problèmes aux Limites nonlinéaires, Dunod (1969)

20. E. N. Lorenz: Tellus, 21, 289 (1969)

21. B. B. Mandelbrot: J. Fluid Mech. 62, 331 (1974)

22. S. A. Orszag: Statistical Theory of Turbulence, to appear in Pro-

ceedings of the 1973 les Houches Summer School of Theoretical Physics.

23. S. A. Orszag: J. Fluid Mech. 41, 363 (1970)

24. A. Pouquet, M. Lesieur, J. C. Andre: High Reynolds Number Two-dimensional Turbulence Using a Stochastic Model, preprint Observatoire de Nice, submitted to J. Fluid Mech. (1973)

25. P. L. Sulem and U. Frisch: Bounds on energy flux for finite energy turbulence, preprint (1974)

26. H. Swann: Trans. Amer. Math. Soc. 157, 373 (1971)

A NUMERICAL PROCEDURE IN THE HODOGRAPH PLANE FOR THE STUDY OF TRANSONIC FLOW PAST WING PROFILES (*)

Bruno GABUTTI
Laboratorio di Analisi Numerica del CNR - Università - PAVIA
Giuseppe GEYMONAT and Silvio NOCILLA
Politecnico - Torino - ITALY

1. THE PROBLEM

The determination of the aerodynamic field (subsonic and transonic) around symmetric smooth wing profiles in asymptotic uniform flow with zero incidence in the hodograph plane, can be reduced to the following free boundary value problem.

Problem: find $(\psi, a, \mathcal{D})$ such that (see figure 1)

(1.1) $\mathcal{D} = \{(\vartheta, \sigma) ; \vartheta \in]\vartheta_F, \vartheta_A[\text{ and } \sigma > g(\vartheta)\} \setminus \{(0, \sigma) ; \sigma \geq \sigma_\infty\}$

with $\vartheta_A \in]0, \frac{\pi}{2}[$, $\vartheta_F \in [-\frac{\pi}{2}, 0[$ and $g(\vartheta)$ defined, regular on: $\vartheta \in]\vartheta_F, \vartheta_A[$.

(1.2) ψ is defined and regular on $\bar{\mathcal{D}}$ and:

$$\psi_{\sigma\sigma} + k(\sigma)\psi_{\vartheta\vartheta} = 0$$

(1.3) $\psi(0, \sigma) = 0$ on the half line: $\{(0, \sigma); \sigma > \sigma_\infty\}$

(1.4) $\psi(\vartheta, \sigma) = 0$ on : $\{(\vartheta, g(\vartheta)); \vartheta \in]\vartheta_F, \vartheta_A[\}$

(1.5) $\psi_\sigma - k(\sigma) \frac{dg}{d\vartheta} \psi_\vartheta = a \cdot r(\vartheta)\, q(\sigma)$ on $(\vartheta, g(\vartheta))$;

(1.6) $\lim_{\vartheta \to \vartheta_F^+} g(\vartheta) = \lim_{\vartheta \to \vartheta_A^-} g(\vartheta) = +\infty$;

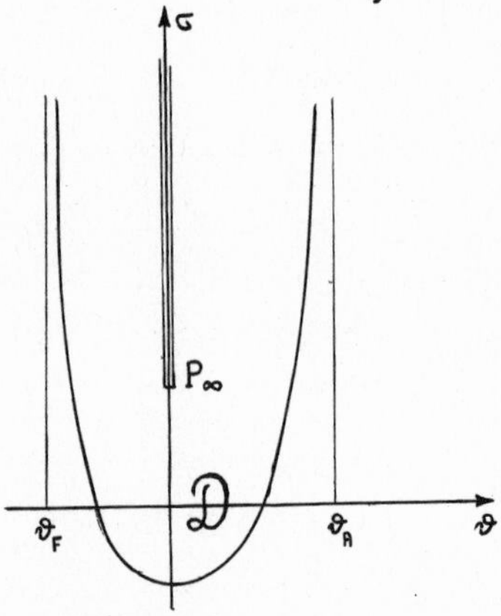

Figure 1

(*) Research partially supported by the CNR (Comitato per la Matematica).

(1.7) $\psi \to 0$ for $\sigma \to +\infty$ and $(\vartheta, \sigma) \in \mathcal{D}$;

(1.8) $a \cdot \psi(\vartheta, \sigma) > 0$ on $\mathcal{D}$;

(1.9) ψ is <u>singular</u> in $P = (0, \sigma_\infty)$ and $z = \psi - \psi_{-\frac{1}{2}}$ is <u>regular</u> in P_∞,

where $\psi_{-\frac{1}{2}}(\vartheta, \sigma)$, $q(\sigma)$, $r(\vartheta)$ and $k(\sigma)$ are given functions; $\psi(\vartheta, \sigma)$, $g(\vartheta)$ are unknown functions and a is an unknown real number.

For an exact formulation of (1.9) and the physical meaning of the problem, see [1] .

2. THE ALGORITHM

In order to solve numerically the problem, the following remarks are useful.

i)It easy to see that the conditions of the problem do not change when we put $-\psi$ and $-a$ instead of ψ and a; we can so find a solution (ψ, a, $\mathcal{D}$) of type:

(2.1) $a > 0$ and $\psi(\vartheta, \sigma) > 0$ on $\mathcal{D}$.

ii)One can prove the followings asymptotics behaviours

(2.2) $\psi(\vartheta, \sigma) \sim h_A(\vartheta) \exp\left(-\frac{\pi}{\vartheta_A}\sqrt{k(\infty)}\,\sigma\right)$ for $\sigma \to +\infty$ and $0 < \vartheta < \vartheta_A$

(2.3) $g(\vartheta) = -D \log(\vartheta_A - \vartheta) - D \log \mu_A + o(1)$ for $\vartheta \to \vartheta_A$

where $k(\infty) = \lim k(\sigma)$ and $D = \vartheta_A/((\pi - \vartheta_A)\ k(\infty)^{-\frac{1}{2}})$ are knowns constants, and $h_A(\vartheta)$, μ_A are unknowns and connected with the unknown constant a by the formula :

(2.4) $h_A(\vartheta) = a\, r(\vartheta_A) B\, (1 - \vartheta_A/\pi) \sin(\pi\vartheta/\vartheta_A)/(\mu\, k(\infty))$

where B, defined by $q(\sigma) \sim B \exp(-\sqrt{k(\infty)}\ )$, is a constant that can be explicitely computed.

The analogous formulas are true for $\vartheta_F < \vartheta < 0$.

iii) For $\sigma < 0$, near the axis $\sigma = 0$, we can assume $k(\sigma) = (\gamma + 1)\sigma$ and it is known that the following problems are well-posed:

<u>Problem I</u> : given the functions $\tau(\vartheta)$, $\nu(\vartheta)$ on $\vartheta_F \le \vartheta \le \vartheta_A$, find $\psi_I(\vartheta, \sigma)$ such that:

$$(2.5)\quad \begin{cases} (\psi_I)_{\sigma\sigma} + (\gamma+1)\sigma\,(\psi_I)_{\vartheta\vartheta} = 0 \\ (\psi_I)_\sigma(\vartheta, 0) = \nu(\vartheta) \qquad \psi_I(\vartheta, 0) = \tau(\vartheta) \end{cases}$$

Problem II : let Γ a curve in the half-plane $\sigma<0$ never tangent to a characteristic which starts and ends on the axis $\sigma=0$ and let ω a given function on Γ ; find $\psi_{II}(\vartheta,\sigma)$ such that:

$$(2.6)\quad \begin{cases} (\psi_{II})_{\sigma\sigma} + (\gamma+1)\sigma\,(\psi_{II})_{\vartheta\vartheta} = 0 \\ \psi_{II} = 0 \quad \text{on } \Gamma, \qquad (\psi_{II})_{\sigma} = \omega \quad \text{on } \Gamma \end{cases}$$

Now an iterative procedure for the solution of the problem considered in the N.1 can be written:

Step 1. Solve the following fixed boundary value problem on the half-plane $\sigma>0$:

Let $\Gamma_+^{(k)}$ be the profile of equation $= g^{(k)}(\vartheta)$ asymptotic to ϑ_A , ϑ_F and let $\nu^{(k)}(\vartheta)$ be a given function on the segment $\{(\vartheta,0)\ ;\ \vartheta_F^{(k)} \le \vartheta \le \vartheta_A^{(k)}\}$; find $\psi_+^{(k)}(\vartheta,\sigma)$ which satisfies (1.1), (1.2), (1.3), (1.4), (1.6), (1.7), (1.9), (2.1) and:

$$\psi_{+\sigma}^{(k)} = \nu^{(k)}(\vartheta) \qquad \text{on } \sigma=0$$

Step 2. Solve the problem (2.5) with $\tau(\vartheta) = \psi_+^{(k)}(\vartheta, 0)$ and $\nu(\vartheta) = \psi_{+\sigma}^{(k)}(\vartheta,0) = \nu^{(k)}(\vartheta)$. Then $\psi^{(k)}(\vartheta,\sigma)$ is known on $\mathfrak{D}^{(k)}$.

Step 3. Find the curve $\Gamma_\wedge^{(k)}$ such that $\psi_I^{(k)}(\vartheta,\sigma)=0$ and find the number $a^{(k+1)}$ by the equation:

$$\psi_I^{(k)}(\vartheta_{min}^{(k)}, \sigma_{min}^{(k)}) = a^{(k+1)}\, r(\vartheta_{min}^{(k)})\, q(\sigma_{min}^{(k)})$$

where $(\vartheta_{min}^{(k)}, \sigma_{min}^{(k)})$ is the minimum point of $\Gamma_\wedge^{(k)}$.

Step 4. Solve the initial value problem:

$$\psi_{-\sigma}^{(k)}\left(1+k(g^{(k+1)}(\vartheta))\left(\frac{dg^{(k+1)}}{d\vartheta}\right)^2\right) =$$

$$= a^{(k+1)} r(\vartheta)\, q(g^{(k)}(\vartheta))$$

$$g^{(k+1)}(\vartheta_{min}) = \sigma_{min}^{(k)}$$

on $\vartheta_F < \vartheta < \vartheta_A$; indeed one uses different explicit formulas for $\sigma<0$ and $\sigma>0$.

Step 5. Solve the problem (2.6) with ω given by:

$$\omega = a^{(k+1)} r(\vartheta) q(g^{(k+1)}(\vartheta))/(1+(\gamma+1)g^{(k+1)}(\vartheta)(dg^{(k+1)}/d\vartheta)^2;$$

on the curve $\Gamma_{\wedge}^{(k+1)}$ obtained in the step 4 for $\sigma < 0$. In this way we can find $\nu^{(k+1)}(\vartheta)$ on the segment $\left\{(\vartheta,0);\ \vartheta_F^{(k+1)} < \vartheta < \vartheta_A^{(k+1)}\right\}$ and so restart from the step 1.

Once obtained the function Γ with some elementary computations we find the local Mach number on the profile.

3. SOME REMARKS ON THE IMPLEMENTATION OF THE ALGORITHM

In practice we have used the procedure obtained by the discretizati on of each step of the algorithm.

We shall now briefly describe some features of the implementation of our algorithm.

Step1. Obviously we have discretized the problem only for $\sigma \leq \sigma_N < +\infty$, but we can use the asymptotic behaviours (2.2), (2.3) to extend the computed formulas to leading edges. Actually, until now, we have extrapolated the profile $\Gamma^{(k)}$ with a trial and error procedure. Let us only point out that we need some accurated smoothing formulas to compute the profile.

Steps 2 and 5. We have used the Tricomi's explicit formulas; however, like is well known, such formulas are of integral type with a singular kernel. For the computation we have used the quadrature formulas of Gauss type with Gegenbauer polynomials (see [2]).

Step 3. For the determination of the curve $\Gamma_{\wedge}^{(k)}$ we used a simple and rapid dicothomic algoritm.

4. NUMERICAL EXPERIMENTS

We are carrying out many computational experiments for different type of symmetric and double symmetric arcs airfoils. For instance, in figure 2 we show the local Mach number for a circular profile.

BIBLIOGRAPHY

[1] S.Nocilla, G.Geymonat, B.Gabutti, Transonic flows past wing profiles: a new direct hodograph method. To be published ia a volume offered to the Professor C. Ferrari.

[2] B. Gabutti, Sulla risoluzione numerica dell'equazione di Tricomi nel piano iperbolico. To appear in Rend. Sem. Mat. Modena.

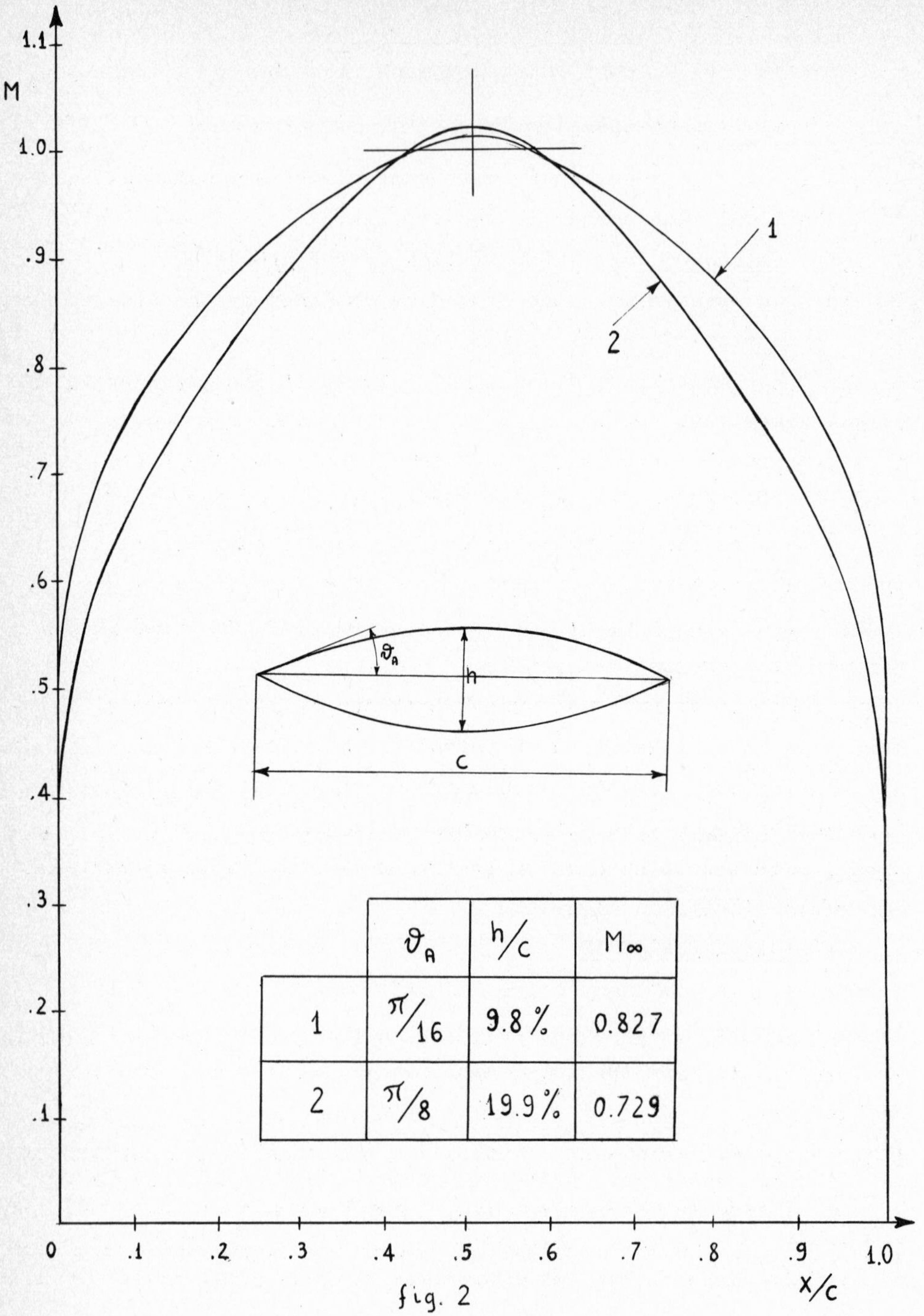

	ϑ_A	h/c	M_∞
1	$\pi/16$	9.8 %	0.827
2	$\pi/8$	19.9 %	0.729

fig. 2

DISCRETISATION OF THE VELOCITY - SPACE IN KINETIC THEORY OF GASES

by Renée GATIGNOL

Mécanique Théorique . Université Pierre et Marie Curie . Paris

The fundamental equation in kinetic theory of gases is the integrodifferential equation of Boltzmann verified by the velocity distribution fonction $f(\vec{\xi}, \vec{r}, t)$:

$$(1) \quad \frac{\partial f}{\partial t} + \vec{\xi} \cdot \frac{\partial f}{\partial \vec{r}} + \vec{X} \cdot \frac{\partial f}{\partial \vec{\xi}} = \iint (f' f'_1 - f f_1)\, g\, b\, db\, d\varepsilon\, d\vec{\xi}_1$$

The numerical treatment of problems of rarefied gas dynamics where the equation (1) intervenes is an important procedure to obtain satisfying results. Following the works of Broadwell, [1], [2], Harris , [6], [7], Godunov and Sultangazin, [5] , it seems to us interesting to describe in a different way the microscopic character of the gas. We propose a method of discretisation in which the velocity space is discretised, the time and space variables staying continuous.

1. DESCRIPTION OF THE MODEL.

The gas is composed of identical particles. The velocities of these particles are restricted to a given finite set of p vectors : $\vec{u}_1$, $\vec{u}_2$, ... $\vec{u}_p$. $N_i = N_i(\vec{r},t)$ denotes the number density of particles with velocity $\vec{u}_i$ at the point $\vec{r}$ and at the time t.

Only binary collisions are considered. Let $\vec{u}_i$ and $\vec{u}_j$ be the velocities of two molecules before an encounter ; after the encounter these molecules have the velocities $\vec{u}_k$ and $\vec{u}_\ell$. These last two vectors must belong to the original set and satisfy the two relations expressing the conservation of momentum and the conservation of energy :

$$(2) \quad \begin{cases} \vec{u}_i + \vec{u}_j = \vec{u}_k + \vec{u}_\ell \\ |\vec{u}_i|^2 + |\vec{u}_j|^2 = |\vec{u}_k|^2 + |\vec{u}_\ell|^2 \end{cases}$$

There is always the trivial solution where the final velocities are the same as the initial velocities, but there can be other solutions.

A "transition probability" $A_{ij}^{k\ell}$ is associated with the collision $\vec{u}_i,\vec{u}_j \to \vec{u}_k,\vec{u}_\ell$: $A_{ij}^{k\ell}\ N_i\ N_j$ is the number of collisions $\vec{u}_i,\vec{u}_j \to \vec{u}_k,\vec{u}_\ell$ per unit time and unit volume; $\frac{1}{2} \sum_{j,k,\ell} A_{ij}^{k\ell} N_j$ is the collision frequency of a molecule with the velocity $\vec{u}_i$. The dynamics of the collision is expressed by the coefficient $A_{ij}^{k\ell}$. If the molecules have a finite radius of action we can write :

$$(3) \quad A_{ij}^{k\ell} = S\, |\vec{u}_i - \vec{u}_j|\, \alpha_{ij}^{k\ell}$$

where S is the effective collision cross-section, and where $\alpha_{ij}^{k\ell}$ is the probability that the pair $\vec{u}_i$, $\vec{u}_j$ gives the pair $\vec{u}_k$, $\vec{u}_\ell$ after collision. As suggested by the properties of similar quantities in statistical mechanics it is inferred that $A_{ij}^{k\ell}$ satisfies the microreversibility principle :

$$A_{ij}^{k\ell} = A_{k\ell}^{ij}$$

The Boltzmann equation is replaced by a system of p non linear partial differential equations

$$(4)\quad \begin{cases} \dfrac{\partial N_i}{\partial t} + \vec{u}_i \ \nabla \vec{N}_i = \dfrac{1}{2} \displaystyle\sum_{j,k,\ell} A_{k\ell}^{ij} (N_k N_\ell - N_i N_j) \\ \\ i = 1,2, \ldots , p \end{cases}$$

We write this system in the form :

$$(5)\qquad \frac{\partial N}{\partial t} + \mathcal{A} N = \mathcal{F}(N,N)$$

N represents $(N_1, N_2, \ldots , N_p)$. $\mathcal{F}(U , V)$ is a bilinear symetric operator. For two vectors $U = (U_1 , U_2 , \ldots , U_p)$ and $V = (V_1, V_2, \ldots , V_p)$ we shall denote their scalar product by $(U , V) = \sum_i U_i V_i$

2. THE MACROSCOPIC STATE VARIABLES.

For this model we define the summational invariants as the functions $\phi(\vec{u})$ associated with quantities that are conserved in an encounter. For this model, $\phi(\vec{u})$ has p components and satisfies the condition :

$$A_{ij}^{k\ell} (\phi_i + \phi_j - \phi_k - \phi_\ell) = 0 \qquad\qquad i , j , k , \ell$$

Here we have adopted the convention $A_{ij}^{k\ell} = 0$ for an unrealizable collision. In particular when $\phi_i = 1$, $\vec{u}_i$ or $|\vec{u}_i|^2$, ϕ is a summational invariant. The summational invariants form a linear space F of dimension q , $(1 \leqslant q \leqslant p)$.

We introduce orthonormal bases in F and in R^p

In F : $V^1 , V^2 , \ldots , V^q$

In R^p : $V^1 , V^2 , \ldots , V^q , W^{q+1} , \ldots , W^p$.

and we write

$$(6)\qquad N = \sum_{i=1}^{q} a_i V^i + \sum_{j=q+1}^{p} b_j W^j$$

$$a_i = (N , V^i) \quad i = 1, \ldots , q \ ; \qquad b_j = (N , W^j) \quad j = q + 1,\ldots,p.$$

As in classical kinetic theory, for each function $\phi(\vec{u})$ we define its mean value and we can write a transport equation for each mean value. If the function $\phi(\vec{u})$ is a summational invariant the transport equation is a conservation equation [3]; we obtain q and only q independent conservation equations. They may be written

$$\frac{\partial}{\partial t} (N , V^i) + (\mathcal{A} N , V^i) = 0 \qquad i = 1 , \ldots , q$$

or

$$(7)\quad \begin{cases} \frac{\partial}{\partial t} a_i + \vec{\nabla} . \vec{h}_i (a_1 , \ldots , a_q , b_{q+1} , \ldots , b_p) = 0 \\ i = 1 , 2 , \ldots , q \end{cases}$$

For the gas of interest here we can consider two types of description : first, a microscopic description corresponding to the knowledge of the densities N_i or equivalently to the knowledge of the quantities a_i and b_j ; and second , a macroscopic description corresponding to a more limited knowledge of the quantities a_i only. The quantities a_i are called macroscopic state variables of the gas. In classical kinetic theory these variables are the density n , the mean velocity $\vec{u}$ and the temperature T. In these models there are n , $\vec{u}$, T among the quantities a_i.

If the gas is uniform (the densities do not depend on $\vec{r}$) Boltzmann's H function can never increase [3] because

$$\frac{dH}{dt} = (\text{Log } N, \quad (N,N) = \frac{1}{8} \sum_{i,j,k,\ell} A_{ij}^{k\ell} (N_i N_j - N_k N_\ell) \text{ Log } \frac{N_i N_j}{N_k N_\ell} \leqslant 0$$

(Log N represents the sequence Log N_1 , Log N_2, ... , Log N_p). This is Boltzmann's H theorem. The limit state of the gas is the maxwellian state and in such a state the p densities N_i are dependent only on the macroscopic state variables. We can demonstrate that for a gas having a given microscopic state N there exists one and only one associated maxwellian state $N^{(o)}$; the microscopic state $N^{(o)}$ and the microscopic state N correspond to the same values of the macroscopic state variables.

We have demonstrated [3] that the system of kinetic equations (4) possesses the essential properties of the Boltzmann equation. In particular we can apply the Chapman Enskog method. It is easy to write the corresponding Euler equations but we cannot obtain the explicit Navier Stokes equations from the general model. However we have succeeded in deriving the Navier Stokes equations from more restricted models [4].

3. EXAMPLES.

Some models of gas with a discrete velocity distribution are met in the literature ; the number of permitted velocities is always very small and eight is a maximum. A model with two velocities is presented by Carleman. Broadwell has introduced two space models : the velocities have the same magnitude and their directions are regularly distributed in space. In the first model, the velocities are the vectors that join the center of a hexahedron to its six vertices ; in the second model, the velocities are the vectors that join the center of a cube to its eight vertices.

Here we propose a coplanar regular model (fig 1) with 2r velocities $\vec{u}_i$, $i = 1 , 2 , \ldots , 2r$, of magnitude c such that

$$\begin{cases} \vec{u}_i + \vec{u}_{i+r} = 0 \\ \vec{u}_{i+1} \text{ is inferred from } \vec{u}_i \text{ by a positive rotation through an angle } \frac{\pi}{r} . \end{cases}$$

Two quantities bearing two indices, which are congruent modulo 2r, are equal. The interesting binary collisions are the following ones :

$$\vec{u}_i \ , \ \vec{u}_{i+r} \rightarrow \vec{u}_j \ , \ \vec{u}_{j+r} \qquad\qquad i \ , \ j \ .$$

We suppose that the molecules having the velocities $\vec{u}_i$, $\vec{u}_{i+r}$ give, after collision, the molecules having the velocities $\vec{u}_{i+\ell}$, $\vec{u}_{i+\ell+r}$ with probability α_ℓ ; we must have $\alpha_\ell = \alpha_{-\ell}$. The equations describing the evolution of the medium are :

$$\begin{cases} \dfrac{\partial N_i}{\partial t} + \vec{u}_i \cdot \vec{\nabla} N_i = 2 \ c \ S \ \sum\limits_{\ell=1}^{2} (\alpha_{-\ell} N_{i+\ell} \ N_{i+\ell+r} - \alpha_\ell N_i \ N_{i+r}) \\ i = 1 \ , \ 2 \ , \ \dots \ , \ 2r \end{cases}$$

We may suppose that the collision of the molecules having the velocities $\vec{u}_i$, $\vec{u}_{i+r}$ is "isotropic" and we take $\alpha_\ell = \frac{1}{r}$, $\ell = 1 \ , \ \dots \ , \ 2r$. So are the coplanar models with four or six velocities used by Harris , [6],[7]. From the thorough study of the elastic shock of two spheres, it is possible to imagine a model with distinct α_ℓ, namely : $\alpha_0 = 1 - \cos\frac{\pi}{4r} + \sin\frac{\pi}{4r}$, $\alpha_\ell = [\cos\frac{\ell\pi}{2r} + \cos\frac{(r-\ell)\pi}{2r}] \sin\frac{\pi}{4r}$ for $\ell = 1, 2, \dots , r-1$.

In the previous models all velocities have the same magnitude. The second of (2) that expresses the conservation of energy is always satisfied ; the temperature is connected with the density and the velocity $\vec{u}$. It is possible to imagine some models in which the velocities have various magnitudes. To be useful a model should admit an explicit computation of the collisions. Simple models are constructed from the superposition of two or more previous models.

We give one example of space model with several magnitudes for the velocities. It is obtained from the suitable superposition of the two space models of Broadwell; we take the velocities $\vec{u}_i^{(1)}$, $i = 1,\dots,6$ of the six velocity model to be orthogonal to the faces of the cube defining the model with the eight velocities $\vec{u}_i^{(2)}$, $i=1,\dots,8$, (fig 2). There is not nontrivial collision between two molecules with the velocities $\vec{u}_i^{(1)}$ and $\vec{u}_j^{(2)}$ except in the case where $|\vec{u}_i^{(1)}| = \frac{2\sqrt{3}}{3} |\vec{u}_i^{(2)}|$ or in the case where $|\vec{u}_i^{(1)}| = \frac{\sqrt{3}}{3} |\vec{u}_i^{(2)}|$. If $|\vec{u}_i^{(1)}| = \frac{2\sqrt{3}}{3} |\vec{u}_i^{(2)}|$ the nontrivial collisions are :

$$\vec{u}_1^{(1)} \ , \ \vec{u}_3^{(1)} \rightarrow \vec{u}_2^{(1)} \ , \ \vec{u}_4^{(1)}$$
$$\vec{u}_1^{(2)} \ , \ \vec{u}_4^{(2)} \rightarrow \vec{u}_2^{(2)} \ , \ \vec{u}_3^{(2)}$$
$$\vec{u}_1^{(2)} \ , \ \vec{u}_8^{(2)} \rightarrow \vec{u}_2^{(2)} \ , \ \vec{u}_7^{(2)}$$
$$\vec{u}_2^{(1)} \ , \ \vec{u}_1^{(2)} \rightarrow \vec{u}_3^{(1)} \ , \ \vec{u}_4^{(2)}$$

and so forth. In this model with fourteen velocities, the summational invariants are the classical invariants corresponding to the conservation of mass, momentum and energy during an encounter. The dimension of F is 5.

If $|\vec{u}_i^{(1)}| = |\vec{u}_i^{(2)}|$ the two space models are still coupled, the responsible collisions being the following ones : $\vec{u}_2^{(1)}$, $\vec{u}_5^{(1)} \rightarrow \vec{u}_1^{(2)}$, $\vec{u}_8^{(2)}$ and so forth.

We go back to the first model with fourteen velocities (that is, the one with

$|\vec{u}_i^{(1)}| = \frac{2\sqrt{3}}{3}\ |\vec{u}_i^{(2)}|$) and we consider the models deduced from this one by similarity of ratio $(\frac{2\sqrt{3}}{3})^{\alpha}$, whose centre is the centre of the model. The superposition of such models with α equal to successive integers gives some other examples of space models with different magnitudes. If $|\vec{u}_i^{(1)}| = \frac{\sqrt{3}}{3}|\vec{u}_i^{(2)}|$ we construct other similar models.

4. CONCLUSION.

In another connection, using a regular model with six coplanar velocities, we have studied the problem of shock structure. In fact, we study the propagation of shock waves in the direction of one of the basic velocities. It is easy to obtain the shock structure profile for the macroscopic density from the exact equations. The thickness of the shock is of the order of the mean free path. It is more difficult to obtain the shock profile from the Navier Stokes equations associated with this model. The two shock profiles are slightly different. The thickness of the shock for the macroscopic density as obtained from the Navier Stokes equations is smaller than the thickness obtained from the exact equations. Similar results have been found by other methods.

Using the space model with six velocities of Broadwell, Godunov and Sultangazin studied the same problem of the shock wave structure. They obtained similar results. As a continuation of a theoretical study of Temam on the Carleman's model, [8], Mrs Pelissier and Miss Malecot, using the model with six velocities of Broadwell, are calculating the solution of the problem of the shock tube with the fractional steps method. We also point out the work of Broadwell on the Couette flow and the Rayleigh problem and the satisfying results that he has obtained with the space model with eight velocities. It now seems evident that this method of discretisation of velocities can be applied to a broad range of problems. These first results are encouraging to endeavor some careful studies ; among other problems it must be possible to solve three-dimensional problems of rarefied gas dynamics by means of computations on machines using the classical procedure of discretisation for the variables t and $\vec{r}$ that enter in the equations (4).

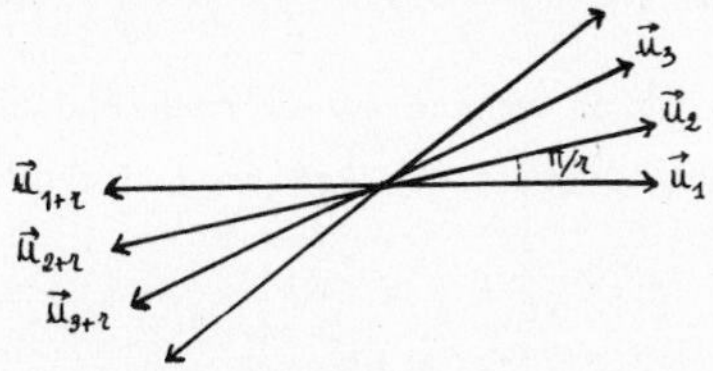

Fig. 1

Modèle à 14 vitesses : $|\vec{u}_i^{(1)}| = \frac{2\sqrt{3}}{3}|\vec{u}_i^{(2)}|$ Hom. $\frac{2\sqrt{3}}{3}$

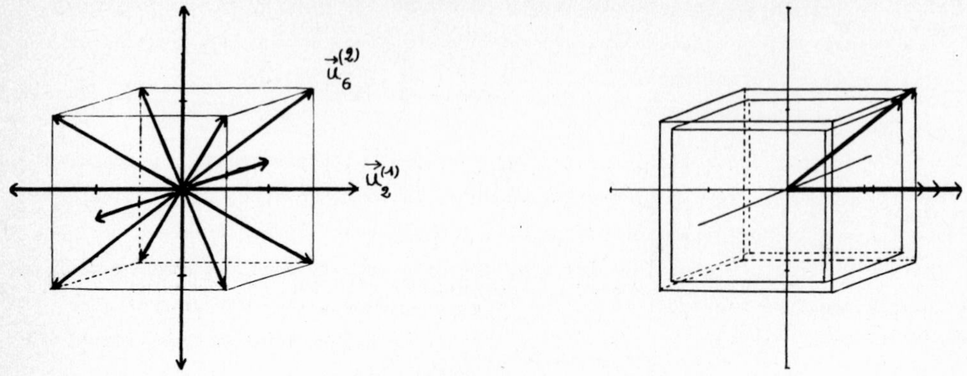

Fig. 2

References

1 - BROADWELL , J.E. Shock Structure in a simple discrete velocity gas. The Physics of Fluids, 7, 1964, n°8, p. 1243.

2 - BROADWELL , J.E. Study of rarefied shear flow by the discrete velocity method. Journal of Fluid Mechanics, 19, 1964, n°3, p. 401.

3 - GATIGNOL , R. Théorie cinétique d'un gaz à répartition discrète de vitesses. Zeitschrift für Flugwissenschaften, 18, 1970, Heft 2/3, p. 93.

4 - GATIGNOL , R. Contribution à la théorie cinétique des gaz à répartition discrète de vitesses. Thèse d'état, Paris, 20 juin 1973.

5 - GODUNOV , S.K. and SULTANGAZIN , U.M. On discrete models of the kinetic Boltzmann equation. Russian Mathematical Surveys, 26, 1971, n°3, p.1.

6 - HARRIS , S. Approach to equilibrium in a moderately dense discrete velocity gas. The Physics of Fluids, 9, 1966, n°7, p. 1328.

7 - HARRIS , S. Proof for a discrete velocity gas that successive derivatives for Boltzmann's H function alternate in sign. Journal of Mathematical Physics, 8, 1967, n°12, p. 2407.

8 - TEMAM , R. Sur la résolution exacte et approchée d'un problème hyperbolique non linéaire de T. Carleman. Archive for Rational Mechanics and Analysis, 35, 1969, n°5, p. 351.

ON THE SOLUTION OF A MIXED PROBLEM FOR THE WAVE EQUATION

A. A. Gladkov

The mixed problem for the wave equation

$$-\frac{\partial^2 u}{\partial t^2} + \frac{\partial^2 u}{\partial y^2} + \frac{\partial^2 u}{\partial z^2} = 0 \tag{1}$$

is formulated so as to find the solution of this equation in the region bounded with time - like and space - like surfaces. On the space - like surface initial Cauchy conditions are given, on the time - like surface one boundary condition is sufficient - the condition on the function or on its conormal derivative.

To be definite we suppose that the initial conditions are zero and that the time - like surface is smooth.

It is well known [1] that the solution of the point $P_0(t_0, y_0, z_0)$ is formally given by Hadamard's formula

$$u(P_0) = -\frac{1}{2\pi}\,\overline{\iint\left(u_1\frac{\partial v}{\partial N_1} - v\frac{\partial u_1}{\partial N_1}\right)dS} \tag{2}$$

where $u_1, \frac{\partial u_1}{\partial N_1}$ are the boundary values of the function and its conormal derivative,

$$v = \left\{(t-t_0)^2 - \left[(y-y_0)^2 + (z-z_0)^2\right]\right\}^{-\frac{1}{2}}$$

It is also known [2] that the solution of equation (1) can be presented in the form of wave potential of a single layer

$$u(P_0) = \iint_S \mu v\, dS \tag{3}$$

and of wave potential of a double layer

$$u(P_0) = \overline{\iint_S \nu \frac{\partial v}{\partial N_1} dS} \tag{4}$$

By analogy with the Newtonian surface layers [4] we can obtain the integral equation for the density of a double layer

$$u_+(P_0) = \pi\nu(P_0) + \overline{\iint_S \nu(P)\frac{\partial v}{\partial N_1} dS} \tag{5}$$

where $u_+(P_0)$ is the limiting value of wave potential obtained when we let P in the region go to P_0 on the surface.

On the opposite side of surface we have:

$$u_-(P_0) = -\pi \nu(P_0) + \iint_S \nu(P) \frac{\partial v}{\partial N_1} dS \qquad (6)$$

The relations (5), (6) are analogous to the Fredholm integral equations in the theory of Newtonian surface layers.

We now make some remarks:

1) Green's formula used in obtaining the relation (2) suggests continuity of the function and its derivatives. But it can be shown that this formula can be applied also in the cases when, on characteristic surfaces in the region, the tangential derivative is continuous but the normal derivative is discontinuous.
2) The domain of influence of elementary surface singularity occupies all the interior of the corresponding characteristic cone.

These remarks allow us to consider complex surfaces in space and time.

1. From the theorem of uniqueness it follows that if the boundary value of the function or its conormal derivative are given, on part of the time - like surface, then the solution is defined in the region for which this part of the surface defines the domain of dependence. From this one can conclude that if some singularities are situated "below the surface" and they create any distribution of potential or its conormal derivative "over the surface", we can receive in sum zero solution in the domain "over the surface" for which this surface defines domain of dependence, if we choose the density of wave surface layer in such manner that it gives boundary value "over the surface" equal to that created by singularities but with opposite sign. This property of cancelling the perturbation by the layer when the proper boundary conditions on the surface are posed can be called the "property of screening". Using in this case the double layer provides identical conormal derivatives on both sides of surface.

The property of screening gives the possibility to make corrections in finding the domains of influence and dependence and in this way allows us to define the areas of posing initial and boundary conditions. In particular, for the surface just considered it is the same to take into consideration the fields of perturbations created by singularities "over the surface" and the proper boundary conditions or not to consider there the influence both of singularities and surface layer.

2. In this connection the question arises how to explain the discrepancy between the domains of influence and dependence obtained with Hadamard's formula and with Fermat's principle. To be definite consider the time - like surface in the form of circular cylinder with the axis along the t axis. On the surface we place the wave surface layer. Let us create the perturbations on the outer surface of the cylinder. The total conormal derivative of potential is posed equal to zero on this cylinder. In consequence of screening the perturbations do not propagate into the cylinder and can go on the surface only between the neighbouring points inside the angle $\vartheta = \pm \frac{\pi}{4}$ formed by generatons of the cylinder. It follows from this that on the surface ahead of the spirals going from the source of perturbation with angles ϑ there will be

no perturbations even though the source lies in the characteristic cone that defines the domain of dependence for points on the cylinder surface ahead of the spirals. So the property of screening can be applied to explain the discrepancy indicated before.

3. For the wave equation in three - dimensional physical space

$$-\frac{\partial^2 u}{\partial t^2} + \frac{\partial^2 u}{\partial x^2} + \frac{\partial^2 u}{\partial y^2} + \frac{\partial^2 u}{\partial z^2} = 0 \qquad (7)$$

the procedure of Hadamard leads to Kirchhoff's formula [1]

$$u(P_0,t) = -\frac{1}{4\pi} \iint \left(\frac{\partial u_1}{\partial n} \frac{1}{r} - \frac{1}{r} \frac{\partial r}{\partial n} \frac{\partial u_1}{\partial t} + u_1 \frac{\partial \frac{1}{r}}{\partial n} \right) dS \qquad (8)$$

where n is normal to surface S,

$P_0 = P_0(x_0, y_0, z_0)$,

$r^2 = (x-x_0)^2 + (y-y_0)^2 + (z-z_0)^2$

By analogy with Newtonian surface layers in this case we may also introduce the wave of single, double and logarithmic layers with potentials (retarded potentials)

$$u(P_0,t) = \frac{1}{4\pi} \iint \frac{1}{r} \mu(P, t-r)\, dS \qquad (9)$$

$$u_0(P_0,t) = \frac{1}{4\pi} \iint \left(\frac{\partial}{\partial n} \frac{1}{r} \right) \nu(P, t-r)\, dS \qquad (10)$$

$$u_0(P_0,t) = \frac{1}{4\pi} \iint r \left(\frac{\partial}{\partial n} \frac{1}{r} \right) \sigma(P, t-r)\, dS \qquad (11)$$

But for the density of the double layer in this case the following integro - differential equation is obtained, which can also be derived in another way [3]

$$u_+(P_0,t) = -\frac{1}{2}\nu(P_0,t) + \frac{1}{4\pi} \iint \left(\frac{\partial}{\partial n} \frac{1}{r} \right) \left[\nu(P, t-r) + r\nu_t(P, t-r) \right] dS \qquad (12)$$

Differentiation of the expression (9) for the density of a single layer with respect to normal to the surface gives also equation of the type (12).

4. The equations (5) and (12) may be solved numerically. For this purpose, the surface may be divided, for example, into elements with constant density of double layer. The derivative with respect to time in equation (12) is replaced by finite difference and then ν is sucessively defined from algebraic equations with given step in t. After finding density of double layer it is not difficult to compute the potential in given points. Thus may be solved for instance the Neumann's problem for the wave equation. As an illustration, in fig. 1, the dependence of potential on time in forward critical point of rigid sphere is given when stepwise sound pulse is incident on the sphere.

BIBLIOGRAPHY

1. Hadamard J., Lectures on Cauchy's problem in linear partial differential equation. N. - Y. Dover publ. (1953)

2. Heaslet M. A., Lomax H., The use of source - sink and doublet distributions extended to the solution of boundary - value problems in supersonic flow. NACA Rep. N 900, (1948)

3. Fulks W., Guenther R. B., Hyperbolic potential theory. Arch. Ration. Mech. and Analysis. v. 49, N 2, (1972)

4. Sretenskii, A. N., Theory of Newtonian potential (Russian) M. - L. Gostekhizdat (1946)

HIGHLY STRETCHED MESHES AS FUNCTIONALS OF SOLUTIONS

by

D. O. Gough

Institute of Astronomy
and Department of Applied Mathematics and
Theoretical Physics, University of Cambridge

E. A. Spiegel

Astronomy Department
Columbia University

and

Juri Toomre

Joint Institute for Laboratory Astrophysics
and Department of Astro-Geophysics
University of Colorado

1. INTRODUCTION

Having chosen the number of mesh points for a representation by finite differences of a differential system, one still has the freedom to decide on their distribution. When the characteristic scale of variation of the solutions does not vary much with position it is common practice to choose a uniform mesh. However when boundary layers are present it is expedient to distribute the points differently, and compute on a stretched mesh. Subject to difficulties in computation, it is presumably best to aim at choosing a distribution of mesh points that minimizes the truncation error introduced by differencing. This paper describes a preliminary attempt to effect such a choice.

First we outline the general principles behind the stretching procedure and then we derive two explicit recipes. Finally we present a sample numerical solution of equations used by Roberts (1966) and Gough, Spiegel & Toomre (1974) to model thermal convection at high Rayleigh number, a problem in which various boundary layers are encountered.

2. GENERAL STRETCHING PROCEDURE

To fix ideas we restrict attention here to two-point boundary-value problems posed in [a,b]. The differential system can be written:

$$\frac{dy_i}{dx} = f_i(y_j, x) \qquad , \qquad i,j = 1,2,\dots I \tag{1}$$

$$B_k[y_i(a), y_j(b)] = 0 \qquad , \tag{2}$$

which is to be represented by finite differences. We now introduce a new independent variable ξ defined in [0,1] with respect to which the mesh points ξ_n (n=1,...N) are evenly spaced, and seek a monotonic nonsingular transformation $\xi(x)$ which in some sense minimizes the errors introduced by differencing. In terms of this variable, equation (1) becomes

$$\frac{dy_i}{d\xi} = \frac{dx}{d\xi} \quad f_i[y_j, x(\xi)] \qquad . \tag{3}$$

The next step is to decide upon a measure of the error. No choice is obvious, though measures come to mind such as the Euclidean norm

$$E_E = \left\{ \sum_{i=1}^{I} r_i^{-2} \sum_{n=1}^{N} \left| Y_{in} - y_i(\xi_n) \right|^2 \right\}^{1/2}$$

where Y_{in} is the solution of the difference equations representing (3) and r_i is the oscillation of y_i in [0,1]. A priori estimates of E_E generally involve sums over i and n of products of derivatives of y_i of order up to $p + 1$, where p is the order of accuracy of the difference representation. We do not attempt to make precise estimates for E_E here but, guided by the above remarks, consider instead the measures

$$E_m \equiv \int_0^1 \left[\sum_{i=1}^{I} \left(r_i^{-1} \frac{d^m y_i}{d\xi^m} \right)^2 + \left(\frac{\omega}{b-a} \frac{d^m x}{d\xi^m} \right)^2 \right] d\xi \quad , \tag{4}$$

where ω is a constant weight factor and we are assuming y_i to be real. Note that in the transformed equations (3), x appears as a dependent variable which will eventually be determined by minimizing E_m. It is therefore treated like the other dependent variables and included in E_m. Otherwise a boundary layer in y_i might simply be transformed into a boundary layer in $x(\xi)$ and no improvement would have been gained. [Indeed, since $dx/d\xi$ appears in every transformed differential equation, we usually weighted it more strongly than any particular y_i by choosing the value of ω to be I. Our case for doing this was strengthened, for us, by the fact that the y_i and x then appeared similarly smooth with respect to ξ, whereas with $\omega = 1$ there were places where $dx/d\xi$ changed more abruptly than the other functions.]

All the computations we have performed so far used centered second-order accuracy differences to represent the derivatives with respect to ξ. Though the most natural error estimate of the type (4) for such differencing is E_3, we have tried only E_1 and E_2, which are considerably simpler to use.

3. FIRST-DERIVATIVE STRETCHING

Here we derive the monotonic stretching transformation between x and ξ that minimizes E_1. Since the solutions $y_i[x(\xi)]$ of the differential system (3) and (2) are independent of the form of $x(\xi)$, it is convenient to derive the minimizing transformation with x as the independent variable. Thus we minimize

$$E_1 = \int_a^b \left(\frac{d\xi}{dx} \right)^{-1} \sum_{i=1}^{I+1} r_i^{-2} f_i^2 dx \quad , \tag{5}$$

where $f_{I+1} = 1$ and $r_{I+1} = (b-a)/\omega$, amongst all monotonic functions $\xi(x)$ satisfying

$$\int_a^b \frac{d\xi}{dx} dx = 1 \quad . \tag{6}$$

A constraint like (6) is clearly necessary, for otherwise the smallest E_1 is zero with $\xi(x)$ unbounded; the choice of the unit range for ξ is arbitrary.

Since it is not straightforward to impose the monotonic constraint at the outset we proceed simply by enlarging the class of admissible functions $\xi(x)$ to all those satisfying (6). Then E_1 is stationary, and is in fact a strong minimum, when

$$\frac{dx}{d\xi} = \lambda^{1/2} \left(\sum_{i=1}^{I+1} r_i^{-2} f_i^{-2} \right)^{-1/2} \tag{7}$$

with λ constant, which yields a monotonic transformation. The Lagrange multiplier λ is determined by the relation between the ranges of the variables x and

ξ. Regarding now ξ as the independent variable, we use

$$\int_0^1 \frac{dx}{d\xi}\, d\xi = b - a \qquad (8)$$

in place of (6) and obtain

$$\lambda^{1/2} = (b-a)\left[\int_0^1 \left(\sum_{i=1}^{I+1} r_i^{-2} f_i^2\right)^{-1/2} d\xi\right]^{-1} \quad . \qquad (9)$$

Equations (3), (7) and (9) subject to the boundary conditions (2) and $x = a$ at $\xi = 0$ (or $x = b$ at $\xi = 1$) provide the basis for constructing the optimized difference equations to represent the original problem.

4. SECOND-DERIVATIVE STRETCHING

As with first derivative stretching we minimize E_2 among all functions $\xi(x)$ satisfying (6). The Euler equation, with ξ as independent variable, is

$$\sum_{i=1}^{I+1} r_i^{-2}[f_i^2(2\phi\phi_{\xi\xi} - \phi_\xi^2) + 4f_i g_i \phi^2 \phi_\xi + (2f_i h_i - g_i^2)\phi^4] - \lambda = 0 \qquad (10)$$

where

$$\phi = x_\xi \quad , \qquad (11)$$

$g_i = df_i/dx$, $h_i = dg_i/dx$ and once again λ is the (constant) Lagrange multiplier. The subscript ξ denotes differentiation with respect to ξ. This is a second order differential equation for $\phi(\xi)$ with an eigenvalue λ. Its solution determines $x(\xi)$, which can be shown to be monotonic, through equation (11) subject to the boundary condition

$$x(0) = a \quad \text{or} \quad x(1) = b \quad . \qquad (12)$$

The boundary conditions which ϕ must satisfy are those that minimize E_2: that is, they are the natural boundary conditions of the variational problem determined by the requirement that the integrated terms in the first variation of E_2 vanish. They are

$$\sum_{i=1}^{I+1} r_i^{-2} f_i (f_i \phi_\xi + g_i \phi^2) = 0 \quad \text{at} \quad \xi = 0,1 \quad . \qquad (13)$$

Equations (3), (10) and (11) subject to the boundary conditions (2), (12) and (13) and the integral constraint (8) are now the equations from which the optimized difference representation is constructed. We have not proved that the stationary values of E_2 are necessarily minima, except for a particular simple class of equations (1). However, in practice we never found the value of E_2 computed with stretching functions different from the extremal solutions to be lower than the stationary value.

5. COMPUTATIONAL METHOD

Centered second-order accuracy difference equations were used to represent the differential system, including any additional differential equations resulting from the stretching procedures. When second-derivative stretching was used, equation (10) was written conveniently as a pair of first order equations in the same form as (3). The nonlinear difference equations were solved by Newton-Raphson iteration (cf. Henrici, 1962) which involves inverting a block bidiagonal matrix with blocks of order JxJ, where J is the order of the total system.

The simplest way to organize the numerical program is to solve all the equations together as a single system, and this we did when we originally tested the procedure. But since the time taken to invert a matrix is proportional to the cube of its order, it is faster to split the system into two. First the difference equations for the basic equations (3) were solved, using an estimated $x(\xi)$. Then the stretching equations were solved, and the basic system was solved again on the new mesh. This method was found to be not only faster but also more stable, especially when the boundary layers were very thin.

The equations for thermal convection which we use for illustration are displayed in the next section. They contain two dimensionless physical parameters, the Rayleigh number R and the Prandtl number σ which specify the physical system they model. The solutions exhibit boundary layers when R is large or σ is small. Solutions for extreme values of the parameters were obtained by computing a sequence of solutions with either R or σ varying. Trial solutions for the Newton-Raphson iterations were constructed by linear projection from two neighboring solutions. Steps of half a decade in R could successfully be taken up to values of about 10^{20} using 600 mesh points. With such steps, the solution on the trial mesh was quite adequate for subsequent use in the mesh stretching equations. Adding a second iteration of mesh stretching followed by solving the basic equations on the new mesh had only a very slight effect on the mesh and made no perceptible improvement to the solution.

6. AN EXAMPLE: STRETCHING FOR THERMAL CONVECTION

The differential system mentioned above which models thermal convection in a Boussinesq fluid between two rigid horizontal isothermal planes can be written (Roberts, 1966; Gough, Spiegel & Toomre, 1974)

$$
\begin{aligned}
W(\partial_x^{\,2} - a^2)^2 W &= Ra^2 W\Theta + \frac{C}{\sigma}\,\partial_x[W^2(\partial_x^{\,2} - a^2)W] \\
\Theta(\partial_x^{\,2} - a^2)\Theta &= -(N - W\Theta)W\Theta + C\,\partial_x(W\Theta^2)
\end{aligned}
\qquad (14)
$$

where W is the vertical component of velocity and Θ is the temperature fluctuation; ∂_x represents differentiation with respect to the independent variable x. The constants a and C are geometrical parameters describing an assumed horizontal structure of the flow. These, together with R and σ, must all be specified. The constant N is the Nusselt number and is an eigenvalue of the problem. Equations (14) are solved subject to the boundary conditions

$$W = \partial_x W = \Theta = 0 \qquad \text{at} \qquad x = 0,1 \qquad (15)$$

and the integral constraint

$$\int_0^1 W\Theta dx = N - 1 \quad . \qquad (16)$$

In Figure 1 below we show a solution, plotted with respect to the stretched variable ξ, for $R = 10^6$, $\sigma = 0.025$, $a = 2$ and $C = 0.408$. It was computed with 301 mesh points using first derivative stretching. Shown also is the original independent variable $x(\xi)$; in order that its shape can easily be seen a solution requiring only mild stretching has been selected. The Nusselt number is 5.27; its reciprocal approximates the thickness of the thermal boundary layer near $\xi=0$. Near $\xi = 1$ are two nested boundary layers: a thermal boundary layer slightly thinner than the boundary layer near $\xi = 0$ and a thinner viscous layer which is evident in the vorticity field (not shown). The same solution W and Θ is plotted in Figure 2, this time against the original independent variable x.

As R is increased and σ is decreased the stretching becomes more extreme. For example when $R = 10^{12}$ and $\sigma = 10^{-5}$, 30 % of the mesh points lie within a viscous boundary layer near $x = 1$ of thickness 10^{-4}. Other solutions of the system (14)-(16) exist which have internal boundary layers in addition to those at the extremes of x, and these are also well resolved by the above procedures.

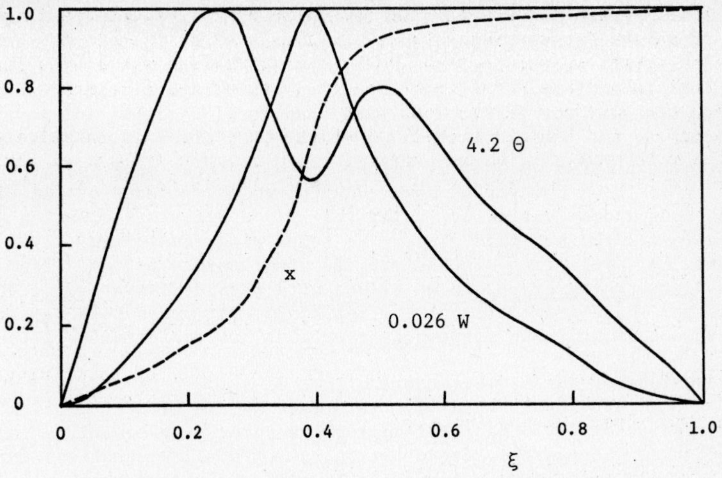

Figure 1. Convection solution on a stretched mesh.

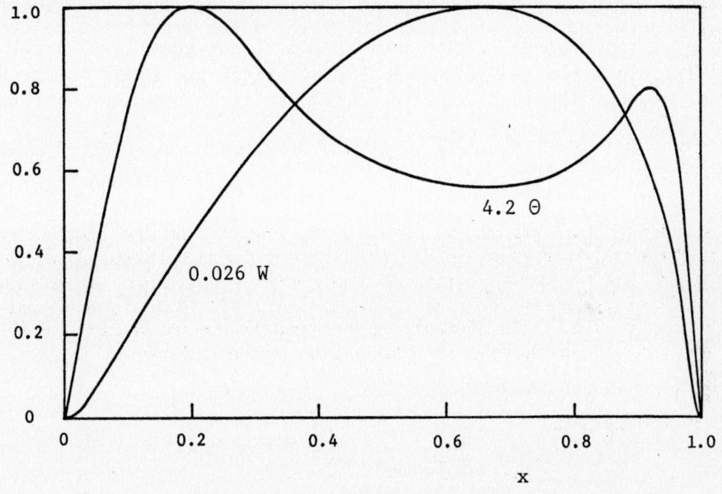

Figure 2. The same solution as in Figure 1, but plotted against its original independent variable x.

7. DISCUSSION

The mesh stretching procedures outlined in this paper have been implemented to utilize efficiently the grid in the numerical solution by differences of nonlinear ordinary differential equations with two-point boundary conditions. In the past, explicitly defined analytic transformations $x(\xi)$ have been used to advantage by Roberts (1970) and ourselves, for example, but these rely on some prior knowledge of the structure of the solutions. Furthermore, it is quite complicated to implement them when one has to deal with nested boundary layers of very different thicknesses and internal boundary layers whose positions are not known in advance. The implicit techniques described here can be automated, and the complexity of the program is independent of the number and structure of the boundary layers. However, one must pay the price of solving additional differential equations. But when the order of the basic system (1) is high, that price is relatively low.

The error estimates we have minimized are not direct measures of the truncation error as is E_E. However we have constructed an estimate of E_E [with $x(\xi)$ included as a dependent variable] using for y_i a numerical solution computed with many more mesh points than Y_{in}. We found with the system (14)-(16) that the E_1 and E_2 minimizing schemes yielded lower values for E_E than any explicitly defined stretching procedure we tried, with second-derivative stretching being slightly better than first-derivative stretching. Of course this may reflect our inability to choose suitable stretching functions, or the unsuitability of the measure E_E.

The advantage of the higher accuracy that we suspect is achieved by the minimization of E_2 rather than E_1 is offset by the greater difficulty in putting it into practice. The first-derivative stretching equations can be integrated by quadrature whereas the second-derivative stretching equations must be solved by iteration which in extreme cases fails to converge unless the trial is very good.

Error estimate minimization can also be achieved in several dimensions. The obvious generalization of the procedure outlined here for use with elliptic equations leads to a mesh which is not rectangular. This gives rise to new problems in differencing which we have not faced. We have, however, stretched rectangular meshes to some advantage in two dimensions (x,y) using y-averaged solutions of the basic differential system for x stretching and x-averages for y stretching. Such stretching preserves the rectangular property of the mesh, but is useful only when all the boundary layers lie along coordinate lines and have uniform thickness.

Much of this work was done while D.O.G. and J.T. held NAS-NRC Resident Research Associateships at the Goddard Institute for Space Studies, New York and were visiting members of the Courant Institute of Mathematical Sciences, New York University. Support from the National Science Foundation under grants NSF GP-32336X and GA-43007 is acknowledged.

REFERENCES

Gough, D.O., Spiegel, E.A. & Toomre, J. 1974 Modal equations for cellular convection. J. Fluid Mech. (in press).

Henrici, P. 1962 Discrete Variable Methods in Ordinary Differential Equations, p. 366, Wiley.

Roberts, G.O. 1970 Computational meshes for boundary layer problems. Proc. 2nd Intern. Conf. Numer. Meth. Fluid Dyn. (Holt, M., ed.),171-8, Springer.

Roberts, P.H. 1966 On non-linear Bénard convection. Non-Equilibrium Thermodynamics, Variational Techniques, and Stability (Donnelly, R., Hermann, R. and Prigogine, I., eds.), 125-62, Univ. Chicago Press.

A NUMERICAL STUDY OF THE THERMAL SPIN-UP OF A STRATIFIED FLUID IN A RAPIDLY ROTATING CYLINDER

Iwao Harada and Norihiko Ozaki

Atomic Energy Research Laboratory, Hitachi Ltd.,
Ozenji, Kawasaki, Kanagawa, Japan

ABSTRACT

We have studied the thermal spin-up (heat-up) process of a Boussinesq fluid in a rapidly rotating cylinder for the case in which the sidewall is conducting or insulating. We use the finite difference method to solve the Navier-Stokes equations. The time variations of the meridional circulation, the zonal velocity (thermal wind in the meteorology), and of the total kinetic energy are discussed in detail. The effects of the Ekman layer, the Stewartson's $E^{1/3}$ layer and of the inner circulation are clarified.

I. INTRODUCTION

If a rotating container is heated from the surface, the thermal convection is produced by the temperature gradient on the surface. The transient process (heat-up) has been investigated by Veronis (1967). The problem is very similar to the spin-up of a stratified fluid studied by Holton (1965), Walin (1969), and Sakurai (1969). They neglected the centrifugal force in comparison with gravity. On the other hand, Barcilon and Pedlosky (1967) and Homsy and Hudson (1969) have investigated the centrifugally driven thermal convection in a rotating cylinder at the steady state.

The object of this paper is to study the thermal spin-up of a stratified fluid, where the Boussinesq approximation is applicable, in a rapidly rotating cylinder. We assume that the centrifugal force is much greater than gravity. The finite difference approximations employed by Williams (1967), who calculated the thermal convection in a slowly rotating annulus, are used to solve the time-dependent Navier-Stokes equations to the axisymmetric flow as an initial value problem.

II. BASIC EQUATIONS

Let us consider a fluid contained in a rotating cylinder with a constant angular velocity Ω, and of radius L and height H. The fluid is thermally driven away from a state of solid body rotation by an imposed vertical temperature gradient Δ T; the top disk is heated impulsively and the bottom disk is retained at the initial temperature. The sidewall is thermally conducting (case I) and that is insulating (case II).

The nondimensional basic equations in the rotating cylindrical co-ordinates (r, φ, z) shown in Figure 1, are as follows;

$$\frac{\partial \zeta}{\partial t} + \epsilon J(\zeta/r) = -r\frac{\partial \theta}{\partial z} - 2\frac{\partial v}{\partial z}\left(1+\epsilon\frac{v}{r}\right) + E\mathcal{L}\zeta \tag{1}$$

$$\frac{\partial v}{\partial t} + \frac{\epsilon}{r}J(v) = \frac{1}{r}\frac{\partial \psi}{\partial r}\left(2+\epsilon\frac{v}{r}\right) + E\mathcal{L}v \tag{2}$$

$$\frac{\partial \theta}{\partial t} + \frac{\epsilon}{r}J(\theta) = E/\sigma_P \; \nabla^2\theta \tag{3}$$

$$-\zeta = \frac{\partial}{\partial r}\frac{1}{r}\frac{\partial \psi}{\partial r} + \frac{1}{r}\frac{\partial^2 \psi}{\partial r^2} \tag{4}$$

$$\nabla^2 = \frac{1}{r}\frac{\partial}{\partial r}r\frac{\partial}{\partial r} + \frac{\partial^2}{\partial z^2} \qquad \mathcal{L} = \nabla^2 - \frac{1}{r^2} \tag{5}$$

The nondimensional numbers in the above equations are

$$E = (\mu/\rho)/\Omega L^2 \qquad \text{(Ekman)}$$

$$\epsilon = \alpha \Delta T \qquad \text{(Thermal Rossby)}$$

$$\sigma_P = \mu C_p / \kappa \qquad \text{(Prandtl)}$$

where $J(\ ) = \frac{\partial \psi}{\partial r}\frac{\partial (\ )}{\partial z} - \frac{\partial \psi}{\partial z}\frac{\partial (\ )}{\partial r}$ represents the convective term by a Jacobian form, ζ is the vorticity, θ the temperature, ψ the stream function of the meridional circulation, u the radial velocity, v the zonal velocity, w the axial velocity, t the time. Here, κ is the thermal diffusivity, ρ the density, μ the viscosity, Cp the specific heat at constant pressure, α the coefficient of thermal expansion. In the non-dimensionalization of quantities, we use the characteristic length L, time Ω^{-1} and velocity $\epsilon \Omega L$.

Boundary conditions for equations (1) ~ (4) are as follows;

$$\psi = \frac{\partial \psi}{\partial n} = v = 0 \quad \text{at} \quad \begin{cases} z = (\frac{1}{2} \pm \frac{1}{2})\Lambda & (0 \le r \le 1) \\ r = 1 & (0 \le z \le \Lambda) \end{cases}$$

$$\theta = (\tfrac{1}{2} \pm \tfrac{1}{2})\Lambda \quad \text{at} \quad z = (\tfrac{1}{2} \pm \tfrac{1}{2})\Lambda \quad (0 \le r \le 1)$$

$$\left.\begin{array}{ll} \theta = z & \text{for case (I)} \\ \frac{\partial \theta}{\partial r} = 0 & \text{for case (II)} \end{array}\right\} \quad \text{at} \quad r = 1 \quad (0 < z < \Lambda)$$

where $\partial \psi / \partial n$ is the normal derivative and $\Lambda = H/L$ is an aspect ratio. The initial conditions are

$$\psi = v = \zeta = \theta = 0 \qquad (0 \le r \le 1,\ 0 \le z \le \Lambda)$$

The important integral quantity is the kinetic energy. We define the total kinetic energy E_T as the sum of the meridional kinetic energy E_S and the zonal kinetic energy E_v, viz.,

$$E_T = \frac{1}{2}\langle v^2 + \psi \frac{\zeta}{r} \rangle, \quad E_S = \frac{1}{2}\langle \psi \frac{\zeta}{r} \rangle, \quad E_v = \frac{1}{2}\langle v^2 \rangle$$

where $\langle\ \rangle$ denotes the volume integral over the r, z cross section. These quantities represent the amount of heat energy which is transformed to the kinetic energy.

III. FINITE DIFFERENCE EQUATIONS

In the derivation of the finite difference equations, we use the difference and averaging operators in the notation of Shuman (1962)

$$\delta_x \phi = [\phi(x + \tfrac{1}{2}\Delta x) - \phi(x - \tfrac{1}{2}\Delta x)]/\Delta x, \quad \bar{\phi}^x = [\phi(x + \tfrac{1}{2}\Delta x) + \phi(x - \tfrac{1}{2}\Delta x)]/2$$

where ϕ represents one of the field variables, and Δx is the grid interval of one of the coordinates x. Finite difference equations corresponding to (1) ~ (4) are as follows;

$$\delta_t \bar{\zeta}^t + \epsilon J_A(\zeta / r) = -\delta_z r \bar{\theta}^{zrr} + 2\delta_z \bar{v}^z + \epsilon \frac{1}{r}\delta_z(\overline{v^2})^z + E\mathcal{L}(\zeta) \tag{6}$$

$$\delta_t \bar{v}^t + \epsilon/r\, J_A(v) = \frac{1}{r}\delta_z \bar{\psi}^z (2 + \epsilon \frac{v}{r}) + E\mathcal{L}(v) \tag{7}$$

$$\delta_t \bar{\theta}^t + \epsilon/r\, J_A(\theta) = E/\sigma_P [\frac{1}{r}\delta_r(r\delta_r \theta) + \delta_{zz}\theta] \tag{8}$$

$$-\zeta = \delta_r(\frac{1}{r}\delta_r \psi) + \frac{1}{r}\delta_{zz}\psi \tag{9}$$

where

$$\mathcal{L}(\phi) = [\delta_r \frac{1}{r}\delta_r(r\phi) + \delta_{zz}\phi]/ag \tag{10}$$

$$\left.\begin{aligned} J_2(\phi) &= \delta_z(\overline{\phi}^z \delta_r \overline{\psi}^{rz}) - \delta_r(\overline{\phi}^r \delta_z \overline{\psi}^{rz}) \\ J_4(\phi) &= \delta_r(\overline{\psi \delta_z \overline{\phi}^z}^r) - \delta_z(\overline{\psi \delta_r \overline{\phi}^r})^z \\ J_A(\phi) &= \frac{1}{3} J_2(\phi) + \frac{2}{3} J_4(\phi) \end{aligned}\right\} \quad (11)$$

and $J_A(\phi)$ is the scheme given by Arakawa (1966).

Finite difference equations (6) ~ (9) are the same as those given by Williams (1967) except the centrifugally driven buoyancy term on the right-hand side of the vorticity equation (6). In these equations, we use the central time differencing to avoid numerical instability. The convective, buoyancy, Coriolis terms are evaluated at the central time $n\Delta t$, and the diffusion terms with the subscript 'lag' are evaluated at non-central time $(n-1)\Delta t$.

IV. NUMERICAL RESULTS AND DISCUSSION

Two cases are computed corresponding to the set of values shown in TABLE I. In each case we use an aspect ratio of 1.0, and the time mesh $\Delta t = 6.5 \times 10^{-3}$ based on the limit of the numerical stability condition $\Delta t < \min.(h^2/8E, \sigma_P h^2/8E)$, where $h = \min.(\Delta r, \Delta z)$. The Poisson equation (9) is solved by the method of successive over relaxation (SOR) with the relaxation factor $\omega = 1.80$ at each time step.

TABLE I

The set of values used for computations

Case	Surface condition (disks, sidewall)	Number of grid interval in r, z	E	ϵ	σ_P
I	conducting, conducting	40 x 40	10^{-2}	5×10^{-2}	1.0
II	conducting, insulating	40 x 40	10^{-2}	5×10^{-2}	1.0

Figure 2 shows the time variations of the kinetic energies for the case (I). There are four time scales predicted by the theoretical analysis. In these computations, the time scales for the Ekman layer, the Stewartson's $E^{-1/3}$ layer, spin-up and for diffusion are 1, 5, 10 and 100, respectively. For the sake of comparison, we mark the time scales of order unity and of $E^{-1/3}$. We can see that the increase of the total kinetic energy is born out by the meridional motion until $t \sim 1$. Thereafter, the zonal motion takes over and the zonal and the total kinetic energies increase until $t \sim 3.0$. Three nondimensional times 1.3, 2.6 and 3.9 (3.8) are chosen to show how the Ekman, transient Stewartson and Stewartson layers are formed.

Figures 3(a)-(d) show contur plots of streamlines ψ for the meridional circulation field at $t = 1.3$, 2.6 and 3.9 (3.8).

Figure 3(a) shows that a strong circulation generated on the sidewall at initial stages develops into a direct circulating flow from the top of the hot disk, across the interior, to downward the bottom of the cold disk. The Ekman layers on the top and bottom disks act as a source and a sink, respectively. The strength of the source is greater than that of the sink according to the temperature gradients. This aspect is seen by the denser streamlines in the upper half region of the flow field. The zonal component of the Coriolis force which is caused by the radial component of the velocity balances $\frac{\partial v}{\partial t}$ in the interior region. The axial velocity of the initial Stewartson layer (of thickness $E^{1/3}$) is large corresponding to the strong Ekman suction.

Figure 3(b) shows the transient Stewartson layer at $t = 2.6$. The strength of the Ekman suction becomes weak in comparison with that at $t = 1.3$ because of the increased average temperature. The axial velocity on the sidewall decreases according to the weakened Ekman suction. The streamlines are pushed to the sidewall, and tend to become parallel to the axis of rotation.

When the thermal diffusion layer develops, the zonal velocity is produced by the thermal wind relation. The time derivative of v does not balance the zonal component of the Coriolis force but begins to balance that of the viscous shear.

Figure 3(c) shows the streamlines at $t = 3.9$. The transient Stewartson layer fully develops, and the viscous force balances the term of $\frac{\partial v}{\partial t}$. The relic of the initial impulsive circulation disappears from the interior region. The secondary flow induced by the main circulation agrees qualititatively with the theoretical analysis (Homsy and Hudson, 1969).

The streamlines at $t = 3.8$ in the case (II) shown in Figure 3(d) corresponds to the formation time of the Stewartson layer. In this case, the circulation at the initial stages is mainly controlled by the Ekman layer rather than by the sidewall layer as in the case (I). The stream lines of the meridional velocity are very similar, excepting the fact that the absolute value is small, to those in the case (II). The main circulation includes two small circulations in the upper and lower regions. This is because the heat energy is not supplied from the insulating sidewall but only from the top disk. The temperature in the Stewartson layer is, therefore, governed by the axially one dimensional heat equation.

Figure 4(a) shows contur plots of the zonal velocity at $t = 1.3$ in the case (I). We can see the thermal wind effect which induces zonal velocity by comparing Figure 4 (a) with Figure 3(a). The thermal wind, initially induced on the sidewall, grows up in the sidewall layer at $t = 3.9$ and becomes symmetric with respect to $z = \Lambda/2$ [Figure 4(b)]. This is due to the fact that the angular momentum at the upper region is competitive to that at the lower one as the temperature gradient on the sidewall is nearly uniform at this stage.

Figure 5 shows the comparison of analytic results by boundary layer approximation (Homsy and Hudson, 1969) and numerical results at $t = 3.9$ in the case (I). The numerical result is different from the analytic result for the zonal velocity [Figure 5(a)]. One reason is that the zonal velocity requires diffusion time scale E^{-1} to reach the steady state. The axial velocities of both results [Figure 5(b)] show a good agreement.

V. REFERENCES

Arakawa, A. J. Comput. Phys. 1, 119 (1966)

Barcilon, V., and Pedlosky, J. J. Fluid Mech. 29, 1 (1967)

Holton, J. R. J. Atoms. Soc. 22, 402 (1965)

Homsy, G, M., and Hudson, J.L. J. Fluid Mech. 35, 33 (1969)

Sakurai, T. J. Fluid Mech. 37, 689 (1969)

Shuman, F.G. Proc. Int. Symp. on Numerical Weather Prediction. Tokyo, 1960 Meteor. Soc. of Japan (1962)

Stewartson, K. J. Fluid Mech. 3, 17 (1957)

Veronis, G. Tellus 19, 326 (1967)

Veronis, G. Tellus 19, 620 (1967)

Walin, G. J. Fluid Mech. 36, 289 (1969)

Williams, G.P. J. Atoms. Soc. 24, 144 (1967)

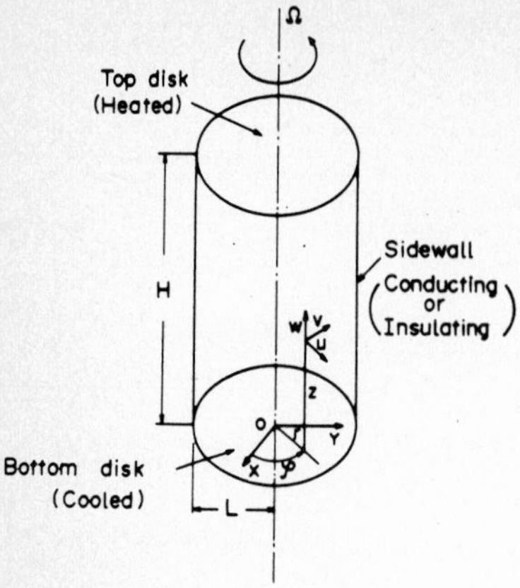

A model for computations

FIG.1

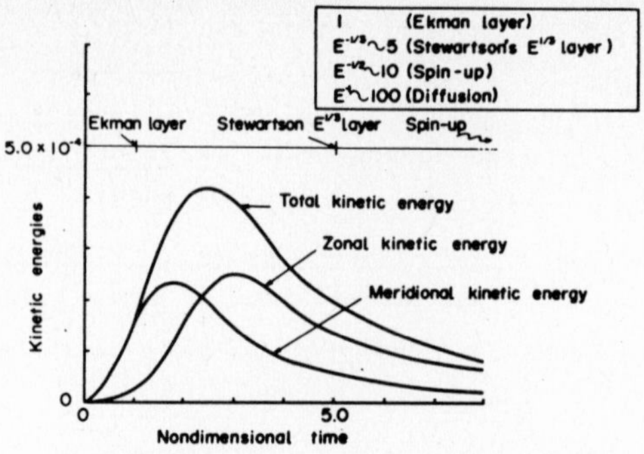

Kinetic energies during the spin-up (conducting sidewall)
($E = 10^{-2}$, $\epsilon = 5\times10^{-2}$, $\sigma_0 = 1.0$)

FIG.2

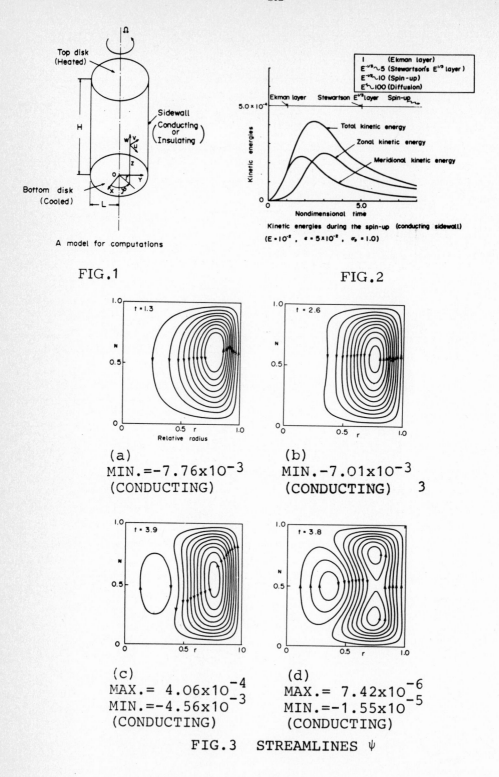

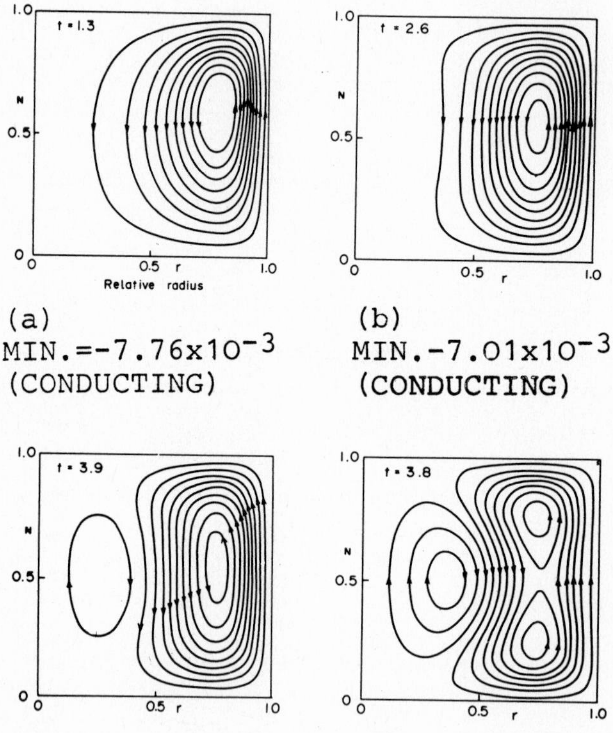

(a)
MIN.=-7.76×10^{-3}
(CONDUCTING)

(b)
MIN.-7.01×10^{-3}
(CONDUCTING)

(c)
MAX.= 4.06×10^{-4}
MIN.=-4.56×10^{-3}
(CONDUCTING)

(d)
MAX.= 7.42×10^{-6}
MIN.=-1.55×10^{-5}
(CONDUCTING)

FIG.3 STREAMLINES ψ

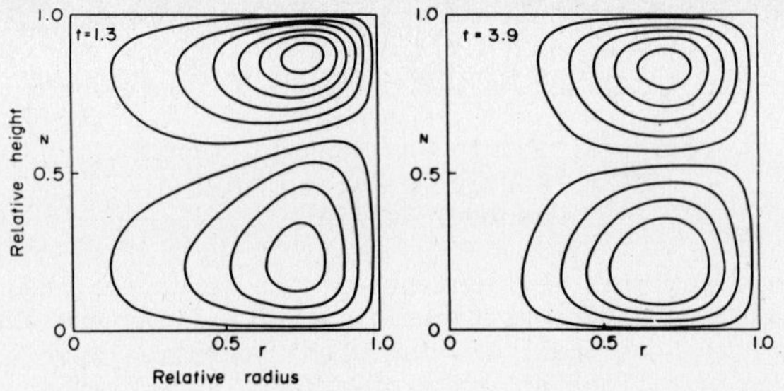

(a) $v_{max.} = 3.83 \times 10^{-2}$, $v_{min.} = -2.52 \times 10^{-2}$ (b) $v_{max.} = 6.29 \times 10^{-2}$, $v_{min.} = -5.99 \times 10^{-2}$

Zonal velocity (Conducting sidewall)

Fig.4

(upper;positive,lower;negative)

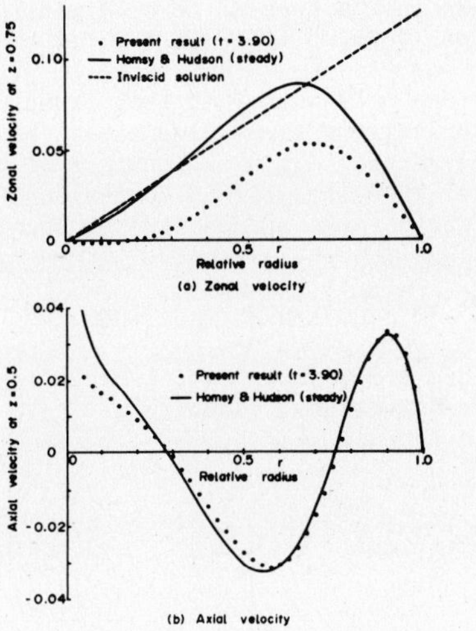

A comparison of the analytic solutions and the numerical results

FIG.5

SOLUTION OF THE THREE-DIMENSIONAL COMPRESSIBLE, LAMINAR, AND TURBULENT BOUNDARY-LAYER EQUATIONS WITH COMPARISONS TO EXPERIMENTAL DATA

Julius E. Harris and Dana J. Morris
NASA Langley Research Center
Hampton, Virginia

Current design and analysis requirements of the aerospace industry necessitate development of accurate and efficient techniques for solving the three-dimensional compressible, laminar, transitional, and turbulent boundary-layer equations for flows over general configurations at angle-of-attack. The further development of large-storage, high-speed digital computer systems has allowed application of algorithms for solving the complete system of nonlinear partial differential equations governing two- and three-dimensional boundary layer flow. Numerical experimentation and detailed experimental turbulent boundary-layer research has resulted in the calibration/verification of mean field turbulence models (eddy viscosity/mixing length) for two-dimensional boundary-layer flows which are sufficiently accurate for application over a wide range of flow and boundary conditions (see Refs. 1 to 3). The numerical techniques developed for two-dimensional flows can be directly applied, with minor modifications, to general three-dimensional flows; however, questions concerning the extension of turbulence models developed from two-dimensional boundary-layer data to three-dimensional flows still remain to be answered (Refs. 4 to 6). The goal of developing a general three-dimensional solution technique is further complicated by three numerical problem areas: selection of (1) coordinate system and (2) variable system, plus the problem of (3) inviscid flow field data (Ref. 7). For general configurations the problems associated with obtaining accurate inviscid flow field data is by far the more difficult of the three problem areas; however, substantial progress has been made in the area of three-dimensional boundary layer flows over the past few years (see Refs. 8 to 12).

In the present paper one of a number of techniques under development at NASA Langley Research Center for solving three-dimensional boundary-layer flows is presented together with advantages and disadvantages of the procedure. A simple mean field turbulence model is discussed and comparisons are shown for a test case where numerical results, for a range of nodal-point spacings, are compared with experimental data.

GOVERNING EQUATIONS

The governing equations are written as follows (see Fig. 1 for coordinate system; overbar designates dimensional quantity):

Continuity

$$\frac{\partial}{\partial\xi}(\bar{h}_2\bar{h}_3\bar{\rho}\bar{u}) + \frac{\partial}{\partial\eta}(\bar{h}_1\bar{h}_3\bar{\rho}\bar{v}) + \frac{\partial}{\partial\zeta}(\bar{h}_1\bar{h}_2\bar{\rho}\bar{w}) = 0 \tag{1}$$

ξ - momentum

$$\frac{\bar{u}}{\bar{h}_1}\frac{\partial\bar{u}}{\partial\xi} + \frac{\bar{v}}{\bar{h}_2}\frac{\partial\bar{u}}{\partial\eta} + \frac{\bar{w}}{\bar{h}_3}\frac{\partial\bar{u}}{\partial\zeta} + \frac{\bar{u}\bar{v}}{\bar{h}_1\bar{h}_2}\frac{\partial\bar{h}_1}{\partial\eta} - \frac{\bar{v}^2}{\bar{h}_1\bar{h}_2}\frac{\partial\bar{h}_2}{\partial\xi} = -\frac{1}{\bar{\rho}\bar{h}_1}\frac{\partial\bar{p}}{\partial\xi} + \frac{1}{\bar{\rho}\bar{h}_3}\frac{\partial}{\partial\zeta}\left(\frac{\bar{\mu}_{eff}}{\bar{h}_3}\frac{\partial\bar{u}}{\partial\zeta}\right) \tag{2}$$

η - momentum

$$\frac{\bar{u}}{\bar{h}_1}\frac{\partial\bar{v}}{\partial\xi} + \frac{\bar{v}}{\bar{h}_2}\frac{\partial\bar{v}}{\partial\eta} + \frac{\bar{w}}{\bar{h}_3}\frac{\partial\bar{v}}{\partial\zeta} - \frac{\bar{u}^2}{\bar{h}_1\bar{h}_2}\frac{\partial\bar{h}_1}{\partial\eta} + \frac{\bar{u}\bar{v}}{\bar{h}_1\bar{h}_2}\frac{\partial\bar{h}_2}{\partial\xi} = -\frac{1}{\bar{\rho}\bar{h}_2}\frac{\partial\bar{p}}{\partial\eta} + \frac{1}{\bar{\rho}\bar{h}_3}\frac{\partial}{\partial\zeta}\left(\frac{\bar{\mu}_{eff}}{\bar{h}_3}\frac{\partial\bar{u}}{\partial\zeta}\right) \tag{3}$$

Energy

$$\frac{\bar{u}}{\bar{h}_1}\frac{\partial \bar{T}}{\partial \xi} + \frac{\bar{v}}{\bar{h}_2}\frac{\partial \bar{T}}{\partial \eta} + \frac{\bar{w}}{\bar{h}_3}\frac{\partial \bar{T}}{\partial \zeta} = \frac{1}{\bar{\rho}\bar{C}_p}\left\{\frac{\bar{u}}{\bar{h}_1}\frac{\partial \bar{p}}{\partial \xi} + \frac{\bar{v}}{\bar{h}_2}\frac{\partial \bar{p}}{\partial \eta} + \frac{\bar{\mu}_{eff}}{\bar{h}_3^2}\left[\left(\frac{\partial \bar{u}}{\partial \zeta}\right)^2 + \left(\frac{\partial \bar{v}}{\partial \zeta}\right)^2\right] + \frac{1}{\bar{h}_3}\frac{\partial}{\partial \zeta}\left(\frac{\bar{K}_{eff}}{\bar{h}_3}\frac{\partial \bar{T}}{\partial \zeta}\right)\right\} \qquad (4)$$

The system of equations is closed with the perfect gas equation of state and Sutherland's law for molecular viscosity.

The effective viscosity, $\bar{\mu}_{eff}$ and conductivity, $\bar{K}_{eff}$ are defined as follows:

$$\bar{\mu}_{eff} = \bar{\mu}\left(1 + \frac{\bar{\epsilon}}{\bar{\mu}}\Gamma\right) \qquad (5)$$

$$\bar{K}_{eff} = \frac{\bar{C}_p\bar{\mu}}{\sigma}\left(1 + \frac{\bar{\epsilon}}{\bar{\mu}}\frac{\sigma}{\sigma_t}\Gamma\right) \qquad (6)$$

Where $\bar{\mu}$, $\bar{\epsilon}$, σ, and σ_t represent the molecular viscosity, eddy viscosity, Prandtl number, and static turbulent Prandtl number, respectively. The streamwise intermittency, Γ (Ref. 3), models the flow in the transitional region and is a function of (ξ,η); $0 \leq \Gamma \leq 1$. In the present analysis the beginning and end of transition are empirically specified in (ξ,η) coordinates. The eddy viscosity model is assumed to be a continuous scalar function independent of the coordinate direction; that is (Refs. 5 and 13),

$$\bar{\epsilon} = \bar{\rho}\bar{M}_\ell^2\left\{\left(\frac{1}{\bar{h}_3}\right)^2\left[\left(\frac{\partial \bar{u}}{\partial \zeta}\right)^2 + \left(\frac{\partial \bar{v}}{\partial \zeta}\right)^2\right]\right\}^{1/2} \qquad (7)$$

where

$$\frac{\bar{M}_\ell}{\bar{x}_{3,e}} = K_2 \tanh\left[K_1/K_2\left(\frac{\bar{x}_3}{\bar{x}_{3,e}}\right)\right] D \qquad (8)$$

$$D = 1 - \exp\left(-\frac{\bar{x}_3}{\bar{A}}\right) \qquad (9)$$

$$\bar{A} = K_3\left(\frac{\bar{\mu}}{\bar{\rho}}\right)_w\left(\frac{\bar{\tau}}{\bar{\rho}}\right)_w^{-1/2} \qquad (10)$$

For the results presented in the present paper K_1, K_2, K_3, and σ_t were assigned values of 0.435, 0.09, 26.0, and 0.95, respectively. The shear stress, τ (Eq. (10)), is the total wall value, $(\bar{\tau}_\xi^2 + \bar{\tau}_\eta^2)^{1/2}$; $\bar{x}_3$ is the physical coordinate normal to the wall. It should be noted that the constants used in the present analysis, while valid over a wide range of flow conditions, must be modified for certain classes of flow and wall boundary conditions (see Ref. 7, for example); furthermore, for some classes of flow the invariant model is highly suspect.

TRANSFORMATION

Equations (1) to (10) are cast into nondimensional form (Ref 8, nondimensional variables) and a similarity-type transform is defined for the normal coordinate and velocity as follows:

$$\bar{h}_3 = \tilde{\xi}\,\frac{\rho_e}{\rho}\,\frac{\bar{L}}{\sqrt{R_e}}\,h_3 \tag{11}$$

$$\bar{w} = \frac{1}{\tilde{\xi}\sqrt{R_e}}\,\frac{\rho_e}{\rho}\,\bar{U}_\infty\,w \tag{12}$$

where for a sharp cone, $\tilde{\xi} = \sqrt{\xi}$. The scale factors h_2 and h_3 are defined such that the resulting equations will be of the Crocco-type; that is, $h_2 = h_2(\xi,\eta)$. We further assume

$$\zeta = (1 - F)^{1/2} \tag{13}$$

where $F=u/ue$.
The continuity and ξ - momentum equations are combined to form the shear equation where the shear parameter, ϕ, is defined as

$$\phi = -\frac{1}{h_3}\,\frac{\mu}{T}\,(1 + \frac{\epsilon\Gamma}{\mu}) \tag{14}$$

This results in F being replaced by ϕ as a new dependent variable and removes $H=w/u_e$ from the remaining equations. Consequently, the governing system of equations reduces to three coupled, nonlinear partial differential equations in $\theta = \frac{T}{T_e}$, $G = \frac{v}{v_e}$, and ϕ, and an explicit relationship for H. The primary advantage of the Crocco-type transformation is that the solution domain is bounded between definite limits, $0\leq\zeta\leq 1$. The one restriction placed on the system is that F must increase monotonically from zero at the body surface to the specified edge value and must not exceed 1. Edge vorticity and streamline-swallowing effects are not considered in the present paper.

The governing equations then assume the following form:

$$\frac{\partial^2\omega}{\partial\zeta^2} + \alpha_1\frac{\partial\omega}{\partial\zeta} + \alpha_2\omega + \alpha_3 + \alpha_4\frac{\partial\omega}{\partial\xi} + \alpha_5\frac{\partial\omega}{\partial\eta} = 0 \tag{15}$$

where ω represents θ, G, and ϕ. The main disadvantage of this procedure is that the form of the coefficients, α_i required for linearization is rather complex; however, the resulting difference equations are linear and of tridiagonal form.

The boundary conditions on Equation (15) are as follows:

$$\zeta = 0\begin{cases}\theta = 1\\ G = 1\,;\\ \phi = 0\end{cases}\quad \zeta = 1\begin{cases}\theta = \theta_w \quad\text{or}\quad \frac{\partial\theta}{\partial\zeta} = f(\xi,\eta)\\ G = 0\\ \left.\frac{\partial\phi}{\partial\zeta}\right|_w = -\left(a_1 H + a_2\frac{\bar{\ell}\theta}{\phi} + \phi\right)_w\end{cases} \tag{16}$$

where the a_i are functions of geometry and inviscid edge conditions and $\bar{\ell} = \frac{\mu}{T}(1 + \frac{\epsilon}{\mu}\Gamma)$. The wall boundary condition on ϕ presents somewhat of a problem since ϕ_w is unknown; however, the wall-derivative relationship can be directly incorporated into the iterative solution procedure.

SOLUTION TECHNIQUE

Equation (15) is solved in an iterative mode using a marching, implicit finite difference technique suggested by Dwyer (Ref. 14) and modified by Krause (Ref. 15). The method is unconditionally stable (conditional stability for reverse cross flow) and second order accurate for equally spaced nodal points. For turbulent flows at least two to three points must be located in the viscous sublayer; consequently, it is necessary to use a variable nodal-point distribution in the ζ -direction because of the chosen $\zeta = \zeta(F)$ relationship (see Eq. (13)). Current studies indicate that for turbulent flows it would be possible to determine an optimum functional relationship in place of Equation (13) such that equally spaced nodal points could be used in the ζ -direction. For the present study a geometric progression was assumed; that is, $\Delta\zeta_{n-1} = K\Delta\zeta_n$, n=2, 3,...NE-1 (Ref. 3). The difference quotients used in the present analysis as well as the treatment of the ϕ_w boundary conditions are the same as that in Reference 8 with the exception that a Crank-Nicolson differencing is used at the maximum η station (Ref. 11). In order to obtain the solution for $\xi,\eta\neq 0$ two initial orthogonal data planes must be specified; $\xi=0, 0\leq\eta\leq\eta_{max}$ and $0\leq\xi\leq\xi_{max}$, $\eta=0$. These data planes are generated in the present procedure in much the same way as in Ref. 10.

RESULTS AND DISCUSSION

The numerical procedure and turbulence model have been applied to a number of test cases; however, for the present paper only one test case will be presented. The body is a 12.5° semiapex angle, sharp, right-circular cone at 15.75° angle of attack. The free stream Mach number, total pressure, and total temperature were 1.8, 1.724×10^5 N/m^2. and 2.94×10^2 °K, respectively. Transition was assumed to be initiated and completed in the region $.03 \leq \bar{x}_1/\bar{L} \leq .08$ ($\bar{L}$ = 105.6 cm). The wall temperature was near adiabatic. No experimental data were input into the viscous flow solution. The inviscid pressure distribution $p_e = p_e(\xi,\eta)$ was obtained from a numerical solution of the inviscid flow equations. Experimental profile data were available for $\frac{\bar{x}_1}{\bar{L}} = .85$ (Ref. 16).

The numerical results for F, G, and ϕ are compared with experimental data for circumferential locations of 0°, 45°, 90°, and 135° in Figure 2. Solutions were obtained for N (number of nodal points in ζ -direction) = 301, 201, 101, 61, and 21 for K=1.02 in order to determine the effect of nodal spacing on the resultant solution. For all essential purposes the numerical results for N=301 and 201 were identical; those for N=101 were within one-half of one percent of the converged values. As can be seen from Figure 2 the agreement in very good for 301 points and in general good for 21 points. A comparison of the numerical results for $C_{f,\infty}$, $\bar{\tau}/(1/2\rho_\infty \bar{U}_\infty^2)$, is presented in Figure 3. In the turbulent region of flow the percent difference between N=61 and N=21 in relation to N=301 was approximately one percent and three percent, respectively (Fig. 3(a)). $C_{f,\infty}$ is presented as a function of the circumferential angle Φ in Figure 3(b) for N values of 301, 61, and 21. Figure 3 indicates that for engineering calculations as few as 21 points could be used to obtain within three--percent accuracy. Comparisons of surface streamline direction ω_s, $\tan^{-1}\frac{\bar{\tau}\eta}{\bar{\tau}\xi}$, for N=301 and 21 are presented in Figure 4 with experimental data. The agreement with experimental data is good considering the fact that the inviscid pressure distribution was obtained from numerical solution of the inviscid equations and not from experimentally measured data. The important points that should be noted, other than the fact that agreement between numerical results and experimental data was in general good, is that the turbulence model is satisfactory for the type of flow considered (see also Ref. 9) and that the Crocco-type transformation allows as few as 21 points to be used across the boundary layer. The numerical technique is accurate and efficient; approximately .002 sec per nodal point was required on a CDC 7600 computer system. Studies currently underway to accelerate the convergence rate by special treatment of the ϕ_w boundary condition as well as the introduction of more general shear transformations to replace Equation (13) indicate that it may be possible to even further reduce the required number of nodal points as well as increase the computational speed of the technique.

CONCLUDING REMARKS

Solutions of the compressible, three-dimensional laminar, transitional, and turbulent boundary layer equations have been obtained and compare favorably with experimental data. The agreement with experimental data indicates that the procedure yields accurate results for as few as 21 nodal points in the plane normal to the surface. The eddy viscosity model has been shown to be adequate for the class of flow considered; however, previous experience indicates that caution should be exercised in extending the concept to more demanding flows.

REFERENCES

1. Proceedings: Computation of Turbulent Boundary Layers – 1968 AFOSR-IFP-Stanford Conference. Volume 1. Methods, Predictions, Evaluation and Flow Structures [Editors: Kline, S. V.; Morkovin, M. V.; Sovran, G.; and Cockrell, D. J.]

2. Bertram, Mitchel H. (Editor): Compressible Turbulent Boundary Layers. A symposium held at Langley Research Center, Hampton, Virginia, December 10-11, 1968. NASA SP-216.

3. Harris, Julius E.: Numerical Solution of the Equations for Compressible Laminar, Transitional, and Turbulent Boundary Layers and Comparisons With Experimental Data. NASA TR R-368, August 1971.

4. Bradshaw, P.: Effects of Streamline Curvature on Turbulent Flow. AGARD-AG-169, August 1973.

5. Nash, John F.; and Patel, Virendra C.: Three-Dimensional Turbulent Boundary Layers. SBC Technical Books, Scientific and Business Consultants, Inc. Atlanta, Georgia, USA, 1972.

6. Bradshaw, P.: The Strategy of Calculation Methods for Complex Turbulent Flows. I. C. Aero Report 73-05, August 1973.

7. Cebeci, Tuncer; Kaups, Kalle; Mosimskis, G. J.; and Rehn, J. A.: Some Problems of the Calculation of Three-Dimensional Boundary-Layer Flows on General Configurations. NASA CR-2285, July 1973.

8. McGowan, J. J., III; and Davis, R. T.: Development of Numerical Method to Solve the Three-Dimensional Compressible Laminar Boundary-Layer Equations With Application to Elliptical Cones at Angle of Attack. ARL 70-0341, December 1970.

9. Adams, John C., Jr.: Analysis of the Three-Dimensional Compressible Turbulent Boundary Layer on a Sharp Cone at Incidence in Supersonic and Hypersonic Flow. AEDC-TR-72-66, June 1972.

10. Popinski, Zenon; and Davis, R. T.: Three-Dimensional Compressible Laminar Boundary Layers on Sharp and Blunt Circular Cones at Angle of Attack. NASA CR-112316, January, 1973.

11. Blottner, F. G.; and Ellis, M. A.: Three-Dimensional, Incompressible Boundary Layer on Blunt Bodies. Sandia Laboratories, SLA-73 0366, April 1973.

12. Wang, K. C.: Three-Dimensional Laminar Boundary Layer Over Body of Revolution at Incidence. Part VI. General Methods and Results of the Case of High Incidence. AFOSR-TR-73-1045, May 1973.

13. Hunt, James L.; Bushnell, Dennis M.; and Beckwith, Ivan E.: The Compressible Turbulent Boundary Layer on a Blunt Swept Slab With and Without Leading - Edge Blowing. NASA TN D-6203, March 1971.

14. Dwyer, H. A.: Solution of a Three-Dimensional Boundary-Layer Flow With Separation. AIAA Jour., Vol. 6, No. 7, 1968.

15. Krause, E.: Comment on Solution of a Three-Dimensional Boundary-Layer Flow With Separation. AIAA Jour. Vol. 7, No. 3, 1969.

16. Rainbird, William John: Turbulent Boundary-Layer Growth and Separation on a Yawed Cone. AIAA Jour. Vol. 6, No. 12, December 1968.

(1) laminar, (2) transitional, (3) turbulent

$$ds^{-2} = \bar{h}_1^{\,2}d\xi^2 + \bar{h}_2^{\,2}d\eta^2 + \bar{h}_3^{\,2}d\zeta^2$$

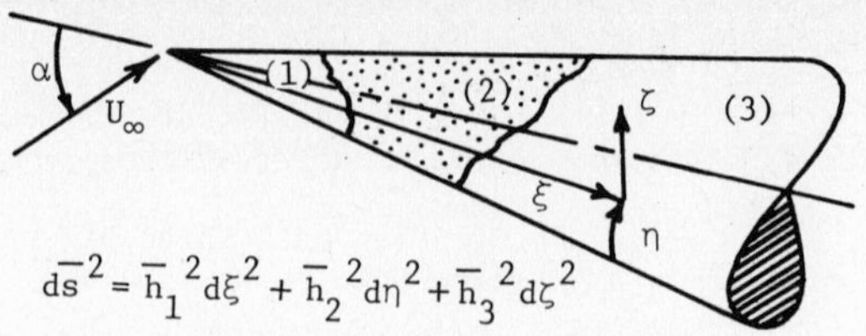

Fig. 1 - Coordinate system

$M_\infty = 1.8$, $R_{\infty,L} = 25 \times 10^6$, $\gamma = 1.4$

EXP	DATA	NUM.	RESULTS
O	u/Q_e	N	
□	v/Q_e	301	————
△	T/T_e	21	– – –

$$Q_e = \sqrt{u_e^{\,2} + v_e^{\,2}}$$

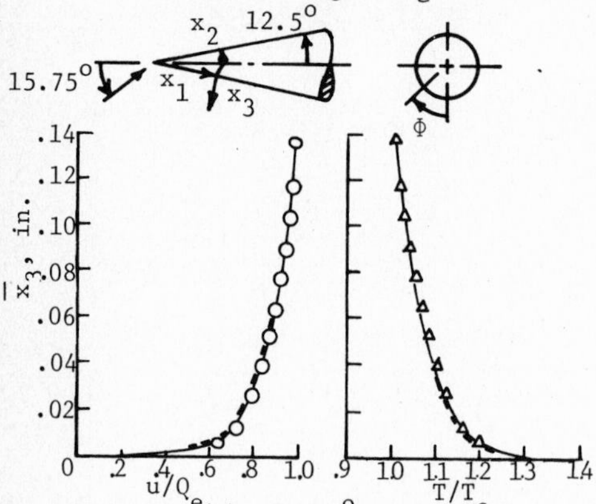

Fig. 2 - Profile comparisons

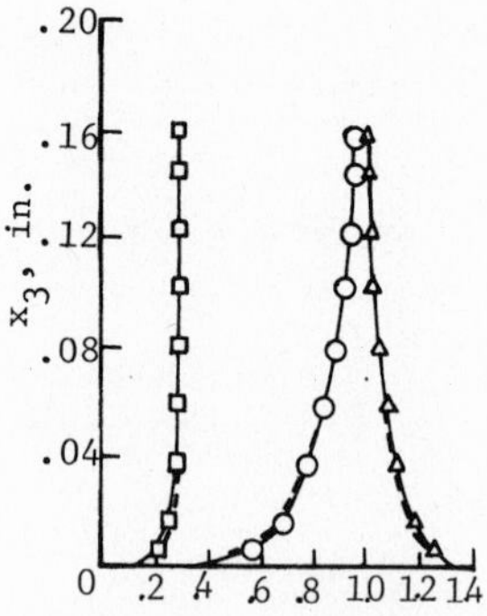

(a) $\Phi = 0^\circ$

(b) $\Phi = 45^\circ$

Fig. 2 - Continued

(c) $\Phi = 90^\circ$

Fig. 2 - Continued

(d) $\Phi = 135^\circ$

Fig. 2 - Concluded

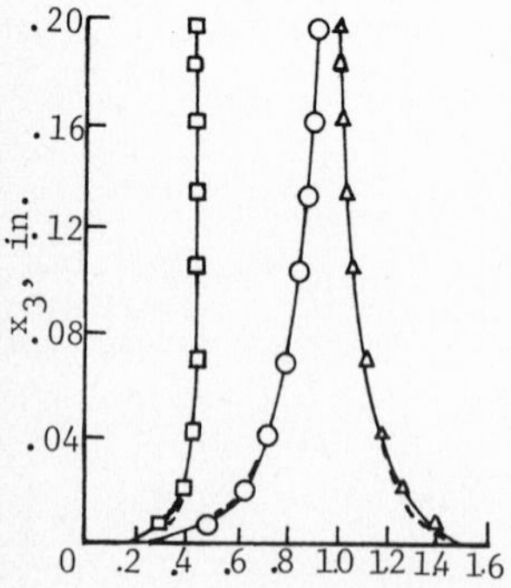

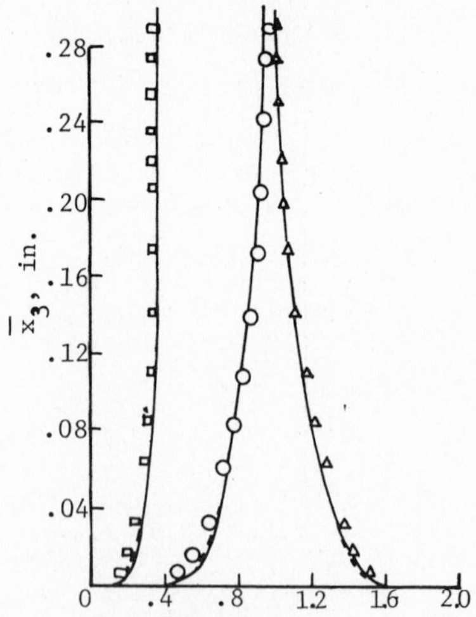

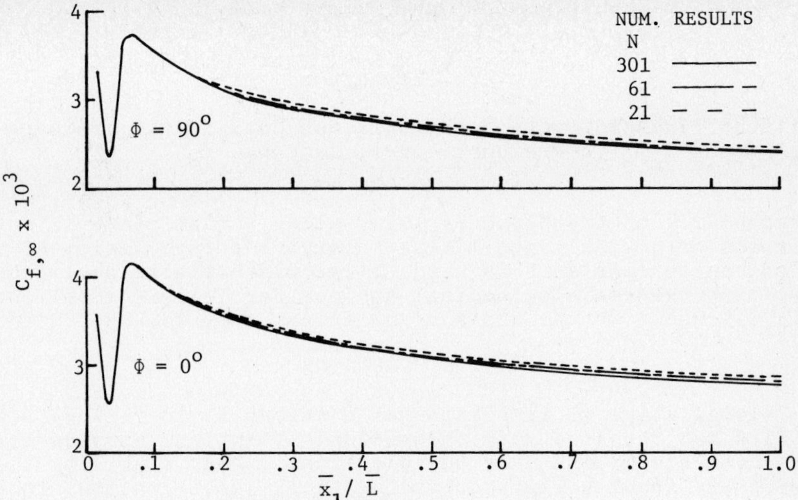

(a) $\Phi = 0^\circ;\ 90^\circ$

Fig. 3 - Effect of nodal spacing on skin friction.

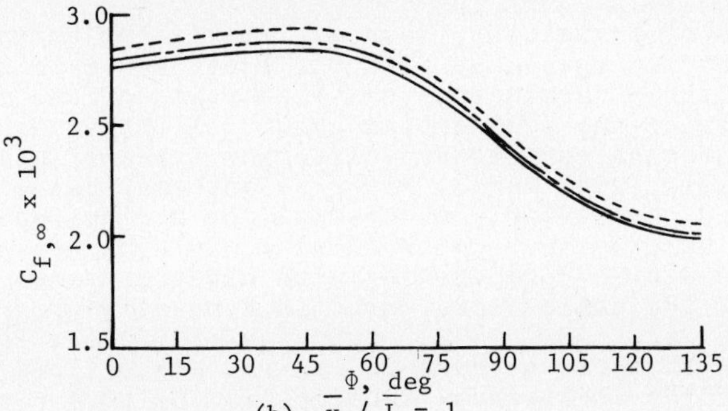

(b) $\overline{x}_1/\overline{L} = 1.$

Fig. 3 - Concluded

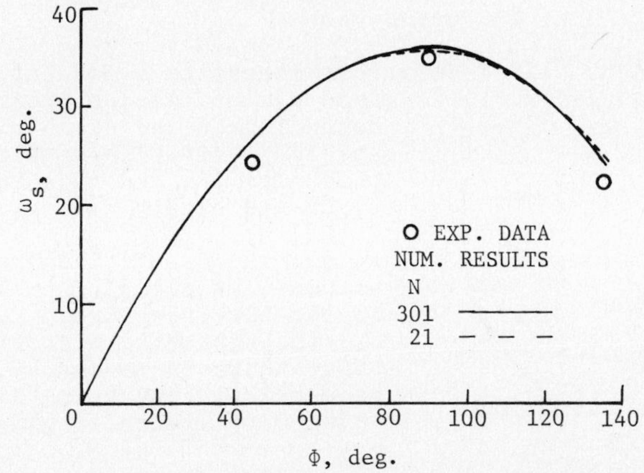

Fig. 4 - Surface flow direction, $\overline{x}_1/\overline{L} = 1.$

ON FINITE AMPLITUDES OF PERIODIC DISTURBANCES OF THE BOUNDARY LAYER ALONG A FLAT PLATE

Th. Herbert

Institut für Angewandte Mathematik der Universität Freiburg
78 Freiburg i.Br., Germany

The stability of the boundary layer along a flat plate is investigated by means of a weakly non-linear theory. A reformulation of the method of Landau constants is used to calculate finite amplitudes of periodic disturbances. The neutral surface for Tollmien-Schlichting waves is discussed on the basis of numerical results.

1. INTRODUCTION

The initial stage of transition of a steady laminar flow to the turbulent flow has received much theoretical attention. Considerable success has been accomplished by the classical linear stability theory, which recently found a rigorous mathematical justification. In this theory instability appears as the exponential growth of an initially very small periodic disturbance for certain parameters describing the basic flow and the disturbance itself.

In extensive studies of transition, however, non-linear terms have to be taken into account. In agreement with the experiments this will lead to generation of higher harmonics, to a distortion of the mean flow and to a limitation of the exponential growth of the amplitudes, as well as to strong interactions between different types of disturbances (which will be disregarded here). Rigorous mathematical work on non-linear stability was particularly successful for steady, space-periodic disturbances. Here we are concerned with disturbances which are periodic in space and time, propagating with a certain phase velocity in the direction of the basic flow. For this type some progress has been made by Joseph and Sattinger (1972), Iudovich (1971) and Iooss (1972). The application of these methods is restricted to flows, where separation of the variables is allowed, finally leading to a one-dimensional problem. Unfortunately the boundary layer flow, which is of exceptional practical importance in fluid mechanics, does not belong to this class.

2. THE BASIC FLOW

The velocity component $\bar{U}$ in free-stream direction x for the boundary layer along a flat plate (fig. 1) is given by the Blasius profile $\bar{U}(\eta)$ depending on the similarity variable $\eta = y/\delta^*$. The displacement thickness δ^*, and as a consequence $\bar{U}$, change slowly with x.

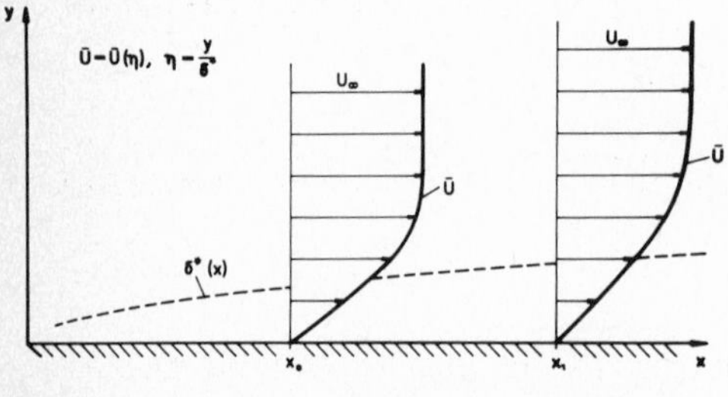

Fig. 1 Blasius boundary layer

All the quantities are reduced to non-dimensional form by introducing the free-stream velocity U_∞, the kinematic viscosity ν and, since there is no characteristic geometrical length in this problem, δ^* as a reference length.

Following the successful method of the linear theory, we approximate the boundary layer locally, i.e. near a fixed position x_0 by a parallel flow. The Reynolds number $Re = U_\infty \delta^*/\nu$ at x_o is then regarded as the parameter for a local analysis of stability. In order to obtain a set of ordinary differential equations for all the components contributing to the disturbed flow field, we assume in addition only a slight dependence on x for the quantities characterizing the disturbance.

3. LANDAU'S SERIES FOR THE AMPLITUDE

The time-dependence $A \sim \exp(a_o t)$ used in the linear theory provides no quantitative information on the complex amplitude A(t). Thus the non-linear analysis centres about an extended equation

$$(1) \qquad dA/dt = A(a_o + a_1|A|^2 + \ldots),$$

which was already proposed by Landau (1944) and studied anew by Stuart (1960), for Tollmien-Schlichting waves in a plane Poiseuille flow. As an infinite series, equation (1) was derived by Watson (1960) from the Navier-Stokes equations, whereby he represented the solution by Fourier series in x and expanded the coefficients in the amplitude A. The linear theory yields a_o as an eigenvalue of the Orr-Sommerfeld problem. According to Watson, the Landau constants $a_1, \ldots$ are determined as long as a_{or} (index r indicates the real part) is sufficiently small. Meanwhile a_1 has been calculated for various parallel flows. Preliminary results obtained by Itoh for the boundary layer were published by Tani (1973).

All these results correspond to reasonable expectations, but the quantitative confirmation is still pending. This confirmation would be of particular importance in view of the open questions concerning the convergence of (1) and the validity of results obtained in the 3rd order of approximation. In the main, experimental data are available for the boundary layer, but there again the comparison is rendered difficult by Watson's restriction. It may be mentioned here that this restriction is most serious when studying interactions of different waves.

In order to avoid this restriction, the method of Landau constants has been presented by Herbert (1974) in a revised form for generally three-dimensional periodic disturbances. Considering only equilibrium states characterized by steady mean flow quantities, the Landau constants are then determined for practically any parameters. In the special case of the non-linear behaviour of a Tollmien-Schlichting wave one obtains the condition

$$(2) \qquad d|A|^2/dt = 0,$$

which was suggested by Reynolds and Potter (1967) and applied by Davey and Nguyen (1971) to the pipe flow. From (1) follows the real equation

$$(3) \qquad d|A|^2/dt = 2|A|^2(a_{or} + a_{1r}|A|^2 + \ldots).$$

The first solution $|A_o|^2 = 0$ of (2), (3) represents the basic flow and is of minor interest. Retaining only the terms up to the 3rd order, one obtains

$$(4) \qquad |A_1|^2 = -a_{or}/a_{1r}$$

as a non trivial steady solution, which is of physical importance only if a_{or} and a_{1r} have contrary signs.

4. BASIC EQUATIONS

During the calculation of the Landau constants a_k, $k \geq o$, a set of ordinary differential equations has to be solved step by step. The most essential of these is the Orr-Sommerfeld equation in the form

$$(5)\qquad \left[\frac{1}{Re}(D^2 - \bar{\alpha}^2)^2 - (i\bar{\alpha}\bar{U} + \bar{a})(D^2 - \bar{\alpha}^2) + i\bar{\alpha}D^2\bar{U}\right]\varphi = f$$

with the homogeneous boundary conditions

$$(6)\qquad \varphi = D\varphi = 0 \quad \text{at } y = 0,\ \ y \to \infty$$

where $D = d/dy$; $\bar{a}$ and $\bar{\alpha}$ are constants, and $\varphi = \varphi(y)$ is an unknown periodic component of the disturbed flow. For a given Reynolds number Re and wave-number α the following cases appear:

(a) $\bar{\alpha} = \alpha$, $\bar{a} = a_o$, $f \equiv 0$. This is the well known non-selfadjoint eigenvalue problem of the linear theory, which yields a_o as a solution of the characteristic equation $F(Re,\alpha,a_o) = 0$ and the eigenfunction φ.

(b) $f = f(y)$, $F(Re, \bar{\alpha}, \bar{a}) \neq 0$, the inhomogeneous two-point boundary value problem, where f essentially depends on the previously calculated solutions

(c) $\bar{\alpha} = \alpha$, $\bar{a} = a_o$, $f = a_\kappa \cdot g(y) + h(y)$, the resonance case with $F(Re,\alpha,a_o) = 0$. Here g and h again contain, in the main, previous results; a_κ is one of the free Landau constants a_k, $k > o$. In order to solve for this case, the adjoint homogeneous problem has to be solved first. a_κ is then determined by the condition of solvability.

Another set of second order differential equations yields the mean-flow distortion. It should be noticed that the assumption of a quasi-parallel flow leads to serious difficulties in calculating the components without periodicity in the x-direction. These were overcome by modified equations approximately maintaining the growth of the boundary layer thickness.

5. NUMERICAL TREATMENT

The difficulties associated with solving case (a) of the problem (5), (6) are well known, and various asymptotic and numerical methods are in use now. A survey has been given by Gersting and Jankowski (1972) recently. The problems mainly arise from the fact that the highest derivative in (5) is multiplied by the small value 1/Re. Besides this, in the present analysis the functions f in (5) are successively formed from previously evaluated numerical solutions and their derivatives. This may lead to serious error propagation and thus requires high accuracy in the computation.

The method used here may be regarded as an extension to the cases (b) and (c), and modification, concerning the difference-scheme, of the "method of near-orthonormalized integration", described meanwhile by Gersting and Jankowski (1972). Experience has shown that by this means, single precision arithmetic provides sufficient accuracy in the results - also for the much more extensive studies of interactions - thus saving computing time and storage.

6. THE NEUTRAL SURFACE

Detailed computations of a_o, a_1 were carried out for Tollmien-Schlichting waves. The equilibrium states obtained in this 3rd order of approximation are discussed in the following.

The basic flow described by $|A_o| = 0$ is stable outside and unstable inside the neutral curve C_o of the linear theory, which is shown in fig. 2 with the critical Reynolds number and wave-number. According to (4) and to the signs of a_{or} and a_{1r}, significant values $|A_1|^2 \geq 0$ result only outside the shaded area. The stability of the equilibrium state related to $|A_1|$ is indicated in fig. 2.

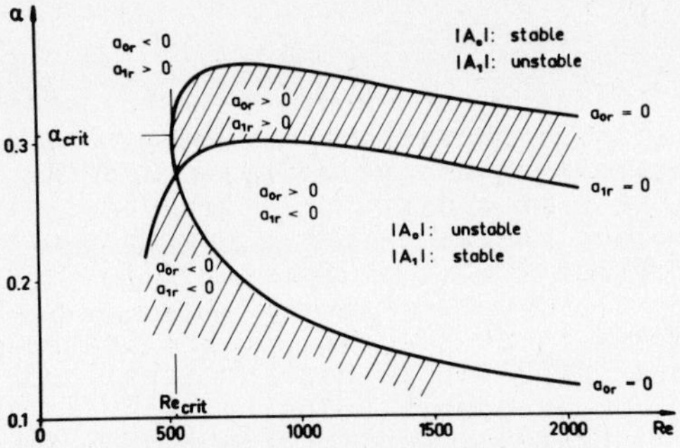

Fig. 2 Neutral curve C_o (a_{or}=0) and the curve a_{1r} = 0

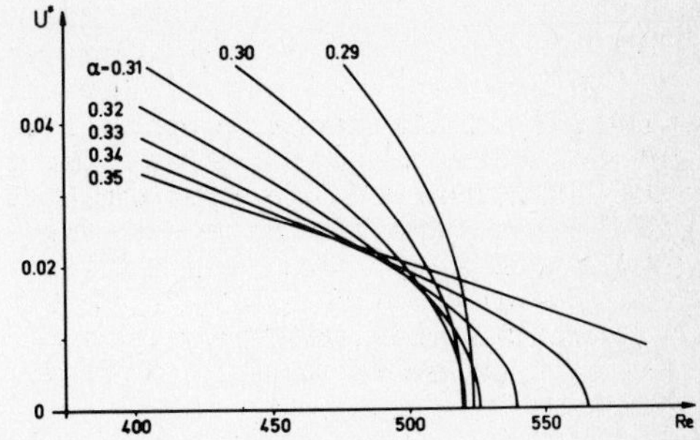

Fig. 3 Cross-sections at α=const. near Re_{crit} of C_o

Since $|A_1|$ as yet has only formal importance, we use for the following discussion the maximum value U* of the root-mean-square velocity u*(y) derived from $|A_1|$, which can be measured directly. The relation

(7) $\quad C(Re, \alpha, U^*) = 0$

is the implicit representation of a spatial neutral surface for Tollmien-Schlichting waves of finite amplitude, cutting the plane U* = 0 along C_o = C(Re,α,0) = 0. Various cross-sections U*(Re,α) of this neutral surface are shown in the figures 3 - 5.

Fig. 3 indicates a moderate decrease of Re_{crit} (given by the envelope) and an increase of α_{crit} with growing amplitude U*. Thus the boundary layer shows subcritical instability. Experimental data for comparison are not yet available. The slight shift in Re_{crit} from 519 at U* = 0 to 514 at .01 and to 492 at .02, however, may be one of the reasons, why measurements agree well with the results of the linear theory, in spite of using finite disturbances in the experiment.

In the region of amplified linear disturbances above the lower branch α_I(Re) of the neutral curve C_o a stable equilibrium state exists at values of about 4% of the free-stream velocity in a wide range of Reynolds numbers and wave-numbers (fig. 4). The growth of the curve Re=800 is due to a_{1r} approaching zero.

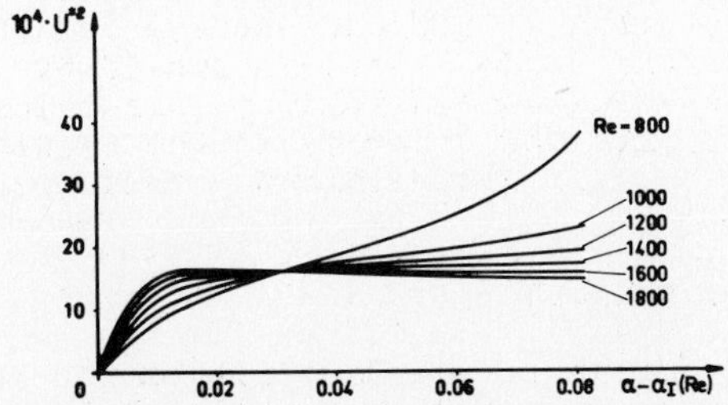

Fig. 4 Cross-sections at Re=const. above the lower branch α_I

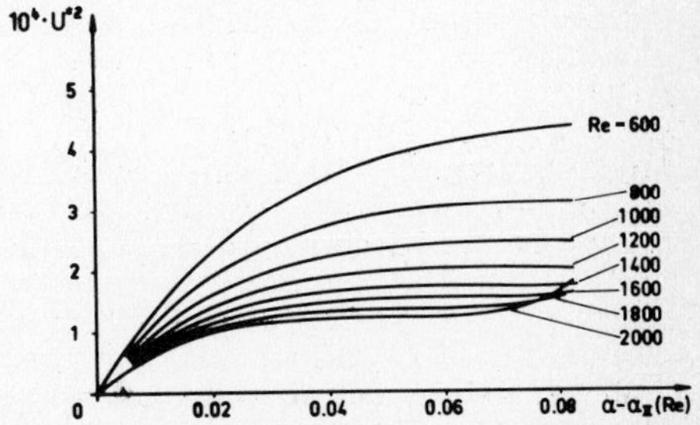

Fig. 5 Cross-sections at Re=const. above the upper branch α_{II}

The unstable equilibrium state near the upper branch (fig. 5) is obtained at smaller values of U*; they are approximately 1 - 2% of the free-stream velocity. The sudden growth of the curves at higher wave-numbers, which can be seen most noticably for Re=2000, occurs also at other Reynolds numbers.

7. COMPARISON WITH EXPERIMENTS

In the measurements of Klebanoff, Tidstrom and Sargent (1962) oscillations of a fixed frequency f and initial amplitude u'_r were generated at a reference position x_r. The root-mean-square velocity u' was recorded at various downstream positions x at a distance $\bar{y}$ from the wall. The parameters indicated in fig. 5 place the wave-number near the upper branch of C_o at Reynolds numbers of about 2000.

A calculated cross-section U* = U*(x,f) at f=const. of the neutral surface is shown in fig. 5. The curve $u^*(\bar{y})$ represents the local root-mean-square velocity at $\bar{y}$ for the unstable equilibrium state. In the dashed curve the values of $u'(\bar{y})$ recorded with u'_r = .0008 are plotted against x. The disturbance develops as a Tollmien-Schlichting wave throughout the entire region and passes a flat maximum of the amplitude. The value u'_{max} = u* = .0039 supplies a real comparative value for the numerical result u* = .0086. Other comparisons, in particular for the stable equilibrium state near the lower branch of C_o, lead to similar situations.

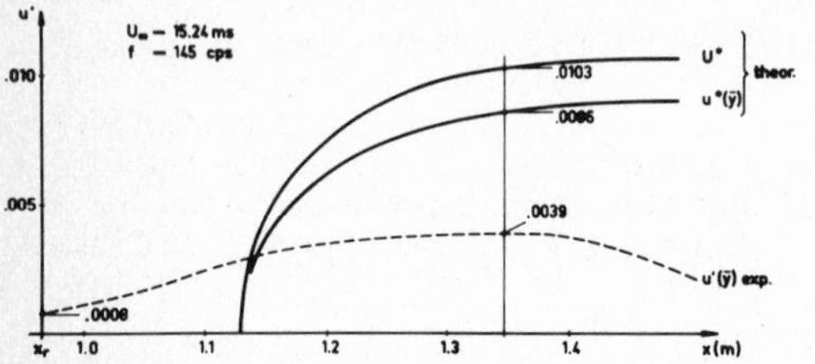

Fig. 6 Comparison between numerical and experimental data

Qualitatively, the observed behaviour of the disturbances agrees well with the theoretical results. The difference of the values may be partially due to the restriction of the 3rd order theory to "sufficiently" small amplitudes. The main cause, however, seems to be the separation of variables. This implies the assumption, that the disturbances, in spite of their phase velocity, are able to reach the local equilibrium state, which therefore should not change much in the downstream direction. In contrast to this, fig. 6 shows a considerable variation of U* with x.

For an improved analysis it is necessary, therefore, to take not only the growing thickness of the boundary layer into account, but also the much more important downstream variations of the disturbance itself. It seems promising to apply the method of multiple scales here, which was recently used in a linear stability analysis of the boundary layer by Bouthier (1973).

8. HIGHER APPROXIMATIONS

Finally, the previous numerical results will be used for a more general consideration of the neutral surface. In fig. 7 a cross-section U^{*2} (Re,α) at Re=const. is shown qualitatively. α_I and α_{II} indicate the lower and upper branch of the neutral curve, and α_o the zero of a_{1r}. The calculations yield the curves referred to as (I) and (II). In consequence of (4), curve (I) approaches infinity at α_o, since for a_{1r} = 0 the 3rd order terms have no effect on the exponential growth of the amplitude. A complete nonlinear theory may, on no account, give such an unbounded growth of the disturbance. Therefore, at a higher approximation we expect a curve like (A) instead of (I). Moreover, taking the rise of (II) at large α into account, we can well suppose that a closed curve (B) will result. The change between stable and unstable equilibrium at fixed α would be perfectly feasible.

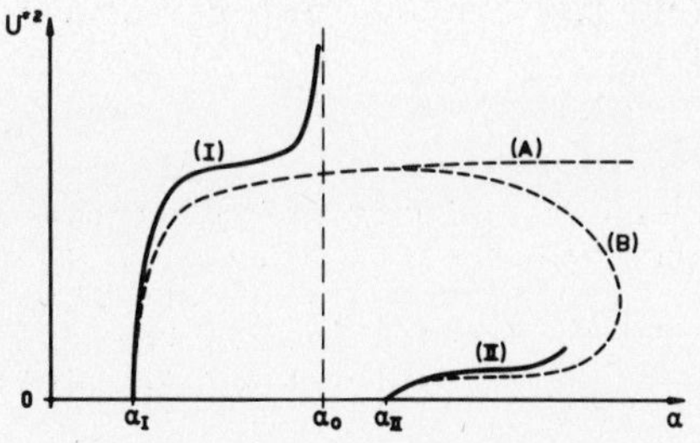

Fig. 7 Cross-section of the neutral surface at Re=const.

If we extend this consideration to smaller Reynolds numbers, we obtain the picture of a connected neutral surface, enclosing all the amplified disturbances of finite amplitude. All disturbances beyond this surface would be damped out. To consolidate this picture by a numerical investigation, we require in addition at least one further Landau constant.

Such an extended investigation could also be of use in estimating the validity of the 3rd order approximation. However, the higher order constants a_k, $k > 1$, are not determined uniquely. This results in basic limitations of the scope of the method of Landau constants.

Meanwhile an alternative way has been found to calculate the neutral surface, thereby avoiding the series expansions in the amplitude. It is based on the homogeneous system of coupled equations suggested long ago by Heisenberg (1924). The representation (7) of the neutral surface appears immediately as the characteristic equation. This method is being applied in current work to various parallel flows.

Altogether, the present investigation has shown the possibilities, as well as certain limitations existing in applications of the method of Landau constants. Moreover it has led to improved methods to be used in future work.

9. REFERENCES

Bouthier, M., Journal de Mécanique 12, 75-95 (1973)
Davey, A. and Nguyen, H.P.F., J.Fluid Mech. 45, 701-720 (1971)
Gersting, J.M. and Jankowski, D.F., Int. J. Num. Meth. Eng. 4, 195-206 (1972)
Heisenberg, W., Ann. d. Physik 74, 577-627 (1924)
Herbert, T., Dissertation, Universität Karlsruhe (1974)
Iooss, G., Arch. Rational Mech. Anal. 47, 301-329 (1972)
Iudovich, V.I., Prikl. Mat. Mek. 35, 638-655 (1971)
Joseph, D.D. and Sattinger, D.H., Arch. Rational Mech. Anal. 45, 79-109 (1972)
Klebanoff, P.S., Tidstrom, K.D. and Sargent, L.M., J. Fluid Mech. 12, 1-34 (1962)
Landau, L., C. R. Acad. Sci. URSS 44, 311-314 (1944)
Reynolds, W.C. and Potter, M.C., J. Fluid Mech. 27, 465-492 (1967)
Stuart, J.T., J. Fluid Mech. 9, 353-370 (1960)
Tani, I., ZAMM 53, T25-T32 (1973)
Watson, J., J. Fluid Mech. 9, 371-389 (1960)

ASYMPTOTIC SOLUTION FOR SUPERSONIC VISCOUS FLOW PAST A COMPRESSION CORNER*

R. Jenson, O. R. Burggraf,
and D. P. Rizzetta
Department of Aeronautical Engineering
The Ohio State University

I. INTRODUCTION

When a boundary layer on a plane surface encounters a downstream compressive disturbance of sufficient magnitude the flow is forced to separate from the wall. The separation point is commonly observed to lie at a rather large distance ahead of the disturbance, which contradicts the inherent nature of Prandtl's boundary layer theory that no upstream influence can occur. Crocco and Lees (1952) have shown that coupling the pressure of the external inviscid flow to the displacement thickness of the boundary layer permits upstream influence to be consistent with the boundary layer equations. This concept led to the extremely useful integral methods of Lees and Reeves (1964) and others. However the correct mathematical structure of such a viscous-interacting flow was only recently developed by Stewartson and Williams (1969) in their paper on self-induced separation. In an independent paper, the identical asymptotic structure was developed by Messiter (1970) for incompressible flow near a trailing edge. In the present paper, a numerical solution of the Stewartson-Williams problem is presented for the case of separation produced in supersonic flow past a compression corner. The flow through separation ahead of the corner agrees with the Stewartson and Williams solution, but the reattachment of the separated flow onto the ramp is of different nature than the downstream flow in their case.

II. TRIPLE DECK STRUCTURE

When the ramp angle is of order $(Re^{-1/4})$, then as Reynolds number $Re \to \infty$, a "triple-deck" structure develops similar to that found by Stewartson and Williams (1969) for self-induced separation. In fact, except for a modified boundary condition on the wall, the Stewartson and Williams analysis can be taken over to the present problem virtually intact. The length of the interaction is $O(Re^{-3/8})$ and there are three distinct vertical scalings in which different physical processes dominate (See Figure 1). The X- length scale $O(Re^{-3/8})$ is necessary to match the solutions in each region, producing a consistent asymptotic structure. The "main deck" has a vertical scale $O(Re^{-1/2})$ and consists of the fluid in the upstream boundary layer passing through the short $O(Re^{-3/8})$ region near the corner in which the interaction is completed. Because the interaction region is so short, the dominant

* The research reported here was sponsored by The Office of Naval Research, United States Navy, under Contract No. N00014-67-A-0232-0014.

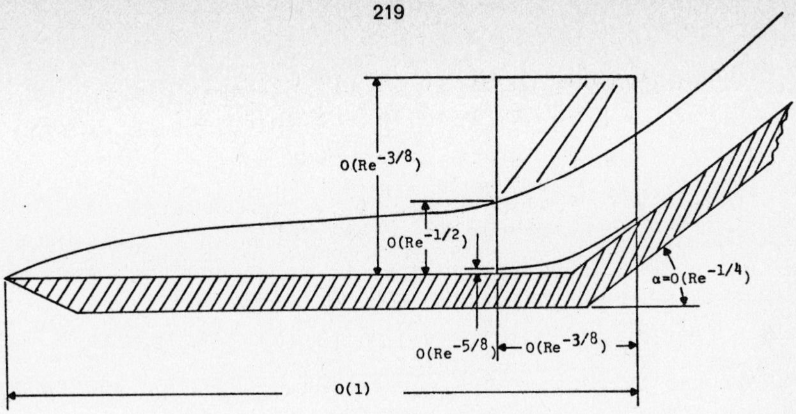

Figure 1. Triple Deck Structure

mechanism in the main deck is an inviscid turning of the flow, and the first order equations in this region are those of a simple displacement of the streamlines. The "upper deck" has a vertical scale $O(Re^{-3/8})$ and consists of fluid which, through outward propagating Mach waves, is disturbed by the flow in the interaction region. Here the dominant process is irrotational as well as inviscid, so that the first-order equations are just the Prandtl-Glauert equations. The "lower deck" or sublayer, has a height $O(Re^{-5/8})$ and the dominant process here is a viscous, incompressible flow. After a suitable rescaling, the equations in this inner region are the familiar incompressible boundary layer equations, and the boundary conditions at the wall are the usual boundary layer wall conditions. However the outer edge condition comes, not from matching to an outer irrotational inviscid flow, but from matching to the rotational inviscid "main deck," which is matched in turn to an irrotational inviscid "upper deck." When the details of the matching are carried out the result to first order is that the "main deck" assumes the passive role of merely transmitting streamline displacements from the "lower deck" to the "upper deck" and pressure perturbations from the "upper deck" to the "lower deck." The "upper deck" provides a relation between the pressure and the streamline displacement at the edge of the inner layer. Hence the whole problem reduces to the solution of the incompressible boundary layer equations (with unusual boundary conditions) in the lower deck.

The resulting first-order equations are:

$$\frac{\partial U}{\partial X} + \frac{\partial V}{\partial Y} = 0 \qquad (1)$$

$$U\frac{\partial U}{\partial X} + V\frac{\partial U}{\partial Y} = -\frac{dP}{dX} + \frac{\partial^2 U}{\partial Y^2} \qquad (2)$$

with boundary conditions

$$U = V = 0 \text{ on } \begin{bmatrix} Y = 0, X < 0 \\ Y = \alpha X, X > 0 \end{bmatrix}$$
$$U \to Y \text{ as } X \to -\infty$$
$$U \to Y - \alpha X \text{ as } X \to \infty \tag{3}$$
$$\lim_{Y\to\infty}[U-Y] = -\int_{-\infty}^{X} P(\xi)d\xi$$

The variables used above are related to corresponding lower case variables in section 6 of Stewartson and Williams, and

$$\alpha = \left\{[M_\infty^2/(M_\infty^2-1)^{1/2}]^{1/2}(R_e \nu_w)^{-1/2}[U_0'(0)]^{1/2}/\ M_0'(0)\right\}R_e^{1/4}\ \alpha^* \tag{4}$$

where α^* is the tangent of the corner angle. The assumption has been made that the corner angle is $O(Re^{-1/4})$. For smaller angles, separation does not occur. On the other hand, if the angle is larger than $O(Re^{-1/4})$ it might be surmised that the "self-induced separation" solution of Stewartson and Williams would be displaced upstream of the corner to a distance greater than $O(Re^{-3/8})$. Unlike the Stewartson and Williams problem, as long as reattachment occurs within the $Re^{-3/8}$ length scaling the downstream conditions are known and no uniqueness problems arise due to reversed flow as $X\to\infty$.

III. TRANSPOSITION OF THE INNER LAYER PROBLEM

To facilitate treatment of the problem, the geometry was simplified by using Prandtl's transposition theorem. The variables introduced for this purpose are

$$Z = \begin{Bmatrix} Y \text{ for } X \leq 0 \\ Y - \alpha X \text{ for } X \geq 0 \end{Bmatrix} \tag{5}$$

$$W = \begin{Bmatrix} V \text{ for } X < 0 \\ V - \alpha U \text{ for } X > 0 \end{Bmatrix} \tag{6}$$

The continuity and momentum equations are invariant under this change of variables.

The pressure is temporarily eliminated from the problem by differentiating the momentum equation (2) with respect to Y. In terms of the non-dimensional shear stress

$$\tau = \frac{\partial U}{\partial Y} = \frac{\partial U}{\partial Z} \tag{7}$$

this results in the shear transport equation

$$\frac{\partial \tau}{\partial t} + U\frac{\partial \tau}{\partial X} + W\frac{\partial \tau}{\partial Z} = \frac{\partial^2 \tau}{\partial Z^2} \tag{8}$$

subject to the conditions

$$\tau \to 1 \text{ for } X \to \pm\infty$$
$$\tau \to 1 \text{ for } Z \to \infty \tag{9}$$

We have found it convenient to use the device of calculating an unsteady flow to achieve the desired steady state solution. The velocity components are obtained from τ via (7) and (1).

For the remaining boundary condition on τ, we use the compatibility condition, which requires special treatment here because of the discontinuity introduced by the transposition theorem. Differentiating the pressure integral in Eq. (3) with respect to X, holding Z fixed, we obtain

$$P(X) = -\left[\frac{\partial U}{\partial X}\right]_{Z\to\infty} + \alpha H(X) \tag{10}$$

where H(X) is the Heavyside step function; i.e., H(X) takes the value 0 for X<0 or 1 for X>0. Requiring P to be continuous, and noting that U is the Z-integral of τ, Eq. (10) yields the following jump condition at X = 0:

$$\int_0^\infty [\tau_X(0^+,\eta) - \tau_X(0^-,\eta)]d\eta = \alpha \tag{11}$$

The subscript X here implies the X-derivative of τ.

The pressure gradient now is evaluated by differentiating P(X) in Eq. (10), and then is substituted into the momentum equation (2). Evaluating Eq. (2) at Z = 0 gives the compatibility condition:

$$\left[\frac{\partial \tau}{\partial Z}\right]_{Z=0} = -\frac{\partial^2}{\partial X^2}\int_0^\infty \tau(X,\eta)d\eta \ , \quad X \neq 0 \tag{12}$$

The boundary value problem for the compression corner is now specified by Eqs. (8), (9), (11) and (12). It is interesting to note that in this form α appears only in Eq. (11), the corner jump condition.

IV. NUMERICAL ALGORITHM

At the interior points the unsteady equation of motion (8) is used to work in time from an initial state of uniform shear flow to a final steady solution. Both time-explicit and partially time-implicit schemes have been used, but the partially time-implicit scheme was found to be more efficient. The scheme used was

$$\frac{\partial \tau}{\partial t} + \Delta t\left[\hat{U}\frac{\partial}{\partial X}\left(\frac{\partial \tau}{\partial t}\right) + \hat{W}\frac{\partial}{\partial Z}\left(\frac{\partial \tau}{\partial t}\right) - \frac{\partial^2}{\partial Z^2}\left(\frac{\partial \tau}{\partial t}\right)\right] = -\left[\hat{U}\frac{\partial \hat{\tau}}{\partial X} + \hat{W}\frac{\partial \hat{\tau}}{\partial Z} - \frac{\partial^2 \hat{\tau}}{\partial Z^2}\right] \tag{13}$$

Where the circumflex indicates that a variable is calculated at the previous time step. All first X and Z derivatives are evaluated by "windward differences" and all second derivatives by centered differences. This scheme requires solution of a tridiagonal matrix equation for the new values of $\partial\tau/\partial t$ at each x station for each time step. At the upstream and downstream boundaries the known asymptotic form of the solution is used to reduce the error incurred by applying these conditions at a finite distance. This asymptotic boundary condition is especially important downstream due to algebraic decay.

The central feature of our numerical procedure is the treatment of the compatibility condition (12). In difference form, Eqs. (11) and (12) can be combined into a single equation by use of the Kronecker delta symbol, δ_m^n; after premultiplying by ΔX we find

$$\Delta X[(\tau_{k,2}-\tau_{k,1})/\Delta Z] = -(\Delta Z/\Delta X)[(\tau_{k-1,1} - 2\tau_{k,1} + \tau_{k+1,1})/2$$
$$+ \sum_{j=2}^{N}(\tau_{k-1,j} - 2\tau_{k,j} + \tau_{k+1,j})] + \alpha\delta_k^{kc}$$

where the truncation error is of order ΔX at the corner, $(\Delta X)^2$ elsewhere. Here $\tau_{k,j}$ refers to τ at the kth X-station and the jth Z-station, j = 1 refers to Z = 0, and k = k_c refers to the corner. We now regard this equation as an implicit algorithm (in the X direction) for determining the wall values of τ. Thus we rearrange the terms to obtain the tri-diagonal matrix equation

$$a\tau_{k-1,1} + \tau_{k,1} + a\tau_{k+1,1} = b_k$$

$$a = 1/2[1 + (\Delta X/\Delta Z)^2]^{-1}$$

$$b_k = \tau_{k,2}/[1 + (\Delta Z/\Delta X)^2]$$

$$+ 2a[\sum_{j=2}^{N} (\tau_{k-1,j} - 2\tau_{k,j} + \tau_{k+1,j}) - \alpha(\Delta X/\Delta Z)\delta_k^{kc}] \qquad (14)$$

The implicit nature of condition (14) allows instantaneous interaction of all the shear profiles, which appears to aid in suppressing downstream-growing eigenfunctions that would otherwise occur.

V. DISCUSSION OF RESULTS

Owing to the scaling of variables used in the asymptotic theory, as in Eq. (4), parameters such as Mach number and Reynolds number have been suppressed. Consequently the reduced boundary value problem for the compression corner, Eq. (1) - (3), contains only one parameter, the reduced corner angle α. Numerical solutions have been obtained over a range of α for various values of the mesh parameters. The results presented here appear to be accurate to about one per cent.

Figure 2 shows the skin-friction distribution on the plate and ramp in terms of the scaled variables of the inner layer ($\partial U/\partial Z$ vs. X). Flow separation and reversal near the corner (X=0) are indicated by negative values of the friction. Incipient separation is indicated for a value of α near 1.65. For the larger values of α, the curves are shaped much like those of Carter (1972), who solved the full Navier-Stokes equations for this problem, although at relatively low Reynolds number.

The wall pressure distribution is shown in Figure 3. For increasing α, the portion of the curve ahead of separation is unchanged except for an upstream displacement, corresponding to the self-induced separation solution of Stewartson and Williams. A pressure plateau is developing for the larger α, and the plateau level appears to be approaching the value 1.8 suggested as the downstream limit for the Stewartson-Williams solution.

A comparison of the asymptotic theory for α=2.5 is made with the experimental data of Lewis, Kubota, and Lees (1968) for α=2.4 in Figure 4. At the conditions (M_∞=4, α^*=10°), the linearized simple-wave boundary condition underpredicts the ramp pressure rise by about 20%. To correct for this effect, the theoretical results shown in Figure 4 were modified by use of the exact simple-wave relation between pressure and flow angle. The ramp pressure then agrees well with the data, but the wall pressure indicates that separation is delayed in the experiment relative to the theory. This is explained by the relatively low Reynolds number of the experiment (Re = 68,000). In fact, the conditions are such that the leading edge of the computation mesh for the asymptotic theory lies ahead of the plate leading edge in the experiment, having the effect of an excessively thick boundary layer in the theory. This condition implies that the boundary layer on the plate has not truly evolved for the low Reynolds number of the experiment, since the region of influence of the corner overrides the region of normal boundary layer development.

VI. REFERENCES

Crocco, L., and Lees, L. 1952. J. Aero. Sci. 19, 649.

Lees, L., and Reeves, B. 1964. AIAA J. 2, 1907-1920.

Stewartson, K., and Williams, P. G. 1969. Proc. Roy. Soc. London A 312, 181.

Messiter, A. 1970. S.I.A.M. Jour. Appl. Math. 18, 241.

Carter, J.E. 1972. Proc. 3rd Int. Conf. on Numerical Methods in Fluid Mech., Lecture Notes in Physics, Vol. 19 Springer-Verlag.

Lewis, J.E., Kubota, T., and Lees, L. 1968. AIAA J. 6, 7-14.

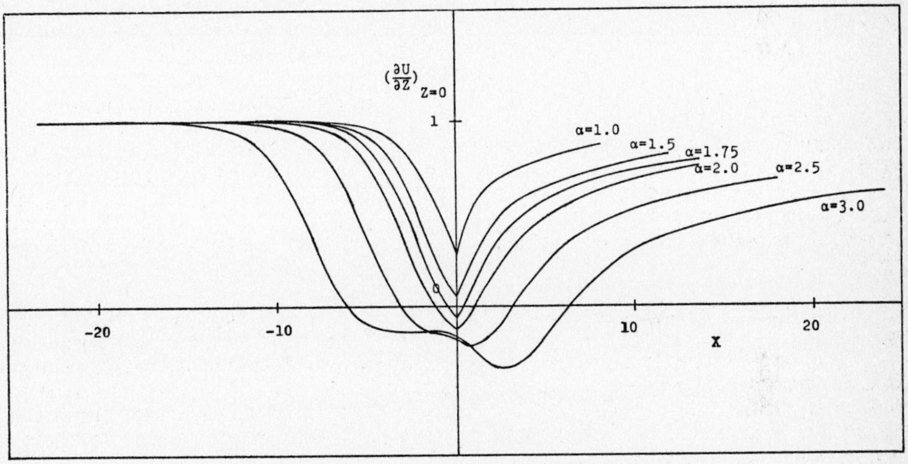

Figure 2. Scaled Wall Shear Distributions

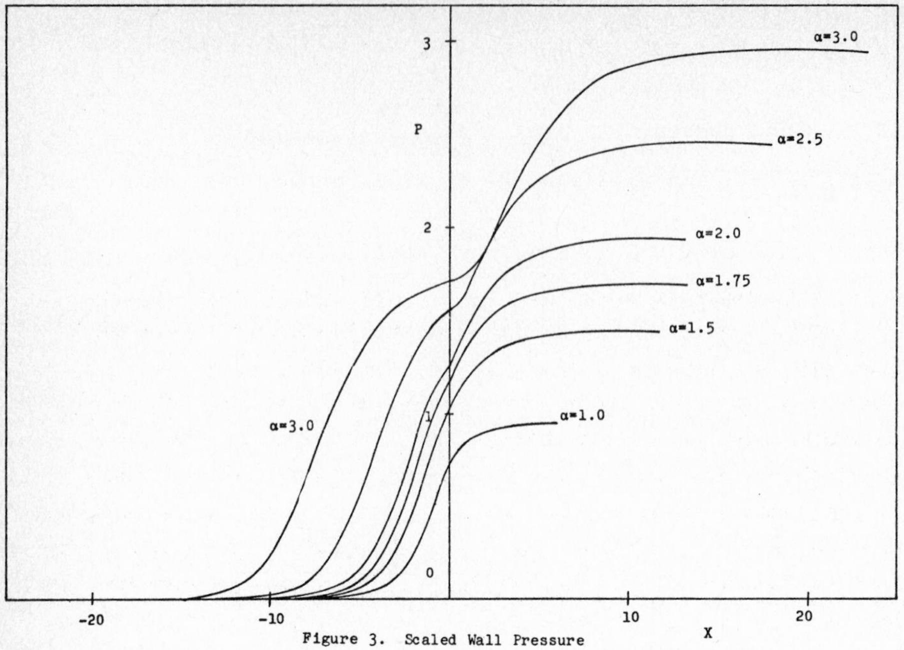

Figure 3. Scaled Wall Pressure

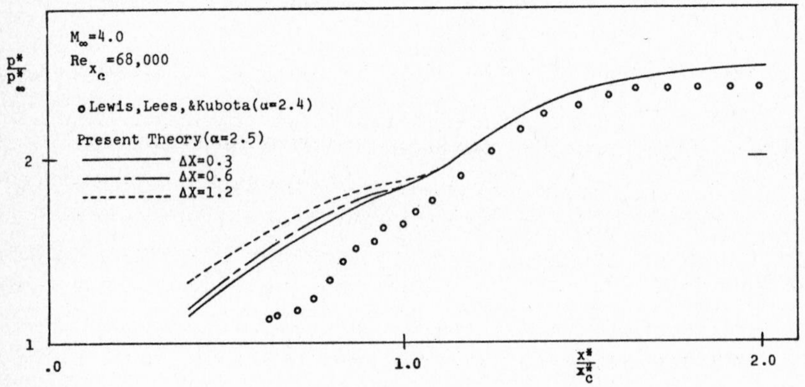

Figure 4. Comparison Between Asymptotic Solution and Experiment

LEADING EDGE SEPARATION FROM NON-CONICAL SLENDER WINGS AT INCIDENCE

by

I. P. Jones

University of East Anglia, Norwich, England.

INTRODUCTION

In this paper we consider the flow past flat plate slender wings at incidence. At the leading edges of such a wing the boundary layers on the upper and lower surfaces meet and coalesce to form a vortex sheet, which spirals into a vortex core situated above the wing. Most of the previous investigations into this problem, for example, Brown and Michael (1955), Smith (1966), Barsby (1973), have utilised the conical flow approximation which is appropriate for, say, a slender delta wing. This approximation breaks down for wings of arbitrary plan-form, and here we describe a numerical method which is suitable for such wings. Another method using a different coordinate system and numerical procedure has been developed by Clark (1974) at the Royal Aircraft Establishment, Farnborough.

The assumptions and approximations that we adopt in our model of the flow are as follows:

(i) The flow is effectively inviscid.

(ii) The vorticity in the fluid is condensed onto vortex sheets which emanate from the leading edges. Each of these sheets is infinite in length and rolls up into a tight spiral core. These infinite sheets are replaced by a finite outer part springing from the leading edge and an isolated potential vortex. Across the trace joining the end of the sheet to the isolated vortex there is a discontinuity of pressure that is independent of its shape. This trace is represented by a cut which renders the physical variables single-valued and joins the isolated vortex to the free end of the finite vortex sheet.

(iii) The secondary separation which is induced under the primary vortex is unimportant.

(iv) The slender body theory of Munk, Jones and Ward (see Ward (1955)) is applicable.

MATHEMATICAL TREATMENT

We take cartesian axes (x, y, z) with the origin at the apex of the wing, as shown in Fig. 1.

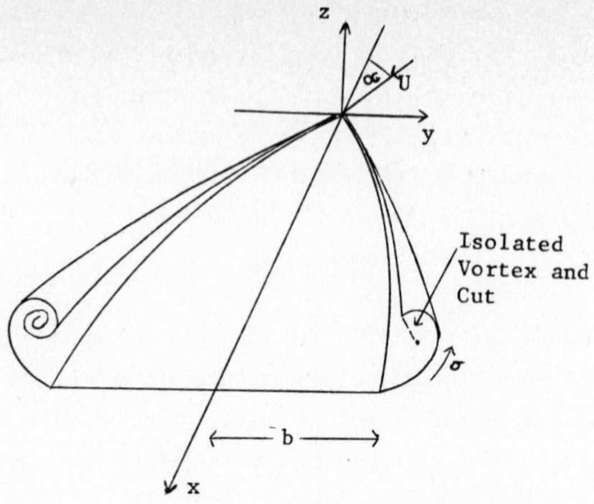

Fig. 1. Slender wing with vortex sheets

Relative to these axes the free stream has velocity components $(U, 0, \alpha U)$ where α is the angle of incidence. A perturbation velocity potential Φ can be defined such that the total potential is $Ux + \Phi$, where the first term is the contribution from the free stream and the second term arises from the disturbance to the flow by the wing. In this case, from slender body theory, Φ satisfies Laplace's equation,

$$\Phi_{yy} + \Phi_{zz} = 0 \quad , \tag{1}$$

in each 'cross-flow' plane, x = constant, perpendicular to the wing. The boundary conditions to be satisfied are that the normal velocity to the wing, Φ_z, is zero on z = 0, $\Phi \sim \alpha U z$ as $z \to \infty$ together with appropriate conditions on the vortex sheets and at the leading edge. Since Φ is a harmonic function the solution of (1) is most easily determined by constructing a complex potential $W(Z) = \Phi + i\Psi$ where $Z = y + iz$ and Ψ is the stream function. The boundary conditions which must be satisfied on a vortex sheet are discussed in detail by Smith (1971). They are that the vortex sheet is a stream surface of the three-dimensional flow, and that it cannot sustain a pressure jump.

Barsby (1973) has shown that an intrinsic coordinate system has advantages over the polar coordinates used by earlier workers, especially for low angles of incidence. Thus we adopt an intrinsic coordinate system (η, ψ) in each cross-flow plane, with $\eta = \sigma/s$, where σ is the arc length of the sheet, s is an arbitrary function of x chosen in such a manner that as the vortex system grows downstream , along with the dimensions of the wing, the angular extent of the vortex sheet is maintained at an approximately constant value, and ψ is the angle made by the tangent to the sheet with the real axis. In this coordinate system the stream-surface condition referred to above may be written as

$$\frac{\Phi_n}{U} = -\sin\psi \left.\frac{dy}{dx}\right|_\eta + \cos\psi \left.\frac{dz}{dx}\right|_\eta \tag{2}$$

and the pressure condition as

$$\frac{d\Delta\Phi}{dx} = \frac{d\Delta\Phi}{d\sigma}\left(\left.\frac{dy}{dx}\right|_\eta \cos\psi + \left.\frac{dz}{dx}\right|_\eta \sin\psi - \frac{\Phi_{\sigma m}}{U}\right) \tag{3}$$

where Φ_n is the normal component of the velocity to the sheet by the cross-flow plane, $\Phi_{\sigma m}$ is the mean velocity tangential to the sheet also in the cross-flow plane, Δ is the difference operator across the vortex sheet (inside minus outside) and $\left.\frac{d}{dx}\right|_\eta$ denotes the derivative with respect to x keeping η constant. $\Phi_{\sigma m}$ and Φ_n are related to the complex velocity potential through the formula

$$\Phi_{\sigma m} - i\Phi_n = \frac{dW}{d\sigma} = \frac{dW}{dZ}\frac{dZ}{d\sigma} .$$

Since the inner part of the vortex sheet is replaced by an isolated vortex and cut it clearly is not possible to satisfy the pressure condition (3) at all points of the vortex sheet. Instead we replace, as is usual, the pressure condition on the inner part of the vortex sheet by an overall force condition, namely that the total force which acts on the isolated vortex and cut is zero. This condition may be expressed as

$$U\frac{d\Gamma}{dx}(Z_v - Z_E) + \Gamma U \frac{dZ_v}{dx} - \overline{\lim_{Z\to Z_v}\Gamma\left(\frac{dW}{dZ} - \frac{\Gamma}{2\pi i(Z-Z_v)}\right)} = 0 , \tag{4}$$

where Γ is the strength of the isolated vortex, Z_v its position, Z_E the position of the free end of the vortex sheet and the bar denotes the complex conjugate.

To the above conditions we add a Kutta condition, namely,

$$\frac{dW}{dZ} \text{ is finite} \tag{5}$$

at the leading edge of the wing.

In the case of conical flow Γ, Z_v and y,z at constant η are all proportional to x, s is chosen as the wing semi-span, and the boundary conditions (2) to (5) may then be written in such a way that they are independent of x.

We must now construct an analytic expression for the complex potential W in each cross-flow plane which satisfies the boundary conditions on the wing and at infinity, together with the conditions (2) to (4) on the vortex sheet and at the isolated vortex, whose locations are not known a priori, and the Kutta condition (5) at the leading edge.

To satisfy the wing boundary condition the slit representing the wing in the cross-flow plane is transformed into a convenient shape. In this case the Joukowski transformation, $Z = \frac{b}{2}(\zeta + 1/\zeta)$, which maps the region outside the wing into the region outside a unit circle, is chosen. Here ζ is the coordinate of a point in the transformed plane and b(x) is the wing semi-span in the cross-flow plane. This transformation is also suitable for non-symmetric flows, for example with a wing at yaw, although only symmetrical ones are considered in this paper.

The velocity field which satisfies the wing boundary conditions and the correct conditions at infinity is then given at the point ζ in the transformed plane by

$$\frac{dW}{d\zeta} = -\frac{i\alpha b U}{2}(1 + 1/\zeta^2) + \frac{\Gamma}{2\pi i}\left\{\frac{1}{\zeta - \zeta_v} - \frac{\bar{\zeta}_v}{\zeta\bar{\zeta}_v - 1} - \frac{1}{\zeta + \bar{\zeta}_v} + \frac{\zeta_v}{\zeta\zeta_v + 1}\right\}$$
$$+ \frac{1}{2\pi i}\int_0^{\sigma_E} -\frac{d\Delta\phi}{d\sigma}\left\{\frac{1}{\zeta - \zeta(\sigma)} - \frac{\bar{\zeta}(\sigma)}{\zeta\bar{\zeta}(\sigma) - 1} - \frac{1}{\zeta + \bar{\zeta}(\sigma)} + \frac{\zeta(\sigma)}{\zeta\zeta(\sigma) + 1}\right\}, \qquad (6)$$

where ζ_v is the position of the vortex in the transformed plane and σ_E is the arc length of the finite part of the vortex sheet.

At the leading edge of the wing, z = b(x), and hence $dZ/d\zeta = 0$. Because of this singularity in the transformation the Kutta condition (5) becomes

$$\frac{dW}{d\zeta} = 0 \qquad (7)$$

In each cross-flow plane, as noted earlier, we use an intrinsic coordinate system (η, ψ) to describe the initially unknown sheet shape. The sheet coordinates are then given by

$$Z = b(x) + \int_0^{\sigma} e^{i\psi}\, d\sigma \quad . \qquad (8)$$

With the complex velocity now given by (6) in the transformed cross-flow plane the equations which express the boundary conditions, and so determine the position and strengths of the isolated vortex and vortex sheets may be reduced to a set of non-linear algebraic equations which can be solved using an iterative procedure. In the finite difference representation the shape of the sheet in each cross-flow plane is defined by the value of ψ at n intermediate points on it, with the sheet strengths defined by $-\frac{1}{U}\frac{d\Delta\Phi}{d\sigma}$, again evaluated at the intermediate points.

The intermediate points are defined in the next paragraph. These values which define the sheet strength and shape, together with the position Z_v and the strength Γ/U of the isolated vortex, give 2n + 3 unknowns, represented by the vector $\underline{X}$, to be determined in each cross-flow plane.

Near the leading edge of the wing the sheet shape can change quite rapidly, especially at low angles of incidence, and for this reason it is desirable to have an accurate representation of the sheet in this region. To achieve this n + 1 pivotal points, which define the location of the vortex sheet in each cross -flow plane, are taken to be equal intervals in t apart, where $\eta = \frac{\lambda t^2(7-t)}{6(1+t)}$ with the leading edge as the first point; λ is an arbitrary constant chosen so that the sheet extends some way around the isolated vortex at the initial station. This scaling ensures that the points on the sheet are closer together near the leading edge. Points on the sheet midway between these pivotal points are defined as the intermediate points. Because of the form of the pressure and stream surface conditions and the necessity to march downstream keeping η constant, the same step length in t, h_t say, must be used for each cross-flow plane.

The method adopted here for advancing the solution downstream through successive cross-flow planes is based on the Crank-Nicolson marching procedure. Initially the solution, $\underline{X}(x_0)$, at some station $x = x_0$ is known because, upstream of this point, all the wings considered are chosen to be triangular and so an available conical solution may be assumed at this point. Using linear extrapolation an estimate is then made of the solution $\underline{X}(x_0+h)$ at a station $x = x_0+h$ downstream. The objective is to satisfy the stream surface and pressure conditions, (2) and (3), at the intermediate points on the sheet, the no force condition, (4), on the vortex and cut, and the Kutta condition, (7), at the leading edge. For this purpose we have 2n + 3 equations to solve for the 2n + 3 unknowns

The streamwise derivatives in equations (2), (3) and (4) are approximated using central difference formulae. Thus

$$\frac{d}{dx} f(x_0+\tfrac{1}{2}h,\eta) = \frac{f(x_0+h,\eta) - f(x_0,\eta)}{h} . \tag{9}$$

To use this central difference formula in the boundary conditions (2), (3), (4) and (7) the quantities appearing therein must be evaluated at $x = x_0+\frac{1}{2}h$ by interpolating between $\underline{X}(x_0)$ and $\underline{X}(x_0+h)$. Since the solution $\underline{X}(x_0+h)$ is found iteratively, the quantities in the plane $x = x_0+\frac{1}{2}h$ must be updated at each iteration. Furthermore to apply this central difference formula (9) to the stream surface (2) and pressure conditions (3) the sheet coordinates and circulations must first be calculated in the physical planes x_0 and x_0+h. These are calculated from $\underline{X}(x_0)$ and $\underline{X}(x_0+h)$ in a similar manner to that described below for the quantities in the cross-flow plane $x_0+\frac{1}{2}h$. In this cross-flow plane it is necessary, for example to calculate ψ and $-\frac{1}{U}\, d\Delta\Phi/d\sigma$ at the pivotal points, to interpolate along the sheet and for this purpose four point Lagrangian interpolation formulae are used. All the integrations which have to be performed in each cross-flow plane, in (6), (8) and the formula for the sheet circulation, namely

$$\Delta\Phi(\sigma) = \Gamma + \int_{\sigma}^{\sigma_E} -\frac{d\Delta\Phi}{d\sigma}\, d\sigma = \Gamma + \int_{t}^{t_E} -\frac{d\Delta\Phi}{d\sigma}\frac{d\sigma}{dt}\, dt \qquad ,$$

have been carried out with respect to t using Simpson's rule so that comparable accuracy with the interpolation formulae is achieved. Care must be taken when evaluating the integrals in (6) for points on the sheet since the integrand is singular and the integral is treated as a Cauchy Principal Value using the method adopted by Barsby (1973).

The 2n + 3 algebraic equations which we have established for the unknown quantities at $x = x_0 + h$ can now be written vectorially in the form $\underline{F}(\underline{X}(x_0+h)) = 0$ This set of equations is solved using the Newton iterative scheme,

$$\underline{X}^{n+1} = \underline{X}^n - J^{-1}\,\underline{F}(\underline{X}^n) \qquad ,$$

where J is the Jacobian, $\partial F_i/\partial X_j$, evaluated at the first iteration, and $\underline{X}^n$ denotes $\underline{X}$ evaluated at the nth iteration. Once the solution has been found the numerical procedure continues marching downstream using $\underline{X}(x_0+h)$ as the known value at the point $x = x_0 + h$. The iterative procedure at each step may be accelerated by recalculating J^{-1} but this is only carried out when the rate of convergence slows down relative to its initial rate of convergence.

EXAMPLES

The method described above has been applied to wings of two different basic shapes. The first, a gothic wing for which b, the wing semi-span, is given by

$$\left.\begin{array}{ll} b = 0.25\, x & 0 \le x < 1.1 \\ b = -0.15125 + 0.525 - 0.125\, x^2 & 1.1 \le x < 2.1 \\ b = 0.4 & x \ge 2.1\,. \end{array}\right\}$$

Figures 2 and 3 show vortex sheet shapes for two different angles of incidence. The value of α and the form assumed for S are shown in the Figures. The overall circulation for these cases is shown in Fig. 4. Also shown in the diagram is the circulation calculated by Smith (1959) using a Brown and Michael model, in which the entire vortex sheet and core is replaced by an isolated vortex and cut. For $\alpha = 0.1$ the results of calculations by Clark (1974) using a polar coordinate representation and a different marching scheme are also shown.

The second example studied is an ogee wing, similar in planform to the Concorde and the Tu 144. This wing is sketched in Fig. 5. Figures 6 and 7 show, at different streamwise stations, the vortex sheet shapes and vortex positions for different incidences, and the corresponding circulation is shown in Fig. 8.

Whilst no comparison with experiment has been carried out for these cases, extensive comparison has been carried out by Smith (1966) for the conical-flow case. These comparisons show that the height of the vortex is predicted correctly and the lateral distance of the vortex from the centre is overestimated, mainly as a result of secondary separation. If the overall lift is calculated, however, the agreement between experiment and theory is very good. These features can also be expected to hold for the non-conical model described in this paper.

ACKNOWLEDGEMENTS

Financial support for this work was provided by the Ministry of Defence under research agreement AT/2162/05.

The author would like to thank Dr. J. E. Barsby of the University of Liverpool, Professor N. Riley of the University of East Anglia and Mr. J. H. B. Smith of the Royal Aircraft Establishment, Farnborough, for their help and guidance.

REFERENCES

Barsby, J. E. Aeronautical Quarterly, 24, 120 - 128, 1973.

Brown, C. E. and Michael, W. H. N.A.C.A. Tech. Note 3430, 1955.

Clark, R. W. Private Communication.

Smith, J. H. B. A.R.C. R & M 3116, 1959.

Smith, J. H. B. R.A.E. Technical Report No.66070, 1966.

Smith, J. H. B. R.A.E. Tech. Memo., Aero, 1368, 1971.

Ward, G. N. Linearized Theory of Steady High Speed Flow (Cambridge University Press, 1955).

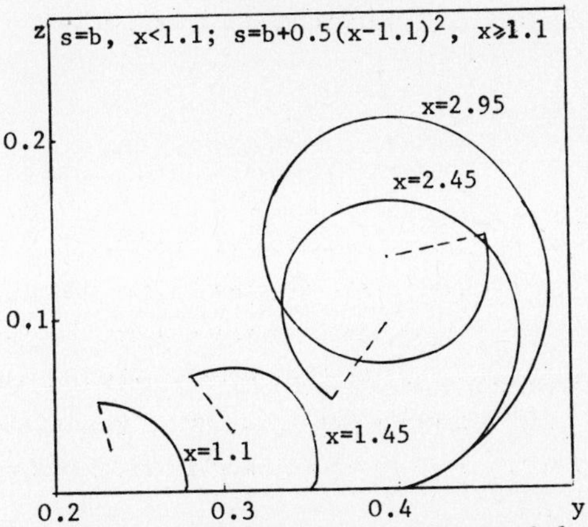

Fig.2 Vortex sheet shapes $\alpha = 0.1$

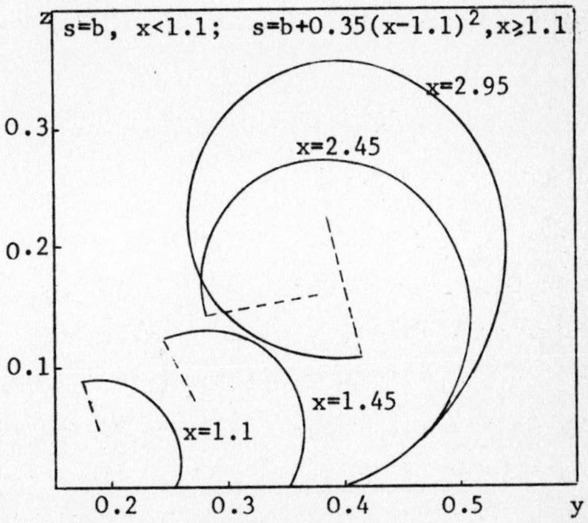

Fig. 3 Vortex sheet shapes, $\alpha = 0.2$

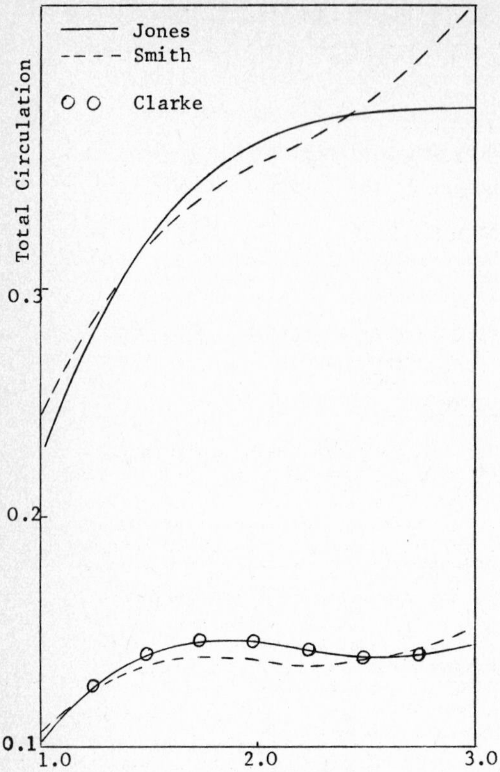

Fig.4 Total Circulation

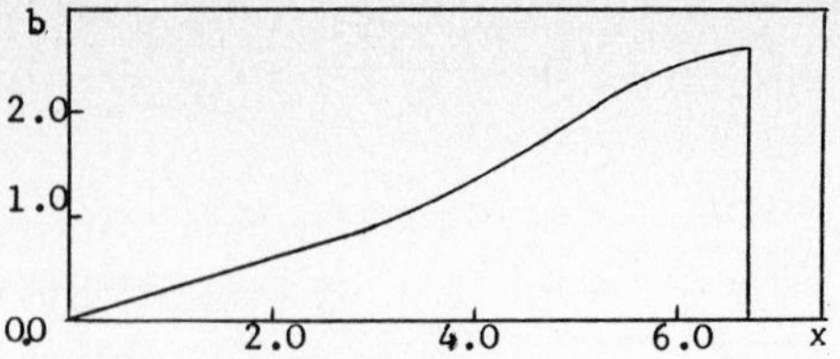

Fig. 5 Ogee wing

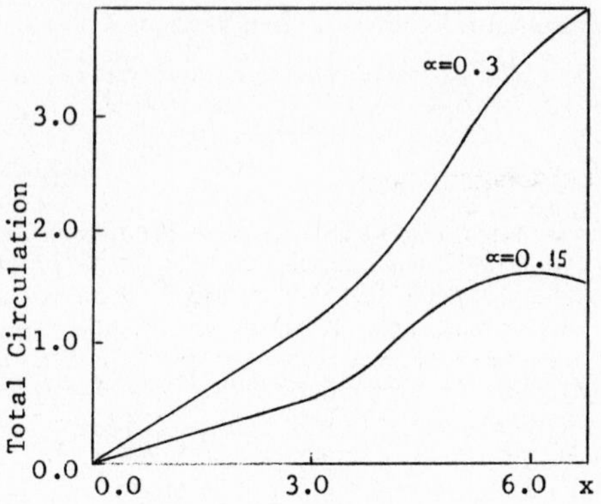

Fig. 8 Total Circulation

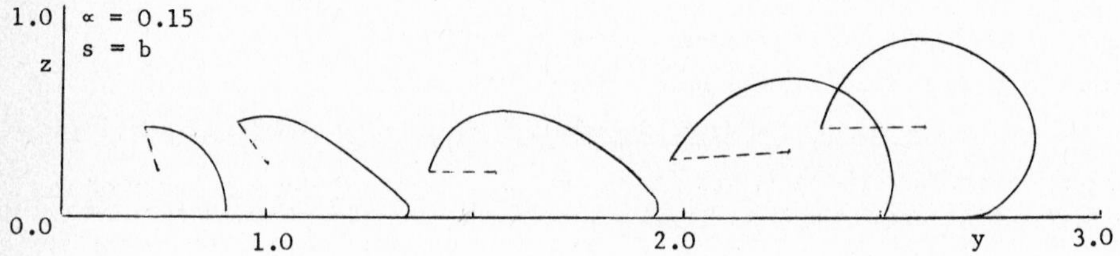

Fig. 6 Vortex sheet shapes

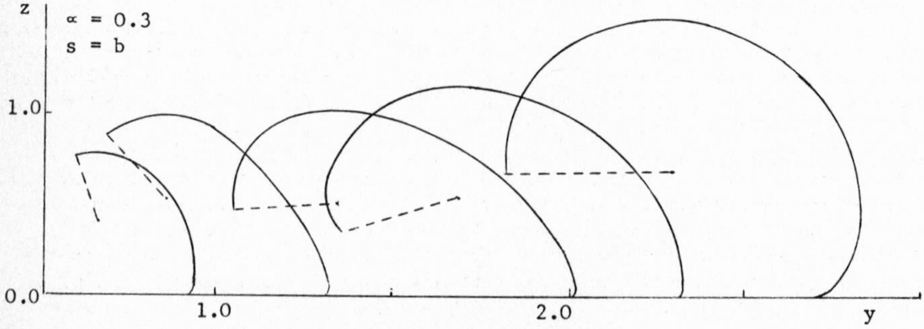

Fig. 7 Vortex sheet shapes

A METHOD OF CALCULATION FOR THE THREE-DIMENSIONAL FLUID FLOW IN TURBOMACHINES

C. Korving
Delft University of Technology
Department of Mathematics
The Netherlands

1. INTRODUCTION

This report presents a consideration of the internal flow in a centrifugal pump impeller. Because of the three-dimensional configuration of the flow the problem to be solved is very complicated, even if it is simplified by using the irrotational flow of an incompressible, inviscid fluid. The various quasi-three-dimensional methods which are used to solve both direct and design problems in turbo machinery are based on the assumption of solving a two-dimensional flow problem on flow surfaces of a known shape. Since these methods have the disadvantage that the accuracy of the results of the calculation depends on the geometry of the impeller and, moreover, it is very difficult to estimate the error involved, a method of solving the direct problem has been developed on the basis of a real three-dimensional approach. The basic equation, which is governed by the Laplace equation can be derived from the steady flow, which occurs in the system of coordinates fixed to the rotor. The purpose of this study is twofold: firstly, to find a method which considerably reduces the computation time required to calculate the blade velocity distribution and, secondly,to use an approximation technique providing results whose accuracy is independent of the geometry of the impeller, the blade shape and the number of blades.

2. FORMULATION OF THE PROBLEM

Since the absolute flow is assumed to be known at the stations far upstream and downstream from the rotor,the latter as far as the meridian component of the velocity is concerned, it is necessary to consider a flow region consisting of three parts: an inlet, a part with rotating blades and an outlet. Because of the periodicity in the geometry, we shall confine ourselves to a region between two successive blades and extensions from these blades in the inlet and outlet regions. The shape of the surface lying between the leading edge of the blade and the entrance plane may be chosen arbitrarily. The boundary conditions at this surface are only subject to the condition of periodicity. The corresponding surface in the outlet part, however, must actually coincide with the vortex plane containing the vortices shed at the trailing edge of the blades. Since the position of this surface is unknown, the shape has initially to be assumed prescribing boundary conditions specified by the condition of continuity of the pressure over the assumed surface and by the condition of identical normal derivatives in corresponding points of the two successive surfaces. After the calculations have been performed, the resulting shape must be checked with the assumed shape to see whether improved approximations are necessary. In view of these considerations the boundary value problem can now be formulated in terms of the Laplace equation, subject to the boundary conditions (Fig. 2.1):

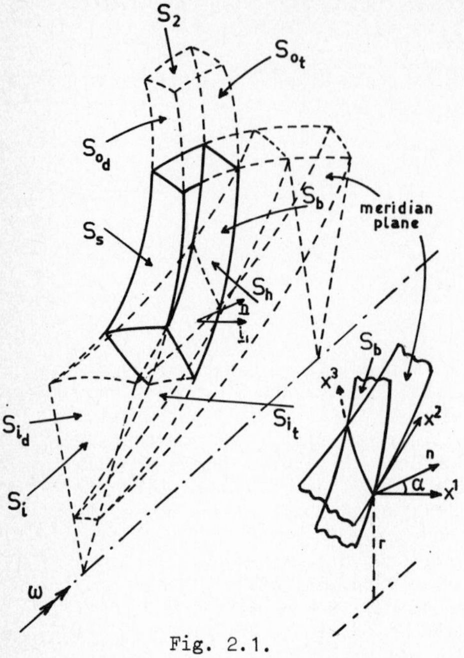

Fig. 2.1.

1) at the blade surfaces S_b, the hub S_h, the shroud S_s and the entrance and discharge stages S_1 and S_2 we have:

$n^i \phi_{,i} = \omega r \cos\alpha$	S_b
$= 0$	S_h and S_s
$= f$	S_1 and S_2

2) at the extensions from the blades S_i and S_o in the inlet and outlet regions respectively:

$$n^i \phi_{,i}\big|_t = -n^i \phi_{,i}\big|_d \;; \quad \phi_t = \phi_d + q(x^3)$$

(the subscrpts t and d indicate the trailing and driving surfaces). The quantity q vanishes at S_i when the tangential component of the velocity is zero ahead of the rotor, whereas this quantity is unknown at S_o, which has to be determined by a process of iteration in order to satisfy the Kutta condition at the trailing edge of the blades.

3. THE METHOD OF SOLUTION

On the one hand, the solution of the above boundary value problem can be based on a direct method of numerical integration of the Laplace equation, while on the other hand it can be based on the method of solving the equivalent variational problem. Regardless of which method is chosen, a direct numerical solution of the problem will involve too much computation time. This difficulty can be overcome by adopting the following approach:

A regular set of coaxial surfaces of revolution is located in the space between the hub and the shroud of the rotor. These are denoted as coordinate surfaces of constant x^3 at which $x^3 = x^3_j$, where j runs from 1 (hub) to n (shroud). The functions ϕ^j and ϕ^j_u of the surface coordinates x^α ($\alpha = 1,2$) are defined, representing the potential and its derivative with respect to x^3 respectively. The potential ϕ at an arbitrary point in a subdomain (j) lying between the surfaces $x^3 = x^3_j$ and $x^3 = x^3_{j+1}$ is assumed to be a polynomial in x^3 with coefficients containing the defined quantities ϕ^j and ϕ^j_u at their boundaries, expressed by

$$\phi^{(j)}(x^1,x^2,x^3) = l_0^{(j)}(x^3)\,\phi^j(x^1,x^2) + l_1^{(j)}\phi^{j+1} + m_0^{(j)}\phi_u^{j} + m_1^{(j)}\phi_u^{j+1} \qquad (3.1)$$

The coefficients of the interpolation polynomicals l_i and m_i are calculated from the condition that the approximated potential and its derivative with respect to x^3 correspond exactly to the quantities ϕ^j and ϕ^j_u at the boundaries $x^3 = x^3_j$ and

$x^3 = x^3_{j+1}$, resulting in polynomials of the third degree. The approximation (3.1) is used to produce a set of partial differential equations, following from the requirement that the Laplace equation is only satisfied on the surfaces $x^3 = x^3_j$. Arranging this system, the result of the collocation method is represented in the set of n partial differential equations of the form:

$$\frac{G_i^{33}}{\delta}\left[\frac{3}{\delta} A_{ij}\phi^j + B_{ij}\phi_u^j\right] + \frac{\partial}{\partial x^\alpha}\left(U_{ij}^{\alpha\beta}\frac{\partial}{\partial x^\beta}\phi^j\right) + 2\frac{\partial}{\partial x^2}\left(U_{ij}^{23}\phi_u^j\right) + \qquad (3.2)$$

$$+ V_{ij}\frac{\partial}{\partial x^2}\phi^j + W_{ij}\phi_u^j = 0 \qquad i,j = 1(1)n \quad \alpha,\beta = 1,2$$

where:

$$(\sqrt{g}\, g^{ij})_k = G_k^{ij} \qquad G_k^{13} = 0$$

$$\delta = x^3_{j+1} - x^3_j$$

$$A_{ij}: \begin{pmatrix} -2 & 2 & & & \\ 1 & -2 & 1 & & \\ & \ddots & \ddots & \ddots & \\ & & 1 & -2 & 1 \\ & & & 2 & -2 \end{pmatrix} ; \quad B_{ij}: \begin{pmatrix} -4 & -2 & & & \\ 1 & 0 & -1 & & \\ & \ddots & \ddots & \ddots & \\ & & 1 & 0 & -1 \\ & & & 2 & 4 \end{pmatrix}$$

$$U_{ij}^{\alpha k} = G_i^{\alpha k}\,\delta_j^i$$

$$V_{ij} = \left(\frac{\partial}{\partial x^3} G^{23}\right)_i \delta_j^i$$

$$W_{ij} = \left\{\left(\frac{\partial}{\partial x^3} G^{33}\right)_i - \frac{\partial}{\partial x^2} G_i^{23}\right\}\delta_j^i$$

combined with n-2 relations, actually resulting from the requirement of the continuity of the second derivative of the potential with respect to x^3 at the intersurfaces $x^3 = x^3_j$ (j=2(1)n-1) :

$$\frac{3}{\delta}\left(\phi^{j+1} - \phi^{j-1}\right) = \phi_u^{j+1} + 4\phi_u^j + \phi_u^{j-1} \qquad (3.3)$$

Next, the complete set of equations in the 2n unknowns ϕ^j and ϕ_u^j is obtained if the boundary conditions at the hub and the shroud are added to the system:

$$G_{(m)}^{23}\frac{\partial}{\partial x^2}\phi^{(m)} + G_{(m)}^{33}\phi_u^{(m)} = 0 \qquad m = 1, n \qquad (3.4)$$

The boundary conditions must be prescribed on the bounding curves Γ_i which enclose the domains at each surface of revolution $x^3 = x^3_i$, the relations being found by the replacement of ϕ by ϕ^i in the remaining boundary conditions. Considering the system of equations (3.2), it can be noted that the interdependency of the various equations is produced by the first two terms. Since these terms are dominant in the equations, it is possible to transform the system of equations into a system

containing dominant terms with respect to one dependent variable in each of the equations. Each equation can then be solved individually this operation being part of a process of successive iteration. The basic operation of the solution procedure is, in fact, similar to the case when a real two-dimensional problem at one surface of revolution has to be solved by using an iteration procedure for only one constant in relation to the Kutta condition.
Before the transformation with respect to the dependent variables can be performed, the bounding curve Γ_i at each surface $x^3 = x^3_i$ must be represented by the same values of the x^1, x^2 coordinates. The initial boundary conditions can then be mixed up in the boundary conditions expressed in the new dependent variables. This can be achieved by the choice of an axisymmetric system of coordinates in which the x^3 coordinate curves are generated by the intersection lines of the mid-surface of the blade with meridian planes x^1 = constant (Fig. 2.1). As a result, the x^3 coordinate is generally not perpendicular to the x^2 coordinate, hence the tensor notation is used. In order to obtain the diagonal formation of the equations (3.2) the eigenvalues λ and eigenfunctions have to be calculated from:

$$A_{ij}\,\phi^j + \frac{\delta}{3}\,B_{ij}\,\phi_u^{*j} - \lambda\,\delta_{ij}\,\phi^j = 0 \qquad i,j = 1(1)\,n \qquad (3.5)$$

combined with the equations (3.3):

$$C_{ij}\,\phi_u^{*j} = \frac{3}{\delta}\,B_{ij}\,\phi^j \qquad i = 2(1)\,n-1, \quad j = 1(1)\,n \qquad (3.6)$$

with

$$\begin{aligned} C_{ij} &= A_{ij} && i \neq j \\ &= 4 && i = j \\ \phi_u^{*m} &= 0 && m = 1, n \end{aligned} \qquad (3.7)$$

The relationship between ϕ_u^{*j} and ϕ_u^j follows from the condition (see eq. 3.6 and 3.3):

$$C_{ij}\,\phi_u^j = C_{ij}\,\phi_u^{*j} \qquad i = 2(1)\,n-1, \quad j = 1(1)\,n \qquad (3.8)$$

the homogeneous boundary conditions for ϕ_u^{*j} (3.7) and the inhomogeneous boundary conditions for ϕ_u^j (3.4). If the eigenfunctions for ϕ^j and ϕ_u^{*j} are expressed in the form

$$E_l^j = C_l^1 \cos(j-1)\beta_l$$

and

$$Eu_l^j = C_l^2 \sin(j-1)\beta_l \qquad (3.9)$$

respectively, the eigenvalues can easily be calculated. Performing the transformation

$$\begin{aligned} \phi^j &= E_l^j\,\psi^l \\ \phi_u^{*j} &= Eu_l^j\,\psi^l \end{aligned} \qquad (3.10)$$

the complete system of 2n equations (3.2), (3.3) and (3.4) can be reduced to a system of n partial differential equations in ψ^l, with the property that ψ^l is dominant in equation m for l = m.

The solution of the partial differential equations is obtained by using the Galerkin method. An approximate solution can then be expressed in the form:

$$\psi^l = \psi_s^l\,p^s(x^1, x^2) \qquad s = 1(1)\,N, \quad l = 1(1)\,n \qquad (3.11)$$

where p^s is a certain system of N functions, in the form of polynomials in the finite element representation, and ψ_s^l are undetermined coefficients. Using the symbol L_i as operator for equation i of the system of n partial differential equations and writing the boundary conditions in the general form

$$F^i = [n^k \phi_{,k}]^i - \bar{f}^i = 0 \qquad \text{on } \Gamma \qquad (3.12)$$

we arrive at the system of equations:

$$\iint_{S_t} L_i p^s dx^1 dx^2 - \int_{\Gamma} \alpha_{ij} F^j p^s d\eta = 0 \qquad (3.13)$$

which serves for the determination of the coefficients ψ_s^l. After integrating the surface integral by parts, the result is represented by the system of linear equations:

$$S^{sr}_{m(m)} \psi_r^{(m)} = T^s_m \qquad \text{no sum on } m \qquad \begin{matrix} m = 1(1)n \\ s, r = 1(1)N \end{matrix} \qquad (3.14)$$

which clearly indicates that n subsystems of N equations now have to be solved by means of successive iteration instead of solving the initial version of a complete system of 2n x N equations.

4. RESULTS

The blade velocity distribution was calculated for fluid flows through an impeller of a given geometry, blade shape and number of blades.
The finite element method was used, taking a regular division of triangular elements at a surface of revolution, and a linear shape function was chosen. The calculations were carried out on an IBM 360/65 computer and the computation time required to complete a run, including the iteration due to a leading edge not following an x^3 coordinate curve, amounted to less than one minute for the case n = 2. Comparing only the results of the calculations for two different positions of the leading edge (see Fig. 4.1), it can be concluded that case 2 gives better results than case 1. This is expressed by the reduction of the backflow at the pressure side of the blade near the hub, and smoothing of the velocity peak at the suction side of the blade near the shroud. In the existing program the boundary conditions are prescribed for the mid-surface of the blade, taking account of the thickness of the blade only as far as the blockage effect is concerned. Calculations carried out for the case n = 3 show that quite accurate results have already been obtained for the case n = 2. If desired, the method may be further refined, especially with respect to the inclusion of the blade thickness, to give more accurate values of the blade surface velocities near the leading edge than can be obtained with the method described.

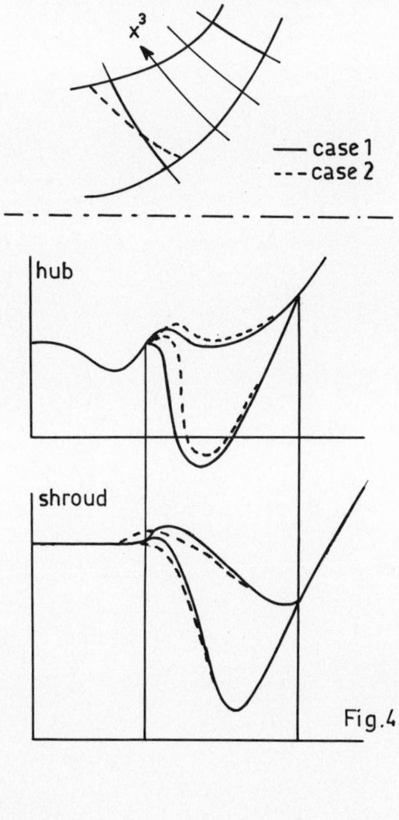

Fig.4.1

5. CONCLUSIONS

The method of solution presented above demonstrates that the complex fluid flows in turbo machines can be analysed by a three-dimensional approach. In view of the low computation time achieved with the existing computer program, various effects such as the flow near the leading edge can be studied more accurately and may possibly be extended to cover the case where a boundary layer thickness is taken into account. Moreover, the use of the collocation method makes it possible to apply a similar method for compressible and rotational flows, which is one of the reasons for preferring the collocation method to the variational method.

6. REFERENCES

Katsanis T. NASA TN - D - 2546, (1964).

Korving C. Dissertation, Delft University of Technology, (to be published in 1974).

Stockman N.O. and Kramer J.L. NASA TN - D - 1562, 1963.

Vavra M.H. Aero-thermodynamics and flow in turbo machines (1960)

SOME HYDRODYNAMIC PROBLEMS FOR MULTICOMPONENT SYSTEMS

Kuznetzov B.G., Zuykov Y.P.

Computer Center, Siberian Branch of the USSR Academy of Sciences, Novosibirsk

The existing mathematical models for multicomponent systems can be approximately divided into two groups: diffusive and multi-velocity models. Each has its own field of application. In diffusive models individual motion of components are characterized by small diffusive additions to the mixture velocities. Therefore their application is difficult if relative phase velocities are high. In multi-velocity models diffusive effects are neglected, therefore they do not work in the diffusive condition of discrete phases. In this connection of interest are multi-velocity models operating within a wide range of relative velocities of components in the presence of diffusive effects.

I. An $(n+1)$-component medium is considered consisting of a continuous phase $(i=0)$, into which descrete phases are embeded in a form of solid particles, drops, bubbles $(i=1,2,\dots,n)$. In the absence of phase conversions the equations of mass conservation are as follows:

$$\frac{\partial}{\partial t}(m_i\alpha_i) + div(m_i\alpha_i v_i) = 0\ ,\quad \sum_{i=0}^{n}\alpha_i = 1\ ,\quad (i=0,1,\dots,n). \qquad (1.1)$$

Here m_i is the substance density, v_i - the average mass velocity of the i-component, α_i - their volume concentration in the mixture.

Impulse equations for components can be presented as follows:

$$\rho_i\frac{dv_i}{dt_i} = \rho_i\mathcal{F} + \mathcal{F}_i + div\,\sigma_i\ ,\ \rho_i = m_i\alpha_i\ ,\ \frac{d}{dt_i} = \frac{\partial}{\partial t} + v_i\cdot grad. \qquad (1.2)$$

Here $\mathcal{F}$ is the mass force; the $\mathcal{F}_i$ are volume forces of interaction of the components, σ_i is the tension tensor for the i-component. By virtue of Newton's third law:

$$\sum_{i=0}^{n}\mathcal{F}_i = 0. \qquad (1.3)$$

Representations of $\mathcal{F}_i$ and σ_i usually cause difficulty in the description of concrete motions of multicomponent media within the framework of the multi-velocity model. In this connection it should be noted that in contrast to one-component system, where the volume forces are given, the multicomponent systems have no sharp distinction between the surface and volume forces of the component interaction: part of the surface forces, which are associated with some tensors σ_i^+, that additively enter into σ_i and satisfy the condition (1.3):

$$\sum_{i=0}^{n} div\,\sigma_i^+ = 0 \qquad (1.4)$$

can be related to the volume forces of interaction. It is clear that the expressions for the volume forces of interaction change, too, but so that the expressions $div\,\sigma_i + \mathcal{F}_i$ remain invariant. Désignating the tension tensor of the mixture by σ, the impulse equation for the mixture

$$\left.\begin{array}{l}\rho \dfrac{dv}{dt} = \rho \mathcal{F} + \operatorname{div}[\sigma - \sum_{i=0}^{n} \rho_i (v - v_i)(v - v_i)] \,, \\ \rho = \sum_{i=0}^{n} \rho_i \,, \quad \rho v = \sum_{i=0}^{n} \rho_i v_i \,, \quad \dfrac{d}{dt} = \dfrac{\partial}{\partial t} + v \cdot \operatorname{grad}\end{array}\right\} \qquad (1.5)$$

must be the consequence of (1.2). Then it is necessary that

$$\operatorname{div} \sum_{i=0}^{n} \sigma_i = \operatorname{div} \sigma .$$

Considering the last relation as the equation for σ_i , its general solution can be written to within the additive members of σ_i^+ satisfying (1.4), e.g. in the form [1] :

$$\sigma_i = c_i \sigma \quad , \quad \rho_i = c_i \rho . \qquad (1.6)$$

If later, in describing the volume forces of the interaction of the components, all the forces are taken into account, including the forces $\operatorname{div} \sigma_i^+$, then expressions (1.6) can be regarded as the most general expression for tension tensors of the components in the most general case of the mixture motion.

If the mixture is a Stokes fluid [2] , the relationship between the tension tensor and deformation is of the form

$$\sigma = [-p + \zeta \operatorname{div}(\sum_{i=0}^{n} \alpha_i v_i)] I + \mu_1 D + \mu_2 D^2 , \qquad (1.7)$$

where ζ, μ_1, μ_2 are functions of the invariants D , characteristics of the components and energy of the micromotions of mixture, respectively.

2. To distinguish the volume forces of interaction explicitly, one can use the following procedure. Equations (2.1) and the expressions σ_i in the form (1.6) can be regarded as general relations which must be true in different special cases, when there is some additional data about $\mathcal{F}$. Here one can always create, or at least, imagine such conditions when there act forces to be determined. Thus, to distinguish explicitly the forces of interaction due to the law of Archimedes, let us consider a special case of the mixture motion when in some point M at the time t , the velocities of all the components are equal and $\operatorname{grad} \alpha_i = 0$, while some environs of the M move as a solid body. An example of this situation is the initial state of the medium at rest with a uniform distribution of the components in the field of the gravity forces. Let us take the local system of coordinates with the origin in M , whose acceleration is that of the above-mentioned environs of M at the time t . Then equations (1.3), (1.5) yield:

$$\rho \mathcal{F}_* = \operatorname{grad} p \,, \quad \frac{\partial}{\partial t}(\rho_i v_i) = \rho_i \mathcal{F}_* - c_i \operatorname{grad} p + \mathcal{F}_{i1} . \qquad (2.1)$$

Here $\mathcal{F}_*$ is the basic vector of the external mass forces and the force of inertia, the $\mathcal{F}_{i1}$ - are the forces of interaction, due to the law of Archimedes. On the other hand, the change of the impulse $\partial/\partial t(\rho_i v_i)$ due to the law of Archimedes is $(\rho_i - \alpha_i \rho)\mathcal{F}_*$, where $-\alpha_i \rho \mathcal{F}_*$ is the buoyancy force. Substituting it into (2.1) we obtain:

$$\mathcal{F}_{i1} = \rho(c_i - \alpha_i)\mathcal{F}_* = \rho(c_i - \alpha_i)(\mathcal{F} - \frac{dv}{dt}) . \qquad (2.2)$$

The average motion of the components with respect to the mixture and to one another results in the appearance of the known interaction forces

$$\mathcal{F}_{i2} = \varphi_i (v_0 - v_i) + \sum_{\kappa=0}^{n} \varphi_{i\kappa} (v_\kappa - v_i) , \qquad (2.3)$$

where the $\varphi_i, \varphi_{i\kappa}$ are dependent on the number, shape, dimensions of the particles and the distance between them, on viscisity of the solution and the energy of the mixture micromotion.

Let us dwell in more detail on the forces of diffusive interaction, which are usually neglected in multi-velocity models. Let us consider a special case of media, the density of whose matter is the same for all components, but there are different from zero gradients of concentration c_i . Let us choose the frame of reference so that the velocity of the mixture as a whole is equal to zero. Under these conditions the individual motions of the components are determined only by the diffusion:

$$v_i = -\lambda \operatorname{grad} c_i . \qquad (2.4)$$

Substituting these expressions into (1.2), taking (1.5), (1.6), (1.7) into account, we find that in order to satisfy the law (2.4), it is necessary to introduce the forces of the diffusive interaction of the components:

$$\mathcal{F}_{i3} = p \operatorname{grad} c_i - \frac{d}{dt}(\lambda \operatorname{grad} c_i) . \qquad (2.5)$$

3. In the multi-velocity model, one can prescribe to each component its own temperature T_i , the inner pressure p_i , the energy e_i . Later for simplicity we consider only the case, when all the components are incompressible. In this case $e_i = e_i(T_i), m_i = m_i(T_i)$. Let us note that besides the inner energy of the components, the mixture, as a whole, has one more significant parameter e , which characterizes the energy of micro-motions of the mixture. When the points of the mixture move with different velocity, the individual particle velocity of descrete phases and the points of the continuous phase can essentially differ from their average characteristics v_i . In other words all the components generally take part in some micro-motions, whose energy e can be significant. These micro-motions are the decisive factor in forming viscosity and diffusion of the mixture.

In the incompressible medium the energy exchange between the components is described by the following equations:

$$\left.\begin{aligned} \rho_i \frac{de_i}{dt_i} &= \vartheta_i (T_0 - T_i) + \operatorname{div}(\lambda \operatorname{grad} \rho_i e_i) , \quad (i = 1,2,\dots,n) \\ \rho_0 \frac{de_0}{dt_0} &= \sum_{i=1}^{n} \vartheta_i (T_i - T_0) + \operatorname{div}(\lambda \operatorname{grad} \rho_0 e_0) + C_e \\ C_e &= \mu_0 D : D + k_1 \mu_0 e R^{-2} \end{aligned}\right\} \qquad (3.1)$$

where ϑ_i are coefficients of the conductive heat exchange. The value C_e characterizes dissipation of the mechanical energy into heat, including dissipation of the energy of micro-motions, which is directly proportional to the molecular viscosity of the continuous phase μ_0, and the micro-motion energy e and inversely propotional to the square of the average distance between the particle surfaces R. Using the theorem of the energy conservation for the evolved volume of the mixture as a whole, we obtain the equations for the micro-motion energy:

$$\rho \frac{de}{dt} = div(\lambda \, grad \, \rho e) + \sigma : D - C_e \, . \qquad (3.2)$$

In case of the linear relationship between stresses and deformations (1.7) the coefficients of viscosity μ and heatconductivity λ can be defined by the relations:

$$\mu = \mu_0 + k_\mu r \rho \sqrt{e} \, , \quad \lambda = \lambda_0 + k_\lambda r \sqrt{e} \, , \quad k_\mu, k_\lambda = const. \qquad (3.3)$$

The scale of micro-motions r satisfies the equation

$$\rho \frac{dr}{dt} = div(\lambda \, grad \, \rho r) + k_r \rho \sqrt{e} \, , \quad k_r = const > 0. \qquad (3.4)$$

Finally, the equation for the quantity of particles under the assumption of concervation of their quantity, is as follows:

$$\frac{dN_i}{dt_i} + div(N_i V_i) = 0 \, , \quad (i = 1, 2, \ldots, n). \qquad (3.5)$$

4. Within the model suggested there are solved the problems of precipitation of suspension of solid particles in the viscous heavy fluid [3] and of motion of the homogeneous particle suspension in the viscous fluid in a plane tube. In the first problem it was found that if one takes account of the diffusive forces of interaction the suspension process becomes longer than without such forces. The problem of the boiling layer was also considered. It was found that the steady boiling layer is provided only for volume concentrations of the particles $\alpha_1 > 0.275$.

In the second problem for simplicity a case is considered, when eigen-densities of the continuous phase and of particles are the same. In this case the problem reduces to the definition of the function $v(x)$, $e(x)$, $r(x)$ satisfying, within $0 < x < 1$, the system of equations:

$$\left. \begin{aligned} & \frac{d}{dx}\left(\mu \frac{dv}{dx}\right) = a = const \, , \quad \frac{d}{dx}\left(\lambda \frac{de}{dx}\right) = be - \frac{\mu}{\rho}\left(\frac{dv}{dx}\right)^2 \\ & \frac{d}{dx}\left(\lambda \frac{dr}{dx}\right) = -c\sqrt{e} \, , \quad b = const, \; c = const, \; \rho = const \end{aligned} \right\} \qquad (4.1)$$

under the conditions:

$$\left.\begin{array}{l} \dfrac{dv}{dx} = \dfrac{de}{dx} = \dfrac{d\tau}{dx} = 0 \ , \quad x = 0 \\[2ex] v = 0 \ , \ e = e^{\circ} = const, \ \tau = \tau^{\circ} = const \ , \ x = 1 \end{array}\right\} \qquad (4.2)$$

The calculations were carried out for the following values of the parameters:

$$k_{\lambda} = k_{\mu} = 1 \ ; \ c = 0.01 \ ; \ e^{\circ} = \tau^{\circ} = 0 \ ; \ \lambda_{o} = 1._{10}{}^{-8} \ ; \mu_{o} = 0.001 \ ; \ \rho = 1000. \qquad (4.3)$$

Fig. 1 shows characteristics of the stream for the plane canal with the smooth walls for $\alpha=-5$ and for two values of the coefficient b : 1) $b = 2$ (continuous lines), 2) $b = 0.1$ (dashed lines). Numbers 1,2,3,4 indicate diagrams of energy ($0.1 \times e$), velocity ($10 \times v$), amplitude of oscillation (τ) and viscosity of mixture ($100 \times \mu$), respectively. Here at α small flow of stream the flow of the mixture is similar to the Poiseuille flow. At large values of α the characteristics of the stream are similar to the average characteris - tics of the turbulent flow. Fig. 2 shows diagrams of the velocity of the mixture (curves 1) and the corresponding velocities of the Poiseuille flow with the same flow of stream -(curves 2) for the values of the parameters: $b = 2$; $\alpha = -500$ (continuous lines); $\alpha = -2500$ (dashed lines).

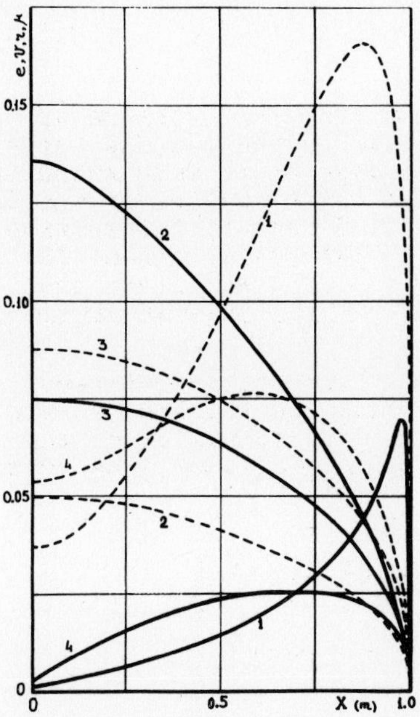

Fig. 1

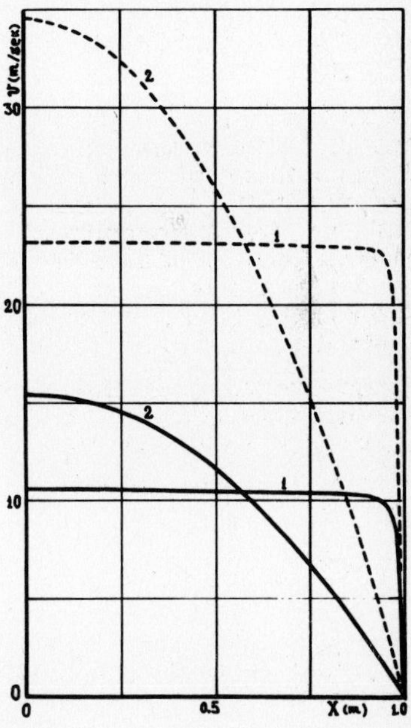

Fig. 2

R E F E R E N C E S

1. Kuznetsov B.G. About the hydrodynamic equations of multiphase systems. Sb. " Tchislennye metody mekhaniki sploshnoi sredy ", t. 4, No 1, Novosibirsk, 1973.

2. Serrin J. Mathematical Principles of Classical Fluid Mechanics. Berlin - Göttingen - Heidelberg, 1959.

3. Zuykov Yu.P., B.G.Kuznetsov. Calculation of motion of the two-component media. PMTF, No 6, 1973.

NUMERICAL SIMULATION OF INTERACTING, THREE-DIMENSIONAL VORTEX FILAMENTS*

A. Leonard
Ames Research Center, NASA, Moffett Field, Calif. 94035

1. INTRODUCTION

Many unsteady three-dimensional fluid flows of interest are characterized by low viscosity and the presence of distinct regions of high vorticity imbedded in an otherwise irrotational flow. Jets, wakes, free-shear layers, vortex rings, and trailing vortices are examples of such flows. The usual source of vorticity in flows of this type is through viscous interaction with a solid boundary and subsequent ejection from the boundary in the form of a free-shear layer or vortex sheet. Generally, this sheet then either breaks into filaments or tubes through a Kelvin-Helmholtz instability or rolls up into a vortex filament. In any case, the vorticity away from the boundary can often be characterized as an intermingling of vortex tubes or vortex filaments. Even in the fine scales of turbulent flows, filament-like or "stringy" structures appear to be dominant (Kuo and Corrsin, 1972).

A numerical method to simulate incompressible flows of this type is presented here which takes advantage of the above-mentioned characteristic in a natural way. The basic idea is to model the vorticity distribution in terms of continuous closed filaments and to track these filaments in a Lagrangian reference frame. Viscous effects are assumed to be important only in the determination of the fine structure within the filament core. Thus each filament moves according to the local fluid velocity appropriately averaged over the filament core. The vorticity distribution within a filament is parameterized by a locally defined, effective core radius determined dynamically by the effects of viscous diffusion and vortex stretching. Similar techniques have been used for two-dimensional flows in which the vorticity is confined to straight tubes parallel to the third dimension (see, e.g., Chorin (1973)).

2. INVISCID MOTION OF A FILAMENT

A vortex filament is taken to be a collection of vortex lines** bunched to form a thin tube that may follow an arbitrary curve in three-dimensional space. Since the vorticity field $\underline{\omega}$ is solenoidal, this curve must be a closed loop. In addition, the integral $\int \underline{\omega} \cdot d\underline{A}$ over any cross section of the filament must be the same anywhere along the filament; this quantity is called the circulation, Γ, of the filament:

$$\Gamma = \int \underline{\omega} \cdot d\underline{A} \tag{1}$$

To develop the necessary equations of motion for the filaments, start with the incompressible vorticity equation

$$D\underline{\omega}/Dt = \underline{\omega} \cdot \nabla\underline{u} + \nu\nabla^2\underline{\omega} \tag{2}$$

where $\underline{u}$ is the velocity field and ν, the kinematic viscosity. Assume that inviscid effects dominate the motion of the filament as a whole and only the internal structure is influenced by the viscosity. Hence, the motion is purely kinematic and the vortex lines follow material lines (Helmholtz, 1858). The inviscid motion of a collection of vortex lines or filament will then be given by a suitable average of the material velocity over the core of the filament. Since $\underline{\omega} = \nabla \times \underline{u}$ and $\nabla \cdot \underline{u} = 0$, $\underline{u}$ is given by

$$\underline{u}(\underline{r},t) = -\frac{1}{4\pi}\int \frac{(\underline{r} - \underline{r}') \times \underline{\omega}(\underline{r}',t)d\underline{r}'}{|\underline{r} - \underline{r}'|^3} + \underline{v}(\underline{r},t) \tag{3}$$

---

*Research performed under a NASA-NRC Senior Research associateship.

**A curve everywhere tangent to a given continuous vector field $\underline{c}$ is a <u>vector line</u> of $\underline{c}$. Vortex lines are vector lines of the vorticity field.

where $\underline{v}$ is a given irrotational contribution (if solid boundaries are present, $\underline{v}$ must be constructed to ensure that $\underline{u}$ is tangent at the solid surface). If $|\underline{r} - \underline{r}'| >> \delta$ (where δ is the local radius of the filament core), then the contribution of $d\underline{r}'$ to the velocity at $\underline{r}$ is essentially constant in $\underline{r}$ over the cross section of the core and may be computed without complications. Averaging over nearby portions of the filament requires more care because these contributions produce large rotational velocities around the center of the core. Only recently has an arbitrary curved filament been treated in a fairly rigorous fashion. Using the methods of matched asymptotic expansions, Widnall, Bliss, and Zalay (1971) and Moore and Saffman (1972) have shown that, in the limit $\delta << R_c$ (R_c is the local radius of curvature of the filament), the contribution to the filament velocity due to any segment containing $\underline{r}$ may be computed by compressing the segment to zero cross section and using an appropriate cutoff in the resulting line integral. Thus if $\underline{u}^*$ denotes this contribution,

$$\underline{u}^*(\underline{r}(s),t) = -\frac{\Gamma}{4\pi} \int_{|s-s'|>\varepsilon} \frac{[\underline{r}(s) - \underline{r}'(s')] \times \frac{\partial \underline{r}'}{\partial s'} ds'}{|\underline{r} - \underline{r}'|^3} \tag{4}$$

where the integration is over the arclength of the segment, s,s' are distance coordinates along the filament, ε and δ are related through

$$\ell n\, \varepsilon = \ell n(\delta/2) + (1/2) - A \tag{5}$$

and A is an $O(1)$ constant that depends on the vorticity distribution within the filament.

3. INTERNAL STRUCTURE

If there is no net axial velocity within the filament, the vorticity distribution there is subject to two important effects - vortex stretching and viscous diffusion. If the filament is undergoing local stretching (extension of the vortex lines as measured with material markers), then by the incompressibility constraint and the fact that vortex lines follow material lines the radius of the filament contracts. Thus if ℓ is the length of a segment of a filament with circular cross section, then $\ell\pi\delta^2$ = const or

$$(d\delta^2/dt)_{\text{inviscid}} = -(\delta^2/\ell)(d\ell/dt) \tag{6}$$

On the other hand, viscous effects tend to spread vortex lines or fatten the filament. For a viscous vortex ring with small cross section, Tung and Ting (1967) and Saffman (1970) found that the distribution of vorticity across the core is Gaussian and that the characteristic radius increases according to

$$(d\delta^2/dt)_{\text{viscous}} = 4\nu \tag{7}$$

Combining the two effects yields, in general (see Marsden (1974)),

$$d\delta^2/dt = 4\nu - (\delta^2/\ell)(d\ell/dt) \tag{8}$$

For the simulations, either of two simplified models for ℓ were used. In the constant-filament-volume model (Moore and Saffman, 1972), ℓ is taken as the total length of the filament and the internal waves in the core are assumed to be efficient in smoothing any variations in δ along the filament. If, on the other hand, effects of these internal waves are assumed negligible, then ℓ is taken as the local length of a small segment - the constant-local-volume model.

4. NUMERICAL PROCEDURES

The core radius of each filament is assumed to be much smaller than the local radius of curvature with no axial velocity in the filament so that the distribution of vorticity is nearly Gaussian in magnitude and the vortex lines are parallel at each cross section normal to the filament tangent vector.

For numerical purposes, the vorticity field is composed of M filaments $(i = 1,2, \ldots, M)$. Each is described by its circulation Γ_i, position vectors of its sequence of node points $\underline{R}_{i,j}(t)(j = 1,2, \ldots, N_i)$ marking the core center and $\delta_{i,j}(t)$, the corresponding filament core radii (characteristic Gaussian widths) for the segment between nodes j and j+1. By use of the arguments cited above, the node points are moved according to the local velocity suitably averaged over the core. For this purpose, the integral in Eq. (3) is split into two parts, one for the contribution from nonadjacent segments and the other for the contribution of the two segments adjacent to the node at which the velocity is being computed. For the former, the filament is approximated as connected straight-line segments extending between each pair of sequential node points. For the latter, a circular arc is fitted through the node point and its two nearest neighbors and the cutoff integral approximation is used to compute the induced velocity due to local curvature.

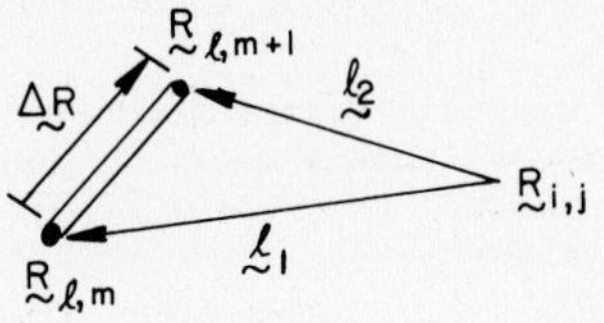

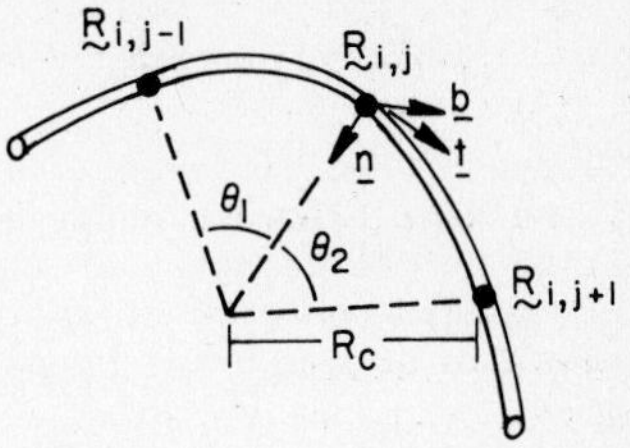

With these assumptions and approximations, the equations of motion for the node position vectors become

$$\frac{d\underline{R}_{i,j}}{dt} = \sum_{\ell} \frac{\Gamma_\ell}{4\pi} \sum_{m} \underline{F}\left(\underline{R}_{i,j}, \underline{R}_{\ell,m}, \underline{R}_{\ell,m+1}\right) + \underline{G}\left(\underline{R}_{i,j-1}, \underline{R}_{i,j}, \underline{R}_{i,j+1}\right) + \underline{V}\left(\underline{R}_{i,j}\right) \tag{9}$$

where $\underline{F}$ represents the contribution of a nonadjacent line segment

$$\underline{F} = \frac{\underline{\ell}_1 \times \Delta\underline{R}}{|\underline{\ell}_1 \times \Delta\underline{R}|^2} \left[\Delta\underline{R} \cdot \left(\frac{\underline{\ell}_2}{|\underline{\ell}_2|} - \frac{\underline{\ell}_1}{|\underline{\ell}_1|}\right)\right] \tag{10}$$

$\underline{G}$ is the local curvature term

$$\underline{G} = \frac{\Gamma \underline{b}}{4\pi R_c} \left[\ln\left(\frac{8R_c}{\delta}\right) - 0.558 - \frac{1}{2} \ln\left(\cot \frac{\theta_1}{4} \cot \frac{\theta_2}{4}\right)\right] \tag{11}$$

and $\underline{V}$ is the irrotational contribution (see figure above for definitions). Note that the local core radius must also be computed as it enters as a parameter in the above formulas. Let $|\Delta\underline{R}_{i,j}| = |\underline{R}_{i,j+1} - \underline{R}_{i,j}|$ and assume that internal waves instantaneously smooth the core diameter; then the dynamical equations for the core radii $\delta_{i,j}$ may be written approximately as

$$\frac{d\delta_{i,j}^2}{dt} = 4\nu - \delta_{i,j}^2 \frac{d}{dt}\left[\ln\left(\sum_{k} |\Delta\underline{R}_{i,k}|\right)\right] \tag{12}$$

On the other hand, if the smoothing due to internal waves is assumed to be negligible, then the summation over k in Eq. (12) must be replaced by k = j.

Equations (9) and (12) define a nonlinear initial value problem for $\underline{R}_{i,j}$ and $\delta_{i,j}$. Leapfrog time differencing is used with an occasional forward Euler step to suppress the weak instability associated with leapfrog differencing (Lilly, 1965). Additional node points are added during the simulations as required by the development of high-filament curvature.

Experimental results (Kambe and Takao, 1971; Oshima, 1974) and early simulation runs show that filament loops in close proximity may, depending on their relative orientation, attract each other to the extent that their vorticity fields are cancelled significantly where they overlap. It is postulated that under these circumstances their respective vortex lines become interwoven until viscous effects destroy the fine structure, thereby leaving no net vorticity in that neighborhood. From this process, a new geometric configuration arises

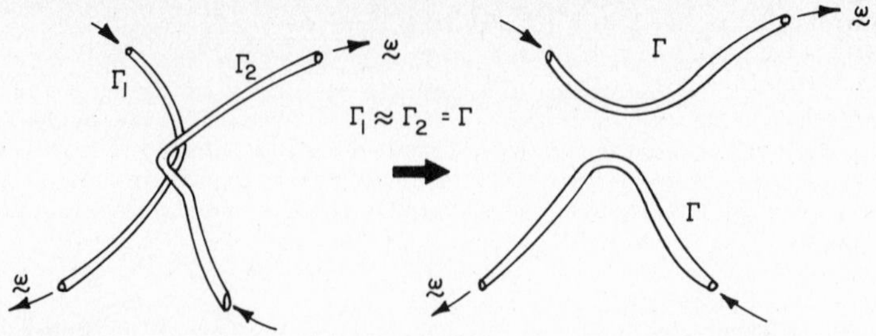

Thus if loops of two different filaments collide, a single filament emerges; but if loops of the same filament undergo this process, two filaments emerge. Such a redefinition of the geometry of the filament is made when the center of one filament segment is within the core of another segment and their respective circulations are nearly equal in strength and oppositely directed.

5. SIMULATION RESULTS

In this section the highlights of some of the simulations are summarized. Detailed quantitative analyses of the results are still being carried out.

Aircraft Trailing Vortices

Figure (1) shows three displays from a simulation of two trailing vortices with separation distance b. The parameters used represent a B-47 aircraft (Crow, 1970). The vortices were initially parallel straight lines but with a small random perturbation having a Kolmogorov energy spectrum $E(k) \sim k^{-5/3}$. The imposed boundary conditions were essentially periodic with period 30b since the summation in Eq. (9) was extended over the two neighboring 30b lengths which were moved as shifted replicas of the center 30b computational domain.

At 44 sec, note that the vortices are joining at two points and nearly joining at a third, indicating that a symmetric mode of wavelength 10b is dominating the growing perturbation in essential agreement with linear stability theory (Crow, 1970). Fourier analyses of the simulation results substantiate this visual observation although other modes of λ = 30b, 15b, and 7.5b are important contributors. It is apparent that ring vortices should form as the two line vortices join, but this feature was not implemented for periodic line vortices.

Perturbed Ring Vortices

The self-induced propulsion of a smoke ring is a familiar phenomenon. Long-wavelength disturbances on a ring vortex are neutrally stable according to inviscid linear stability theory (Thompson, 1883) and have been studied experimentally by Kambe and Takao (1971). Figure 2 is from the simulation of an $n = 2$ disturbance ($\lambda_n = 2\pi R_{ring}/n$) and shows the characteristic oscillatory behavior.

Interacting Ring Vortices

Two ring vortices travelling side by side approximately a diameter apart are mutually attracted to each other and eventually collide and fuse together, forming a single deformed filament (Kambe and Takao, 1971; Oshima, 1974). This process is illustrated in Fig. 3. Experimentally, under certain conditions, the single filament will fission back into two filaments, although this has not yet been observed in the numerical simulations.

6. FURTHER CONSIDERATIONS

The simulations above were performed with several hundred node points or less and therefore were not costly from the standpoint of computer usage. However, as the total number of node points, N, becomes large, the processor time/time step grows asymptotically as N^2 if the procedure described above is used to compute the velocities at each grid point. Typically, one would hope to decrease this growth to

O(N log N) for incompressible flow problems. Two possible modifications or remedies are: (1) For large separations, the integral in Eq. (5) may be expanded with only leading terms retained. As a result, a number of segments in sequence may be treated simultaneously as one contribution. (2) The velocity potential equation $\nabla^2\underline{\psi} = -\underline{\omega}$ may be solved on a relatively coarse grid and the velocities then interpolated at the filament nodes (Christiansen, 1973). The first scheme has been partially implemented for the present calculations.

Further modifications are required if the effects of solid boundaries are to be included. First, to ensure tangency of the velocity field at the boundary, an harmonic contribution must be computed at each time step. Second, vorticity must be generated at the surface in such a way as to enforce the no-slip condition. This may be possible with vortex filaments (Chorin, 1973) or may require coupling with a time-dependent, boundary-layer calculation.

A motion picture film produced from the CRT displays of several of the simulations is available to interested readers on request. The author acknowledges many helpful discussions with Dr. R. S. Rogallo.

REFERENCES

Chorin, A. J. J. Fluid Mech. 57, 785 (1973).
Christiansen, J. P. J. Comp. Phys. 13, 364 (1973).
Crow, S. C. AIAA J. 8, 2172 (1970).
Helmholtz, H. J. Reine Angew. Math. 55, 25 (1858) (Trans. by P. G. Tait, Phil. Mag. 33, 485 (1867)).
Kambe, T., and Takao, T. J. Phys. Soc. Japan 31, 591 (1971).
Kuo, A. Y., and Corrsin, S. J. Fluid Mech. 56, 447 (1972).
Lilly, D. K. Mon. Weather Rev. 93, 11 (1965).
Marsden, J. Bull. Am. Math. Soc. 80, 154 (1974).
Moore, D. W., and Saffman, P. G. Phil. Trans. Roy. Soc. London 272, 403 (1972).
Oshima, Y., private communication (1974).
Saffman, P. G. Studies Appl. Math. 44, 371 (1970).
Thompson, J. J. Motion of Vortex Rings, MacMillan and Co., London (1883).
Tung, C., and Ting, L. Phys. Fluids 10, 901 (1967).
Widnall, S. E., Bliss, D. and Zalay, A. in Aircraft Wake Turbulence and Its Detection, ed. by J. H. Olsen, A Goldburg, and M. Rogers, Plenum Press, New York (1971).

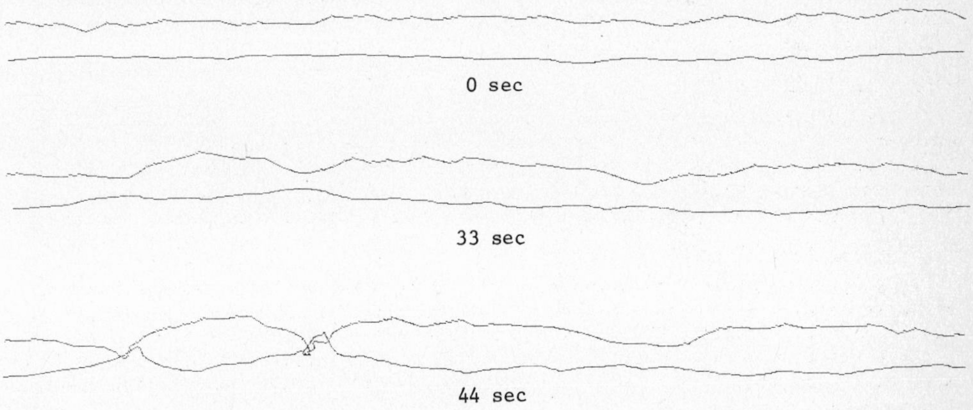

Figure 1.- Trailing vortices; $\Gamma = 980\ ft^2/sec$, $b = 90$ ft, $\delta = 9$ ft, $\nu = 0$, constant local volume.

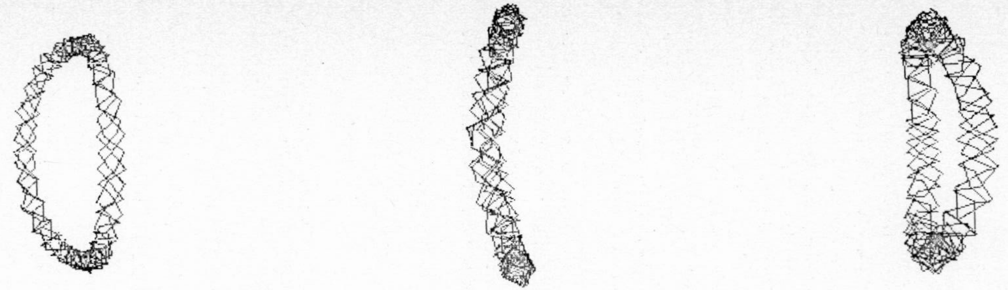

0 sec 6 sec 12 sec

Figure 2.- Ring vortex with n = 2 perturbation; $\Gamma = 20\ cm^2/sec$, $R = 3$ cm, $\delta = 0.5$ cm, $\nu = 0.01\ cm^2/sec$, constant filament volume.

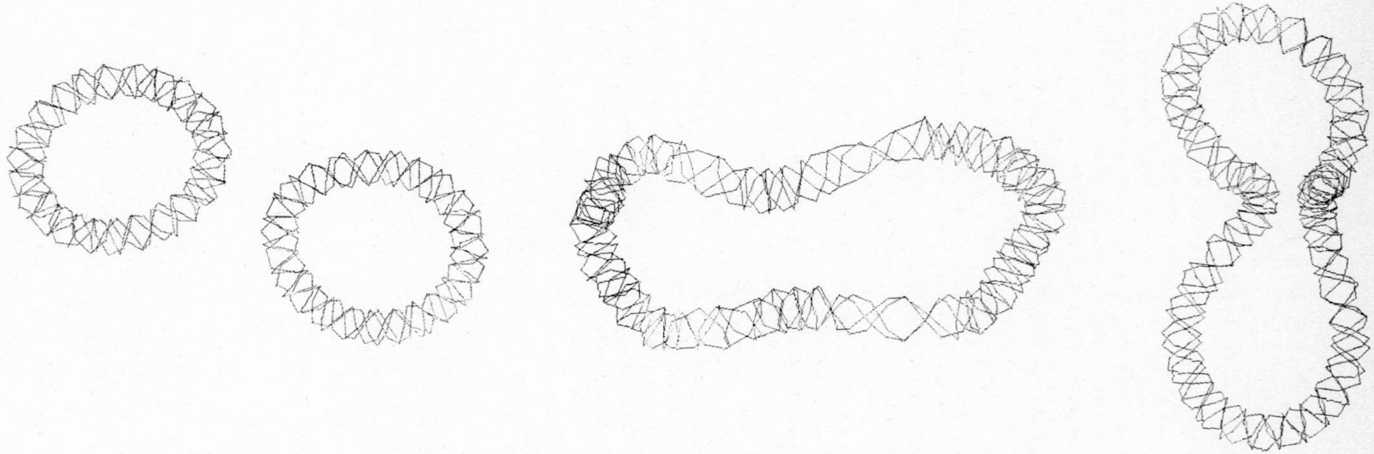

12 sec 16.5 sec 24 sec

Figure 3.- Two-ring vortices whose centers are 12 cm apart initially. Vortex parameters are the same as in Fig. 2.

THE PROBLEM OF SPURIOUS OSCILLATIONS IN THE NUMERICAL SOLUTION OF THE EQUATIONS OF GAS DYNAMICS

A. LERAT * and R. PEYRET **

INTRODUCTION

Generally, spurious oscillations appear in shock waves, in contact discontinuities, and also in compression or rarefaction waves when the equations of gas dynamics are solved numerically by second order schemes. In the case of the scalar equation $u_t + (u^2/2)_x = 0$, a study [1], [2] permitted us to show clearly the important effect of the nonlinear part of the truncation error in the occurence of spurious oscillations. By considering a general class of schemes $\mathcal{S}_\beta^\alpha$ (depending on two parameters α and β), we have shown in [3] that one may define, in the scalar case, a criterion of optimality (no oscillation in any shock profile, minimal spread in the numerical shock structure) for the schemes belonging to this class. These studies were made by analyzing a certain differential equation (the "equivalent equation") whose discretization by the scheme has an order of accuracy higher than the order of accuracy corresponding to the original equation. This very fruitfull method of studying the properties of the finite difference schemes has been independently introduced by Hirt [4] and by Yanenko and Shokin [5] ; similar methods are considered in [6]-[9].

The present work is an extension of [1]-[3] to the study of the dispersive and dissipative properties of the schemes $\mathcal{S}_\beta^\alpha$ in the case of a hyperbolic system of conservation laws. As a consequence of that study it is possible to compare the oscillatory (or not oscillatory) character of the numerical solution given by each of these schemes. Also we show how the addition of an artificial viscosity modifies the "equivalent system" by improving its dissipative properties.

THE CLASS OF SCHEMES $\mathcal{S}_\beta^\alpha$

We consider the problem of solving the nonlinear system

(1) $$w_t + [f(w)]_x = 0 ,$$

with appropriate initial and boundary conditions ; $w(x,t)$ is a vector with m components. The system (1) is hyperbolic, i.e. the matrix $A = f'(w)$ has m real and distinct eigenvalues $\lambda^{(k)}(w)$.

We shall make use of the schemes $\mathcal{S}_\beta^\alpha$ [3] :

(2) $$\begin{cases} \tilde{w}_i = (1-\beta)\, w_i^n + \beta\, w_{i+1}^n - \alpha\sigma\,(f_{i+1}^n - f_i^n) , \\ w_i^{n+1} = w_i^n - \frac{\sigma}{2\alpha}\left[(\alpha-\beta) f_{i+1}^n + (2\beta-1) f_i^n + (1-\alpha-\beta) f_{i-1}^n + \tilde{f}_i - \tilde{f}_{i-1}\right] , \end{cases}$$

where $\sigma = \Delta t/\Delta x$, α and β are two arbitrary parameters ($\alpha \neq 0$). The predictor $\tilde{w}_i$ is an approximation to the solution at $x = (i+\beta)\Delta x$, $t = (n+\alpha)\Delta t$. The notation in (2) are the usual ones : $$f_i^n \equiv f(w_i^n) , \quad \tilde{f}_i \equiv f(\tilde{w}_i) .$$

The schemes $\mathcal{S}_\beta^\alpha$ form a general class of predictor-corrector, 3-point, schemes of second-order accuracy. We note that :

(a) $\mathcal{S}_{1/2}^\alpha$ corresponds to a class of centered schemes, analogous to those introduced by Gourlay and Morris [10] and identical to the schemes studied by Mc Guire and Morris [8]. $\mathcal{S}_{1/2}^{1/2}$ is the "two-step Lax-Wendroff" scheme of Richtmyer [11] ; $\mathcal{S}_{1/2}^1$ has been proposed by Rubin and Burstein [12].

(b) $\mathcal{S}_0^\alpha$ and $\mathcal{S}_1^\alpha$ correspond respectively to the cases $\varepsilon = 0$ and $\varepsilon = 1$ of the noncentered

---

* Conservatoire National des Arts et Métiers, rue Pinel, Paris 13e

** C.N.R.S., Mécanique Théorique, Université Paris VI, Place Jussieu, Paris 5e et O.N.E.R.A., 92 Chatillon-sous-Bagneux .

schemes considered by Warming, Kutler and Lomax [13]. In particular, $\mathcal{S}_0^1$ and $\mathcal{S}_1^1$ are the Mac Cormack schemes [14].
In the linear case, all the schemes $\mathcal{S}_\beta^\alpha$ are identical with the Lax-Wendroff scheme [15] ; consequently the stability criterion is $\sigma \,\mathrm{Max}|\lambda^{(k)}| \leq 1$.

THE EQUIVALENT SYSTEM (AT THIRD ORDER)

For studying the properties of the $\mathcal{S}_\beta^\alpha$, we consider the system of equations discretized by the schemes with a truncation error of third order (the "equivalent system"). As usually, this system is obtained by Taylor expansions of a function w interpolating the discrete approximation $\{w_i^n\}$. This function w must be sufficiently differenciable ; for Δt and Δx fixed, such a function exists, even in the case where the exact solution of (1) is discontinuous, because it is known that the discrete approximation $\{w_i^n\}$ is only able to represent a discontinuity by spreading it over at least one mesh width.
The system discretized by (2) to third order is [16] :

$$(3)\quad w_t + [f(w)]_x = \frac{\Delta x^2}{6}\Big\{[A(\sigma^2A^2-I)w_x]_{xx} + \sigma^2[\mathcal{B}(f_x,f_x)-\mathcal{B}(Af_x,w_x)]_x + \frac{3}{2\alpha}[\mathcal{B}(\varphi_x^0,\varphi_x^1)]_x\Big\},$$

where I is the unit matrix, $A = f'(w)$, $\varphi^0 = (1-\beta)w + \alpha\sigma f(w)$, $\varphi^1 = \beta w - \alpha\sigma f(w)$ and $\mathcal{B}$ is the symmetrical bilinear transformation $\mathcal{B} = f''(w)$. The system (3) may be written as :

$$(4)\qquad w_t + [f(w)]_x = \Delta x^2\Big[\mathcal{E}_1(w)w_{xxx} + \mathcal{E}_2(w,w_x)w_{xx} + \mathcal{E}_3(w,w_x)\Big]$$

which generalizes the scalar equivalent equation of [3]. Here :

$$(5)\quad \mathcal{E}_1(w) = \frac{1}{6}A(\sigma^2A^2 - I) \equiv P(A)\ ,$$

$$(6)\quad \mathcal{E}_2(w,w_x) = \frac{1}{2}\Big\{\sigma^2\big(\tfrac{1}{3}B(w_x)A^2 + \tfrac{2}{3}AB(w_x)A + A^2B(w_x) + (\tfrac{2}{3}-\alpha)B(Aw_x)A + \tfrac{1}{3}AB(Aw_x)\big) + \frac{\sigma}{2}(2\beta-1)\,[B(w_x)A + B(Aw_x)] + \big(\beta\,\tfrac{1-\beta}{\alpha} - 1\big)B(w_x)\Big\}\ ,$$

where $B(Y)$ is the matrix associated with the linear transformation $X \to \mathcal{B}(Y,X)$, i.e. $B(Y)X \equiv \mathcal{B}(Y,X) \equiv f''(w)(Y,X)$; $B(w_x) = A_x$. The form of $\mathcal{E}_3(w,w_x)$ is very complicated but this term is unnecessary [1] for the present study.
Mac Guire and Morris [8], who have recently studied the centered schemes $\mathcal{S}_{1/2}^\alpha$ by a similar method, made certain approximations in order to get valuable qualitative informations on the dissipative properties of the schemes. As in the scalar [3], our aim here is to study the quantitative effect of the choice of the parameters α and β on the oscillatory character of the numerical results.

PROPERTIES OF SCHEMES $\mathcal{S}_\beta^\alpha$

Let $\ell^{(k)}$ and $r^{(k)}$ be respectively the left and right eigenvectors associated with $\lambda^{(k)}$; by left-multiplying (4) by $\ell^{(k)}$, we get

$$(7)\qquad \ell^{(k)}(w_t + \lambda^{(k)}w_x) = \Delta x^2\Big[\ell^{(k)}\mathcal{E}_1(w)w_{xxx} + \ell^{(k)}\mathcal{E}_2(w,w_x)w_{xx} + \ell^{(k)}\mathcal{E}_3(w,w_x)\Big].$$

The exact compatibility equation along the characteristic $dx/dt = \lambda^{(k)}$ is obtained by setting the right-hand side of (7) equal to zero. If $\ell^{(k)}\mathcal{E}_1 = 0$, the term $\Delta x^2\mathcal{E}_1 w_{xxx}$ produces no dispersion and if $\ell^{(k)}\mathcal{E}_2 = 0$, the term $\Delta x^2\mathcal{E}_2 w_{xx}$ produces no dissipation along this characteristic (extension of the property K of [5]).

We now introduce the following definitions : along a characteristic line $dx/dt = \lambda^{(k)}$ the schemes $\mathcal{S}_\beta^\alpha$ are called <u>k-dispersive to second order</u> if the scalar coefficient

$$(8)\qquad \ell^{(k)}\mathcal{E}_1(w)\,r^{(k)} \neq 0\ ,$$

and <u>k-dissipative to second order</u> if the scalar coefficient

here $\ell^{(k)}$ and $r^{(k)}$ are such that :
$$\ell^{(k)}\mathcal{E}_2(w,w_x)\,r^{(k)} > 0\ ,\qquad \ell^{(k)}\cdot r^{(k)} = 1\ .$$

From (5), we get easily :

(9)
$$\begin{cases}\ell^{(k)}\mathcal{E}_1(w)\,r^{(k)} = E_1(\eta^{(k)})\ ,\\ E_1(\eta^{(k)}) = \frac{1}{6\sigma}\,\eta^{(k)}(\eta^{(k)2}-1)\ ,\quad \eta^{(k)} = \sigma\lambda^{(k)}.\end{cases}$$

From (6), we get, if w is a k-simple wave :

(10)
$$\begin{cases}\ell^{(k)}\mathcal{E}_2(w,w_x)\,r^{(k)} = -E_2(\eta^{(k)})\,\lambda^{(k)}_x\ ,\\ E_2(\eta^{(k)}) = E_2(\eta^{(k)};\alpha,\beta) = -\frac{1}{2}\left[(3-\alpha)\eta^{(k)2}+(2\beta-1)\eta^{(k)}+\beta\frac{1-\beta}{\alpha}-1\right]\end{cases}$$

The relation (10) is obtained as follows : first, the differentiation of $Ar^{(k)}=\lambda^{(k)}r^{(k)}$ with respect to x and multiplication by $\ell^{(k)}$ gives $\ell^{(k)}B(w_x)r^{(k)} = \lambda^{(k)}_x$. Now let $w(x,t)$ be a k-simple wave and $v(w)$ any k-Riemann invariant of (1) [17] ; by differencing $v(w(x,t)) = \text{const.}$ with respect to x, we find that the derivative w_x of the solution w of (4) is proportional to $r^{(k)}$ to second order ; then the relation (10) can be deduced.

Consequently, we may state the following results :

(a) The schemes $\mathcal{S}^\alpha_\beta$ are k-dispersive to 2nd order along a characteristic $dx/dt=\lambda^{(k)}$ at points where $E_1(\eta^{(k)}) \neq 0$.

(b.1) If the k-th characteristic field is genuinely nonlinear [17] ($\text{grad}_w\lambda^{(k)}\cdot r^{(k)} \neq 0$ for any w) and if w is a k-simple wave, we have $\lambda^{(k)}_x = \text{grad}_w\lambda^{(k)}\cdot r^{(k)} \neq 0$. If $\lambda^{(k)}_x < 0$ the wave is a k-compression wave, and if $\lambda^{(k)}_x > 0$ it is a k-rarefaction wave. Thus, the schemes $\mathcal{S}^\alpha_\beta$ are k-dissipative to 2nd order along a characteristic $dx/dt=\lambda^{(k)}$ of a k-compression wave (resp. k-rarefaction wave) at points where $E_2(\eta^{(k)}) > 0$ [resp. $E_2(\eta^{(k)}) < 0$].

(b.2) If the k-th characteristic field is degenerate ($\text{grad}_w\lambda^{(k)}\cdot r^{(k)} = 0$ for any w) and if two nearby different states have the same k-Riemann invariants, these states are connected by a k-contact discontinuity (theorem 8.7, [17]) ; in this case, (10) again holds and furthermore $\lambda^{(k)}_x = 0$ since $\lambda^{(k)}$ is one of the k-invariants. Therefore, the $\mathcal{S}^\alpha_\beta$ have zero dissipation at 2nd order along the characteristic associated with a k-contact discontinuity.

(c) Case of a shock wave. Recalling that the numerical representation of a discontinuity is spread over several meshes we make the approximation that a k-shock can be identified with a k-compression wave, at least when the shock is not too strong. Hence, we shall apply the previous results in that case.

Remark : a scheme which is not k-dissipative to 2nd order is not necessarily unstable because the presence of the term of higher order, namely

(11)
$$T_4 = \frac{1}{8}\Delta x^3\,\sigma A^2(\sigma^2A^2 - I)\,w_{xxxx}$$

which introduces a third order dissipation when the linear stability criterion is satisfied.

We note that the coefficients $E_1(\eta^{(k)})$ and $E_2(\eta^{(k)})\lambda^{(k)}_x$ are similar to those obtained in the scalar case [3] with $\lambda^{(k)}$ instead of u. Therefore, the discussion given in [3] can be extended to the present problem. σE_1 and E_2 are shown in fig. 1.

ARTIFICIAL VISCOSITY

If one introduces an artificial viscosity of Lax-Wendroff type [15] in the Lapidus form (cf. [11]) the schemes $\mathcal{S}^\alpha_\beta$ become $\bar{\mathcal{S}}^\alpha_\beta$ and may be written as :

(12)
$$\begin{cases}\tilde{w}_i = (1-\beta)w^n_i + \beta w^n_{i+1} - \alpha\left[(\sigma+g^+_2)(f^n_{i+1}-f^n_i) + g^+_1(w^n_{i+1}-w^n_i)\right]\\ w^{n+1}_i = w^n_i - \frac{\sigma}{2}\left[\frac{\alpha-\beta}{\alpha}f^n_{i+1} + \frac{2\beta-1}{\alpha}f^n_i + \frac{1-\alpha-\beta}{\alpha}f^n_{i-1} + \frac{1}{\alpha}(\tilde{f}_i-\tilde{f}_{i-1}) + g^-_0(w^n_i-w^n_{i-1}) - g^+_0(w^n_{i+1}-w^n_i)\right]\end{cases}$$

where $g^\pm_0, g^+_1, g^+_2$ are defined as in [11]. Then, the equivalent system (4) is modified by the addition of the term $\chi\frac{\Delta x^2}{2}(Q(w,w_x)w_x)_x$ where $\chi = \text{const.} > 0$ is the viscosity coefficient and $\chi Q(w,w_x)$ is the interpolation polynomial $g(A)$ (cf. [11]), the coefficients of which depend on $\lambda^{(k)}$ and $|\lambda^{(k)}_x|$. The matrix $\mathcal{E}_2(w,w_x)$ is replaced by

$\overline{\mathcal{E}}_2(w,w_x)$ so that, for a k-simple wave :

(13) $$\ell^{(k)}\overline{\mathcal{E}}_2(w,w_x)\,r^{(k)} = -\,E_2(\eta^{(k)})\,\lambda_x^{(k)} + \chi\,|\lambda_x^{(k)}|\,.$$

Therefore, if $\chi > \underset{\eta^{(k)}\in[-1,1]}{\mathrm{Max}} |E_2(\eta^{(k)};\alpha,\beta)|$, the $\overline{\mathcal{S}_\beta^\alpha}$ are always k-dissipative except for a contact discontinuity ($\lambda_x^{(k)} = 0$) .

APPLICATION TO GAS DYNAMICS

We treat here the case of the unsteady one-dimensional equations of gas dynamics : $w = (\rho, \rho u, e)^T$ where ρ is the density, u the velocity and e the total energy. The matrix A has three eigenvalues : $\lambda^{(1)} = u-c$, $\lambda^{(2)} = u$, $\lambda^{(3)} = u+c$ where c is the sound velocity. The characteristic fields associated respectively with $\lambda^{(1)}$ and $\lambda^{(3)}$ are genuinely nonlinear, the one associated with $\lambda^{(2)}$ is degenerate.

In order to illustrate the results obtained above, we consider the problem of the flow in a shock tube. Fig. 2-4 show results given by four particular schemes $\mathcal{S}_\beta^\alpha$, say :

$$\mathcal{S}^{I} : \alpha = 1,\ \beta = 0 \quad , \quad \mathcal{S}^{II} : \alpha = \tfrac{1}{4},\ \beta = \tfrac{1}{2} \quad , \quad \mathcal{S}^{III} : \alpha = \tfrac{1}{2},\ \beta = \tfrac{1}{2} \quad , \quad \mathcal{S}^{IV} : \alpha = 1+\tfrac{\sqrt{5}}{2},\ \beta = \tfrac{1}{2}.$$

For the scalar case [3], $\mathcal{S}^{IV}$ is the optimal scheme in the sense defined in the introduction, that is $E_2(\eta;\alpha,\beta) \geq 0$ for any $\eta \in [-1,1]$ and $\underset{\eta\in[-1,1]}{\mathrm{Max}} E_2(\eta;\alpha,\beta)$ minimal.

Fig. 2 corresponds to the first stage of the flow : the rarefaction wave is associated with $\lambda^{(1)}$ and the shock with $\lambda^{(3)}$.
Fig. 3 shows the second stage : the shock is reflected and it then corresponds to $\lambda^{(1)}$.
The contact discontinuity in this stage is shown in fig. 4. All these results were obtained with $\Delta x = 1/500$ and $\sigma \, \underset{i}{\mathrm{Max}} (|u_i^n| + c_i^n) = 1$.

By considering the values of E_1 and E_2 (fig. 1) for the different schemes $\mathcal{S}_\beta^\alpha$, we may explain the character of the numerical solutions. It is important to point out that all the schemes $\mathcal{S}_\beta^\alpha$ have the same $\mathcal{E}_1(w)$, therefore produce the same dispersive error when the $\eta^{(k)}$ and the σ considered are the same. The values of the $\eta^{(k)}$ associated with each wave considered here are given in fig. 2-4.
In the rarefaction wave (fig. 2), the dispersion is nearly at a maximum. As is known, the oscillations appear behind the wave. Also, the graph of E_2 shows that $\mathcal{S}^{II}$ is the most dissipative scheme in this case ($E_2^{II} = E_2(\eta^{(1)};\frac{1}{4},\frac{1}{2}) < 0$) and that the numerical solution is the least oscillatory. More precisely :

$$E_2^{II} = E_2(\eta^{(1)};\tfrac{1}{4},\tfrac{1}{2}) < E_2^{III} = E_2(\eta^{(1)};\tfrac{1}{2},\tfrac{1}{2}) \lesssim E_2^{I} = E_2(\eta^{(1)};1,0) < E_2^{IV}(\eta^{(1)};1+\tfrac{\sqrt{5}}{2},\tfrac{1}{2})$$

in the region considered $(-0.69 < \eta^{(1)} < -0.35)$. Note that $E_2^{IV} > 0$ yields a "weak nonlinear instability" which increases the oscillations due to dispersion ; we recall that $\mathcal{S}^{IV}$, although non-dissipative here, remains stable due to the effect of T_4 [Eq. (11)] .
In the case of the shock of fig. 2 ($0.63 < \eta^{(3)} < 1.$) , we have

$$E_2^{II} < E_2^{III} < 0 < E_2^{IV} < E_2^{I}$$

and we note (i) the absence of oscillations in dissipative cases ($E_2 > 0$), (i.i) the presence of oscillations and the relative importance of their amplitude in non-dissipative cases ($E_2 < 0$) .
In the case of the reflected shock of fig. 3 ($-0.8 < \eta^{(1)} < -0.5$) ,

$$E_2^{II} < E_2^{III} \lesssim E_2^{I} < 0 < E_2^{IV} \,.$$

Here, the important dispersive effect induces relatively large oscillations even when $\mathcal{S}^{IV}$ is used, this being the only scheme which is dissipative in the present situation.
Finally, in fig. 4, we show the importance of the oscillations in a contact discontinuity : the dissipation of schemes $\mathcal{S}_\beta^\alpha$ is zero in such a wave (These oscillations are the same for all the four schemes).

CONCLUSION

One of the purposes of the present study was to show how the analysis of the equivalent system corresponding to the general class of schemes $\mathcal{S}^{\alpha}_{\beta}$ is useful for determining the relative merits of these schemes ; hence, it is possible to find a scheme well-suited to a specific problem. The scheme $\mathcal{S}^{IV}$ is the best among all $\mathcal{S}^{\alpha}_{\beta}$ for representing compression or shock waves even if it may give oscillations when the dispersion is too large. A similar conclusion ($\alpha \# 2$) was obtained in [8] from numerical experiments with $\mathcal{S}^{\alpha}_{1/2}$. On the other hand, this advantage becomes a disadvantage in a rarefaction wave ; but the oscillations are damped with time because the gradients tend to decrease in such a wave (compare fig. 2 and fig. 3). Moreover, since the couple $(\alpha = 1 + \frac{\sqrt{5}}{2}, \beta = \frac{1}{2})$ minimizes the $\underset{\eta \in [-1,1]}{\text{Max}} E_2$ with the constraint $E_2 \geqslant 0$, the scheme $\mathcal{S}^{IV}$ is, among the $\mathcal{S}^{\alpha}_{\beta}$ which are dissipative in compression and shock waves, also the best one to represent rarefaction waves.

REFERENCES

[1] Lerat A., Peyret R. C.R. Acad. Sc. Paris, série A, 276, 759-762 (1973)
[2] Lerat A., Peyret R., Computers and Fluids, to appear
[3] Lerat A., Peyret R. C.R. Acad. Sc. Paris, série A, 277, 363-366 (1973)
[4] Hirt C.W., J. Computational Phys., 2, 339-355 (1968)
[5] Yanenko N.N, Shokin Y.I., Phys. Fluids, 12, II 28-II 33 (1969)
[6] Cheng S.I., Phys. Fluids, 12, II 34-II 41 (1969)
[7] Roache P.J., J. Computational Phys., 10, 169-184 (1972)
[8] Mc Guire G.R., Morris J.L., J. Computational Phys., 11, 531-549 (1973)
[9] Warming R.F., Hyett B.J., J. Computational Phys., 14, 159-179 (1974)
[10] Gourlay A.R., Morris J.L., Math. Comp., 22, 28-39 (1968)
[11] Richtmyer R.D., Morton K.W., Diff.Meth.for initial value probl., Intersc.(1967)
[12] Rubin E.L., Burstein S.Z., J. Computational Phys., 2, 178-196 (1967)
[13] Warming R.F., Kutler P., Lomax H., A.I.A.A. Journal, 11, 182-196 (1973)
[14] Mc Cormack R.W., Lect. Notes in Phys., 8, 151-163 (1971)
[15] Lax P.D., Wendroff B., Comm.Pure Appl. Math., 13, 217-237 (1960)
[16] Lerat A., Peyret R., to be published
[17] Lax P.D. Comm.Pure Appl. Math, 10, 537-566 (1957)

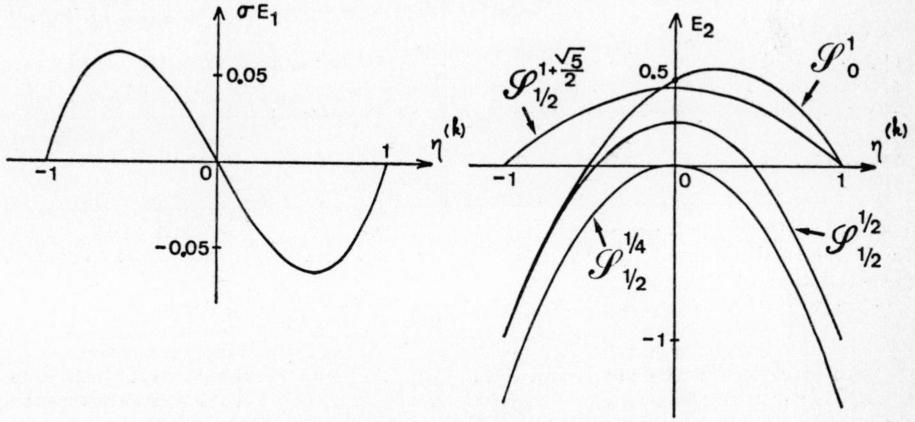

Fig. 1 : σE_1 and E_2 versus $\eta^{(k)}$

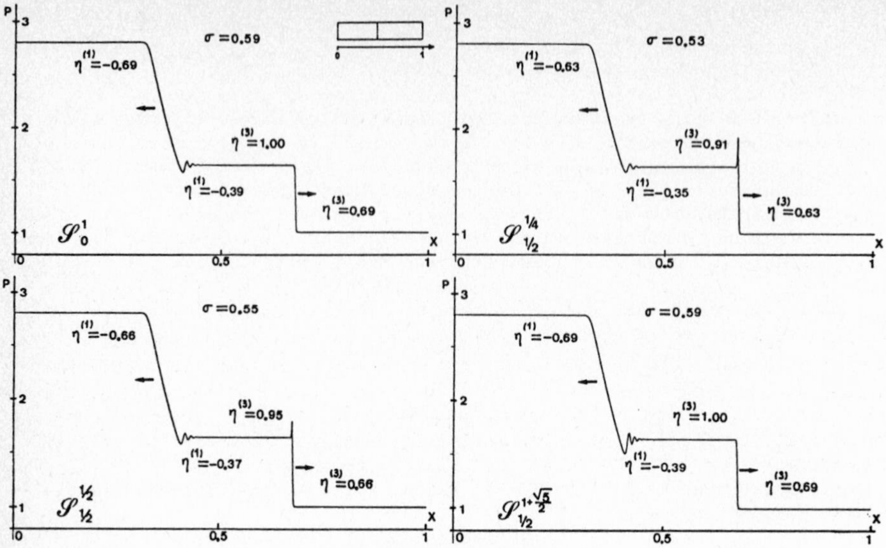

Fig. 2 : Pressure versus space in a shock tube

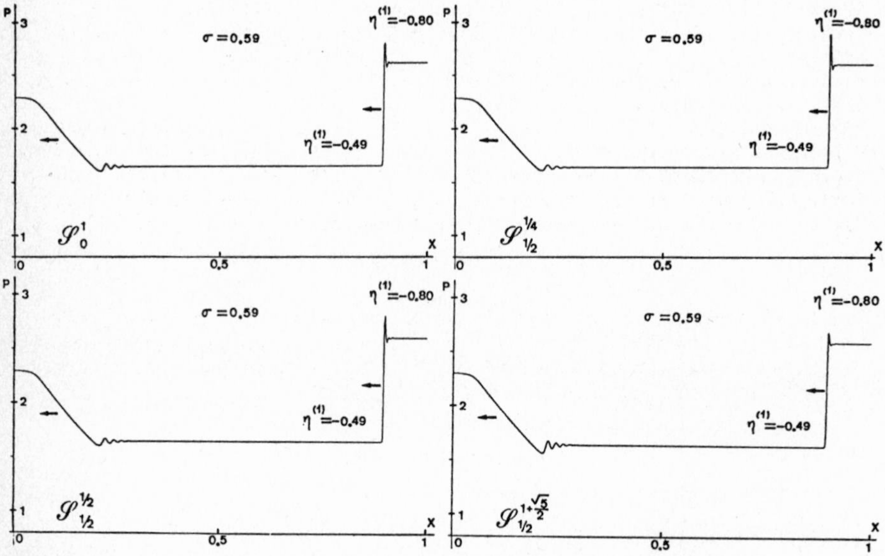

Fig. 3 : Pressure versus space in a shock tube after reflection of the shock

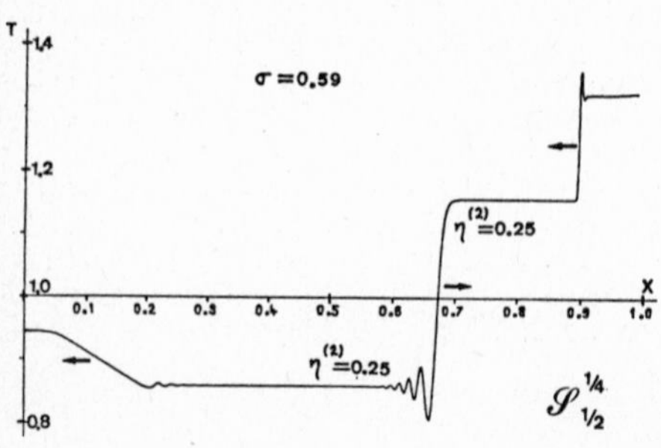

Fig. 4 : Temperature versus space in a shock tube after reflection of the shock

ROTATING THIN ELLIPTIC CYLINDER IN A PARALLEL VISCOUS FLUID FLOW

Hans J. Lugt and Samuel Ohring

Naval Ship Research and Development Center, Bethesda, Maryland 20034

INTRODUCTION

The numerical computation of flow fields around bodies rotating in a parallel stream poses a much greater challenge than computing only the translational motion of bodies. This is largely due to the positioning of the grid of the finite-difference or finite-element scheme. For noncircular cylinders and for non-axisymmetric bodies it is of great advantage to keep the grid fixed with the rotating body in order to obtain sufficiently high accuracy near the body surface where the largest vorticity gradients occur in a homogeneous fluid. However, the changing position of the grid, which rotates relative to the free stream, introduces other inaccuracies which also affect the quality of the solution very close to the solid surface. These sources of numerical inaccuracies require special handling of the boundary conditions.

In the literature only a few studies deal with numerical methods for flows past rotating bodies. One paper (Thoman and Szewczyk, 1966) presents solutions for flows past rotating circular cylinders in a parallel stream. Another paper (Clark, 1972) gives solutions for the motion past spinning axisymmetric bodies under a non-zero angle of attack. In both cases the above mentioned grid problem did not occur because of the circular geometries of the bodies.

The implications of rotation on the numerical analysis of the solution of the Navier-Stokes equations were studied for a specific physical problem: the laminar flow of an incompressible fluid about an abruptly started thin elliptic cylinder translating with constant speed U while rotating with constant angular velocity Ω. The initial condition consists of the potential-flow solution (Lamb, 1945) and a vorticity sheet at the body surface which is due to the adherence of the fluid to the surface.

For this problem numerical techniques were studied and preliminary results presented by the authors (1973). The physical results for the nonrotating case were discussed by Lugt and Haussling (1974). In this paper the investigation is focused on the numerical difficulties due to rotation.

NUMERICAL ANALYSIS

The flow problem was solved by constructing solutions of the Navier-Stokes equations in the stream function-vorticity formulation. After testing a number of finite-difference schemes reported in Lugt and Ohring (1973), the following combination appeared to be very efficient: the DuFort-Frankel scheme for solving the vorticity equation and the fast Fourier analysis/cyclic reduction technique of Hockney for the Poisson equation. It should be stressed here that without the use of a fast Poisson solver like that of Hockney, Buneman, and others a problem such as the present one could not have been tackled successfully. A comparison with SOR (Lugt and Ohring, 1973) shows that with equal accuracy the ratio of computer time is larger than 1:50 in favor of Hockney's method. This comparison is based on the (η,θ) grid of (97 x 96) where the elliptic coordinates used in this study are defined by $x + iy = a \cosh (\eta + i\theta)$ with $a > 0$. The basic equations for the dimensionless stream function ψ and the dimensionless vorticity component ω normal to the (η,θ)-plane are then

$$\frac{\partial\omega}{\partial t} + \frac{1}{h^2}\left[-\frac{\partial}{\partial\eta}\left(\frac{\partial\psi}{\partial\theta}\,\omega\right) + \frac{\partial}{\partial\theta}\left(\frac{\partial\psi}{\partial\eta}\,\omega\right)\right] = \frac{2}{Re}\nabla^2\omega \;, \tag{1}$$

$$\nabla^2\psi = \omega \;. \tag{2}$$

In these equations, t is the dimensionless time defined by t^*U/a with time t^*, $Re = 2aU/\nu$ is the Reynolds number, ν the kinematic viscosity, and U the constant velocity of the free stream. The characteristic length and velocity scales in the dimensionless quantities are a and U. The coefficient h is defined by $h^2 = \cosh^2\eta - \cos^2\theta$. The following data are useful for the description of the grid network (Lugt and Ohring, 1973): With $\eta = \eta_1$ representing the body contour (Fig. 1) the grid is defined by $[\eta_1 + (i-1)\Delta\eta, (j - \frac{1}{2})\Delta\theta]$ where $i = 1,\ldots 97$; $j = 1,\ldots 96$ with $\eta_1 = 0.1$ and $\Delta\eta = 0.04$. Hence, the outer boundary is about 12 plate lengths away from the body center. Except for the initial phase, where smaller Δt's were used, $\Delta t = 0.0025$ for $t > 0.186$. It may be pointed out that Δt is not only limited by the numerical instability but also by the relation $\Delta t < 2\pi Ro/96$, whichever is smaller. Here, the Rossby number is defined by $Ro = U/\Omega a$.

The main computational difficulties due to the rotation of the body enter through the boundary conditions. In the grid fixed to the body the boundary conditions are:

$$\eta = \eta_1: \quad \psi = 0, \quad \partial\psi/\partial\eta = 0, \tag{3}$$

$\eta = \eta_{97} < \infty$; $0 \le \theta - \alpha \le \pi/2$; $3\pi/2 \le \theta - \alpha \le 2\pi$; (upstream half):

$$\partial\psi/\partial\eta = h\sin(\theta-\alpha) + \frac{1}{Ro}\cosh\eta\,\sinh\eta, \quad \omega = \frac{2}{Ro}\;, \tag{4}$$

$\pi/2 < \theta - \alpha < 3\pi/2$ (downstream half):

$$\frac{\partial\omega}{\partial t} + \frac{1}{U}(\vec{U}\cdot\nabla)\omega = 0\;, \tag{5a}$$

$$\left[\frac{\partial\vec{v}}{\partial t} + \frac{1}{U}(\vec{U}\cdot\nabla)\vec{v} - 2\frac{a}{U}(\vec{\Omega}\times\vec{v}) - \nabla\frac{1}{2Ro^2}(\cosh^2\eta-\sin^2\theta)\right]_\theta = 0\;, \tag{5b}$$

where $\vec{v}$ is the velocity vector, and the subscript θ denotes the θ-component of the bracketed expression. If $\theta - \alpha < 0$, 2π must be added repeatedly until $\theta - \alpha \geq 0$. The angle of attack is $\alpha = \Omega t^* = t/Ro$. The surface vorticity $\omega_{1,j}$ is computed by

$$\omega_{1,j} = \frac{1}{4h_{1,j}^2(\Delta\eta)^2}\,(\psi_{2,j} + 4\psi_{3,j} - \psi_{4,j})\;. \tag{6}$$

Equations (4) through (6) are discussed in Lugt and Haussling (1974) for the non-rotating case.

Although the reference frame fixed with the body is of advantage for obtaining accurate values of the vorticity at and near the body surface, the overall ψ-field is not accurate due to the large values of ψ near the outer boundary. This can be remedied without loosing the advantages of a grid fixed with the body by the transformation

$$\psi = \psi^* + \frac{1}{2Ro}(\cosh^2\eta - \sin^2\theta), \quad \omega = \omega^* + 2/Ro\;. \tag{7}$$

In solving the Poisson equation (2) at t=0, that is for the potential flow ψ_P, Hockney's method gives the following percentage errors for $Ro = 2$, $\eta_1 = 0.1$, $\Delta\eta = 0.04$, grid (97 x 96): for ψ^* of $\nabla^2\psi^* = 0$, $|(\psi_P^* - \psi^*)/\psi_P^*|\cdot 100 < 0.01\%$ except immediately near $\psi^* = 0$, where the error is of the order 1%. However, when solving

$\nabla^2\psi = 2/Ro$ the error is extremely high for ψ: in the large area near the body, where ψ is very small, the error is 500% on the average, otherwise near the body about 50%. The reason for this difference is that the error term of the finite-difference Laplace operator is much larger for the untransformed function ψ than for ψ^*.

The usage of a grid rotating relative to the main stream introduces a numerical difficulty caused by the outer boundary conditions (5). The discontinuities in $\partial\psi^*/\partial\eta$ and ω^* at $\eta = \eta_{97}$, $\theta-\alpha = \pi/2$ and $3\pi/2$, which grow with advancing time and which were irrelevant for the calculation of nonrotating bodies, generate a disturbance every time an outer boundary grid point switches from the downstream to the upstream half of the outer boundary. (The influence of the switch from the upstream to the downstream half is negligible.) These disturbances are felt immediately throughout the fluid and cause large errors in the surface vorticity. This numerical phenomenon gets aggravated at about $\alpha = 5\pi/2$ in Fig. 2 and soon afterwards renders the solution meaningless (for $\alpha > 3\pi$). It may be mentioned that by that time the vorticity has not yet reached the outer boundary.

This difficulty has been overcome by smoothing out the discontinuities. The transition was done linearly over the outer-boundary arcs $\pi/2 < (\theta-\alpha) \leq 3\pi/4$; $5\pi/4 \leq \theta-\alpha < 3\pi/2$. The continuation of the solution in this manner is marked in Fig. 2 by circles. The computation could then be continued even when the vorticity crossed the outer boundary.

RESULTS

The initial-boundary value problem was solved for the two initial positions $\alpha_0 = 0°$ (in Fig. 2, $\alpha_0 = \pi$) and $\alpha_0 = 90°$. The C_D, C_L, and C_M-coefficients, defined by $C_D = \text{Drag}/\frac{\rho}{2}U^2 a \cosh \eta_1$, $C_L = \text{Lift}/\frac{\rho}{2}U^2 a \cosh \eta_1$, and $C_M = \text{Torque}/\frac{\rho}{2}U^2 a^2 \cosh^2\eta_1$ with ρ the fluid density, are plotted against α (or t) in Fig. 2 with a phase shift of $\pi/2$ between both solutions. The transient period caused by the abrupt start of the body is surprisingly short. From $\alpha = 3\pi/2$ onward the curves almost fall together. For part of the C_M-curves the actual values are plotted using dots and circles so that the amount of scattering can be observed. The average values of the C_L-curve are negative as expected since they represent the Magnus force on the body. The C_M-curve shows an almost autorotating behavior of the body. This means that the positive and negative values of C_M almost balance over one revolution and that the kinetic energy of the free stream is almost sufficient for self-sustained spinning. Autorotation for such low Reynolds numbers is not surprising since in wind-tunnel experiments autorotation has been observed for Re as low as 100 (Smith, 1971).

In Fig. 3 a sequence of constant ψ and ω patterns for half a revolution is displayed. The reference frame is chosen such that it is fixed with the center of the body in the flow direction. In distinction to the nonrotating-body problem (Lugt and Haussling, 1974) the frequency of vortex shedding is determined by the rate of rotation. Fig. 2 reveals, however, that the maxima and minima of C_D, C_L, and C_M do not coincide with the vertical and horizontal positions of the body but shortly after. The detailed description and explanation, for which a computer-generated movie will be exploited, will be published later.

ACKNOWLEDGMENT

The authors would like to thank Mr. H.J. Haussling of the Naval Ship Research and Development Center for suggesting the transformation (7).

REFERENCES

Clark, B.L., AIAA paper 72-112 (1972).

Lamb, H., Hydrodynamics. Dover Publications, New York, 1945. 6th ed., p. 86.

Lugt, H.J. and Haussling, H.J., Forthcoming in Journ. Fluid Mechanics (1974).

Lugt, H.J. and Ohring, S., Proceedings International Conference on Numerical Methods in Fluid Dynamics, University of Southampton, England, Sept. 26-28, 1973.

Smith, E.H., Journ. Fluid Mechanics 50, 513-534 (1971).

Thoman, D.C. and Szewczyk, A.A., University of Notre Dame, Dept. of Mechanical Engineering, Tech. Rep. 66-14, July 1966.

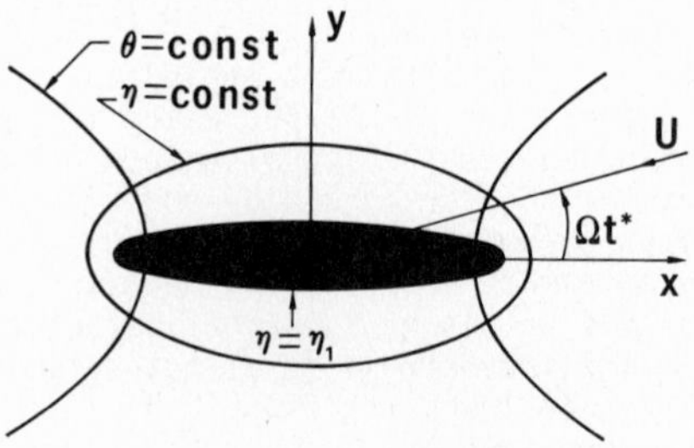

Fig. 1: Elliptic coordinate system.

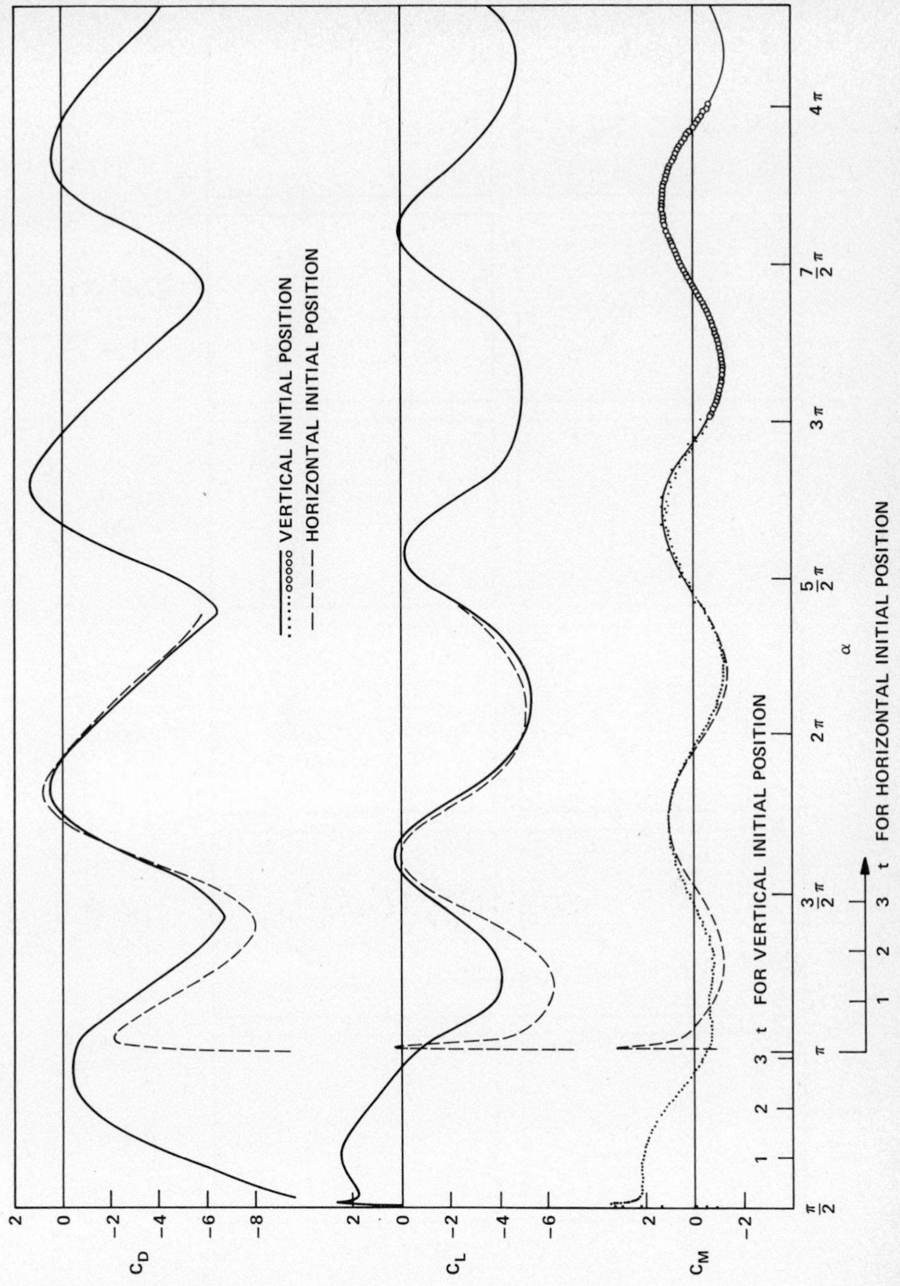

Fig. 2: Drag, lift, and moment coefficients versus α (or time) for Re = 200, Ro = 2, η_1 = 0.1.

$t = 10.46$

$\alpha = 2\pi + \frac{2\pi}{6}$
$t = 11.51$

$\alpha = 2\pi + \frac{\pi}{2}$
$t = 12.56$

$\alpha = 2\pi + \frac{4\pi}{6}$
$t = 13.61$

$\alpha = 2\pi + \frac{5\pi}{6}$
$t = 14.66$

$\alpha = 3\pi$
$t = 15.71$

ψ ω

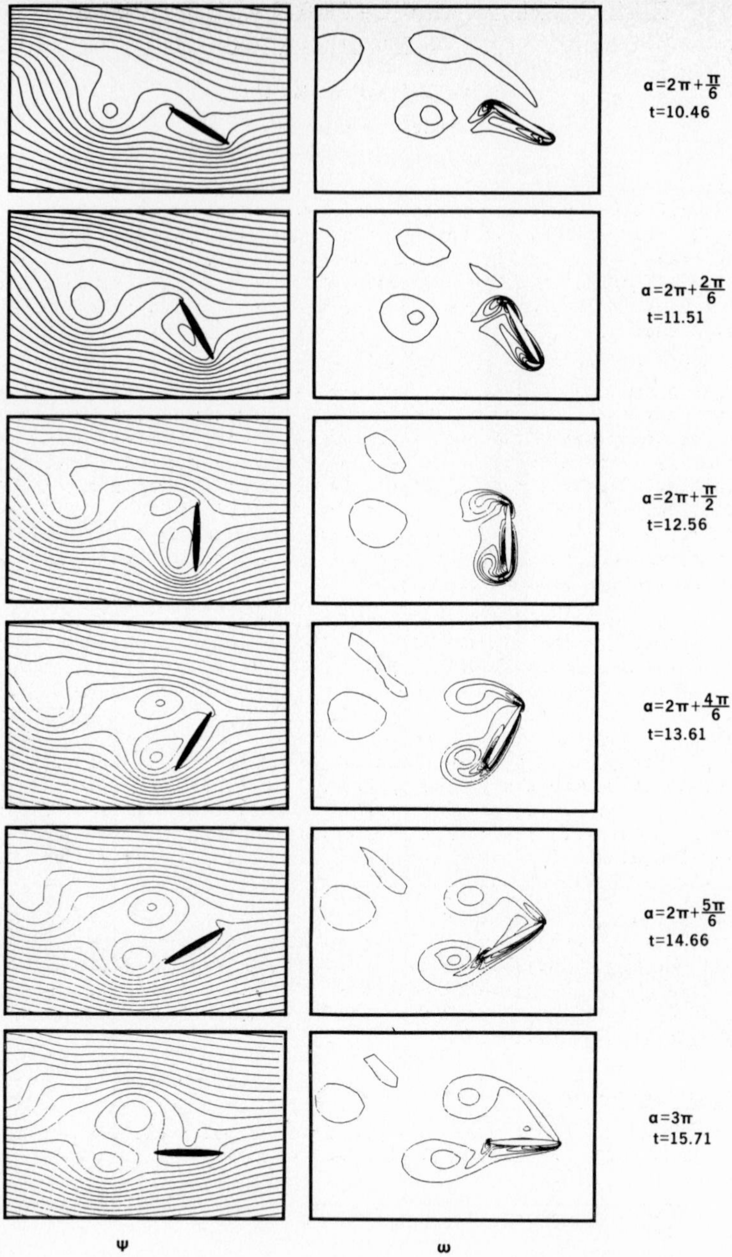

Fig. 3: Sequence of streamlines and equal-vorticity lines for Re = 200, Ro = 2, $\eta_1 = 0.1$, half a revolution between $2\pi + \frac{\pi}{6} \leq \alpha \leq 3\pi$.

COMPUTATIONAL MAGNETOHYDRODYNAMICS OF TIME-DEPENDENT FLOWS IN TOROIDAL SYSTEMS

H. C. Lui and C. K. Chu

Columbia University

Introduction

Nearly all the important fusion experiments today are toroidal in geometry, e.g., tokamaks and toroidal pinches. While numerous theoretical studies are carried out at present for the equilibrium and stability of such devices, and many one-dimensional computations have been made to simulate their time-evolution, there are as yet virtually no two-dimensional calculations for initial-value problems in toroidal geometry.

This paper presents one such study, with preliminary but strongly encouraging results. The plasma is represented by a single fluid with finite electrical and thermal conductivities, but no viscosity and no particle-kinetic effects (no waves, turbulence, or trapping). The motion is studied in the poloidal plane (R-z or r-θ plane of fig. 1), and there is no variation in the ignorable toroidal (φ) direction. A plasma initially at rest and at relatively low temperature is driven by two prescribed functions of time: the current in the toroidal windings $I_c(t)$, and the total flux enclosed by the torus $\Psi(t)$. Currents and magnetic fields are then induced in the plasma, and the plasma moves. Naturally, all the currents, magnetic fields and velocities are three-dimensional vectors.

This study gives the gross motion of the plasma in the poloidal plane, complete with pinching, bouncing, and toroidal column shift. Our results have been compared to the experimental results of the SP-1 screw pinch at Jutphaas, and the agreement is remarkably close. The method is currently being extended to belt pinches, i.e., toroidal devices with non-circular cross-section. The method is not immediately applicable to low density devices, such as tokamaks, since in these machines the pinching motion is unimportant while diffusion and particle-kinetic effects are dominantly important. The method as it stands also cannot be used for non-axisymmetric devices, nor does it treat instabilities with toroidal variation (These latter instabilities are often important in practice, but as of now we rely on theoretical analysis for their treatment). Axisymmetric instability modes, are of course, automatically accounted for in our study.

In the course of this research, we have benefited greatly from discussions with numerous colleagues at Courant Institute, at Princeton Plasma Physics Laboratory, and at Los Alamos Scientific Laboratory. The work was supported by the USAEC under contracts AT(11-1)-3086 and AT(11-1)-2456. The writing was completed while one author (CKC) was visiting professor at the University of Paris XI, and he is greatful for their hospitality.

Equations and Boundary Conditions

The differential equations of single fluid magnetohydrodynamics with finite electrical and thermal conductivities are well-known (here written in Gaussian units):

$$\frac{\partial \rho}{\partial t} + \operatorname{div} \rho \underline{u} = 0 \tag{1a}$$

$$\rho\left(\frac{\partial \underline{u}}{\partial t} + \underline{u}\cdot\nabla\, \underline{u}\right) = -\nabla p + \underline{j} \times \underline{B}/c \tag{1b}$$

$$\rho\, c_v\left(\frac{\partial T}{\partial t} + \underline{u}\cdot\nabla T\right) = -\, p \operatorname{div} \underline{u} + j^2/\sigma + \operatorname{div} \varkappa \nabla T \tag{1c}$$

$$\frac{1}{c}\frac{\partial \underline{B}}{\partial t} = - \text{curl } \underline{E} \tag{1d}$$

$$\underline{j} = \frac{c}{4\pi} \text{curl } \underline{B} \tag{1e}$$

$$\underline{E} = \underline{j}/\sigma - \underline{u} \times \underline{B}/c \tag{1f}$$

$$\text{div } \underline{B} = 0 \tag{1g}$$

The basic geometry and coordinate systems for this problem are shown in fig.1. These equations will be written in r, θ, φ coordinates, although $R = R_o + r \cos\theta$ will be retained for brevity. Since $\partial(\ )/\partial\varphi = 0$ by hypothesis, and by (1g), we can introduce a flux function ψ, as well as the quantity $\chi = RB_\varphi$. These two scalar functions completely describe all the magnetic field components, and by (1e), also all the current components:

$$B_r = -\frac{1}{Rr}\frac{\partial\psi}{\partial\theta}, \quad B_\theta = \frac{1}{R}\frac{\partial\psi}{\partial r}, \quad B_\varphi = \chi/R \tag{2a) - (2c}$$

$$j_r = -\frac{c}{4\pi Rr}\frac{\partial\chi}{\partial\theta}, \quad j_\theta = \frac{c}{4\pi R}\frac{\partial\chi}{\partial r} \tag{2d), (2e}$$

$$j_\varphi = -\frac{c}{4\pi}\left[\frac{1}{r}\frac{\partial}{\partial r}\left(\frac{r}{R}\frac{\partial\psi}{\partial r}\right) + \frac{1}{r^2}\frac{\partial}{\partial\theta}\left(\frac{1}{R}\frac{\partial\psi}{\partial\theta}\right)\right] \tag{2f}$$

Substituting these first into (1f) and then into (1d), we get

$$\frac{\partial\psi}{\partial t} + u_r\frac{\partial\psi}{\partial r} + \frac{u_\theta}{r}\frac{\partial\psi}{\partial\theta} = \frac{c^2}{4\pi\sigma}\left[\frac{R}{r}\frac{\partial}{\partial r}\left(\frac{r}{R}\frac{\partial\psi}{\partial r}\right) + \frac{R}{r}\frac{\partial}{\partial\theta}\left(\frac{1}{Rr}\frac{\partial\psi}{\partial\theta}\right)\right] \tag{3}$$

and a similar looking equation for χ. Moreover, in these coordinates, the energy equation (1c) for T is also of this form. These constitute a group of three parabolic equations for ψ, χ, and T; they are coupled to the fluid motion and fluid density, since ρ, u_r, u_θ, u_φ are contained in the coefficients.

At the same time, the mass and momentum conservation equations (1a),(1b) form an essentially hyperbolic system, with T appearing in the coefficients and $\underline{j}$ and $\underline{B}$ appearing in the forcing functions. We set $\underline{W} = (\rho, \rho u_r, \rho u_\theta, \rho u_\varphi)$; then (1a) and (1b) in r,θ coordinates become

$$\frac{\partial\underline{W}}{\partial t} + A(\underline{W},T)\frac{1}{r}\frac{\partial}{\partial r}(r\underline{W}) + B(\underline{W},T)\frac{1}{r}\frac{\partial\underline{W}}{\partial\theta} = S(\underline{W},r,\theta,T) + F(\underline{j},\underline{B}) \tag{4}$$

Here S consists of all the terms containing no derivatives of W and arising from the curvature effects, while $F(\underline{j},\underline{B})$ consists of all the $\underline{j} \times \underline{B}$ terms in the three directions. A and B are 4×4 matrices containing only the components of $\underline{W}$ and the temperature T.

We now describe the boundary conditions for the three parabolic equations. The outer shell is assumed to be ideally conducting, and is appropriately cut with radial and circumferential slots to permit the fields to penetrate into the plasma. The magnetic boundary condition on the circular shell $r = a$ (in the cases done so far, but again, the shell shape can be arbitrary) is $\underline{B}\cdot\underline{n} = 0$, or equivalently,

$$\psi = \psi_w(t) \text{ , a pure function of time.}$$

It is easily seen by applying Stokes theorem that the boundary value ψ_w is just equal to the flux in the"hole"of the torus, Ψ, which is the difference of the external transformer flux and the flux induced by the toroidal plasma current I_φ.

One may prefer to prescribe $I_\varphi(t)$ instead of $\Psi(t)$. In this case, we must assume $\psi_w(t)$ and perform a simple cut-and-try procedure. There is no significant increase in difficulty.

At the same time, the boundary condition for χ is obtained by applying Ampere's law fromthe coil current $I_c(t)$:

$$\chi_w = (RB_\varphi)_w = 2\pi N\, I_c(t) / c$$

where N is the total number of turns of the coil carrying I_c.

The boundary condition for the temperature is $T_w = T_o$, some constant, or $\partial T/\partial r = 0$, etc. In any case, there is very little plasma at the wall so that the solution is relatively insensitive to the precise condition employed.

For the set of hyperbolic equations, we can mathematically only prescribe $u_r = 0$ at $r = a$. However, as is common in numerical computations of hyperbolic equations, the difference scheme may require additional extraneous boundary conditions; in that case, we set as usual the normal derivatives or differences of u_θ, u_φ, and ρ to zero when needed.

The initial conditions of the problem is a plasma at rest with no currents and no fields, and a slight temperature, which in turns gives us enough electrical conductivity σ to start up the problem.

Strictly speaking, we must take into account the fact that a vacuum can develop as the plasma pinches inward. This case results in two difficulties. First, when the vacuum region is produced, we must solve a different set of equations for the vacuum, and match across the interface (to be determined) to the plasma --- the so called sharp boundary model. Second, when the density becomes low, the Alfven speed becomes enormous, which cases instability in explicit schemes and large inaccuracies in implicit schemes, unless one takes very small time steps. To avoid these two difficulties, we have at present simply stipulated a low density cut-off, or $\rho \geq \rho_{min} = 0.15\rho_{initial}$. Physically, this is approximately representative of particles diffusing from the walls, and our model is thus a special simplified form of a diffuse profile.

Numerical Methods

The systems of three parabolic equations and that of the four hyperbolic equations are split into two groups. Let us say that at $t = n\Delta t$, all the variables have been calculated. We take $\underline{W}^n = (\rho^n, (\rho u_r)^n, (\rho u_\theta)^n, (\rho u_\varphi)^n)$, and find an intermediate or predictor value $\underline{W}^*$ according to equation (5a) below. We then substitute $\underline{W}^*$ into the coefficients for the three parabolic equations, and calculate ψ^{n+1}, χ^{n+1}, T^{n+1} by the standard alternating direction implicit method for parabolic equations. These are then used in the coefficients for the hyperbolic equations to calculate $\underline{W}^{n+1}$. No further iterations are used.

Implicit schemes are also used to solve the hyperbolic system in order to insure stability when ρ becomes low. Our scheme is in essence a Crank-Nicolson scheme, with a predictor step for the nonlinear coefficients. Let

$$H_\theta\, \underline{W}_{i,j} \equiv (\underline{W}_{i,j+1} - \underline{W}_{i,j-1}) / r_{i,j}\, \Delta\theta$$

$$H_r\, \underline{W}_{i,j} \equiv (r_{i+1}\underline{W}_{i+1,j} - r_{i-1}\, \underline{W}_{i-1,j}) / r_{i,j}\, \Delta r\ .$$

Also, let

$$S^n_{i,j} = S\,(\underline{W}^n_{i,j}\,,\ r,\ \theta,\ T^n\,)$$

$$S^*_{i,j} = S\,(\underline{W}^*_{i,j}\,,\ r,\ \theta,\ T^{n+1}\,)$$

and use the same notation for A^n, A^*, B^n, B^*. The approximation for (4) is written in two main steps, a predictor step and a corrector step, similar to that used in ref. 1:

$$\underline{W}^*_{i,j} = \frac{1}{4}\,(\underline{W}^n_{i+1,j} + \underline{W}^n_{i-1,j} + \underline{W}^n_{i,j+1} + \underline{W}^n_{i,j-1})$$

$$- \frac{\Delta t}{2}\,(A^n_{i,j}H_r + B^n_{i,j}\,H_\theta)\,\underline{W}^n_{i,j} + \Delta t\,(S^n_{i,j} + F^n_{i,j}) \qquad (5a)$$

$$(I + \frac{\Delta t}{4} B^{*}_{i,j} H_{\theta}) (I + \frac{\Delta t}{4} A^{*}_{i,j} H_{r}) \underline{W}^{n+1}_{i,j}$$
$$= (I - \frac{\Delta t}{4} B^{n}_{i,j} H_{\theta}) (I - \frac{\Delta t}{4} A^{n}_{i,j} H_{r}) \underline{W}^{n}_{i,j}$$
$$+ \frac{\Delta t}{2} (S^{*}_{i,j} + S^{n}_{i,j} + F^{n+1}_{i,j} + F^{n}_{i,j}) \qquad (5b)$$

The implicit equation (5b) is then solved in two fractional steps (D'jakonov splitting) as follows:

$$(I + \frac{\Delta t}{4} B^{*}_{i,j} H_{\theta}) \underline{\tilde{W}}_{i,j} = \text{right-hand side of (5b)}$$
$$(I + \frac{\Delta t}{4} A^{*}_{i,j} H_{r}) \underline{W}^{n+1}_{i,j} = \underline{\tilde{W}}_{i,j} \qquad (6)$$

Results and Discussion

The method has been applied to simulate an actual experiment, the SP-1 screw pinch at Jutphaas, Holland[2]. Fig. 2 gives the main parameters and driving currents. Fig. 3 shows the pinching, bounce, and second pinching of the plasma column, as well as the toroidal shift of the column, as seen from the density and temperature profiles at various instants of time. The axis of the machine is to the left of the coordinate system, i.e., negative r corresponds to the direction toward the axis of the machine. The stronger magnetic fields at the smaller radii R give greater magnetic pressures, hence the profiles are not symmetrical with respect to $r = 0$ or $R = R_o$. The computation breaks down (so far) after the second pinch, since near $r = 0$, there is large B, small ρ, as well as small cell size $r\Delta\theta$, so that the Alfven spped is high and the Courant condition is exceeded strongly there.

Fig. 4 is a comparison of B_{φ} and B_{θ} as functions of time and position along the plane of symmetry $\theta = 0^{o}$ and $\theta = 180^{o}$, as computed and as actually measured. The agreement is very encouraging. Two observations are in order: (1) Our rise time of the fields is not exactly the same as the experimental value, and we have compared the B-profiles at exactly the same time instants; the agreement would have been even better had we compared slightly different time instants. (2) The measured B_{θ}-profiles have a flat center region, indicating little current there, and more nearly a sharp boundary plasma, than the profiles calculated. This is probably due to the dissipativeness of the numerical scheme, or to the dissipative coefficients used.

Fig. 5 is a comparison of the toroidal field B_{φ} at $r = 0$ as a function of time. The closeness of the calculated and experimental results indicate that at these small time scales, magnetohydrodynamics indeed is an adequate description of the various phenomena, and the neglecting of the precise transport mechanisms does not cause too much error. Needless to say, for long-time behavior and equilibrium calculations, more detailed transport mechanisms must be included in the model.

References

1. A.R. Gourlay and J. Ll. Morris, Math. Comp. 22, 549 (1968).
2. C. Bobeldijk, L.H. Th. Rietjens, P.C.T. van der Laan, T. Th.de Bats, Equilibrium and stability of a toroidal screw pinch. FOM Rijnhuizen Report 66-32, 1966.

   C. Bobeldijk, The toroidal screw pinch. FOM Rijnhuizen Report 68-45, 1968.

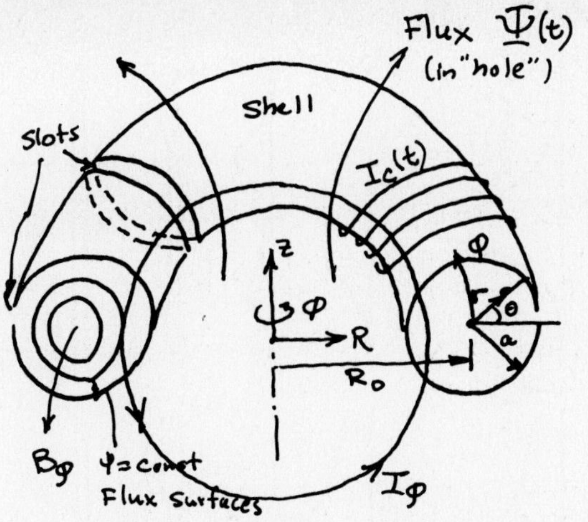

Fig. 1 Toroidal Screw Pinch and Coordinate Systems

Gas : hydrogen

ρ_{init} : 0.4×10^{-8} gm/cm^3

T_{init} : 1 ev

B_φ bias : 100 G

B_φ max applied : 6 kG

Minor radius a : 6 cm

Major radius R_o : 36 cm

σ : Spitzer value

χ : 2×10^6 erg/cm sec °K

ρ_{min} : 0.15 ρ_{init}

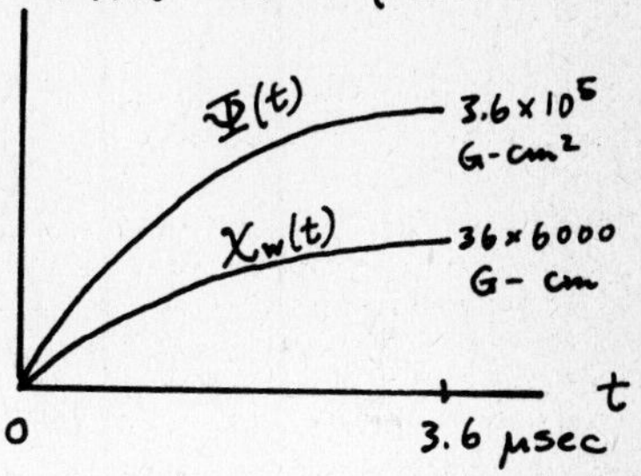

Fig. 2. Input Data

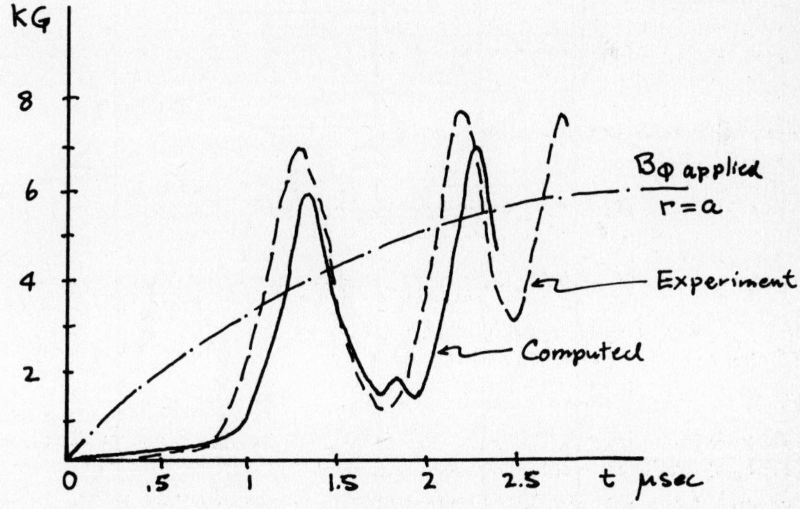

Fig. 5 $B_\varphi(t)$ at $r = 0$

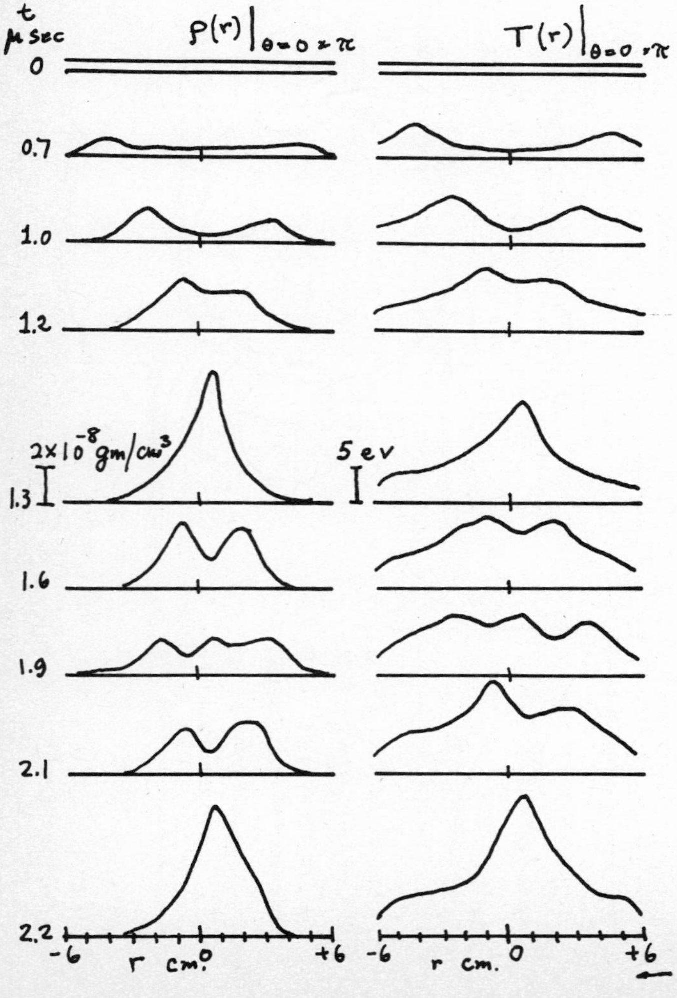

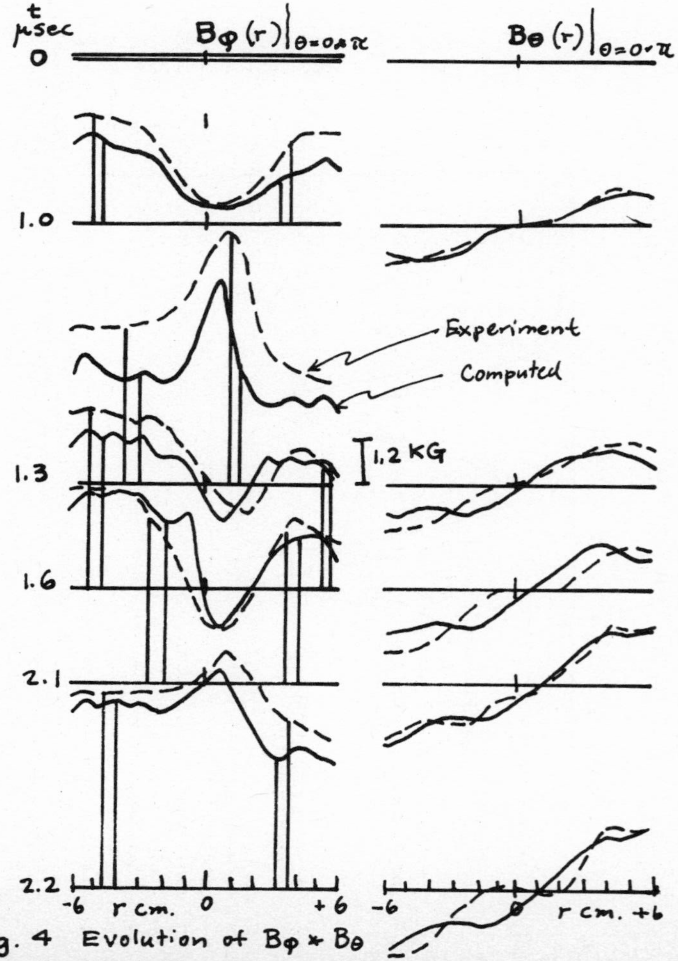

Fig. 4 Evolution of Bφ & Bθ

Fig. 3. Evolution of ρ & T

COMPUTATIONAL ASPECTS OF A VISCOUS INCOMPRESSIBLE FLUID

Guillermo Marshall
Universidad de Buenos Aires
Argentina

INTRODUCTION

The purpose of the present work is to study the solution of the Navier-Stokes system of partial differential equations describing viscous incompressible flows in two dimensions. The method of solution is based on the perturbation of the Poisson type equation. The resulting regular system is integrated by means of difference schemes. Some results are presented for the problem of the steady flow in a square cavity, for different values of the Reynolds number (the square cavity flow problem has been studied, inter alios, by Pan and Acrivos, Burgraff, Greenspan, Fortin, Temam and Peyret and Bourcier and Francois). The method of solution briefly called the ε-method (for the Navier-Stokes system written in transport vorticity form) has been introduced by Marshall [7] and Marshall and van Spiegel. In the present work especial attention is paid to the finite difference methods used to solve the Navier-Stokes system in association with the ε-method.
The finite difference approximations can be divided into integer and fractional time step methods. The latter are methods of construction of economic difference schemes. The first results on the application of these methods to the problems of mathematical physics were obtained by Douglas, Peaceman and Rachford (see for example Mitchell, Martchouk and Yannénko). These methods suppose a consistent approximation of the differential system in each of the fractional steps. Closely related to these methods is the method of disintegration or splitting or local one dimensional method. This method was formulated for the first time by Soviet mathematicians, Yannenko, Martchouk, Sammarski, D'Yakonov and others. In this method the intermediate fractional steps are not necessarily stable and consistent with the differential equation. Consistency and stability are satisfied on the whole (on the integer time step). We have applied these methods to the solution of the Navier-Stokes system, in what we have called the LODE and LODI schemes.
Further, Martchouk introduced a series of numerical algorithms for the method of disintegration of the equations of dynamical atmospheric processes. This method is based on the disintegration of the equations of weather forecasting, disintegration that is made to correspond to physical feasible stages of atmospheric processes; in this sense, it can be regarded as a 'natural' method of approximation. The first stage of this process is the transport of the substance along the trajectory, the second is the dynamic adaptation of the field and the third is the turbulent diffusion. We have applied these ideas to the solution of the Navier-Stokes system, in what we have called the PE and PI schemes.
Concluding, in the present work, the highlights of each scheme are analysed and a comparison of their perfomance is made in connection with a specific problem: the steady state square cavity flow problem.

FORMULATION OF THE PROBLEM

Let us consider the associated perturbed problem (cf [7] and [9]), approaching the Navier-Stokes system of equations describing the laminar flow of a viscous incompressible fluid in two dimensions (dimensionless form)

$$\left\{\begin{array}{l} \dfrac{\partial \omega}{\partial t} + U_i \dfrac{\partial \omega}{\partial x_i} = \dfrac{1}{R} \dfrac{\partial^2 \omega}{\partial x_i \partial x_i} \\[2ex] \varepsilon \dfrac{\partial \phi}{\partial t} = \dfrac{\partial^2 \phi}{\partial x_i \partial x_i} + \omega \\[2ex] i = 1,2. \end{array}\right. \qquad (1)$$

Where U_i are the velocity components: $U_1=\partial\phi/\partial x_2$ and $U_2=-\partial\phi/\partial x_1$, ϕ is the stream function, ω is the vorticity: $\omega=\partial U_2/\partial x_1-\partial U_1/\partial x_2$, R is the Reynolds number: $R=U_0D/\upsilon$, U_0 is a reference velocity, D is a reference length and υ is the kinematic viscosity, t is the time, x_i are the space coordinates, i=1,2. The term $\varepsilon\,\partial\phi/\partial t$ has no physical meaning; it is used only for regularizing system (1).

We are seeking functions $W=\{\omega,\phi\}$ defined in the cylinder $G=\Omega\times[0,T]$ (where Ω is a bounded two dimensional region with boundary Γ, $T>0$), satisfying (1) together with the following initial and boundary conditions:

$W(x_i,0)=W_0(x_i)$, $x_i\in\Omega$, ($\phi_0(x_i)$ arbitrary chosen),

$W(x_i,t)=W(x_i)$, $x_i\in\Gamma$, $t>0$.

For ε fixed (1) has a unique solution, furthermore, it is conjectured that when ε tends to zero the solution of problem (1) tends to the solution of the Navier-Stokes equations. This type of perturbation method is inspired in the works of Chorin, Lions, Temam and Yannenko. System (1) constitutes a mixed Cauchy problem. For $\varepsilon=0$ we shall obtain a real evolutionary problem, for $\varepsilon\neq0$, we shall obtain a pseudo-evolutionary problem i.e. a problem in which the intermediate time step solutions do not represent solutions of the real evolutionary problem. A full account of the influence of the parameter ε is given in references [7] and [9]; in the present work we shall consider it as constant and equal to unity.

We shall study numerical solutions of (1) using the steady state square cavity flow problem as a test problem. In this case we are seeking steady state solutions of (1) (t tending to infinite and ε arbitrary). The steady state square cavity flow problem deals with the steady flow in a two dimensional square cavity where the motion of the viscous fluid is driven by the uniform translation of the top wall. The mathematical formulation of this problem as well as the discretization problem can be found for example in [5].

CENTRED AND NON-CENTRED EXPLICIT SCHEMES

A consistent explicit integer time step scheme approaching (1) follows

$$\frac{W^{n+1}-W^n}{\Delta t}+A\,W^n=B\,W^n+F^n+O(\Delta t) \qquad (2)$$

where W is defined as $W=\{\omega,\phi\}$ and A, B and F as

$$A=\begin{bmatrix} U_i\dfrac{\partial}{\partial x_i} & 0 \\ 0 & 0\end{bmatrix},\quad B=\begin{bmatrix}\dfrac{1}{R}\dfrac{\partial^2}{\partial x_i\partial x_i} & 0 \\ 0 & \dfrac{1}{\varepsilon}\dfrac{\partial^2}{\partial x_i\partial x_i}\end{bmatrix},\quad F=\begin{bmatrix}0\\ \dfrac{1}{\varepsilon}\omega\end{bmatrix}$$

For simplicity we have only discretized the time variable. As usual W^n represents the value of the numerical solution at time $t=n\,\Delta t$; t is the time step. If in the approximation of the term AW^n we use centred differences we obtain the so called CES scheme (centred explicit scheme). However if we use non-centred differences according to the sign of the velocities (upstream scheme), we obtain the so called NES scheme (non-centred explicit scheme). The former is a second order scheme in space while the latter is a first order scheme. In both schemes the dissipative term, i.e. BW^n, is approached by standard second order differences.

A linear stability analysis shows that the CES and NES schemes are stable if

$$\Delta t\leq\varepsilon\Delta x^2/4,\quad \Delta t\leq R\,\Delta x^2/4 \quad\text{and}\quad \Delta x\leq 2/(R|U_i|) \qquad \text{(CES scheme)},$$

$$\Delta t\leq\varepsilon\Delta x^2/4,\quad\text{and}\quad \Delta t\leq\left[(|U_1|+|U_2|)/\Delta x+4/(R\,\Delta x^2)\right]^{-1} \qquad \text{(NES scheme)}$$

where $\Delta x_1=\Delta x_2=\Delta x$ indicates the space step.

The NES scheme introduces a numerical dissipation which is a function of the velocity field and the step size. Several authors have studied this approximation and it is now recognized to be inaccurate for any but very special circumstances. However, at present it seems to be the only way to overcome the perturbations appearing at high Reynolds numbers.

LOCAL ONE DIMENSIONAL SCHEMES

Let us write system (1) as

$$\frac{\partial W}{\partial t} = A\,W + F \tag{3}$$

where now A is the sum of the operators A and B of (2). A split scheme approaching (3) follows

$$\begin{cases} \dfrac{W^{n+1/2} - W^n}{\Delta t} = \alpha A_1 W^{n+1/2} + \beta A_1 W^n + F_1^{n+1/2} \\[2ex] \dfrac{W^{n+1} - W^{n+1/2}}{\Delta t} = \alpha A_2 W^{n+1} + \beta A_2 W^{n+1/2} + F_2^{n+1/2} \\[2ex] A = A_1 + A_2,\ F = F_1 + F_2, \end{cases} \tag{4}$$

where $\alpha + \beta = 1$, α and β being weighting coefficients. If $\alpha = 0$ we obtain the so called LODE scheme (local one dimensional explicit scheme). If $\alpha \neq 0$ we obtain the so called LODI scheme (local one dimensional implicit scheme). In both schemes the convective and diffusive terms are approached by centred differences. It can be shown that (4) is a consistent approximation of (3) (cf [8]).
A linear stability analysis shows that for the LODE scheme ($\alpha = 0$), the stability conditions are

$$\Delta t \leq \varepsilon \Delta x^2/2, \qquad \Delta t \leq R\,\Delta x^2/2 \qquad \text{and} \qquad \Delta x \leq 2/(R\,|U_i|)$$

and for the LODI scheme ($\alpha = 0$), the stability condition is

$$\Delta x \leq 2/(R\,|U_i|\,).$$

In actual computations with the LODI scheme we have used $\alpha = \beta = 1/2$.

PREDICTOR (CORRECTOR) SCHEMES

For the sake of simplicity we write system (1) as

$$\begin{cases} \dfrac{\partial \omega}{\partial t} + A\,\omega = B\,\omega \\[2ex] \dfrac{\partial \phi}{\partial t} = D\,\phi + F \end{cases} \tag{5}$$

where A, B, D and F are defined as

$$A = U_i\ \partial/\partial x_i, \quad B = 1/R\ \partial^2/\partial x_i \partial x_i, \quad D = 1/\varepsilon\ \ \partial^2/\partial x_i \partial x_i \quad \text{and } F = 1/\varepsilon\ \omega$$

A disintegrated scheme, in the sense of Martchouk, approaching (5) follows

$$\frac{\omega^{n+1/2} - \omega^n}{\Delta t} + \alpha A\ \omega^{n+1/2} + \beta A\ \omega^n = 0 \tag{6.1}$$

$$\frac{\omega^{n+1} - \omega^{n+1/2}}{\Delta t} = \alpha B\ \omega^{n+1} + \beta B\ \omega^{n+1/2} \tag{6.2}$$

$$\frac{\phi^{n+1} - \phi^{n+1}}{\Delta t} = \alpha D\ \phi^{n+1} + \beta D\ \phi^n + F^{n+1/2} \tag{6.3}$$

Equation (6.1) describes a pure convection process, equation (6.2) a diffusion process and finally equation (6.3) describes the process of adjustment of the field produced by means of the integration of the Poisson type equation. The weighting coefficients can be different for each equation. If the α's are zero we obtain the so called PE scheme (predictor explicit). If they are different from zero we have the PI scheme (predictor implicit). In both schemes the convective terms are treated by non-centred differences according to the sign of the velocities. In the PI scheme we have used $\alpha = 1$ and $\beta = 0$ in (6.1) and $\alpha = \beta = 1/2$ in (6.2) and (6.3). Moreover, due to the implicit character of the PI scheme a further split of (6.1), (6.2) and (6.3) in the x_i directions is necessary, in order to solve them by the tridiagonal algorithm. We observe that the way in which (6.1) has been approximated, in the PI scheme, provides tridiagonal matrices belonging to the Jacobi type, thus ensuring the convergence of the process.

Martchouk proposed a further fractional step (the corrector step) after (6.1), to provide second order accuracy. Since the whole process (predictor plus corrector), is equivalent to a centred scheme one obtains the same penalty on the space step, for stability to hold.

A linear stability analysis for the PE scheme gives

$$\Delta t \leq \varepsilon \Delta x^2/4, \qquad \Delta t \leq R \ \Delta x^2/4 \qquad \text{and} \quad \Delta t \leq \Delta x/(|U_1| + |U_2|)$$

while a linear stability analysis of the PI scheme shows that it is unconditionally stable.

BOUNDARY PROCEDURES AND COMPOUND ITERATION

With the methods described above we were not able to obtain convergent and reasonable accurate solutions for Reynolds number 1000 even using grid sizes as fine as 40x40. To overcome this problem, inspired in the work of Greenspan, we introduced two new procedures. The first is a special treatment, at the boundaries, of the vorticity function, in the manner described by the above-mentioned author. The second is the use of smoothing techniques or second iteration. This amounts to the use of the following procedure after every time step of the preceding schemes

$$W^{n+1} = \alpha \ W^n + \beta \ W^{n+1} \qquad \text{where} \quad \beta = 1 - \alpha \ , \quad 0 \leq \alpha \leq 1.$$

This second iteration can be applied to any particular scheme. We have used it in connection with non-centred schemes (NES, PE and PI). For $\alpha = 0$ we obtain the normal version of the schemes already studied. The whole procedure is called a compound iteration (Ortega and Rheinboldt).

NUMERICAL EXPERIMENTS

The three methods (explicit and implicit version) have been applied to the solution of the square cavity flow problem for different values of the Reynolds number. The precision for obtaining the steady state was fixed between 10^{-4} and 10^{-6}. We present here some results (see [8]). Figures 1 and 2 show the evolution of the residual R_1 as a function of the number of cycles for Reynolds numbers 0, 50 and 100 (CES and NES schemes). The rapidity of convergence to the steady state is a function of the Reynolds number. Moreover, the ripples appearing in fig.1 had disappeared in fig.2, this is because the NES scheme is more dissipative than the CES scheme. In this context fig.3 shows the streamlines for Reynolds number 400 (CES and NES schemes superimposed). Figure 4 shows the influence of the splitting process, the second iteration and the especial treatment at the boundaries. Figures 5 and 6 show streamlines for Reynolds number 1000 obtained with an integer and a fractional time step scheme. Figures 7 and 8 give an overall picture of the solution obtained with the CES scheme for R=400. Finally in table I we present

TABLE I

Reynolds number	ϕ max (other authors)	ϕ max (present work)
400	0.094 [4] , $\Delta x = 1/40$	0.101 (CES), $\Delta x = 1/40$
1000	0.085 [4] , $\Delta x = 1/40$	0.105 (NES), $\Delta x = 1/20$
400	0.102 [2] , $\Delta x = 1/40$	0.101 (CES), $\Delta x = 1/40$
1000	0.103 [5] , $\Delta x = 1/20$	0.105 (NES), $\Delta x = 1/20$

comparable results obtained by other authors for the maximum value of the stream function in the square cavity flow problem.

CONCLUSIONS

The main conclusions in relation to the numerical aspects can be summarized as follows: In general stable solutions for higher Reynolds numbers can only be reached with non-centred differences for the approximation of the convective terms* and with smoothing techniques (compound iteration). Furthermore this fact is true for explicit schemes and implicit schemes as well (integer or fractional time step methods). Explicit schemes on the verge of instability are competitive with implicit schemes. In the higher range of Reynolds numbers and with large time steps splitting methods seem very inaccurate. Obviously in such cases a great truncation error is involved and only qualitative global accuracy may be claimed.

---

*We do not mean that it is impossible to use centred schemes for high Reynolds. It is just the question of satisfying the penalty on the space step which is a function of the Reynolds number. For high Reynolds this restriction is unlikely to be satisfied by presentday computers' capacity, not to speak of the total computing time necessary to reach the steady state with such a grid size.

REFERENCES

[1] Bourcier M. and Francois C., Rech. Aèrosp. 131, (1968).
[2] Burgraff O. R., J. Fluid Mech. 24, (1966).
[3] Chorin A. J., J. Comp. Physics 2, (1967).
[4] Fortin M. Temam R. and Peyret R., Internal Report IRIA, (1970).
[5] Greenspan D., Lectures on the numerical solution of linear, singular and nonlinear differential equations, Prentice Hall, (1968).
[6] Lions J. L., in Mathematics applied to Physics, Springer-Verlag, (1970).
[7] Marshall G., Report Na7, Dept. of Math., T. H. Delft, (1972).
[8] Marshall G., Computational methods in viscous flow problems, Ph. D. thesis, Dept. of Mathematics, T. H. Delft, (1973).
[9] Marshall G. and E. van Spiegel, J. Eng. Math. vol. 7, 2, (1973).
[10] Martchouk G. I., Méthodes numériques pour la prévision du temps, A. Colin, (1970).
[11] Mitchell A. R., Computational methods in partial differential equations, Wiley, (1969).
[12] Ortega J. M. and Rheinboldt W. C., Iterative solutions of nonlinear equations, A. Press, (1970).
[13] Pan F. and Acrivos A., J. Fluid Mech. 28, (1967).
[14] Yannenko N. N., Méthode à pas fractionnaires, A. Colin, (1968).

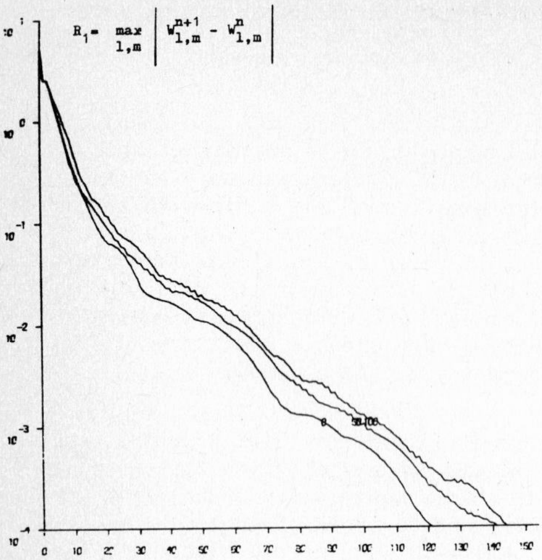

Fig.1. Evolution of the residual R_1 for Reynolds 0, 50 and 100 (CES scheme).

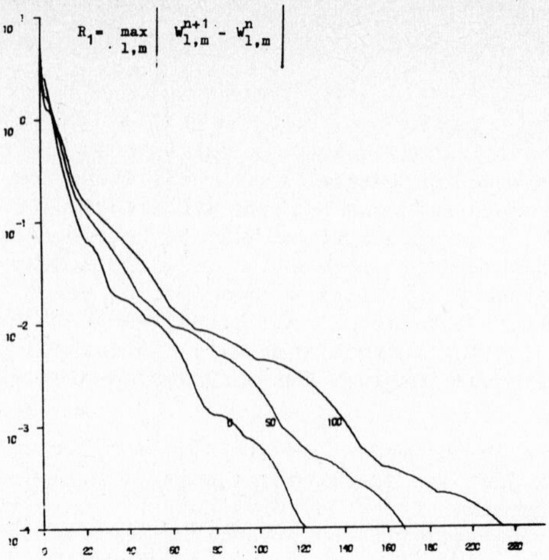

Fig.2. Evolution of the residual R_1 for Reynolds 0, 50 and 100 (NES scheme).

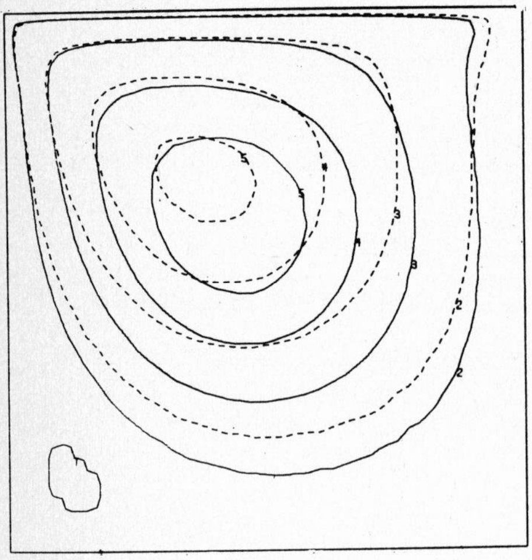

Fig.3. Streamlines for Reynolds number 400 obtained with first and second other schemes.

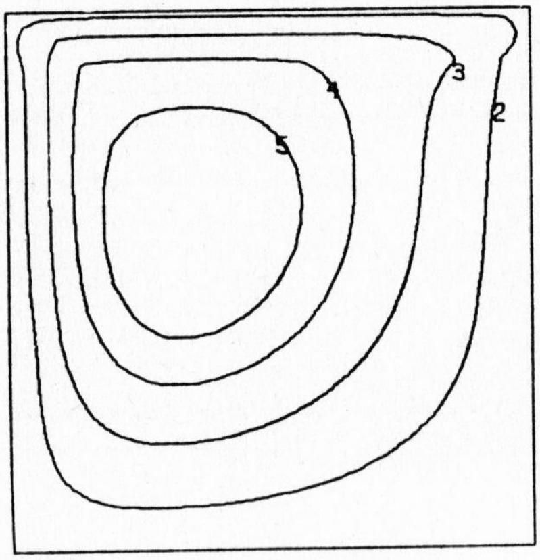

Fig.4. Streamlines for Reynolds number 400 obtained with the PI scheme.

——	CES scheme
---	NES scheme

1	=	0.000
2	=	0.010
3	=	0.040
4	=	0.070
5	=	0.090

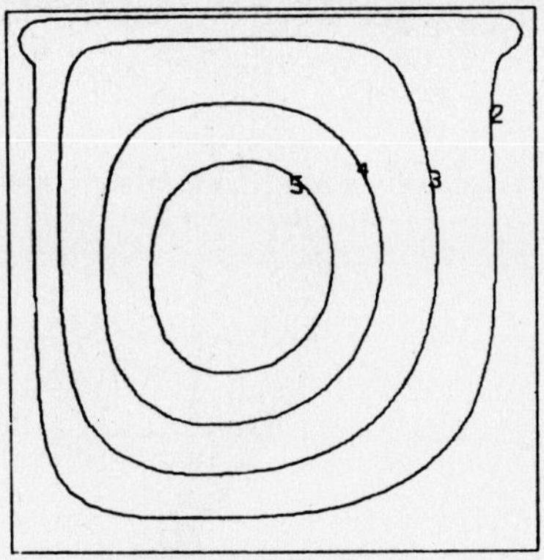

Fig.5. Streamlines for Reynolds number 1000 obtained with the NES scheme.

1 = 0.000
2 = 0.010
3 = 0.040
4 = 0.070
5 = 0.090

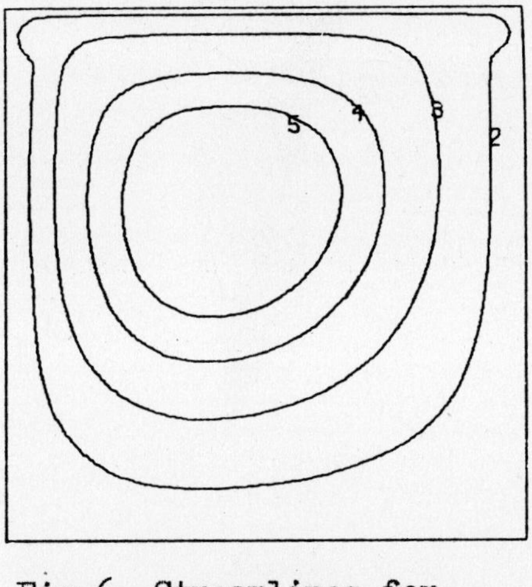

Fig.6. Streamlines for Reynolds number 1000 obtained with the PE scheme.

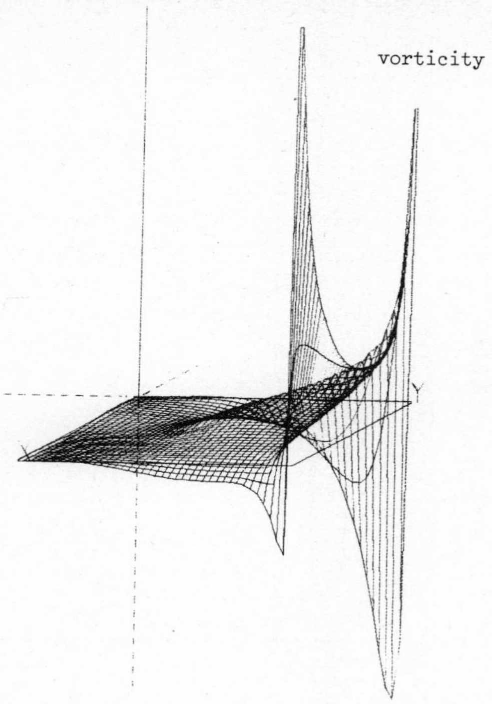

Fig.8. Vorticity function for Reynolds number 400 (40x40 grid size), obtained with the CES scheme.

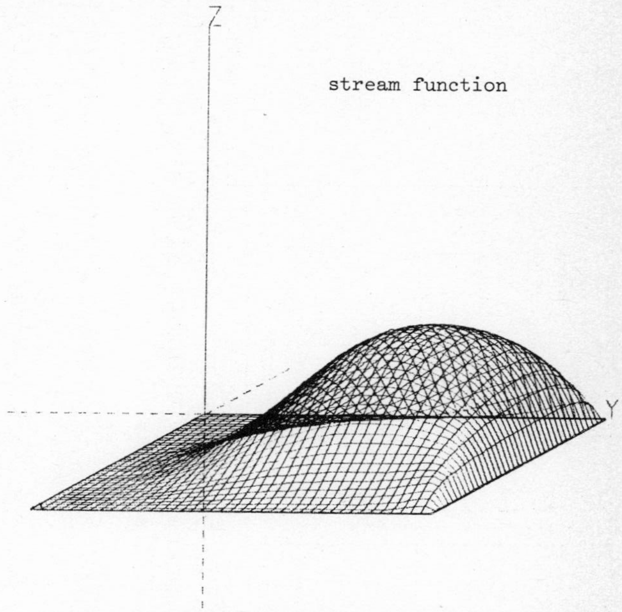

Fig.7. Stream function for Reynolds number 400 (40x40 grid size), obtained with the CES scheme.

DIFFERENCE METHODS FOR TRANSONIC FLOWS ABOUT AIRFOILS *

Arthur A. Mirin
Lawrence Livermore Laboratory
and
Samuel Z. Burstein
Courant Institute of Mathematical Sciences

ABSTRACT

Solutions to the problem of steady transonic flow of an inviscid perfect gas past plane lifting airfoils are computed using a time dependent formulation. The steady solution to the system of hyperbolic partial differential equations is obtained as the asymptotic limit of flow for large times. A coordinate system, which makes use of a conformal map of the interior of the unit circle onto the exterior of the airfoil, is used. The conservation laws are expressed in terms of polar coordinates on the interior of the unit disc and are solved using a second order accurate method. To extend the numerical solution to the boundary of the unit disc the method of characteristics is used.

1. INTRODUCTION

One of the most difficult problems in classical fluid dynamics is the computation of transonic flows. In order to examine questions relating to supercritical flow about airfoils, for supersonic free stream Mach numbers, the equations are cast in a conservative form with no assumption being made on the irrotationality of the flow field. In this regard, the jump condition across shocks will be accurately predicted no matter what their strength. Of course for subsonic free stream Mach numbers the applications of the potential is more secure since one is interested in flows which are shock free or which contain weak shocks that do not cause a separation of the boundary layer. Since there is not much experimental data available for supersonic flows in the transonic region it is hoped that solutions obtained by the present formulation will be a helpful guide which other investigators may compare.

*This research was supported under AEC Contract No. AT(11-1)-3077.

2. PROBLEM FORMULATION

Transonic flow about a two dimensional airfoil with boundary $\mathcal{C}$, with negligible viscosity effects, can be described by a system of four equations in the form

$$w_t + f_x + g_y = 0 \tag{2.1}$$

where the vectors w, f and g are

$$w = \begin{pmatrix} \rho \\ \rho u \\ \rho v \\ E \end{pmatrix}, \quad f = \begin{pmatrix} \rho u \\ \rho u^2+p \\ \rho uv \\ (E+p)u \end{pmatrix}, \quad g = \begin{pmatrix} \rho v \\ \rho vu \\ \rho v^2+p \\ (E+p)v \end{pmatrix} \tag{2.2}$$

Here ρ is the density, p is the pressure, u and v are the x and y components of velocity, and E is the sum of the internal energy, e, and kinetic energy of the fluid, where

$$p = \rho(\gamma-1)e \ . \tag{2.3}$$

In formulating the problem, boundary conditios are required for the system (2.1) on the airfoil. For airfoils of arbitrary shape it is very convenient to conformally map the exterior of the airfoil onto the unit disc with the boundary $\mathcal{C}$ being mapped onto the unit circle; the origin of the unit disc is mapped onto the point at infinity.

Let $z = x + iy$ be the coordinates in the physical plane and $\sigma = re^{i\theta}$ be the coordinates in the unit disc. Denote the mapping function from the unit disc onto the exterior of the airfoil F, and the modulus of the derivative of F as B. We introduce χ, which is related to the argument of the derivative of the mapping function by

$$\chi = \arg\left[\frac{\partial}{\partial\theta} F(re^{i\theta})\right] \tag{2.4}$$

It can be readily seen that for $r = 1$, χ is the slope of the airfoil at the point $F(e^{i\theta})$ and it is assumed that the point $\sigma = 1$ is mapped onto the tail of the airfoil.

Then under (2.4), equation (2.1) becomes

$$w_t + \frac{1}{B}(\sin\chi\, f_r - \cos\chi\, g_r) + \frac{1}{rB}(\cos\chi\, f_\theta + \sin\chi\, g_\theta) = 0 \tag{2.5}$$

Equation (2.5) is subject to the following boundary conditions. On $\mathcal{C}$ ($r = 1$) the normal velocity is zero, i.e.

$$v = u \tan\chi \tag{2.6}$$

At infinity ($r = 0$) it is convenient to describe the boundary conditions in terms of whether the fluid is entering or leaving the flow domain. Take the case where the free stream velocity is subsonic. Since equation (2.5) becomes singular at $r = 0$, this condition is prescribed for $r = r_{min} > 0$; thus the problem is formulated on an annulus $r_{min} \leq r \leq 1$. At $(-\infty)$ the angle of attack α, the Mach number M and the sound speed $c = \sqrt{\gamma p/\rho}$ are prescribed. This allows for the immediate computation of the velocity components. The pressure and density are computed by integrating along the Mach conoid using characteristic compatibility equations. At $(+\infty)$ the angle of attack α is prescribed; all other flow quantities are again computed by application of the method of characteristics.

When the free stream condition is supersonic, all of the flow quantities are prescribed at inflow points on the boundary $r = r_{min}$ while all flow quantities are extrapolated at outflow boundary points by the method of characteristics.

3. DIFFERENCE TECHNIQUE

In this section we describe a finite difference scheme for the solution of equation (2.5) on the interior of the annulus. Advancement of the finite difference solution in the neighborhood of the tail is also discussed.

It is assumed that a mesh $r_i = i\,\Delta r$ and $\theta_j = j\,\Delta\theta$ is formed and that at each mesh point a centered difference formula is used to approximate the spatial derivatives.

The solution is obtained using a two step operator, denoted by D, in which the first step invokes a first order difference operator. The temporary solution obtained is used to construct flux derivatives which can be centered in time as well as space in the second step. It is found that a damping term V is required to stabilize the difference operator D; the numerical solution w then satisfies the difference scheme

$$w^{n+1} = D\cdot w^n + V(D\cdot w^n) \,, \tag{3.1}$$

where V is a third order difference operator.

The coefficients of equation (2.5) become arbitrarily large in the neighborhood of the tail. It is necessary to use a special procedure to advance the numerical solution in the neighborhood of $\sigma = 1$. We construct a cartesian secondary mesh in the neighborhood of the tail. At the beginning of each time step the flow quantities are interpolated onto this secondary mesh. The local coordinates of the mesh do not, in general coincide with the cartesian coordinates in the physical plane. The numerical solution is advanced on all interior points on the secondary mesh using the difference operator D in equation (3.1) but which corresponds to solving the difference approximating to equation (2.1). A first order solution using an algorithm which is similar to the first step of the two step difference operator D is obtained at the tail. This numerical solution is then interpolated back to the mesh (the image of the mesh in the σ-plane) in the physical plane.

Stability of the operator D is considered by writing (2.8) as a quasilinear partial differential equation, i.e.,

$$w_t + Aw_r + Bw_\theta = 0$$

with $A = \alpha f_w + \beta g_w$ and $B = \omega f_w + \tau g_w$. Then the maximum allowable step size is chosen by

$$\Delta t \leq \sigma \inf_{i,j} \left\{ \frac{\Delta r_i}{\rho(A_{ij})}, \frac{\Delta \theta_j}{\rho(B_{ij})} \right\}, \quad 0 < \sigma \leq \frac{1}{\sqrt{2}}. \quad (3.2)$$

The spectral radii ρ can be computed easily since the matrices f_w and g_w can be simultaneously symmetrized. In addition, the maximum allowable time step must satisfy condition (3.2) imposed at the secondary mesh.

4. RESULTS

The transonic model described has been programmed for the CDC 7600 at Lawrence Livermore Laboratory and the CDC 6600 at New York University. The rate at which the difference scheme is solved at each mesh point is 1568 points per second on a 7600 (essentially the same program on a 6600 results in a reduction of this processing rate by a factor of 7). For the computations presented here the finite difference mesh uses 80 points in the θ direction and 21 points in the r direction.

We first present results for a calculation having a subsonic free stream Mach number of .76. The triangular zones shown in Figure 1 are solution points to the pressure coefficient over the airfoil described in Kacprzynski, Ohman, Garabedian and Korn [1] and was obtained from the potential solution method of Bauer, Garab dian and Korn [2]. The solid curve represents the asymptotic state of the transient solution starting from impulsive flow. The pressure distribution shows that a weak shock terminates the supersonic bubble on the upper surface. Figure 2 shows the pressure distribution for this case. A comparison of the pressure distributions over the same airfoil with experimental data (Kacprzynski et al. [1]) is shown in Figure 3.

Figure 4 shows a contour map of the pressure distribution for an asymptotic solution obtained with a supersonic free stream Mach number, $M_\infty = 1.04$, past an NACA 0012 symmetric airfoil. The shock pattern emanating from the tail makes a half angle of approximately 61 degrees. This compares with a two dimensional weak shock solution angle of approximately 60 degrees for the computed Mach number $M = 1.4$ upstream of the shock (the downstream Mach number is computed to be approximately 1.1). The trailing edge angle is 16 degrees for this airfoil. The drag coefficient for this case is computed to be .0.102.

A more detailed discussion of the method will appear in a forthcoming report. The authors would like to thank Dr. Frances Bauer for for the comparison solution used in Figure 1.

REFERENCES

[1] Kacprzynski, J. J., Ohman, L. H., Garabedian, P. R. and Korn, D. G., NRC of Canada Aeronautical Report LR-554, Ottawa, 1971.

[2] Bauer, F., Garabedian, P., and Korn, D., Lecture Notes in Economics and Mathematical Systems, Springer-Verlag, 1972.

Figure Captions

Figure 1. Comparison of the pressure coefficient for the NAE airfoil (Kacprzynski, Ohman, Garabedian and Korn) at $M_\infty = 0.76$, $\alpha = -0.45°$.

Figure 2. Pressure contour map about an NAE airfoil at $M_\infty = 0.76$, $\alpha = -0.45°$.

Figure 3. Comparison of the pressure coefficient for the NAE airfoil; experimental Re = 20 million.

Figure 4. Pressure contour map about an NACA 0012 symmetric airfoil at $M_\infty = 1.04$, $\alpha = 0°$.

Figure 1

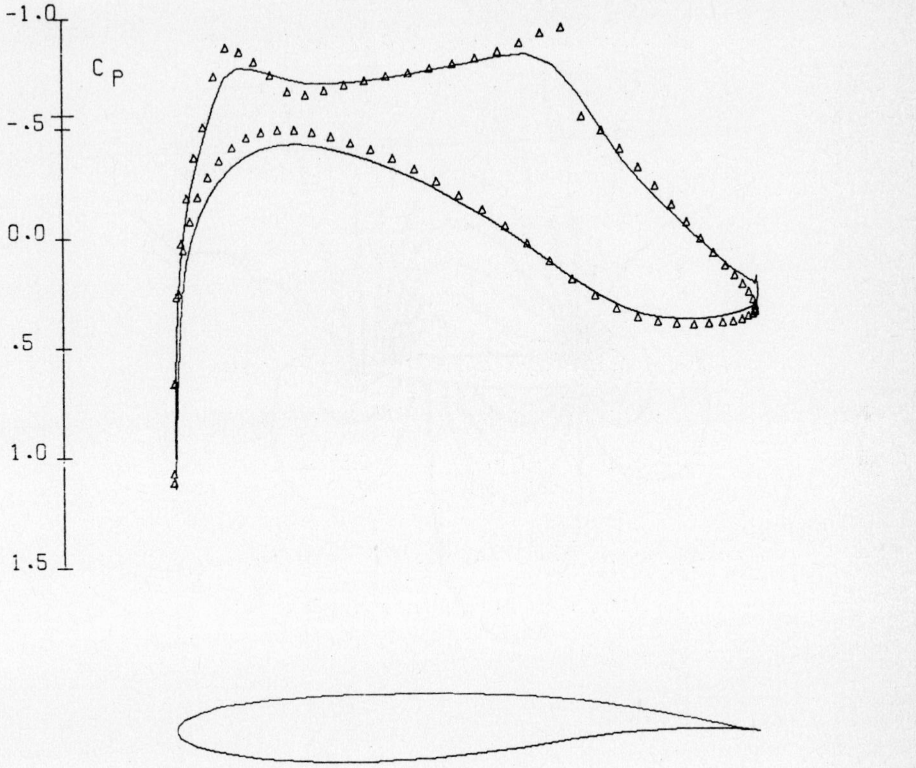

_ TRANSIENT M=.760 ALP= -.45 CL= .497 CD=.0109 T/C= .12

△ POTENTIAL M=.760 ALP= -.58 CL= .497 CD=.0026

Figure 2

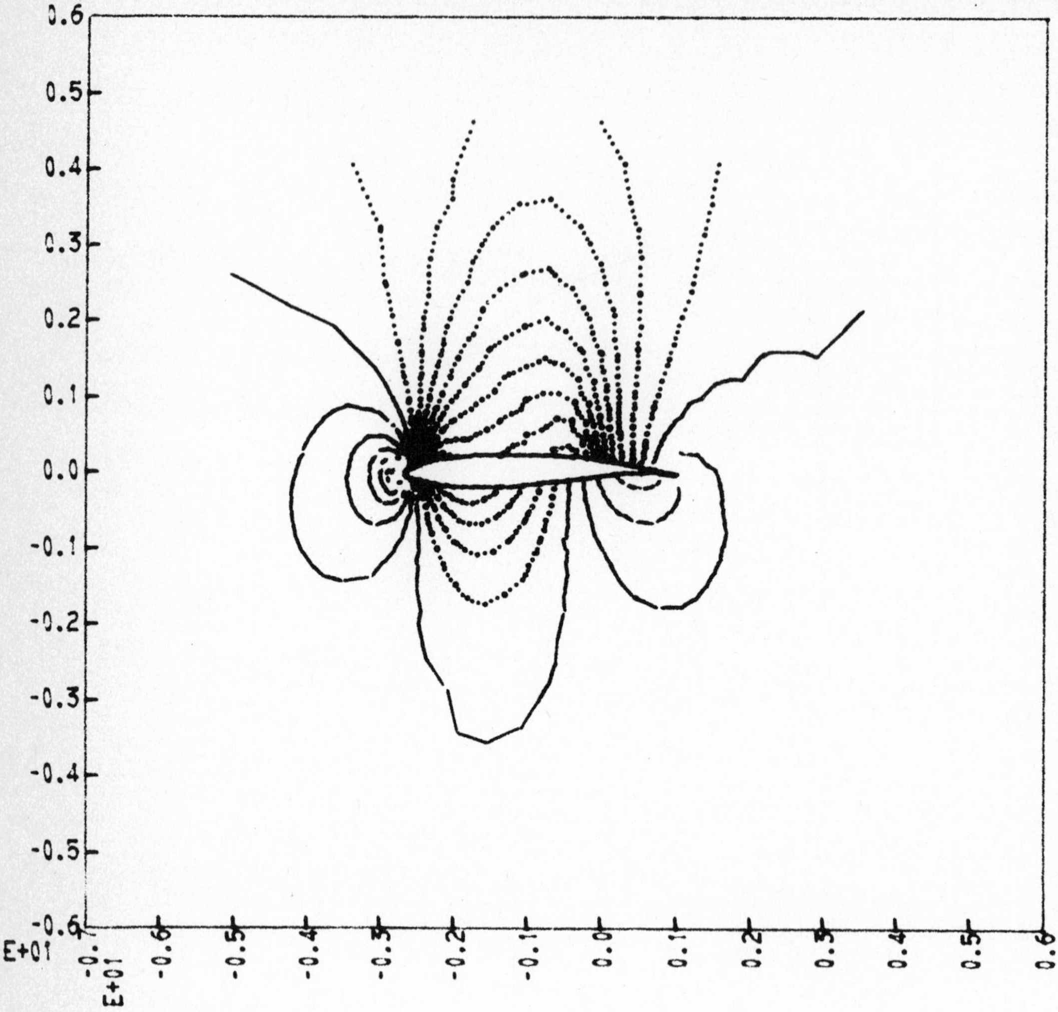
0.6
0.5
0.4
0.3
0.2
0.1
0.0
-0.1
-0.2
-0.3
-0.4
-0.5
-0.6
E+01
-0.
E+01
-0.6
-0.5
-0.4
-0.3
-0.2
-0.1
0.0
0.1
0.2
0.3
0.4
0.5
0.6

Figure 3

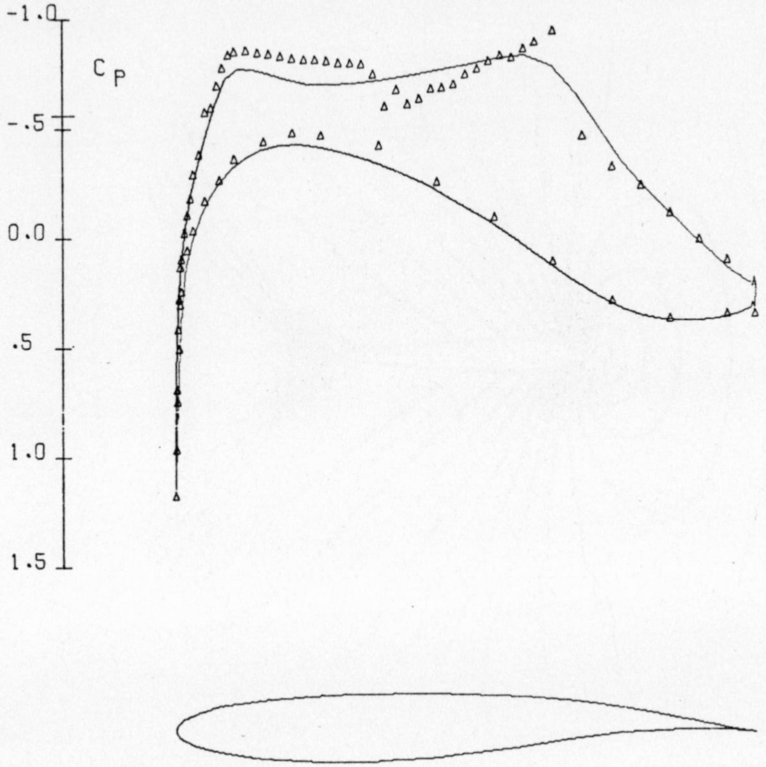

— TRANSIENT M=.760 ALP= -.45 CL= .497 CD=.0109 T/C= .12

△ EXPERIMENT M=.760 ALP= .44 CL= .478 RE=$2\cdot 10^{7}$

Figure 4

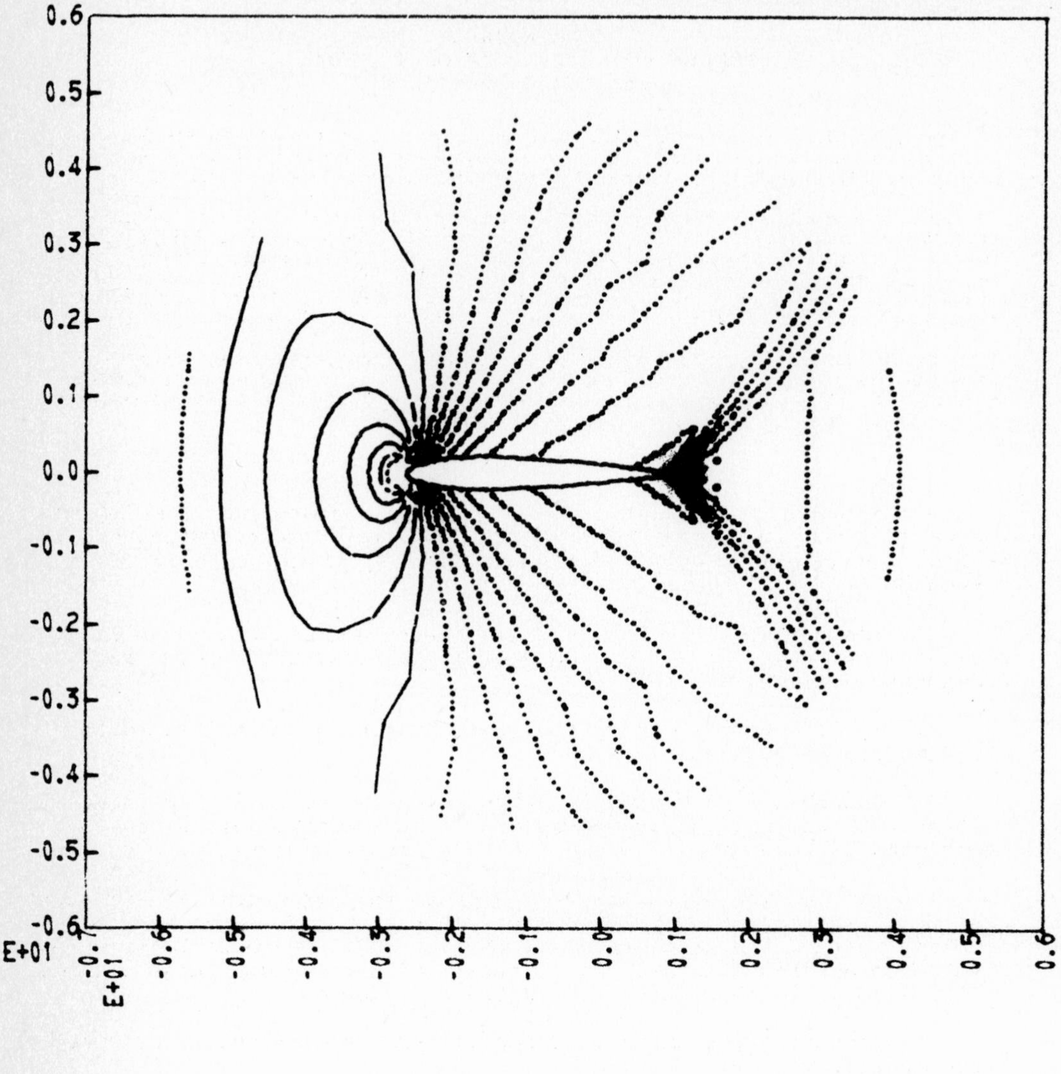
0.6
0.5
0.4
0.3
0.2
0.1
0.0
-0.1
-0.2
-0.3
-0.4
-0.5
-0.6
E+01
-0.7
-0.6
-0.5
-0.4
-0.3
-0.2
-0.1
0.0
0.1
0.2
0.3
0.4
0.5
0.6
E+01

ON THE MATTER OF SHOCK FITTING*

by

Gino Moretti
Polytechnic Institute of New York
Farmingdale, New York

· "For problems in one space variable containing a single strong shock whose position is know at t=0, shock fitting is highly satisfactory; it gives considerably greater accuracy and resolution than the pseudoviscosity methods.

The need for shock-fitting methods in multidimensional problems is rather urgent, for two reasons. First, considerations of economy generally dictate a coarser net than in one-dimensional problems, so that the pseudoviscosity smearing of the shocks and consequent loss of resolution is more severe. At the same time, the shock configurations are generally more intricate; hence, more, rather than less, resolution is required for their accurate description.

In two-dimensional shock-fitting, new types of instability can arise (e.g., in which the shock front tends to buckle and acquire a saw-toothed irregularity). Much work has been done by various people on the development of such methods, but it is my feeling that much more ought to be done".

R. D. Richtmyer, 1972

I strongly agree with the second paragraph above; and I enjoy no longer feeling like the biblical preacher in the desert. Moreover, the rest of the quotation can be modified and supplemented in a positive direction, as follows.

1) Shock fitting is highly satisfactory not only when the shock position is known at t=0 but also when an imbedded shock forms in a previously continuous flow.

2) Satisfactory shock fitting calculations are not limited to one-dimensional problems.

3) Instabilities certainly arise any time the coding is inconsistent with the physical situation which it should simulate; on the other hand, I have not yet found other causes of instability.

4) Not too much work has been done to develop shock-fitting methods; only two years ago, shock smearing has been given a higher status by being renamed "shock capturing"; thus, efforts which could have been devoted to a serious analysis of shock-fitting techniques have been diverted into a new generation of frustrated attempts.

The present paper intends to clarify and support the first three statements above.

There are two ways to fit a shock:

1) by considering it as a boundary between two regions of continuous flow, or

2) by considering it as a locus of discontinuities in a region of otherwise continuous flow.

In the first case (not considered by Richtmyer in his 1972 survey), the computational mesh in each of the regions separated by the shock

*This research was supported by the Office of Naval Research under Contract No. N00014-67-A-0438-0009, Project No. NR 061-135.

can be normalized separately, with the object of mapping each region onto a simple region (a segment, a square or a cube), of which the shock is a boundary, and where an evenly spaced computational grid is defined. Results obtained by using this technique are shown in Figs. 1 through 4. Each problem has some peculiar difficulty.

For the two-dimensional blunt-body flow shown in Fig. 1, the computational region is a square, with the shock on one side. Three-dimensional shock layers about blunt bodies are computed in a similar way, using a cubic computational region with the shock mapped on one face (Moretti and Abbett, 1966; Moretti and Bleich, 1967).

The result of a computation depending on one space-like variable only is shown in Fig. 2. The external heavy lines define the trajectories of two pistons moving in a cylindrical tube. Each discontinuity (shock or contact discontinuity) is assumed to be a boundary between adjacent regions of continuous flow. At each computational step, each region is mapped on a segment of unit length. Here, the treatment of shocks is simpler than in blunt-body problems, because of the dependence on one space-like variable only, but the logic of the computational code is complicated by reflections and interactions of shocks, interactions of shocks and contact discontinuities and formation of imbedded shocks by coalescence of compression waves (Moretti, 1971).

Quasi-one-dimensional flows and two-dimensional or axisymmetrical steady supersonic flows are treated similarly. For example, Fig. 3 shows an underexpanded plume in still air. Here it is interesting to see how different aspects of the flow (formation of imbedded shocks, Mach discs, subsonic cores, interactions of shocks with internal and external contact surfaces) can be treated in a single program, with a perfect blend of all parts (Salas, 1974).

Finally, Fig. 4 shows a cross-section of an aircraft flying at supersonic speed and of its shock layer containing a number of imbedded shocks (heavy lines). To fit the imbedded shocks as boundaries, fictitious continuations of the shocks (broken lines) are necessary. The entire cross-section is subdivided into partial regions, each one of which is mapped onto a square, with sides partially or entirely constituted by shocks (Marconi and Salas, 1973). The complicated topology of this problem lies at the limit of practicality for the technique of fitting shocks as boundaries.

Complicated topology, however, is the only limitation to the use of the technique. As far as accuracy is concerned, the technique has always worked very well. To circumvent topological difficulties, the second technique should be used, which fits the shocks as loci of discontinuities, floating among nodal points of a computational mesh (fixed or variable, but not forced to follow the shocks in their motion). The idea is far from being new; in fact, it is exposed in Richtmyer and Morton, 1967, page 378. For years, however, public opinion has been skeptical about it, on the grounds that computation of floating shocks was essentially unstable. I have been unable to find concrete evidence of attempts to use floating shock-fitting techniques; nor have I found proofs of instability or, at least, valid arguments to support such widespread negative attitudes. Having dealt, on other occasions, with other mythological instabilities such as the "nonlinear instability" which proved to be a natural consequence of incorrect handling of boundary conditions, I decided to attack the problem from the very beginning, avoiding all possible sources of perturbations leading to instabilities, but avoiding also all artificial smoothing or damping of oscillations for being arbitrary, non-physical interventions which can only deface the real physical pattern. On the code I was about to write I tried to

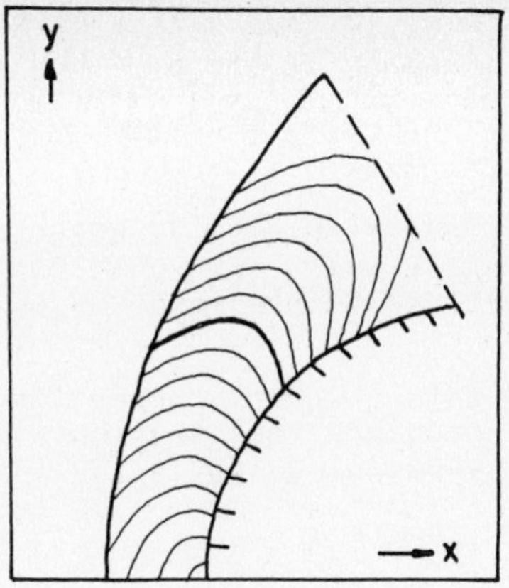

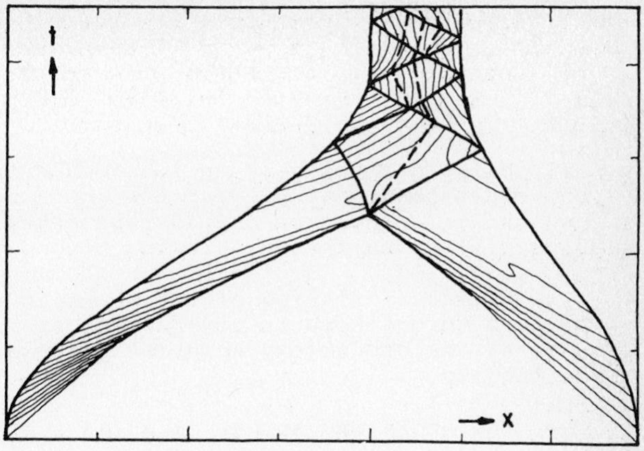

Fig. 1 (left) - Isomachs in a two-dimensional blunt body flow at M=20.

Fig. 2 (right) - Isobars, shocks and contact discontinuities in the flow in a tube between two moving pistons.

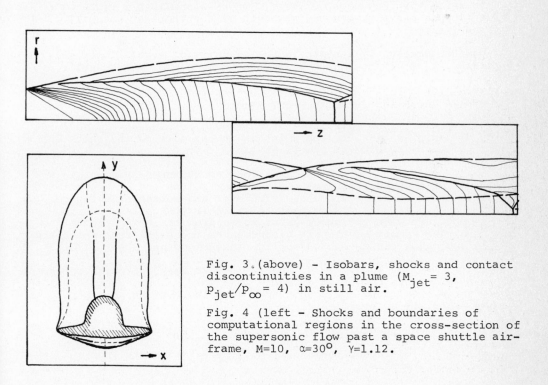

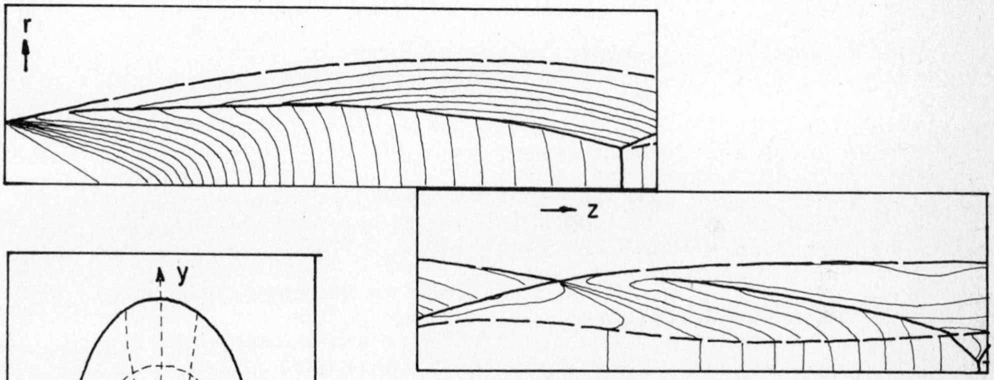

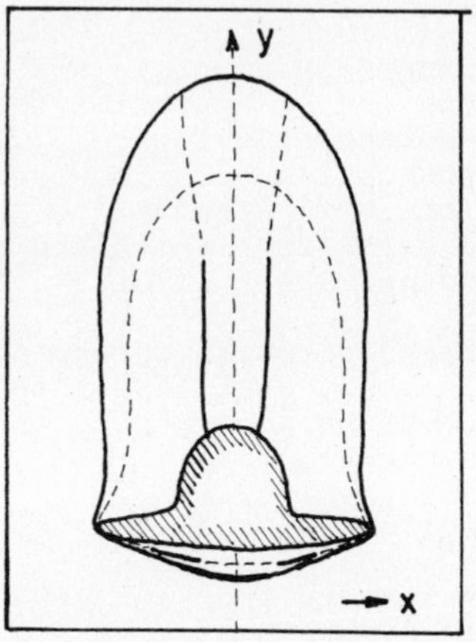

Fig. 3 (above) - Isobars, shocks and contact discontinuities in a plume (M_{jet}= 3, p_{jet}/p_{∞}= 4) in still air.

Fig. 4 (left - Shocks and boundaries of computational regions in the cross-section of the supersonic flow past a space shuttle airframe, M=10, α=30°, γ=1.12.

impose a number of basic rules, whose violation would certainly cause numerical trouble, that is, departure from reality:

1) The location of an incipient imbedded shock must be detected with all possible accuracy.

2) Delayed fitting of an imbedded shock results in wiggles which, on the other hand, may prevent the shock detecting tests from working, and which will propagate and distort other portions of the flow and perhaps pile up somewhere, growing beyond control.

3) Premature fitting of the shock in the region where compression waves tend to coalesce is not harmful at all, provided that the shock behaves as one of the characteristic surfaces coalescing into a finite discontinuity.

4) The shock may appear first as a short arc, with two free end-points.

5) Repeated tests at both ends of such an arc may detect the spreading of the shock and the consequent lengthening of the arc.

6) When the shock reaches a rigid wall, still with a vanishingly small strength, by no means should it be forced to be normal to the wall. In fact, an infinitely weak shock is a characteristic surface across which the flow does not bend. Shocks normal to a rigid wall should appear exactly when the strength of the shock at the wall becomes finite.

7) Conversely, if a shock in the vicinity of a wall loses all its strength, it should no longer be forced to be normal to the wall.

8) In no formula to approximate space-like derivatives should values appear which belong to opposite sides of a shock.

9) Similarly, no time integration should be made at a mesh point if at the initial time, t, the point is at one side of the shock but it is at the opposite side at the final time, $t+\Delta t$.

10) Assuming that all discretizations at ordinary grid points are second order accurate, derivatives at shock points can be approximated by first order accurate formulae (Richtmyer and Morton, 1967, page 382; and Orszag and Israeli, 1974, page 285); their truncation error, however, should not jump when the shock point moves from one mesh interval to the next, lest short-wavelength disturbances are generated, which are one of the major causes of instability (Nieuwland and Spee, 1973).

11) To satisfy condition 8, special discretizing formulae for space derivatives may be needed at ordinary grid points located at less than one mesh interval from the shock; such formulae must be accurate to the second order, if an overall second order accuracy is required.

12) The curvature of the shock may not be negligible, and thus it should not be neglected. Curvatures of opposite signs have opposite effects on the motion of the shock; such effects tend to iron out saw-toothed irregularities.

13) When the shock is weak, but in the process of building up strength, derivatives of certain quantities are extremely large and the speed of propagation of disturbances, relative to the shock, is practically zero. Terms which are crucial for the analysis of the early stages of a shock appear under the form $0.\infty$ and must be rephrased in a

way suitable to discretization.

14) Downstream of the shock, the entropy equation should be discretized in a way which prevents entropy to propagate upstream.

15) The end-point of a shock in the interior of a flow should have a vanishing strength. Its velocity should be found by imposing that the normal Mach number, relative to the shock, equals 1 and by interpolating values at the shock point between values at ordinary mesh points bracketing the shock.

16) Occasional irregularities in the definition of the free shock end-point mentioned above should not affect the other shock points. Consequently, geometrical derivatives along the shock should be approximated by one-sided backward difference formulae.

17) If points on the shock, near its free end-point, show normal relative Mach numbers less than 1, the shock should be truncated to the remaining significant portion.

18) One point which should not be underestimated is that, even if the computation at shock points is performed with the utmost care, the results may still be poor because the shock environment is hard to describe numerically. Consider, for example, a shock travelling downstream in a two-dimensional supersonic region where the steady-state characteristics coalesce inside the flow field (Fig. 5, where AB is a rigid wall and CD is a shock). The gas, moving from left to right supersonically, undergoes an expansion followed by a recompression. Even if the flow is generally unsteady and the shock moves, the flow upstream of the shock is practically steady. Characteristics in the simple wave of such steady flow are shown in Fig. 5. In the simple wave just ahead of the shock no usual approximation of space derivatives by finite difference is accurate; even the usual analysis of truncation error is meaningless (Moretti, 1969). In the equations of motion,

$$f_t = Af_x+Bf_y$$

the time derivative, which has to be practically equal to zero, results as a combination of two space derivatives, each of which is practically infinite. Clearly, the calculation immediately upstream of the shock is difficult and special care must be taken.

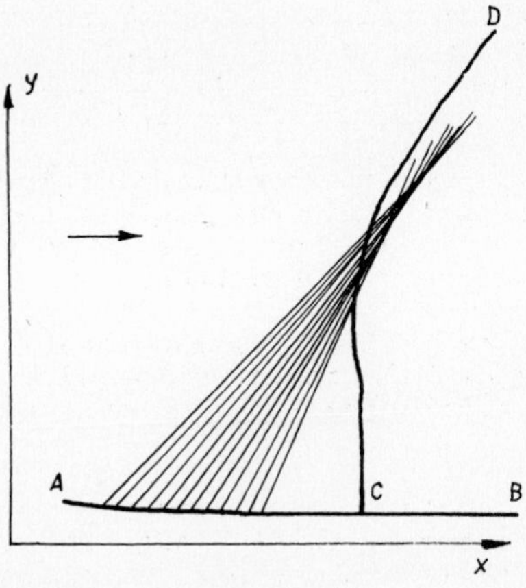

Fig. 5

Description of the discretization process which satisfies all the requirements above exceeds the limits of space for this communication; it will be given in forthcoming Reports of the Polytechnic Institute of New York. Here I will only present a selection of results, showing various stages of evolution of two-dimensional flows above a curved wall (boattail), starting from rest and reaching a given, uniform velocity at infinity. All light curves in Fig. 6 are isobars and the shocks are represented by heavy lines. The shape of the shocks, their motion in time, their final settling into a steady position, and the nature of the surrounding flow field, all seem to prove that the computations are accurate and reliable.

In conclusion, it may be noted that, once all basic rules and discretization formulae have been found, extension of the technique to multiple shock treatment is mostly a matter of bookkeeping.

REFERENCES

Marconi, F. and Salas, M., Computers and Fluids, 1:185 (1973).

Moretti, G., PIBAL Report No. 69-26 (1969).

Moretti, G., PIBAL Report No. 71-25 (1971).

Moretti, G. and Abbett, M., AIAA J., 4, 2136 (1966).

Moretti, G. and Bleich, G., AIAA J., 5, 1557 (1967)

Nieuwland, G.Y. and Spee, B.M., Annual Rev. Fl. Mech., 5:119 (1973).

Orszag, S.A. and Israeli, M., Annual Rev. Fl. Mech., 6:281 (1974).

Richtmyer, R.D., Proc. III Int. Conf. on Num. Meth. in Fl. Mech., Lecture Notes in Physics, 18, Vol. I, 72, Springer-Verlag (1972).

Richtmyer, R.D. and Morton, K.W., Difference methods for initial-value problems, II ed., Interscience Publ. (1967).

Salas, M., AIAA Paper 74-523 (1974).

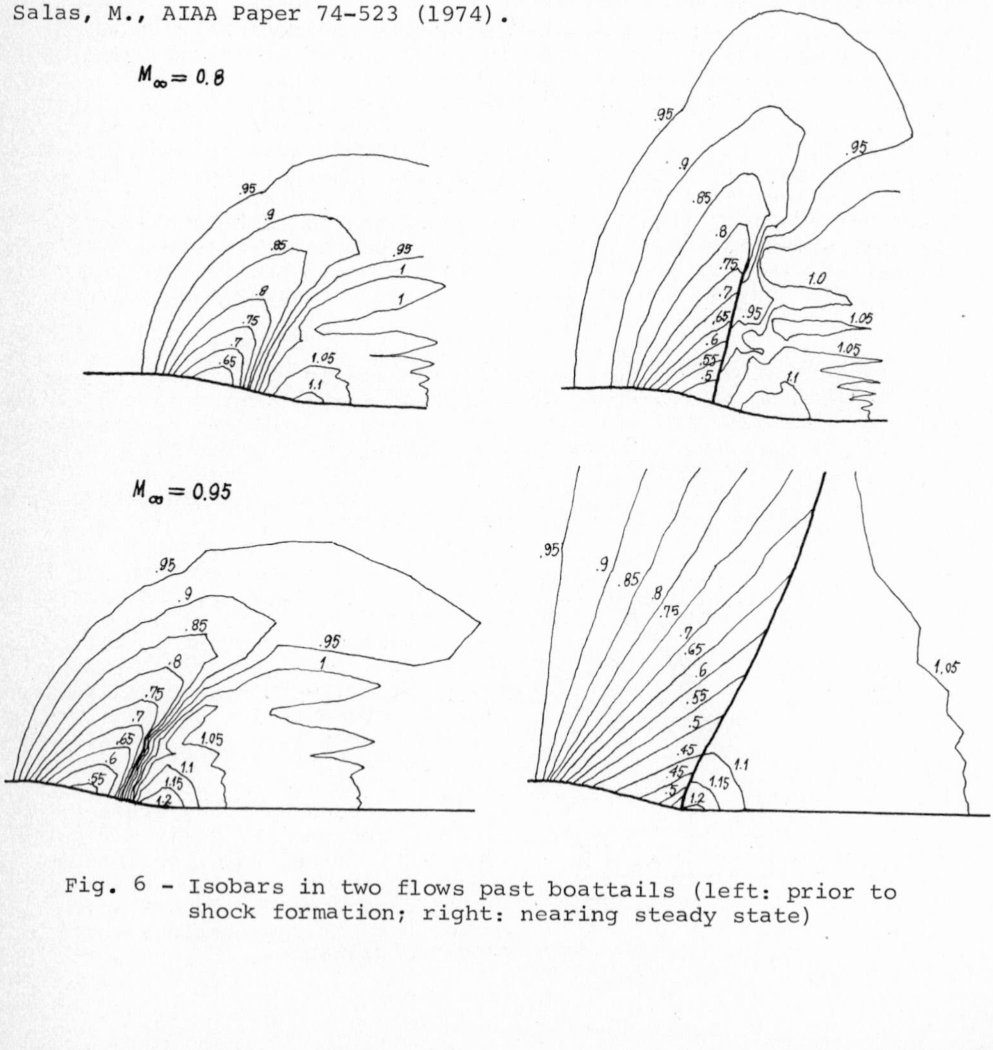

Fig. 6 - Isobars in two flows past boattails (left: prior to shock formation; right: nearing steady state)

ON THE THEORY OF DIFFERENCE SCHEMES FOR GAS DYNAMICS EQUATIONS

(V.I.Paasonen, Yu.I.Shokin, N.N.Yanenko)

(Computing Center of the Siberian Branch of the USSR Academy of Sciences, Novosibirsk State University, Novosibirsk, USSR)

The present paper continues investigations of finite-difference schemes for hyperbolic equations and its subject is closely related to that of papers [1]-[5].

1. Let us consider invariant properties of finite-difference splitting-up schemes. The difference scheme is considered to allow a series of transformations if parabolic form of the first differential approximation (f.d.a.) allows the same series of transformations [2]-[3].

For a system of gas dynamics equations in Eulerian coordinates

$$\frac{\partial w}{\partial t} + \frac{\partial f}{\partial x} + \frac{\partial g}{\partial y} = 0, \qquad (1)$$

where $w = (\rho u, \rho v, \rho, \rho E)'$, $f = (p + \rho u^2, \rho u v, \rho u, \rho u E + u p)'$, $E = \varepsilon + \frac{1}{2}(u^2 + v^2)$, $g = (\rho u v, p + \rho v^2, \rho v, \rho v E + v p)'$ we will consider two finite-difference splitting-up schemes:

$$\frac{w_{ij}^{n+\frac{1}{2}} - w_{ij}^{n}}{\tau} + \frac{f_{i+1j}^{n} - f_{i-1j}^{n}}{2h_1} = \Lambda^1 w_{ij}^{n},$$

$$\frac{w_{ij}^{n+1} - w_{ij}^{n+\frac{1}{2}}}{\tau} + \frac{g_{ij+1}^{n+\frac{1}{2}} - g_{ij-1}^{n+\frac{1}{2}}}{2h_2} = \Lambda^2 w_{ij}^{n+\frac{1}{2}}; \qquad (2)$$

$$\frac{w_{ij}^{n+\frac{1}{2}} - w_{ij}^{n}}{\tau} + \frac{\overset{(1)n}{f}_{i+1j} - \overset{(1)n}{f}_{i-1j}}{2h_1} + \frac{\overset{(1)n}{g}_{ij+1} - \overset{(1)n}{g}_{ij-1}}{2h_2} = L^1 w_{ij}^{n},$$

$$\frac{w_{ij}^{n+1} - w_{ij}^{n+\frac{1}{2}}}{\tau} + \frac{\overset{(2)n+\frac{1}{2}}{f}_{i+1j} - \overset{(2)n+\frac{1}{2}}{f}_{i-1j}}{2h_1} + \frac{\overset{(2)n+\frac{1}{2}}{g}_{ij+1} - \overset{(2)n+\frac{1}{2}}{g}_{ij-1}}{2h_2} = L^2 w_{ij}^{n+\frac{1}{2}}. \qquad (3)$$

Here $f^{(1)} = (p, 0, 0, up)'$, $f^{(2)} = (\rho u^2, \rho u v, \rho u, \rho u E)'$, $g^{(1)} = (0, p, 0, vp)'$, $g^{(2)} = (\rho u v, \rho v^2, \rho v, \rho v E)'$, $\Lambda^k = M^{k(2)}$, $L^k = M^{k(3)}$,

$$M^{k(\ell+1)} = \frac{\tau}{h_1}(T_{\frac{x}{2}} - T_{-\frac{x}{2}})\left[\frac{1}{h_1}\Lambda_{11}^{k(\ell+1)}(T_{\frac{x}{2}} - T_{-\frac{x}{2}}) + \frac{1}{h_2}\Lambda_{12}^{k(\ell+1)}(T_{\frac{y}{2}} - T_{-\frac{y}{2}})\right] +$$

$$+ \frac{\tau}{h_2}(T_{\frac{y}{2}} - T_{-\frac{y}{2}})\left[\frac{1}{h_1}\Lambda_{21}^{k(\ell+1)}(T_{\frac{x}{2}} - T_{-\frac{x}{2}}) + \frac{1}{h_2}\Lambda_{22}^{k(\ell+1)}(T_{\frac{y}{2}} - T_{-\frac{y}{2}})\right],$$

$$\Lambda_{rs}^{k(\ell+1)} = \Lambda_{rs}^{k(\ell+1)}(t, x, y, w, w_t, w_x, w_{tx}, w_{xx})$$

$$(r, s, k, \ell = 1,2),$$

τ, h_1, h_2 are steps of the finite-difference grid along the axes t, x, y respectively, $\Lambda_{rs}^{k(\ell+1)}$ are matrices 4 x 4,

$$T_{\pm\frac{x}{2}}\varphi(x,y) = \varphi(x \pm \frac{h_1}{2}, y), \quad T_{\pm\frac{y}{2}}\varphi(x,y) = \varphi(x, y \pm \frac{h_2}{2}),$$

$$f_{i+\alpha\, j+\beta}^{n+\frac{1}{2}} = f(W_{i+\alpha\, j+\beta}^{n+\frac{1}{2}}), \quad g_{i+\alpha\, j+\beta}^{n+\frac{1}{2}} = g(W_{i+\alpha\, j+\beta}^{n+\frac{1}{2}}),$$

$$f_{i+\alpha\, j+\beta}^{(k)\, n+\frac{1}{2}} = f^{(k)}(W_{i+\alpha\, j+\beta}^{n+\frac{1}{2}}), \quad g_{i+\alpha\, j+\beta}^{(k)\, n+\frac{1}{2}} = g^{(k)}(W_{i+\alpha\, j+\beta}^{n+\frac{1}{2}})$$

$$(\alpha, \beta = -1, 0, 1), \quad k = 1,2.$$

It is assumed that the state equation of gas is of the form: $p = p(\varepsilon, \rho)$. Let us note that scheme (3) is an asymptotic representation of the PIC method [3] .
The following theorem[7] takes the place:
THEOREM 1. 1) If matrix $\Lambda_{rs}^{k(2)} = \| \lambda_{rs}^{k\xi\eta} \|_1^4$ is taken such that

$$\frac{\partial \lambda_{rs}^{k\xi\eta}}{\partial x^\gamma} = \frac{\partial \lambda_{rs}^{k\xi\eta}}{\partial q_{1\gamma}^m} = \frac{\partial \lambda_{rs}^{k\xi\eta}}{\partial q_{\gamma 1}^m} = 0 \quad (k, r, s = 1,2;\ \gamma = 1,2,3;\ m = 1,2,3,4),$$

$$\frac{\partial N^1}{\partial u} = (N_3^1, 0, 0; N_1^1)', \quad \frac{\partial N^1}{\partial v} = (0, N_3^1, 0, N_2^1)', \quad L_6' N^1 = (N_2^1, -N_1^1, 0, 0)', \quad \bar{L}_7 N^1 = N^1,$$

then scheme (2) is invariant with respect to the same series of transformations as the system of equations (1).
2) If matrices $\Lambda_{rs}^{k(3)} = \| M_{rs}^{k\xi\eta} \|_1^4$ are such that

$$\frac{\partial (M_{rs}^{1\xi\eta} + M_{rs}^{2\xi\eta})}{\partial x^\gamma} = \frac{\partial (M_{rs}^{1\xi\eta} + M_{rs}^{2\xi\eta})}{\partial q_{1\gamma}^m} = \frac{\partial (M_{rs}^{1\xi\eta} + M_{rs}^{2\xi\eta})}{\partial q_{\gamma 1}^m} = 0 \quad (k, r, s = 1,2;\ \gamma = 1,2,3;\ m = 1,2,3,4),$$

$$\frac{\partial N^2}{\partial u} = (N_3^2, 0, 0, N_1^2)', \quad \frac{\partial N^2}{\partial v} = (0, N_3^2, 0, N_2^2), \quad L_6' N^2 = (N_2^2, -N_{-1}^2, 0, 0), \quad \bar{L}_7 N^2 = N^2,$$

the difference scheme (3) is invariant with respect to the group of transformations allowed by system 1 .

Here $x^1 = t$, $x^2 = x$, $x^3 = y$, $u^1 = u$, $u^2 = v$, $u^3 = \rho$, $u^4 = p$,

$$q^m_{\gamma 1} = \frac{\partial^2 u^m}{\partial x^\gamma \partial x^1}, \qquad q^m_{1\gamma} = \frac{\partial^2 u^m}{\partial x^1 \partial x^\gamma}, \quad L'_6 = v\frac{\partial}{\partial u} - u\frac{\partial}{\partial v} + \bar{L}_6 ,$$

$$N^k = (N^k_1\, N^k_2\, N^k_3\, N^k_4)' = \frac{\partial}{\partial x}\left[C^k_{11}\frac{\partial w}{\partial x} + C^k_{12}\frac{\partial w}{\partial y}\right] +$$

$$+ \frac{\partial}{\partial y}\left[C^k_{21}\frac{\partial w}{\partial x} + C^k_{22}\frac{\partial w}{\partial y}\right],$$

$$C^1_{rs} = \tau\,(\Lambda^{1(2)}_{rs} + \Lambda^{2(2)}_{rs}) - \frac{\tau}{2}(-1)^{\delta^{s+1}_r} A_r A_s ,$$

$$C^2_{rs} = \tau\,(\Lambda^{1(3)}_{rs} + \Lambda^{2(3)}_{rs}) + \tau\,(\delta^1_r B_2 + \delta^2_r B_4)(\delta^1_s B_1 + \delta^2_s B_3) - \frac{\tau}{2} A_r A_s ,$$

$$A_1 = \frac{df}{dw}, \quad A_2 = \frac{dg}{dw}, \quad B_r = \frac{df^{(r)}}{dw}, \quad B_{r+2} = \frac{dg^{(2)}}{dw}$$

δ^k_ℓ is Cronecker's symbol, operators $\bar{L}_6$ and $\bar{L}_7$ are defined in [8].

In case of a concrete state equation the conditions of invariance of splitting-up difference schemes (2)-(3) can be simplified.

The invariance conditions of the finite difference splitting-up schemes higher order accuracy can be found in a similar way.

The comparison of invariant and noninvariant finite difference schemes was carried out on the example of calculation of the equation (see [3])

$$\frac{\partial u}{\partial t} = \alpha y \frac{\partial u}{\partial x} - \alpha x \frac{\partial u}{\partial y}, \qquad (4)$$

which allows transformation of rotation with the infinitesimal operator $X = y\frac{\partial}{\partial x} - x\frac{\partial}{\partial y}$.

For equation (4) the initial value problem with the initial condition was considered:

$$u(0,x,y) = \begin{cases} 1 - \frac{1}{u_0} r, & r^2 = (x-a)^2 + (y-b)^2 \le u_0^2, \\ 0, & r^2 \ge u_0^2 . \end{cases} \qquad (5)$$

Problem (4)-(5) describes rotation of the circular cone of the height equal to 1 with the radius of the base equal to u_0 around the origin of coordinates with the period $2\pi/\alpha$.

The calculations showed that invariant splitting-up schemes reflected more exactly the peculiarities of the precise solution qualitively and quantitively. Noninvariant finite difference schemes introduce perturbations. Particularly, solutions obtained from such schemes are far from being precise.

Figs. 1 - 4 show lines of the level $u(t,x,y) = const = C$ $(C = 0{,}2;\ 0{,}4;\ 0{,}6;\ 0{,}8)$ of the precise solution (circles) and the difference solution (curve 1 is a solution of the second order noninvariant scheme, curve 2 is that of the second order invariant scheme at $t = 2, 4, 6, 8$. Figs. 5 - 6 show profiles of the solution in the cross-section $x = -2$ at $t = 2$ and $t = 6$, respectively.

2. In numerical calculation of gas dynamic problems we pass from the continuous medium model to the descrete one which must reflect the basic properties of the medium (in particular, respective finite difference analogues of the conservation laws [9]-[10] took place.

Below a concept of algebraic aquivalence of the finite difference schemes is introduced, where the idea of full conservation is used [10] .

Algebraically equivalent difference schemes are such schemes which can be transformed one to another by means of the inverse algebraic transformations. Algebraically equivalent difference equations are determined in a similar way.

One-dimentional gas dynamics equations in Lagrangian variables are usually considered in one of three algebraically equivalent forms:

$$
\begin{array}{lll}
a)\ \dfrac{\partial u}{\partial t} = -\dfrac{\partial p}{\partial x}, & b)\ \dfrac{\partial u}{\partial t} = -\dfrac{\partial p}{\partial x}, & c)\ \dfrac{\partial u}{\partial t} = -\dfrac{\partial p}{\partial x}, \\
\dfrac{\partial v}{\partial t} = \dfrac{\partial u}{\partial x}, & \dfrac{\partial v}{\partial t} = \dfrac{\partial u}{\partial x}, & \dfrac{\partial v}{\partial t} = \dfrac{\partial u}{\partial x}, \\
\dfrac{\partial \varepsilon}{\partial t} = -p\dfrac{\partial u}{\partial x}, & \dfrac{\partial E}{\partial t} = -\dfrac{\partial (up)}{\partial x}, & \dfrac{\partial \varepsilon}{\partial t} + p\dfrac{\partial v}{\partial t} = 0.
\end{array} \qquad (6)
$$

Let us consider three families of finite difference schemes that approximate systems (6a), (6b), (6c) , respectively,

$$
\begin{array}{lll}
a)\ \Delta_0 \varphi_1^1 + \varkappa \Delta_x \varphi_1^4 = 0, & b)\ \Delta_0 \varphi_1^1 + \varkappa \Delta_x \varphi_1^4 = 0, & c)\ \Delta_0 \varphi_1^1 + \varkappa \Delta_x \varphi_1^4 = 0, \\
\Delta_0 \varphi_1^2 - \varkappa \Delta_x \varphi_2^1 = 0, & \Delta_0 \varphi_1^2 - \varkappa \Delta_x \varphi_2^1 = 0, & \Delta_0 \varphi_1^2 - \varkappa \Delta_x \varphi_2^1 = 0, \\
\Delta_0 \varphi_1^3 + \varkappa \varphi_2^4 \Delta_x \varphi_3^1 = 0, & \Delta_0 (\varphi_2^3 + \frac{1}{2} \varphi_4^1 \varphi_5^1) + \varkappa \Delta_x (\varphi_3^4 \varphi_6^1) = 0, & \Delta_0 \varphi_3^3 + \varkappa \varphi_4^4 \Delta_0 \varphi_2^2 = 0.
\end{array} \qquad (7)
$$

Here $\Delta_x = T_x - I$, $\Delta_0 = T_0 - I$, $\varphi_j^i = \sum a_\alpha T_0^{\alpha_0} T_x^{\alpha_1} u$, $A_{ij}^0 = \sum\limits_{\alpha \in \mathcal{J}_{ij}} \alpha_0 a_\alpha$,

$A_{ij}^1 = \sum\limits_{\alpha \in \mathcal{J}_{ij}} \alpha_1 a_\alpha$, $\sum\limits_{\alpha \in \mathcal{J}_{ij}} a_\alpha = 1$, $u^1 = u$, $u^2 = \frac{1}{\rho}$, $u^3 = \varepsilon$, $u^4 = p$, $\mathcal{J}_{ij}$ –

are some finite sets of two-dimentional indices. Let the difference schemes be two-layer ones and all the values be taken on the same time layers. Then $A^0_{11} = A^0_{21} = A^0_{31} = A^0_{32} = A^0_{14} = A^0_{15} = A^0_{33} = A^0_{22} = 0$ and let the schemes considered be of first order approximation. Then in f.d.a. the first two equations coincide and are of the form

$$\frac{\partial u}{\partial t} + \frac{\partial P}{\partial x} + \tau \xi_{11} \frac{\partial}{\partial x}\left(a^2 \frac{\partial u}{\partial x}\right) + h\eta_{11} \frac{\partial^2 P}{\partial x^2} = 0,$$

$$\frac{\partial V}{\partial t} - \frac{\partial u}{\partial x} - \tau \xi_{21} \frac{\partial^2 P}{\partial x^2} - h\eta_{21} \frac{\partial^2 u}{\partial x^2} = 0,$$

and the third equations in f.d.a. of schemes (6a) - (6c) are, respectively, as follows:

$$\frac{\partial \varepsilon}{\partial t} - P\frac{\partial u}{\partial x} + \tau \xi_{31} a^2 \left(\frac{\partial u}{\partial x}\right)^2 + \tau \xi_{32} P \frac{\partial^2 P}{\partial x^2} + h\eta_{31} P \frac{\partial^2 u}{\partial x^2} + h\eta_{32} \frac{\partial P}{\partial x}\frac{\partial u}{\partial x} = 0,$$

$$\frac{\partial E}{\partial t} + \frac{\partial uP}{\partial x} + \tau \xi_{41} \frac{\partial}{\partial x}\left(a^2 u \frac{\partial u}{\partial x}\right) + \tau \xi_{42} \frac{\partial}{\partial x}\left(P\frac{\partial P}{\partial x}\right) + h\eta_{41} \frac{\partial}{\partial x}\left(u \frac{\partial P}{\partial x}\right) + h\eta_{42} \frac{\partial}{\partial x}\left(P\frac{\partial u}{\partial x}\right) = 0,$$

$$\frac{\partial \varepsilon}{\partial t} + P\frac{\partial V}{\partial t} + \tau \xi_{51} a^2 \left(\frac{\partial u}{\partial x}\right)^2 + h\eta_{51} P \frac{\partial^2 u}{\partial x^2} + h\eta_{52} \frac{\partial P}{\partial x}\frac{\partial u}{\partial x} = 0,$$

where a is the speed of sound, ξ_{ij} , η_{ij} are coefficients depending upon A^0_{ij} , A^0_{ij} .

The following statements take place.

THEOREM 2. 2) For scheme (7a) to be algebraically equivalent to the conservative scheme (7b), it is necessary that

$$\xi_{11} - \xi_{31} = 0 , \quad \eta_{32} = \eta_{11} + \eta_{31} , \quad \xi_{32} = 0 .$$

2) For scheme (7b) to be algebraically equivalent to (4.1) it is necessary that

$$\xi_{11} = \xi_{41} , \quad \xi_{42} = 0 , \quad \eta_{11} = \eta_{41} .$$

3) For scheme (7a) to be algebraically equivalent to (7c) it is enough that

$$\varphi^4_2 = \varphi^4_1 , \quad \frac{1}{2}(T_0 + I)\,\varphi^1_1 = T_x \varphi^1_3 .$$

4) For scheme (7a) to be algebraically equivalent to (7c) it is necessary that

$$\xi_{21} = \xi_{32} .$$

5) For scheme (7a) to be algebraically equivalent to (7c) it is necessary that

$$\varphi_2^1 = \varphi_3^1$$

Corollary. If scheme (7a) is algebraically equivalent to scheme (7b) then it is implicit and vice versa.

THEOREM 3. If scheme (7a) has the second order of approximation in X and is algebraically equivalent to scheme (7b), then it has the property K [1]. If, besides, scheme (7a) is also algebraically equivalent to scheme (7b) then in f.d.a. the approximating viscosity enters additively into P.

The statements formulated give criteria for checking the algebraic equivalence of schemes in f.d.a. terms.

4. Let the difference operators

$$a)\ \frac{1}{h}\Delta_{pq} = \frac{1}{h}\sum_{k=-q}^{p}\alpha_k T_h^k, \quad b)\ \frac{1}{h^2}A_d = \frac{1}{h^2}\sum_{k=-d}\lambda_k T_h^k \qquad (8)$$

approximate $\frac{\partial}{\partial x}$ and $\frac{\partial^2}{\partial x^2}$, respectively, in $S = p+q$ and $L = 2d$ orders maximal for the given p, q and d. The operators of these classes will be called accurate.

Definition. The operator Δ has R property if its spectrum projection onto the real axis $\mathrm{Re}\,\delta(\xi)$ is semidefinite in the sign.

THEOREM 4. The accurate operator (8a) has property iff its R measure of asymmetry $\tau = q - p$ satisfies the limit $|\tau| \leq 2$, whereas signs $\mathrm{Re}\,\delta(\xi)$ and τ coincide [4].

THEOREM 5. The accurate operator (8c) is negative definite. In space $\mathcal{L}_2$ let us consider the initial value (Cauchy) problem for the equation with constant coefficients

$$\frac{\partial u}{\partial t} + a\frac{\partial u}{\partial x} + bu = \mu\frac{\partial^2 u}{\partial x^2} + f \ ; \qquad \mu \geq 0$$

and the corresponding scheme with weights

$$G\frac{u^{n+1} - u^n}{\tau} + \left[a\frac{\Delta}{h} + bE - \mu\frac{\Lambda}{h^2}\right]u^n = f^n \qquad (9)$$

where $G = E + \alpha\varkappa\Delta + \tau\beta bE - \gamma\sigma\Lambda$, $\varkappa = \frac{a\tau}{h}$, $\sigma = \frac{\mu\tau}{h^2}$, α, β and γ are the weights corresponding to the convective, source and dissipative terms; $\frac{\Delta}{h}$ and $\frac{\Delta}{h^2}$ are some approximations $\frac{\partial}{\partial x}$ and $\frac{\partial^2}{\partial x^2}$.

THEOREM 6. If the operator Δ has R property and is consistent with the sign of the derivative so that $a\,\mathrm{Re}\,\delta(\xi) \geq 0 \ \ \forall\xi$ and Λ the spectrum of the operator is negative: $\lambda(\xi) \leq 0$ then at $\alpha \geq \frac{1}{2}$, $\gamma \geq \frac{1}{2}$, scheme (9) is unconditionally stable. If viscosity is absent the violation of the condition $a\,\mathrm{Re}\,\delta(\xi) \geq 0$ results in instability of scheme (9) for any fixed α.

According to the theory [4,5,6] , the geometrical criterion of correctness is true.

THEOREM 7. Let the operators $\frac{1}{h}\Delta$ and $\frac{1}{h^2}\Lambda$ be accurate. Then under the conditions for asymmetry measure $\tau = q - p$ of the operator Δ: $a\tau \geq 0$ and $|\tau| \leq 2$ scheme (9) with the weights $\alpha \geq \frac{1}{2}$ and $\gamma \geq \frac{1}{2}$ is conditionally stable. At $\mu = 0$ violation of the geometrical criterion, causes unstability of scheme 9 at any fixed α . The results are true also for the systems in invariant form, when matrix $\frac{1}{h}\Delta$ is diagonal matrix operator, consisting of the approximations $\frac{\partial}{\partial x}$. Besides, the results are true for the schemes of coordinate splitting of multi-dimensional systems. Every fractional step system should be reduced to an invariant.

In a stationary case the following modification is possible, which does not decrease accuracy of the stationary solution and at the same time it simplifies the structure of the scheme and its numerical realization. Let us consider the scheme of type (9), where $\delta \geq 0$ and multipoint operators Δ and Λ are replaced in the upper layer by simpler ones $\bar{\Delta}$ and $\bar{\Lambda}$, i.e.

$$G = E + \alpha\varkappa\bar{\Delta} + \tau\beta\delta E - \gamma\sigma\bar{\Lambda}$$

THEOREM 8. Let operators Δ , Λ , $\bar{\Delta}$, $\bar{\Lambda}$ satisfy the conditions of theorem 6 or 7 and let the inequalities

$$Re[(2\alpha\bar{\delta} - \delta)\delta^{x}] \geq 0 ; \quad a Re(2\alpha\bar{\delta} - \delta) \geq 0 ; \quad \lambda - 2\gamma\bar{\lambda} \geq 0 .$$

be uniformly fulfilled on ξ at $\alpha \geq \alpha_0$ and $\gamma \geq \gamma_0$ for their spectra $\delta(\xi)$, $\lambda(\xi)$, $\bar{\delta}(\xi)$, $\bar{\lambda}(\xi)$ (in the above formula the argument ξ is omitted).

Then at $\alpha \geq \alpha_0$, $\gamma \geq \gamma_0$ scheme (8) is absolutely and strongly stable.

In particular, for symmetric scheme of fourth order accuracy ($\bar{\Delta} = \Delta_{11}$, $\bar{\Lambda} = \Lambda_1$, $\Delta = \Delta_{22}$, $\Lambda = \Lambda_2$) we have $\alpha_0 = \frac{5}{6}$ and $\gamma_0 = \frac{2}{3}$; for an a symmetric scheme of third order accuracy ($a > 0$, $\bar{\Delta} = \Delta_{01}$, $\Delta = \Delta_{12}$) - $\alpha_0 = 7.5 - 4\sqrt{3} \simeq 0.57$, $\gamma_0 = \frac{2}{3}$

Since the schemes with weights considered above are based on operator approximation and are not assumed to use continuous systems, their application in case of variable coefficients and in nonlinear problems has no obstacles. For such problems the criteria of correctness obtained above can be considered as approximate practical recomendations for the choice of scheme's parameters. The schemes of higher accuracy were applied for calculating stationary currents of the viscous incompressible fluid. The splitting-up schemes of Navjer-Stokes equations for the velocity vector and pressure and factorized schemes with complete approximation were tested for the equations in terms of the stream function and velocity vortex.

5. Let us investigate dissipative two-layer schemes $u_t + au_x = 0$ ($a = const$) , which have maximal order accuracy for the given pattern of the difference scheme. Let the coefficients of the scheme

$$Bu^{n+1} + Cu^{n} = 0 , \quad B = \sum_{K=-q}^{p} B_K T_h^K ; \quad C = \sum_{K=-q}^{p} C_K T_h^K \qquad (10)$$

satisfy the system of consistency conditions

$$\sum_{K=-q}^{p} B_K = 1, \quad \sum_{K=-q}^{p} [B_K (K-\varkappa)^{\ell} + C_K K^{\ell}] = (-1)^s \mu(\varkappa) \delta_{2S}^{\ell}; \ \ell = 0,1,\dots,2S, \qquad (11)$$

where $S = p + q$, $\varkappa = \frac{a\tau}{h}$ is Courant's parameter, $\mu(\varkappa)$ is a variable function of $\varkappa$. The following statements [5] are true.

THEOREM 9. The solution of the above system, if it exists, defines a family of schemes (10) of order $L = 2S-1$ with respect to τ and h that depend on the free parameter $\mu(\varkappa)$ and have parabolic (correct) f.d.a. at $\mu > 0 \, (\mu \geq 0)$.

THEOREM 10. Representability of the function μ in the form $\mu(\varkappa) = \alpha(\varkappa) \cdot \prod_{j=-S+1}^{S-1} (\varkappa + j)$ is the necessary and sufficient criterion for a solution of system (11) for any $\varkappa$. If $\alpha(\varkappa) = 0$ for integers $\varkappa$ within $|\varkappa| \leq S$, then schemes (10)-(11) are accurate for these values of $\varkappa$.

THEOREM 11. The necessary and sufficient criterion of dissipativity (strong stability) of (10)-(11) is parabolicity of its f.d.a.

The results are also true if there is a condition of commutativity of matrices $\varkappa$ and $\mu(\varkappa)$.

Let us consider in more detail a scheme of third order accuracy ($S = 2$) of the form

$$B \frac{u^{n+1} - u^n}{\tau} + a \Delta_\alpha u^n = 0, \quad \Delta_\alpha = \frac{1+\alpha(\varkappa)}{2} \Delta_1 + \frac{1-\alpha(\varkappa)}{2} \Delta_{-1}, \Delta_{\pm 1} = \frac{T_h - E}{h}. \qquad (12)$$

In this case $\mu(\varkappa) = \alpha(\varkappa)(\varkappa^2 - 1)\varkappa$. Hence, f.d.a. is correct for $|\varkappa| \leq 1$, if $\alpha(\varkappa) > 0$ at $\varkappa < 0$ and $\alpha(\varkappa) < 0$ at $\varkappa > 0$

Thus, in approximation of the derivative Δ_α of scheme (12) dominates the " backward " difference Δ_{-1} at positive inclination of the characteristic curve, and vice versa " forward " difference dominates Δ_1 at the negative inclination of it. In other words, viscosity of scheme (12) can be considered as the consequence of asymmetric approximation coordinated with the stream direction. Such geometrical interpritation throws light on the approximation character of viscosity and is useful for the extension of the method to quasi-linear systems $u_t + \varphi_x = 0$ ($\varphi = \varphi(u)$) . If we take as a basis the time-centered difference scheme

$$B \frac{u^{n+1} - u^n}{\tau} + \Delta_\alpha \frac{\varphi^{n+1} + \varphi^n}{2} = 0$$

with indefinite operator B and fixed operator Δ_α of structure (12), where $\alpha(\varkappa)$ is a matrix, depending on $\varkappa = \frac{\partial \varphi}{\partial u} \tau / h$, and construct the scheme of higher order accuracy using the error exhaustion method together with the that of indefinite operators, then we obtain third order accurate scheme with respect to τ and h

$$\left[E + \frac{\alpha h}{2} \Delta_0 + \frac{h^2}{6} \Delta_{-1} \Delta_1\right] \frac{u^{n+1} - u^n}{\tau} + \Delta_\alpha \frac{\varphi^{n+1} + \varphi^n}{2} + \frac{\tau^2}{12} \frac{(L\varphi)^{n+1} - (L\varphi)^n}{\tau} = 0, \qquad (13)$$

where

$$L = \Delta_{-1}\left[\left(T_{\frac{x}{2}} \frac{\partial \varphi}{\partial u}\right)\Delta_1\right] \sim \left(\frac{\partial \varphi}{\partial u}\frac{\partial \varphi}{\partial x}\right)_x + O(h^2)$$

F.d.a. of scheme (11) is

$$\frac{\partial u}{\partial t} + \frac{\partial \varphi}{\partial x} = \alpha \frac{h^3}{24} \frac{\partial^2}{\partial x^2}\left(\varphi_{xx} - \frac{\tau}{h^2}\varphi_{tt}\right).$$

The operator of the right hand side is represented as a sum $L_1 u + L_2 u$, where $L_1 u$ does not contain higher derivative and $L_2 u$ coincides with the remaining term of the linear difference scheme at $S = 2$: $L_2 u = -\frac{\alpha(\varkappa)h^4}{24\tau}\varkappa(\varkappa^2-1)\frac{\partial^4 u}{\partial x^4}$. If matrix $M(\varkappa) = \alpha(\varkappa)\varkappa(\varkappa^2-1)$ is nonnegative, scheme (13) has the qualities of a parabolic equation. If, for example, we set $\alpha(\varkappa) = -\alpha_0 \varkappa$, where $\alpha_0 > 0$ is a scalar, then, under Courant's condition, $\varkappa^2 \leq 1$ $M(\varkappa) \geq 0$.

Similar constructions can be made in a multi-dimentional case. The method was applied for calculation of discontituous solution of gas dynamics equations. This method is characterized by large gradients on discontinuities and by insignificant oscillations much smaller than those observed in the methods of second order accuracy.

6. Below on an example of the Cauchy problem for

$$\frac{\partial u}{\partial t} = a \frac{\partial u}{\partial x}, \quad a = const > 0 \qquad (14)$$

the method of constructing the difference schemes of higher order approximation for hyperbolic equations is given which is based on differential corollaries of the initial equations.

The K-th-differential corollary of equation (14) is the equation of the type:

$$\frac{\partial \tau_{(K)}}{\partial t} = a \frac{\partial \tau_{(K)}}{\partial x}, \quad \tau_{(K)} = \frac{\partial^K u}{\partial x^K}.$$

Let

$$u^{n+1}_{(x)} = \sum_\alpha b_\alpha T_x^\alpha u^n(x) = S_h u^n(x)$$

be a stable difference scheme of first order approximation for equation (14). A parabolic form of f.d.a. is :

$$\frac{\partial u}{\partial t} = a\frac{\partial u}{\partial x} + C_2 \frac{\partial^2 u}{\partial x^2}, \quad C_2 = \frac{h^2}{2\tau}\left(\sum_\alpha \alpha^2 b_\alpha - \varkappa^2 a^2\right).$$

If the function $f^n(x)$ approximates with the function $C_2 \tau_{(2)}$ the first order, then the scheme $u^{n+1}(x) = S_h u^n(x) - \tau f^n(x)$ approximates equation (14) with the second order. Let

$f^n(x) = C_2 \tau_{(2)}^{n+1}(x)$ and to find $\tau_{(2)}^{n+1}(x)$ we consider a first order approximation scheme: $\tau_{(2)}^{n+1} = S_h \tau_{(2)}^{n}(x)$, $\tau_{(2)}^{0}(x) = \frac{\Delta_x \Delta_{-x}}{h^2}\varphi(x)$. Then

$$u^{n+1}(x) = S_h u^n(x) - \tau C_2 \tau_{(2)}^{n+1}(x) \qquad (15)$$

approximates equation (14) with the second order.

The following theorems take place.

THEOREM 12. If $S_h = I + \frac{\varkappa a}{2}(T_x - I_{-x}) + \frac{\lambda}{2}\Delta_x \Delta_{-x}$ and $\lambda^2 \leq \varkappa^2 a^2 \leq \lambda$, then the difference scheme (15) of second order approximation is stable in L_2.

THEOREM 13. Let the difference scheme (15) be dissipative of the order $2d$ and have approximation order then the difference scheme

$$u^{n+1}(x) = S_h u^n(x) - \frac{1}{(2d)!}\bar{C}_{2d} S_h^{n+1} (\Delta_x \Delta_{-x})^d u^0(x)$$

where

$$C_{2d} = \frac{h^{2d}}{\tau(2d)!} \quad \bar{C}_{2d} = \frac{h^{2d}}{\tau(2d)!}\left[\sum_\alpha \alpha^{2d} b_\alpha - (\varkappa a)^{2d}\right]$$

is stable and has approximation order $2d$.

REFERENCES

1. Н.Н. Яненко, Ю.И. Шокин. О корректности первых дифференциальных приближений разностных схем. Доклады АН СССР, I82, 4(I968), 776-778.

2. N.N. Yanenko, Yu.I. Shokin. On the group classifications of difference schemes of equations in gas dynamics. Lecture Notes in Physics, vol. 8 (1971), 3-17.

3. N.N.Yanenko, Yu.I.Shokin. Schemas numeriques invariants de groupe pour les equations de la dynamique de gas. Lecture Notes in Physics, vol. 18, part 1, (1973), 174-186.

4. В.И. Паасонен. Абсолютно устойчивые разностные схемы повышенной точности для систем гиперболического типа. Сб. "Численные методы механики сплошной среды", 3(I972), № 3, 82-9I.

5. В.И. Паасонен. Диссипативные неявные схемы с псевдовязкостью высших порядков для гиперболических систем уравнений. Сб. "Численные методы механики сплошной среды", 4(I973), № 4, 44-57.

6. N.N.Anuchina. V.E.Petrenko, Yu.I.Shokin, N.N.Yanenko. On numerical methods of solving gas dynamical problems with large deformations, Fluid Dynamics Transactions, vol. 5, part 1 (1971), 9-32.

7. Ю.И. Шокин, А.И. Урусов. Об инвариантных схемах расщепления. Труды 4-го Всесоюзного семинара по численным методам механики вязкой жидкости, Н., I973.

8. Н.Н. Яненко, Ю.И. Шокин. Групповая классификация неявных разностных схем для системы уравнений газовой динамики. "Численные методы механики сплошной среды, 2, 2(I97I), 85-92.

9. А.Н. Тихонов, А.А. Самарский. Об однородных разностных схемах. Журнал вычислительной математики и матем. физики, I, I(I96I), 5-63.

I0. Ю.П. Попов, А.А. Самарский. Полностью консервативные разностные схемы. Журнал вычислительной математики и матем. физики, 9, 4(I969), 953-958.

NUMERICAL EXPERIMENTS ON FREE SURFACE WATER MOTION WITH BORES

MAURIZIO PANDOLFI - CENTRO STUDI DINAMICA FLUIDI
POLITECNICO DI TORINO - TORINO - ITALY

Introduction

The analogy between the equations describing the one or two dimensional unsteady flow or the two dimensional steady supersonic compressible flow and the water motion with a free surface according to the "shallow water theory" is well known (Ref.1,2). In both problems the equations are non linear and hyperbolic. Until no discontinuities appear in the flows, the integration of such equations is easy. Difficulties are always associated with discontinuities, viz. shock waves in gasdynamics or bores or hydraulic jumps in water motion.

In gasdynamics two different approaches have been commonly used to compute flows with shocks.

According to the first approach, shocks are explicity computed during their evolution on the basis of the Rankine-Hugoniot equations (Ref. 3, 4, 5, 6). In the second one, the Euler equations are rearranged in what is known as the divergence or conservation form and the integration is carried out in such a way that the shocks should appear as sharp transitions, spread over few mesh intervals.

Features of the shock explicit treatment approach are few computational points, fast computations, accurate results and, often, more complicated codes due to the additional logic for the shock calculations.

On the contrary, quite more points are needed when shocks are ignored, computer times are longer, spurious oscillations (numerical wiggles) appear near the shocks and may evolve and propagate through the flow field; the program codes are, of course, simpler.

This paper intends to show how numerical methods of the first type mentioned above, which have been used in gasdynamics successfully, can be applied to problems in the category of shallow water motion.

In my opinion, there are two interesting points in this paper. One is the presentation of some numerical examples supporting the philosophy of treating discontinuities explicitly.

The other is a demonstration of how techniques traditionally confined to problems related to aerospace sciences can be easily applied to the solution of environmental problems.

The numerical examples I present in this paper belong to three classes of physical phenomena:

- unsteady one-dimensional flow in channels
- steady supercritical two-dimensional flow in channels
- steady two-dimensional flow around a blunted obstacle in a supercritical stream.

These examples involve the formation and the evolution of discontinuities such as bores or hydraulic jumps.

The Equations of Motion

The equations of motion according to the shallow water theory are very similar, even simpler, of those for compressible flow in gasdynamics. For example, the Euler equations for unsteady quasi one-dimensional flow in gasdynamics are:

$$P_t + u P_x + \gamma u_x + \gamma u \alpha = 0 \qquad \text{(continuity)}$$
$$(1) \qquad u_t + u u_x + T P_x = 0 \qquad \text{(momentum)}$$
$$S_t + S_x = 0 \qquad \text{(energy)}$$

where $P = \ln p$, $T = \exp(\frac{\gamma - 1}{\gamma} P + \frac{S}{\gamma})$ and $\alpha = (dA/dx)/A$ (A is the cross area of the duct).
All the quantities have been here normalized with respect to reference values: reference lenght l_∞, reference pressure p_∞ and temperature T_∞, reference velocity $V_\infty = \sqrt{p_\infty / \rho_\infty}$ and reference time $t_\infty = l_\infty / V_\infty$.
The corresponding equations for the unsteady flow in a channel with constant depth and variable width are:

$$H_t + u H_x + 2 u_x + 2 u = 0 \qquad \text{(continuity)}$$

(2)

$$u_t + u u_x + h H_x = 0 \qquad \text{(momentum)}$$

where h is the water level, $H = 2\ln h$, the reference velocity is defined as $V_\infty = \sqrt{g\, l_\infty / 2}$, $\alpha = (dA/dx)/A$ (A is the width of the channel).
Only two variables (H, u) describe the flow in hydraulics, instead of three (P, u, S) as in gasdynamics.
The same analogy, shown by Eq.(1) and Eq.(2), holds also for two-dimensional steady or unsteady flows.
If shock waves develop in gasdynamics, they have to be computed during their evolution with an explicit treatment (Ref. 3 and 4), by taking into account the Rankine-Hugoniot equations:

$$\frac{p_1 w_1}{T_1} = \frac{p_2 w_2}{T_2} \qquad \text{(continuity)}$$

(3)

$$p_1 (1 + \frac{w_1^2}{T_1}) = p_2 (1 + \frac{w_2^2}{T_2}) \qquad \text{(momentum)}$$

$$T_1 (1 + \frac{\gamma - 1}{2\gamma} \frac{w_1^2}{T_1}) = T_2 (1 + \frac{\gamma - 1}{2\gamma} \frac{w_2^2}{T_2}) \qquad \text{(energy)}$$

where $w_i = u_i - U_S$ $(i = 1, 2)$.
The labels 1 and 2 refer to the two sides of the shock wave which propagates with the speed U_S. The corresponding equations for the bore, the hydraulic shock, are the following:

$$h_1 w_1 = h_2 w_2 \qquad \text{(continuity)}$$

(4)

$$h_1^2 (1 + \frac{w_1^2}{h_1}) = h_2^2 (1 + \frac{w_2^2}{h_2}) \qquad \text{(momentum)}$$

where $w_i = u_i - U_S$ and U_S indicates the propagation speed of the bore.
Eq.(3) and (4) are very similar also in the case of two-dimensional steady or unsteady shocks.
The close analogy between the two physical phenomena suggests to solve the hydraulic problems with the same numerical procedure used in gasdynamics. If programs for numerical computations in gasdynamics are available, only minor changes and corrections in the codes are needed to solve the corresponding problems in hydraulics.
The general numerical procedure to integrate the partial differential equations in gasdynamics and the methods to compute explicitely the evolution of shock waves are reported in (Ref.3, 4, 5, 6) and have been tested in a large variety of problems.
The integration is carried out by a finite difference method according to the two levels (predictor - corrector) scheme suggested in (Ref.7). The explicit computation of shocks in based on the general philosophy indicated by G. Moretti; the shocks are treated as discontinuities and their evolution is computed by means of the compatibil-

-ity equations along characteristic lines.
Hereafter I present some numerical examples of hydraulic problems involving the development of bores and hydraulic jumps. The results have been represented in the sequence from Fig.1 to Fig.7 by plotting the water isolevel lines.

Unsteady One Dimensional Flow in Channels

Example n°1 (Fig.1)

A semi-infinite channel with constant width and depth is filled with water initially at rest. The channel is bounded at left by a wall. This wall begins to move at the time $t = 0$ from left to right according the law $b \equiv t^3$. According to the theoretical analysis, an imbedded bore is expected to form on the first characteristic at the point denoted by a circle in Fig.1. However the bore has been numerically fitted earlier, but it doesn't grow up and propagates as a very weak shock till a time near to the theoretical one for its formation.
Then it begins to pick up strength.

Example n° 2 (Fig.2)

In this example the wall is moving with a law $b \equiv t^3$. The theoretical analysis predicts the formation of an imbedded bore at the point indicated by a circle in Fig. 2 which is located between the moving wall and the first characteristic.
The shock is numerically fitted very early and, as before, does not pick up strength until the theoretical point for its formation. In fact it propagates at the beginning as a characteristic as it can be observed in Fig.2.
These first two examples have been presented to proof the reliability of the explicit treatment of imbedded bores in very classical problems.

Example n° 3 (Fig.3)

This example deals with a time dependent technique to compute steady flows.
A constant depth channel is placed between two infinite capacity reservoirs with different water levels. The channel width is shaped in a convergent-divergent fashion along the abscissa x.
The steady flow through the channel is subcritical in the convergent portion of the channel till the throat; thereafter, it becomes supercritical, if the level in the downstream reservoir is low enough. However for a particular range of values of this level, a hydraulic jump is expected to take place in the divergent portion of the channel; thereafter, the stream becomes subcritical till it matches, at the end of the channel, the water level in the discharge reservoir.
I can obtain this steady flow as asymptotic result of the following transient (time-dependent technique).
I assume that at the time $t = 0$ the channel is closed by a wall at $x = 1$ and open at $x = 0$, so that the level is every-where equal to the one of the left reservoir and the water is at rest. Suddenly the end wall is removed and a depression wave propagates upstream. During this transient a shock is generated by the coalescence of characteristics going from right to left. First it moves upstream and then downstream, till it gets stabilized in the equilibrium location tipical of the steady state flow. Some numerical oscillations develop after the bore during the transient and reflect some numerical problems. However these perturbations disappear in a short time.

Example n°. 4 (Fig.4)

The rising tide of the ocean can generate bores of remarkable intensity in channels or rivers open to the sea. A beatiful picture of these effects is reportes at pag. 368 of (Ref. 1). This numerical example describes this phenomenon; for the sake of simplicity only one tidal cycle and only the first harmonic of the tide are considered. The smooth tidal wave entering the river tends to generate an imbedded bore; the circle reported in Fig.4 indicates the theoretical point of the bore formation.
The imbedded bore is, numerically fitted earlier but begins to pick up strength only at the proper location. The depression wave of the second half of the tide cycle will decrease then its intensity. More details on these examples are reported in(Ref.8).

Two-Dimensional Supercritical Flow in Channels

The examples, I give here, deal with supercritical streams in channels with constant depth. According to the geometry which describes the channel shape along the abscissa Z , compression or depression waves are generated at the walls. The analogy in these flow between gasdynamics and hydraulics is very often used to interpret the experimental results on water table facilities for simulation of gasdynamical problems.

Example n°. 5 (Fig.5)

A supercritical stream (Froude = 3) is flowing in a channel with uniform transversal distribution of level and velocity at $Z = 0$. The upper wall is shaped so that the channel width is increased gradually to a new value. The bottom wall remains straight. Depression waves are initially generated at the upper wall and are reflected at the bottom one. In the second portion of the channel enlargement compression waves move down from the upper wall: they coalesce till an oblique hydraulic jump is formed.
At $Z = 10$ the hydraulic jump is reflected at the bottom wall and moves towards the upper wall.

Example n°. 6 (Fig.6)

In this example the width of the channel is constant with Z and both the walls are shaped in the same fashion so that the channel axis is displaced and straight after the abscissa $Z = 5$. The oblique hydraulic jump is now generated from the compression waves starting at the bottom wall. A second oblique hydraulic jump tends to be formed by the compression waves given by the upper wall; however these waves are absorbed by the first shock, now reflected on the upper wall, before they can generate the second hydraulic jump.

Two-Dimensional Supercritical Flow Over a Blunted Obstacle (Fig.7)

Gasdynamicists easily recognize in Fig.7 a typical example of supersonic compressible flow over a blunt body followed by a slender after-body. The bow shock is placed ahead of the nose; a pocket of subsonic flow in the front of the shock layer is followed by a supersonic flow with smooth transition through a sonic line.
The computations which allow to obtain these numerical results are done in two steps. First the sub and super-sonic flow region is computed in front of the body by means of a time dependent technique (Ref.5); the equations of the steady two dimensional flow are elliptic and hyperbolic in this region; however they are transformed in hyperbolic equations by introducing the unsteady terms according the well known time dependent technique philosophy and are then integrated with a finite difference method by marching in time.

Once the flow is supersonic, the steady hyperbolic equations of motion are integrated along the abscissa Z (Ref.6). However the results plotted in Fig.7 refer to the computations of a hydraulic example: the supercritical uniform water stream (Froude = = 3.0) over the blunted pile of a bridge. Instead of the bow shock wave we have here the hydraulic jump which wraps the bridge pile. The computational methods are the same I used in the equivalent problem in gasdynamics: the two-dimensional unsteady flow computation in the front according to the time-dipendent technique, and the steady supercritical two-dimensional flow computation on the side of the bridge pile.

REFERENCES

1. J.J. Stoker,: *Water waves*, Interscience Publishers, 1975
2. R. Courant, K.O. Friedrichs,: *Supersonic Flow and Shock Waves*, Interscience Publ. 1967
3. G. Moretti,: *Complicated One-Dimensional Flows*, Polytechnic Institute of Brooklyn, PIBAL Report n° 71-25, September 1971
4. G. Moretti,: *Thoughts and After Thoughts About Shock Computations*, Polytechnic Institute of Brooklyn, PIBAL Report n° 72-37, December 1972
5. G. Moretti, M. Abbett,: *A Time Dependent Computational Methods for Blunt Body Flows*, A.I.A.A. Journal, vol.4, n.12, December 1966
6. G. Moretti, M. Pandolfi,: *Entropy Layers*, Computer and Fluids, vol.1, 1973
7. R.W. Mac Cormack,: *The Effects of Viscosity in Hypervelocity Impact Cratering*, A.I.A.A. 7th Aerospace Sciences Meeting, Paper n° 69-354, 1969
8. M. Pandolfi,: *Numerical Computations of One-dimensional Unsteady Flow in Channels*, Meccanica, AIMETA, December 1973

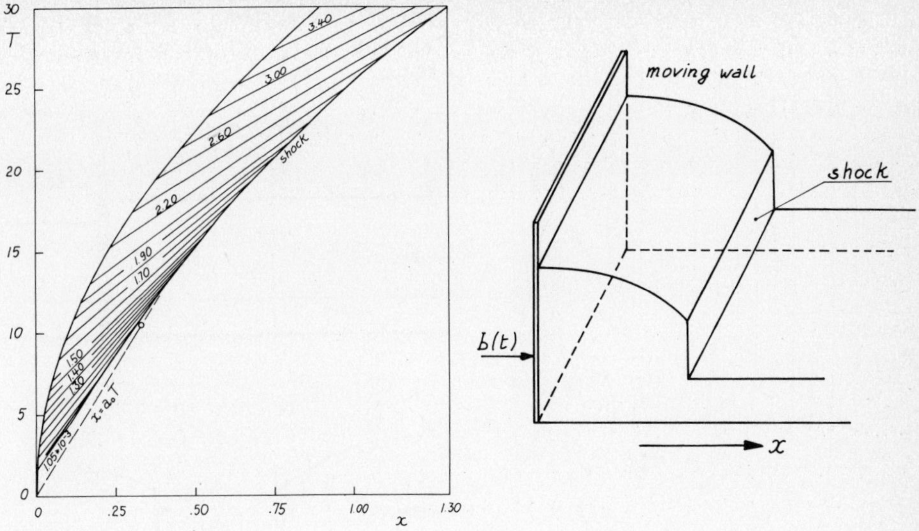
T
x
3.40
3.00
2.60
2.20
1.90
1.70
1.50
1.40
1.30
1.05*10-3
shock
x=a₀T
moving wall
shock
b(t)
x

fig. I

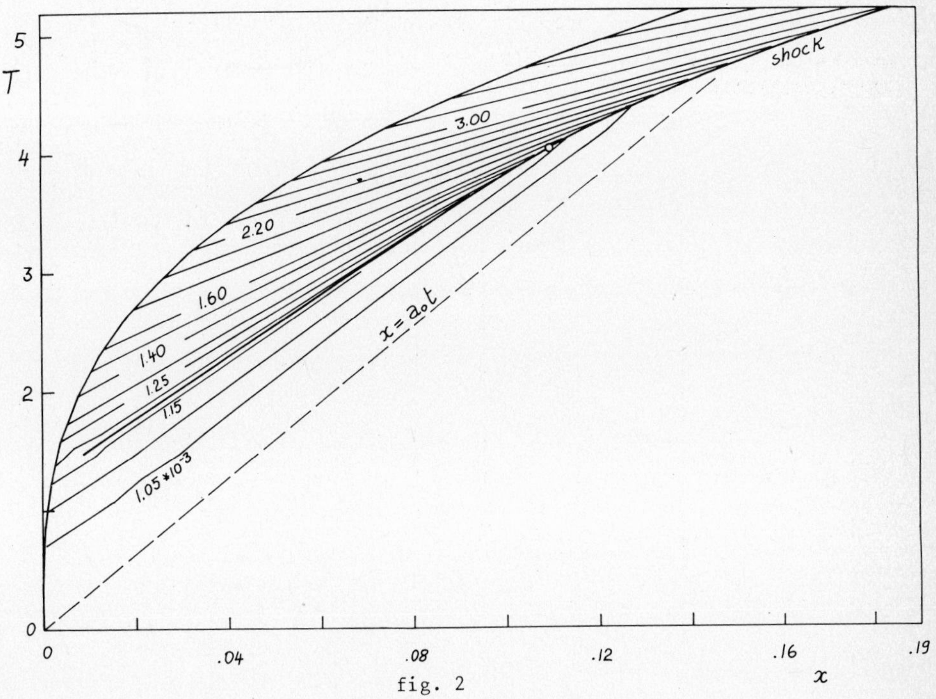
T
3.00
2.20
1.60
1.40
1.25
1.15
1.05*10-3
shock
x=a₀t
x

fig. 2

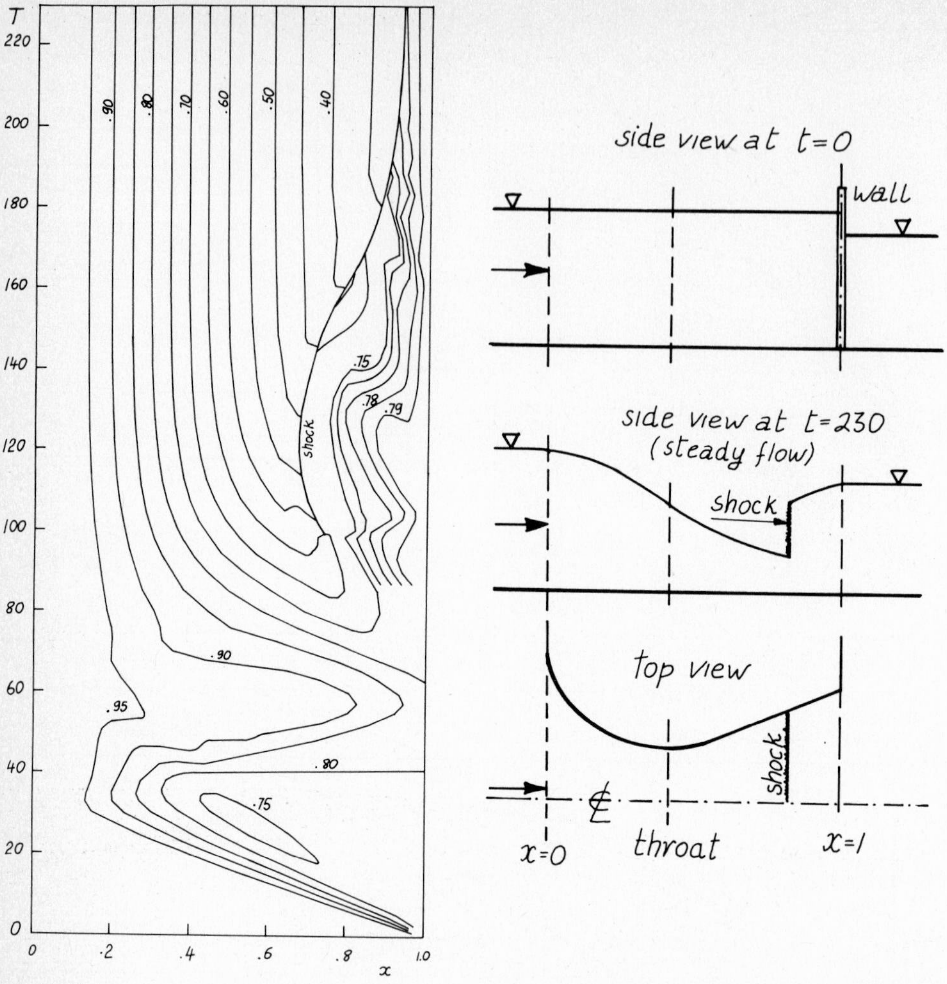
T
220
200
180
160
140
120
100
80
60
40
20
0
0
.2
.4
.6
.8
1.0
x
.90
.80
.70
.60
.50
.40
.75
.78
.79
shock
.90
.95
.80
.75
side view at t=0
wall
side view at t=230
(steady flow)
shock
top view
shock
throat
x=0
x=1

fig. 3

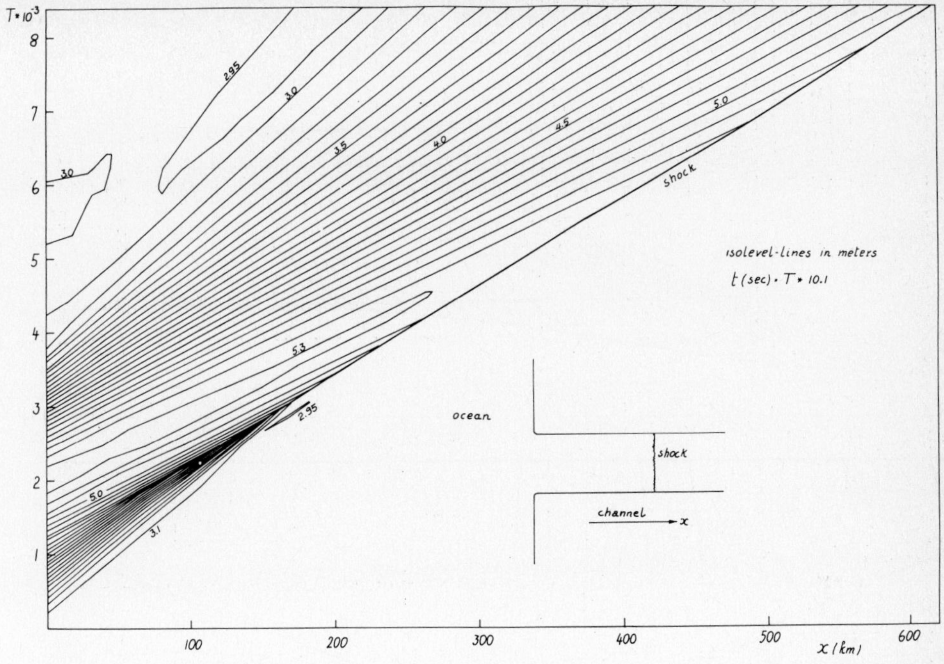
T•10⁻³
isolevel-lines in meters
t(sec) = T * 10.1
ocean
shock
channel
x
x (km)
shock
2.95
3.0
3.5
4.0
4.5
5.0
5.3
3.1

fig. 4

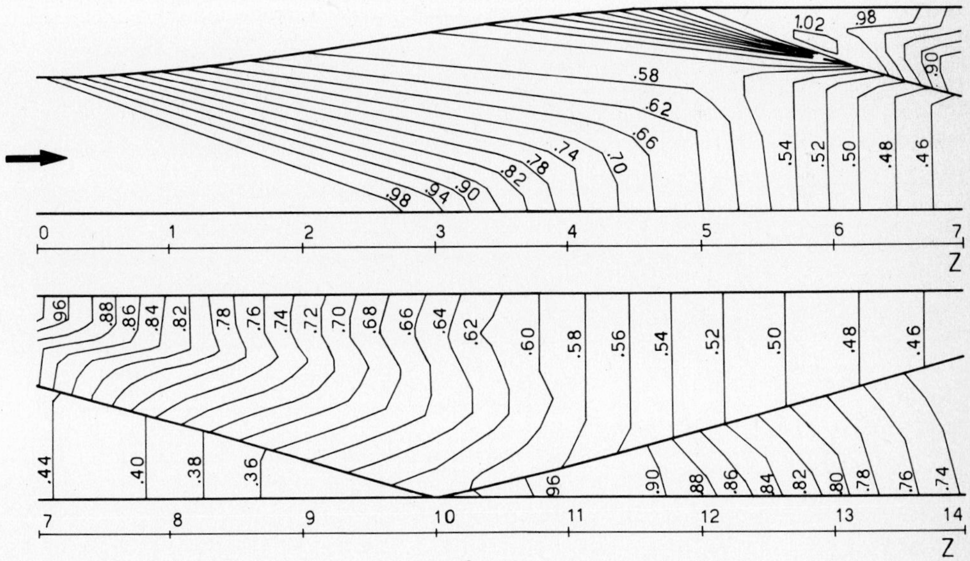
1.02
.98
.58
.62
.66
.70
.74
.78
.82
.90
.94
.98
.54
.52
.50
.48
.46
Z
Z
.44
.40
.38
.36
.96
.88
.86
.84
.82
.80
.78
.76
.74
.72
.70
.68
.66
.64
.62
.60
.58
.56

fig. 5

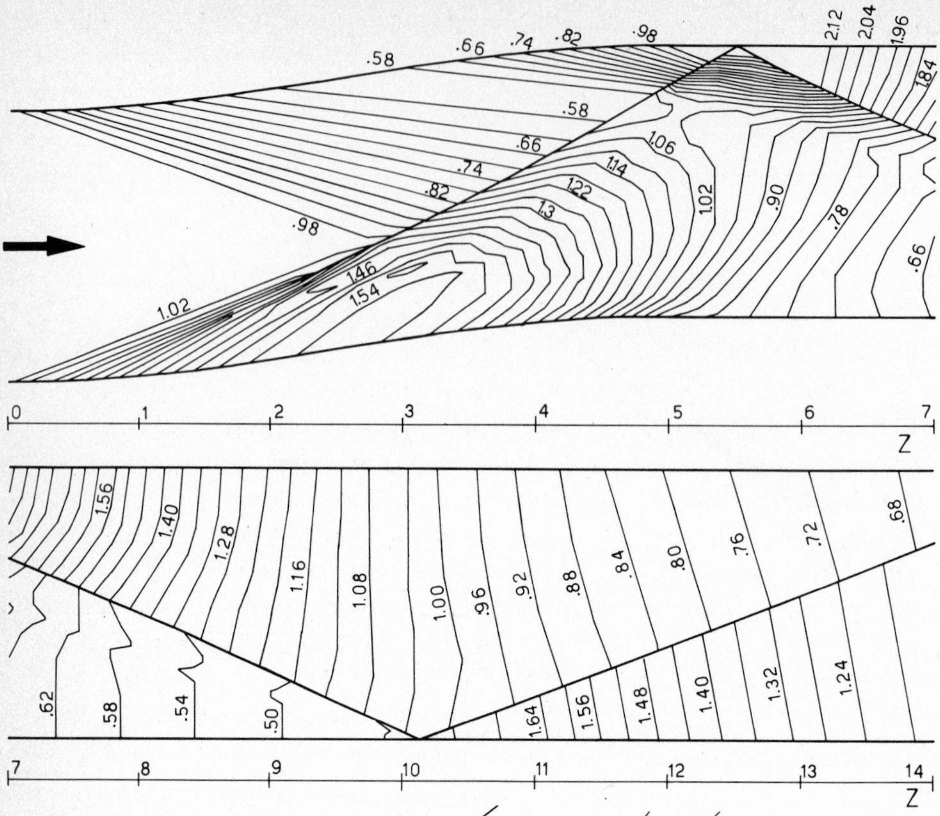

.58
.66
.74
.82
.98
2.12
2.04
1.96
1.84
.58
.66
.74
.82
.98
1.06
1.14
1.22
1.3
1.02
.90
.78
.66
1.46
1.54
1.02
0
1
2
3
4
5
6
7
Z
1.56
1.40
1.28
1.16
1.08
1.00
.96
.92
.88
.84
.80
.76
.72
.68
.62
.58
.54
.50
1.64
1.56
1.48
1.40
1.32
1.24
7
8
9
10
11
12
13
14
Z

fig. 6

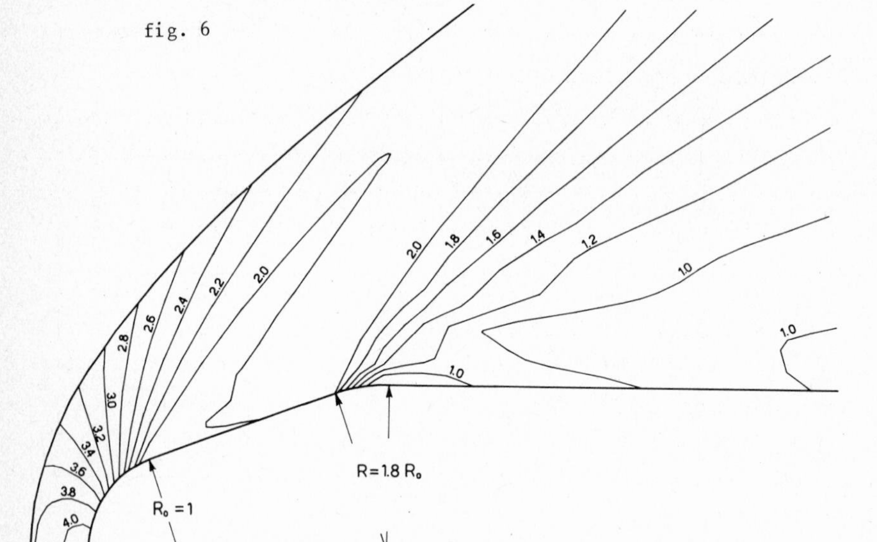

2.0
1.8
1.6
1.4
1.2
1.0
2.2
2.0
2.4
2.6
2.8
3.0
3.2
3.4
3.6
3.8
4.0
1.0
1.0
R=1.8 R₀
R₀ =1
0
1
2
3
4
5
6
7
8
Z

fig. 7

BOUNDARY LAYER CALCULATION BY A HERMITIAN FINITE DIFFERENCE METHOD

Norbert Peters

Institut für Thermo- und Fluiddynamik,
Technische Universität Berlin, Germany

INTRODUCTION

The time-consuming steps in an implicit finite difference method are mainly the following: 1.) evaluation of the coefficients of the linearized differential equations at each grid point, 2.) evaluation of the coefficients of the finite difference equations at each grid point, 3.) solution of the system of difference equations. In boundary layer problems (cf. Blottner [1]), where complex transport mechanisms are considered, step 1 may involve up to 2/3 of the computing time of the whole process. Thus the reduction of the grid points across the boundary layer using higher order approximations should reduce the total computer time considerably. The method proposed in this paper is of fourth order in step size across the boundary layer and leads to the tridiagonal matrix equation as second order methods do.

THE COMPUTATIONAL METHOD

The Hermitian finite difference scheme

The idea of Hermitian finite difference procedures is to replace the differential equation by its finite difference formulation at more than one grid point of an interval of approximation (cf. Collatz [2], p. 164). Consider a single parabolic differential equation of the boundary layer type

$$a \frac{\partial^2 w}{\partial \eta^2} + b \frac{\partial w}{\partial \eta} + c\, w + e \frac{\partial w}{\partial \xi} = d \qquad (1)$$

with the boundary conditions

$\eta = 0$:

$$f \frac{\partial w}{\partial \eta} + g\, w = h \qquad (2)$$

$\eta = \eta_\infty$:

$$w = w_\infty \qquad (3)$$

The dependent variable w may be velocity, temperature, or any concentration in a multicomponent mixture, while the independent variables ξ and η may be similarity variables [1]. The coefficients a to h in the equations may depend on ξ, η and w. Replacing the ξ-derivative at the point ξ_m, η_n by its discrete form

$$\left.\frac{\partial w}{\partial \xi}\right|_{m,n} = \Lambda_m w_{m,n} + \Lambda_{m-1} w_{m-1,n} + \Lambda_{m-2} w_{m-2,n} \cdots \tag{4}$$

where

$$\Lambda_i = \Lambda_i \; (\xi_m, \xi_{m-1}, \xi_{m-2} \; \cdots) \tag{5}$$

one obtaines an ordinary differential equation at each station ξ_m. The terms containing w_{m-1} and w_{m-2}, which are known from previous calculations, may be added to the right hand side, while the term including the still unknwon w_m may be included in the term $c \cdot w$ on the left hand side of eq. (1) to yield an equation of the form

$$a \frac{d^2 w}{d\eta^2} + b \frac{dw}{d\eta} + \bar{c}\; w = \bar{d} \tag{6}$$

According to Zurmühl (cf. [3], p. 478) a Hermitian finite difference formulation using collocation may be derivated using a fourth order interpolation polynomial

$$\begin{aligned} w(\eta) = {} & \frac{1}{2} w_{n+1} \cdot (t^2 + t) + w_n \cdot (1-t^2) \\ & + \frac{1}{2} w_{n-1} \cdot (t^2 - t) + \alpha t \; (1-t^2) \\ & + \beta t^2 \; (1-t^2) \end{aligned} \tag{7}$$

where

$$t = \frac{\eta - \eta_n}{\Delta \eta} \tag{8}$$

which approximates the solution of (6) between w_{n+1}, w_n and w_{n-1} at $t = 1$, $t = 0$ and $t = -1$ respectively.

If the polynomial (7) and its derivatives are introduced into (6) at $t = 1$, $t = 0$, $t = -1$, one obtains three equations

$$\begin{array}{l} \\ t=1 \\ t=0 \\ t=-1 \end{array} \quad \begin{array}{ccccc} w_{n+1} & w_n & w_{n-1} & \alpha & \beta \\ \hline L_{11} & L_{12} & L_{13} & L_{14} & L_{15} \\ L_{21} & L_{22} & L_{23} & L_{24} & L_{25} \\ L_{31} & L_{32} & L_{33} & L_{34} & L_{35} \end{array} \quad \begin{array}{c} \\ = \\ = \\ = \end{array} \quad \begin{array}{c} 1 \\ \hline \bar{d}_{n+1} \\ \bar{d}_n \\ \bar{d}_{n-1} \end{array} \tag{9}$$

where the L's depend on a, b and c at n+1, n and n-1 (see Appendix). In order to eliminate the free parameters α and β, one may multiply these equations with

$$\begin{aligned} \Delta_1 &= L_{24} \cdot L_{35} - L_{25} \cdot L_{34} \\ \Delta_2 &= L_{15} \cdot L_{34} - L_{14} \cdot L_{35} \\ \Delta_3 &= L_{14} \cdot L_{25} - L_{15} \cdot L_{24} \end{aligned} \tag{10}$$

respectively. Adding the rows up, one obtains a finite difference equation of the form

$$A_n\, w_{n+1} + B_n\, w_n + C_n\, w_{n-1} = D_n \tag{11}$$

The finite difference equation at the wall may be derived similarly by considering the differential equation (9) at n = 1 and n = 2 and the boundary condition eq. (2) to yield

$$w_1 = H \cdot w_2 + F \cdot w_3 + h \tag{12}$$

Equation (11) at n = 2 ... n = N-1 represents a tridiagonal matrix which may be solved together with (12) and

$$w_N = w(\eta_\infty) \tag{13}$$

The Newton iteration

In the case of chemically reacting non-equilibrium boundary layers the highly nonlinear reaction rates in the concentration equations need a special treatment. If bimolecular reactions are considered

$$X_1 + X_2 \rightleftharpoons X_3 + X_4 \tag{14}$$

the reaction rate is of second order in the concentrations

$$\Gamma_i = \sum_{k=1}^{NI} \sum_{l=1}^{NI} m_{kl}\, c_k\, c_l \tag{15}$$

where m_{kl} is assumed to be known from the extrapolated values of temperature and density. If third body dissociation-recombination reactions are included

$$X_5 + X_6 + M \rightleftharpoons X_7 + M \tag{16}$$

one may extrapolate the concentration of the third body from previous steps:

$$\Gamma_i = \sum_{k=1}^{NI} n_k\, c_k + \sum_{k=1}^{NI} \sum_{l=1}^{NI} m_{kl}\, c_k\, c_l \tag{17}$$

Through (17) the concentration equations are strongly coupled to each other and are to be solved simultaneously for systems close to chemical equilibrium. Since the reaction rate appears on the right hand side of the concentration equations, the linearization of all the other terms yields a finite difference equation of the form

$$\begin{aligned} F_n(w_{n+1}, w_n, w_{n-1}) &= A_n\, w_{n+1} + B_n\, w_n + C_n\, w_{n-1} \\ &\quad - D_n(w_{n+1}, w_n, w_{n-1}) = 0 \end{aligned} \tag{18}$$

Here w_n may be considered as a vector containing the NI concentrations. A Newton iteration procedure for this system leads to

$$H(w^{\nu})\,(w^{\nu} - w^{\nu+1}) = F(w^{\nu}) \tag{19}$$

where H is a tridiagonal hypermatrix obtained by differentiating $F_n (n = 1,2,...N)$ with respect to w_p $(p = n+1,\ n,\ n-1)$. In view of eq. (17) this differentiation is easily performed for general reaction mechanisms.

Comparison of accuracy

For the heat conduction equation par example

$$a\,\frac{\partial^2 w}{\partial \eta^2} - e\,\frac{\partial w}{\partial \xi} = 0 \tag{20}$$

the truncation error of a second method at ξ_m, η_n is

$$\varepsilon_{m,n} = a \; - \left\{ \frac{\Delta\eta^2}{12} \frac{\partial^4 w}{\partial \eta^4} - \frac{\Delta\eta^4}{360} \frac{\partial^6 w}{\partial \eta^6} \right\}_n - e \left\{ \frac{\Delta\xi}{2} \frac{\partial^2 w}{\partial \xi^2} \right\}_m \tag{21}$$

while the Hermitian fourth order method yields

$$\varepsilon_{m,n} = a \; - \left\{ \frac{\Delta\eta^4}{240} \frac{\partial^6 w}{\partial \eta^6} \right\}_n - e \left\{ \frac{\Delta\xi}{6} \frac{\partial^3 w}{\partial \xi^3} \right\}_m \tag{22}$$

It should be noted that the factor of the sixth derivative is greater in eq. (22) than in eq. (21). This may lead in some particular cases, where the higher order derivatives are much greater than the second and third order ones, to situations where the second order method is more accurate.

AN EXAMPLE CALCULATION

A hydrogen-oxygen laminar diffusion flame in a flat plate boundary layer containing 8 components is presented as an example of the efficiency of the method. The calculation, which considers 20 elementary reactions between H_2, O_2, OH, O, H, HO_2, H_2O and H_2O_2, starts from the quasi-linear frozen flow region at the leading edge of the plate and ends up with the highly nonlinear close-to-equilibrium state far downstream. The hydrogen is injected into the oxygen boundary layer at some distance from the leading edge and very fast chemical reactions leading to ignition can be observed close to the wall. A deflagration wave travelling across the mixed region of the boundary layer consumes the already injected hydrogen. After this a diffusion flame of finite thickness builds up in the boundary layer which approaches local chemical equilibrium. Multicomponent diffusion and thermodiffusion are included in the calculation. With 26 grid points across the boundary layer and 300 steps in flow direction, the whole calculation takes about 20 minutes on a CDC 6400 computer.

APPENDIX

The coefficients of eq. (9) are the following

$L_{11} = \bar{a}_{n+1} + 1.5\,\bar{b}_{n+1} + \bar{c}_{n+1}$, $L_{21} = \bar{a}_n + .5\,\bar{b}_n$

$L_{31} = \bar{a}_{n-1} - .5\,\bar{b}_{n-1}$, $L_{12} = -2\bar{a}_{n+1} - 2\bar{b}_{n+1}$

$L_{22} = -2\bar{a}_n + \bar{c}_n$, $L_{32} = -2\bar{a}_{n-1} + 2\bar{b}_{n-1}$

$L_{13} = \bar{a}_{n+1} + .5\,\bar{b}_{n+1}$, $L_{23} = \bar{a}_n - .5\,\bar{b}_n$

$$L_{33} = \bar{a}_{n-1} - 1.5\,\bar{b}_{n-1} + \bar{c}_{n-1}\,, \qquad L_{14} = -6\bar{a}_{n+1} - 2\bar{b}_{n-1}$$

$$L_{24} = \bar{b}_n \qquad L_{34} = 6\bar{a}_{n-1} - 2\bar{b}_{n-1}$$

$$L_{15} = -10\bar{a}_{n+1} - 2\bar{b}_{n+1} \qquad L_{25} = 2\bar{a}_n$$

$$L_{35} = -10\bar{a}_{n-1} + 2\bar{b}_{n-1}$$

where

$$\bar{a}_n = \frac{a_{m,n}}{(\Delta\eta)^2}\,, \qquad \bar{b}_n = \frac{b_{m,n}}{\Delta\eta}$$

$$\bar{c}_n = c_{m,n} + e_{m,n}\,\Lambda_m$$

$$\bar{d}_n = d_{m,n} - e_{m,n}\,(\Lambda_{m-1}\,w_{m-1,n} + \Lambda_{m-2}\,w_{m-2,n} \ldots)$$

<u>REFERENCES</u>

[1] Blottner, F.G.
<u>AIAA J.</u> 8, 103-205 (1970)

[2] Collatz, L.
The numerical treatment of differential equations, 3. edition, Springer, Berlin/ Heidelberg/ New York (1966)

[3] Zurmühl, R.
Praktische Mathematik für Ingenieure und Physiker, 5. Auflage, Springer, Berlin/Heidelberg/New York (1965)

NUMERICAL SOLUTIONS FOR ATMOSPHERIC BOUNDARY LAYER FLOWS OVER STREET CANYONS

R. Piva, P. Orlandi

Aerodynamic Institute University of Rome

INTRODUCTION

The local dispersion of the pollutants in the lower part of the urban atmosphere is strongly influenced by the interaction between the atmospheric boundary layer, the geometrical building structure and thermal sources distribution in the urban complex. For example the distribution of carbon monoxide from automobiles depends in a complicated way on the effects that adjacent buildings and thermal exhausts have on both the mean streamline pattern and the structure of turbulence. The correct determination of the pollutants level and the possibility to actively control the pollutants dispersion are subjected thus to the knowledge of the entire flow field.

The similarity parameters which control the physical phenomena, Reynolds and Grashof numbers, assume, in these problems, very large values, usually larger than the critical transition values to turbulent regime. Greater accuracies at high Re and Gr numbers and local finer resolutions in computing thechniques along with the introduction of a valid turbulence model are the two sets of problems to be further investigated for an effective numerical simulation of the flows in street canyons. The present work is devoted to obtain stable and sufficiently accurate numerical solutions for large recirculating velocity flow fields but still in prevailing laminar regime, with the purpose of a batter understanding of the flow behaviour in such complicated situations. The physical phenomenon has been schematized, in this preliminary study, with a boundary layer flowing over a wall with a square cavity. A ground level heat source inside the cavity simulates thermal exhausts which produce interacting buoyancy effects.

MATHEMATICAL FORMULATION OF THE PROBLEM

The Navier Stokes equations, simplified with the Boussinesq approximation, have been considered for the numerical simulation of the flow. Their non dimensional and conservative form, in terms of the variables Ψ, Ω, T is given by

$$\frac{\partial \Omega}{\partial t} + \frac{\partial u\Omega}{\partial x} + \frac{\partial w\Omega}{\partial z} = \frac{Gr}{Re^2}\frac{\partial T}{\partial x} + \frac{1}{Re}\left(\frac{\partial^2 \Omega}{\partial x^2} + \frac{\partial^2 \Omega}{\partial z^2}\right) \tag{1}$$

$$\frac{\partial T}{\partial t} + \frac{\partial uT}{\partial x} + \frac{\partial wT}{\partial z} = \frac{1}{Re\ Pr}\left(\frac{\partial^2 T}{\partial x^2} + \frac{\partial^2 T}{\partial z^2}\right) \tag{2}$$

$$\Omega = -\left(\frac{\partial^2 \Psi}{\partial x^2} + \frac{\partial^2 \Psi}{\partial z^2}\right), \quad u = \frac{\partial \Psi}{\partial z}, \quad w = -\frac{\partial \Psi}{\partial x} \tag{3}$$

where $Gr = \dfrac{g\,\beta_o\,L^3(\Theta_H - \Theta_o)}{\nu^2}$, $Pr = \dfrac{\nu}{K}$, $Re = \dfrac{U_\infty L}{\nu}$

with the usual meaning of the symbols.

A uniform undisturbed stream has been assumed as the initial condition for the computation. To solve numerically the problem it is necessary to define a closed boundary, as far as possible from the disturbance, to decrease the influence on the solution of the unavoidable arbitrariness in the choice of the boundary conditions. The boundary conditions were assigned as follows |1|: at the walls

$$u = w = 0 \quad , \quad T = C_\infty \text{ and } T = C_s \tag{4}$$

where C_∞ and C_s are respectively the constant values at infinity and at the source: at the upstream boundary

$$u = u(z) \quad , \quad \frac{\partial}{\partial x}\left(\frac{\partial w}{\partial x}\right) = 0 \quad , \quad T = C_\infty \tag{5}$$

at the upper boundary

$$u = U_\infty \quad , \quad w = 0 \quad , \quad T = C_\infty \tag{6}$$

at the downstream boundary

$$\left(\frac{\partial w}{\partial x}\right) = 0 \quad , \quad \frac{\partial \Omega}{\partial x} = 0 \quad , \quad T = C_\infty \tag{7}$$

or as at the upstream boundary, the less restrictive condition

$$u = u(z) \quad , \quad \frac{\partial}{\partial x}\left(\frac{\partial w}{\partial x}\right) = 0 \; , \; T = C_\infty \tag{8}$$

NUMERICAL SOLUTION

The equations (1), (2), (3) along with initial and boundary conditions have been discretized by a finite difference scheme to obtain the numerical simulation of the flow field and its evolution in time. An implicit alternate direction scheme has been used for the integration of the vorticity and temperature equations, parabolic in type. An iterative over relaxation technique has been used for the ellyptic stream function equation. The non linear space derivatives (convective terms) have been approximated by first order upwind differences |2,3| to preserve numerical stability at large recirculating velocity, without any restriction on the mesh size. The false convection and diffusion, introduced however by the large truncation errors connected with the use of such differences, impose a limit on the mesh size to obtain an acceptable accuracy. To decrease these negative effects, which can completely alterate the solution at high Re and Gr numbers, and to overcome the difficulties connected with the closed boundary, a non uniform grid spacing has been adopted. The step size variability has been introduced analytically by a coordinate stretching transformation in such a way to have in the new variables a constant spacing grid. The purpose of the use of a streched mesh size are summarized as follows:

a) Reduce the truncation errors where they assume their larger values, improving the overall accuracy of the solution. The truncation errors, proportional to the derivatives of the unknown functions and to the step size, can be drastically reduced by a finer grid spacing where larger is the unknown functions variability (e.g. near the walls).

b) Obtain a finer resolution in the flow field regions of major interest, as for example inside the cavity and near the walls, without increasing in a prohibitive way the computer time.

c) To extend the field of computation in order to reduce as much as possible the influence of the inflow, outflow and upper boundaries approximations on the solution of the region of interest.

The transformed equations, complicated by the introduction of new coefficients and terms, may be differenced in a regular constant step size mesh with no deterioration in the order of truncation error. The analytical transformation is given by relations of the type.

$$x = f(k) \quad , \quad z = g(l) \tag{9}$$

where k and l are new coordinates. With the transformation (9), the equations become |4|

$$\frac{\partial \Omega}{\partial t} + \frac{\partial u\Omega}{\partial k}\frac{dk}{dx} + \frac{\partial w\Omega}{\partial l}\frac{dl}{dz} = \frac{Gr}{Re^2}\frac{\partial T}{\partial k}\frac{dk}{dx} + \frac{1}{Re}\left(\frac{\partial^2 \Omega}{\partial k^2}\left(\frac{dk}{dx}\right)^2 + \frac{\partial^2 \Omega}{\partial l^2}\left(\frac{dl}{dz}\right)^2 + \frac{\partial \Omega}{\partial k}\frac{d^2 k}{dx^2} + \frac{\partial \Omega}{\partial l}\frac{d^2 l}{dz^2}\right) \tag{10}$$

$$\frac{\partial T}{\partial t} + \frac{\partial uT}{\partial k}\frac{dk}{dx} + \frac{\partial wT}{\partial l}\frac{dl}{dz} = \frac{1}{Re\ Pr}\left(\frac{\partial^2 T}{\partial k^2}\left(\frac{dk}{dx}\right)^2 + \frac{\partial^2 T}{\partial l^2}\left(\frac{dl}{dz}\right)^2 + \frac{\partial T}{\partial k}\frac{d^2 k}{dx^2} + \frac{\partial T}{\partial z}\frac{d^2 l}{dz^2}\right) \tag{11}$$

$$\frac{\partial^2 \Psi}{\partial k^2}\left(\frac{dk}{dx}\right)^2 + \frac{\partial^2 \Psi}{\partial l^2}\left(\frac{dl}{dz}\right)^2 + \frac{\partial \Psi}{\partial k}\frac{d^2 k}{dx^2} + \frac{\partial \Psi}{\partial l}\frac{d^2 l}{dz^2} = -\Omega \tag{12}$$

RESULTS AND DISCUSSION

Numerical solutions have been obtained for various values of the two non dimensional groups Re and Gr to represent different physical conditions which may be interesting for the atmospheric diffusion problem in the lower atmosphere. In particular it can be noted that at the increase of the Reynolds number the Grashof number effects on the structure of the flow field decrease. The steady state solution for $Gr = 10^6$ $Re = 10^4$ is presented in fig. 1 . In this case the forced convection, induced by the boundary layer flowing over the cavity, prevails over the natural convection induced by the thermal source. The recirculating region generated inside the cavity is stable and its influence on the external flow is very limited (weak interaction). The entire flow field reaches relatively soon the steady state condition without encountering physical instabilities.

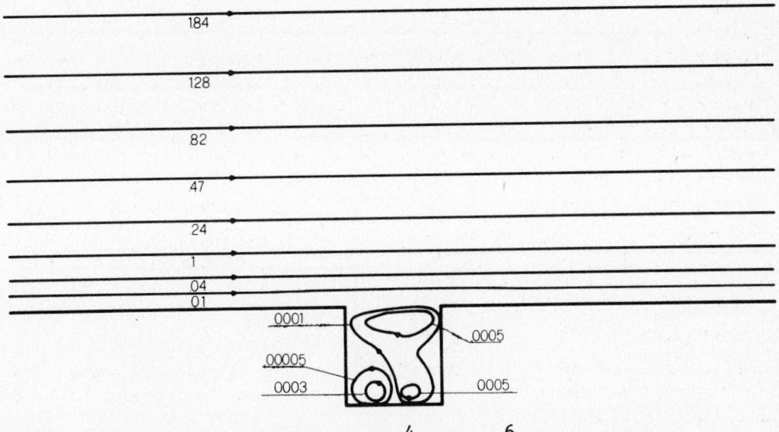

Fig.1: Streamline field for $Re = 10^4$, $Gr = 10^6$ steady state.

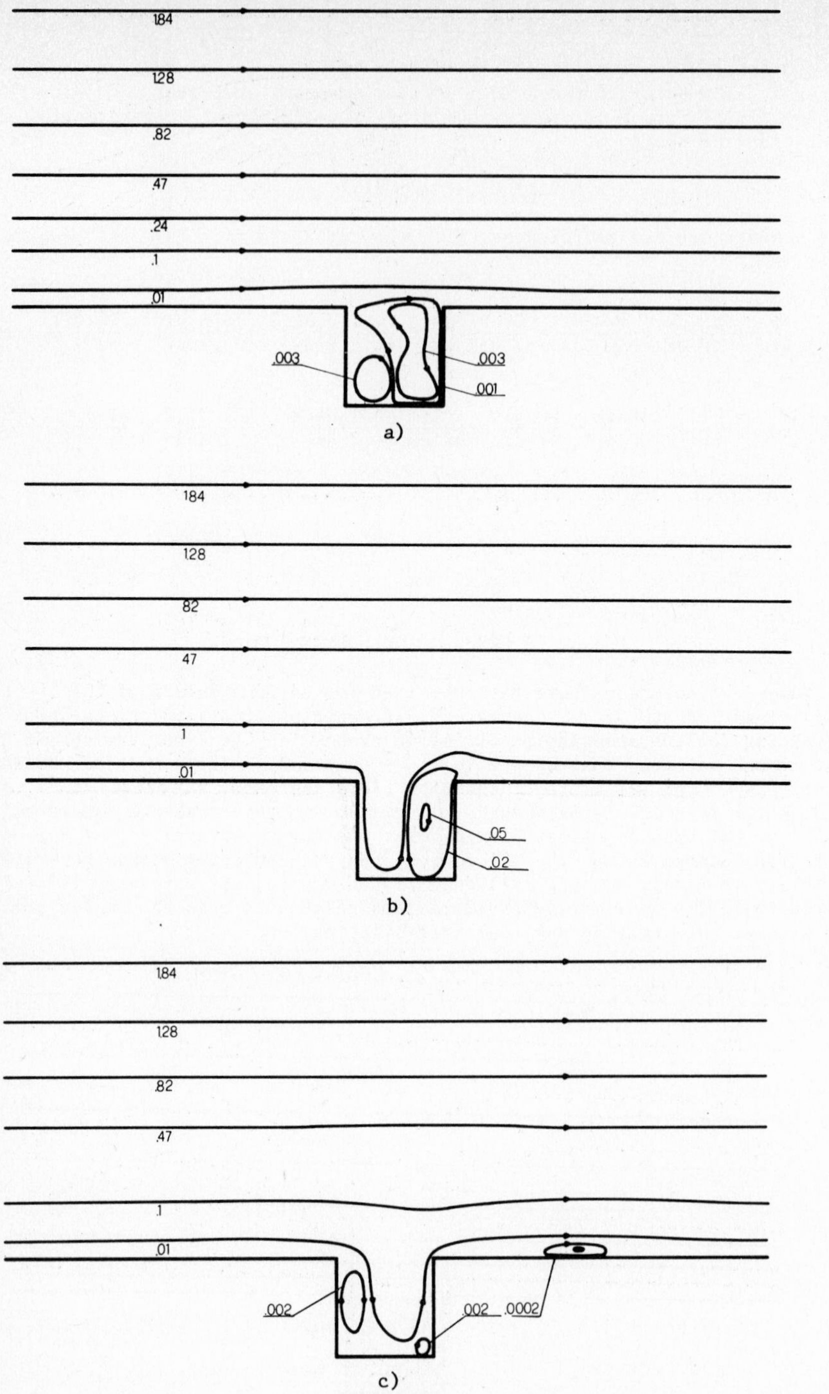

Fig.2: Streamline field for Re = 10^3, Gr = 10^6, a) t = 4, b) t = 10, c) t = 23

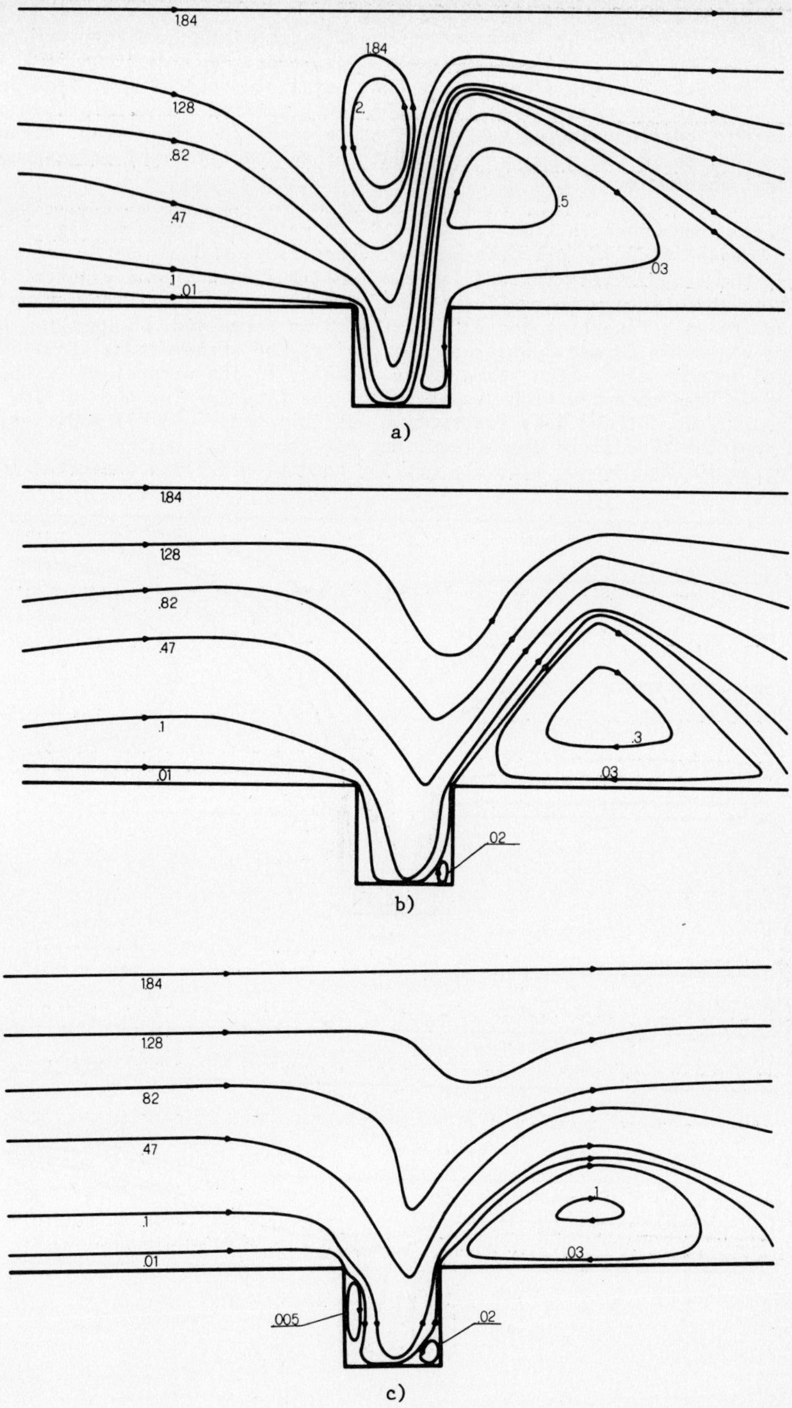

Fig.3: Streamline field for $Re = 10^2$, $Gr = 10^6$, a) t = 2, b) t = 5, c) t = 10

Different sequences in time for Gr = 10^6 Re = 10^3 are shown in fig. 2. In this case the thermal source initiates a double recirculating region, typical of natural convection flows, inside the cavity. In an early stage of the flow field evolution the recirculating flow remains confined inside the cavity, developing later to influence the main stream flow. The lower part of the main stream penetrates the cavity, before reaching the steady state condition. The steady state is not reached easily, as in the previous case, but through oscillations corresponding to the physical phenomena.

Different sequences in time for Re = 10^2 Gr = 10^6 are shown in fig. 3. The thermally induced flow now prevails on the forced flow, disturbing the entire flow field from the early stages (strong interaction). The main stream penetrates the cavity while the upstream thermally induced vortex, which initially goes up to the upper boundary as a "starting vortex", periodically forms and disappears. A large separation region is formed downstream the cavity. The steady state condition is not reached in this case which seems to be unsteady in its nature. All the presented case have been obtained with conditions of the type (8) for the outflow boundary. The application of the less restrictive outflow conditions (7) modifies, in a sensible way, the flow field downstream, but not the cavity region. The two final stages for Re 10^3 and Re 10^2 with the outflow condition (7) are presented in fig.4.

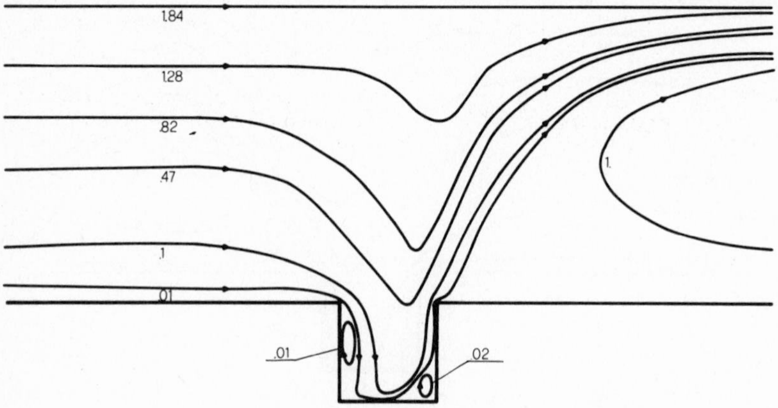

a)

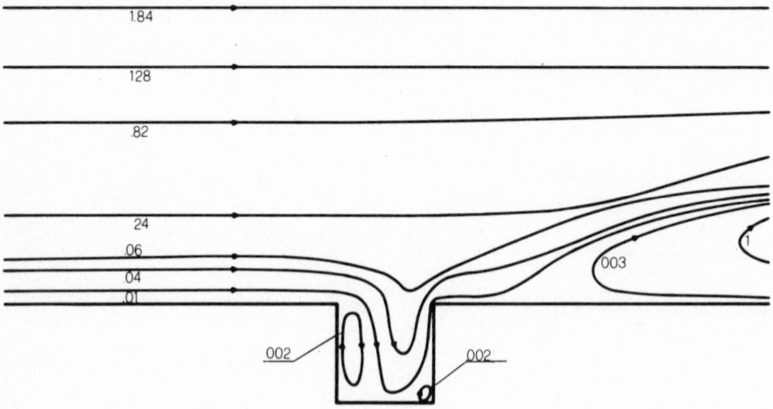

b)

Fig.4: Streamline field, a) Re = 10^2, t = 10, b) Re = 10^3, t = 23; Gr = 10^6 boundary conditions (7)

The comparison with the previous results proove the small influence of the outflow boundary conditions, if applied sufficiently far from the flow disturbance. More restricting seems to be the upper boundary conditions which, impeding the ascensional motion of the vortices, may greatly influence the flow field. As a conclusion, a further investigation must be devoted to the boundary conditions in such open flows. The obtained results look interesting for the pollution problem in street canyons. To different wind and thermal exhaust conditions correspond completely different flow fields: from the case in fig. 1, in which the only diffusion mechanism can "clean" the air, to the case in fig. 3 in which, in addition to diffusion, a strong transport through convection phenomena can drastically reduce the pollutants levels.

REFERENCES

1 - ROACHE, P.J. Computational Fluid Dynamics, Hermosa Publichers Albuquerque U.S.A. (1972)

2 - TORRANCE, K.E. J. Res. NBS, 72B, 4 (1968)

3 - PIVA, R., ORLANDI, P. "Numerical Solutions for high Grashof numbers thermally induced flows in the lower atmosphere" Presented at the 7th AIAA Plasma and Fluid Dynamics Conference June 74.

4 - ORLANDI, P., PIVA, R. "Non uniform mesh effects on numerical solutions for recirculating flows" IAR-10-73 University of Rome (1973).

THE LEADING-EDGE EFFECT IN UNSTEADY FREE-CONVECTION FLOWS

by N. Riley

University of East Anglia, Norwich, England.

INTRODUCTION

In this paper we discuss a class of unsteady flows associated with the free convection boundary layer on a semi-infinite vertical flat plate. The prototype problem in this class is that in which the plate temperature, initially the same as the ambient atmosphere, is raised to a new uniform value. The problem is analogous to that in which a semi-infinite flat plate is moved impulsively in its own plane with uniform speed (see Stewartson (1973)). In that situation the boundary layer grows as if the plate were infinite in extent until it becomes aware of the presence of the leading edge. The leading-edge signal propagates with the free-stream speed. For the free-convection problem Brown and Riley (1973) have discussed the eigensolutions which herald the leading-edge effect under the assumption that, as in the forced convection problem, disturbances travel in the boundary layer with the maximum fluid speed which is now located within the boundary layer. Unlike the forced convection flow there are, at present, no satisfactory numerical solutions of the free convection problem.

Our aim in the present paper is to solve a linear free-convection problem which exhibits many of the features of the non-linear problem described above. Thus we superpose upon the established steady free-convection boundary layer a small amplitude disturbance. The analogue for forced flows is a disturbance to the boundary layer on a flat plate induced by a disturbance to the free stream. This has been considered, for example, by Ackerberg and Phillips (1972), and Brown and Stewartson (1973). The latter authors show that the eigensolutions in the asymptotic solution far downstream have a structure similar to those for the impulsively moved plate, and are associated with the leading edge and not local conditions. The same is true for the free convection problem where the eigensolutions exhibit the same form as those for the non-linear free convection problem described above. To test the conjectures made for the non-linear problem this linear problem has been solved, in two particular cases, by complementing asymptotic solutions with a numerical solution of the coupled partial differential equations. The conjectures concerning the leading-edge effect, which manifests itself through the eigensolutions in both the non-linear and linear problems, are substantiated.

ANALYTICAL DESCRIPTION AND THE EIGENSOLUTIONS

A semi-infinite plate whose undisturbed constant temperature is T_w^* is situated in a fluid at rest, at temperature $T_\infty^* < T_w^*$. The plate is placed vertically in a gravitational field $-g^*\underline{i}$ where $\underline{i}$ is the unit vector in the x*-direction along the plate; the co-ordinate y* is measured perpendicular to the plate. The corresponding components of velocity are $\partial\psi^*/\partial y^*$, $-\partial\psi^*/\partial x^*$, where ψ^* is the stream function. If L* is a typical length and $U^* = \{g^*\beta L^*(T_w^* - T_\infty^*)\}^{\frac{1}{2}}$ a typical velocity, then with the Grashof number $Gr = g^*\beta L^{*3}(T_w^* - T_\infty^*)/\nu^2 \gg 1$, we introduce dimensionless variables by scaling x* with L*, y* with $L^* Gr^{-\frac{1}{4}}$, t* with L^*/U^* ψ^* with $U^* L^* Gr^{-\frac{1}{4}}$ and write $T^* = T_\infty^* + (T_w^* - T_\infty^*)\Theta$.

We suppose that a small-amplitude time-dependent perturbation, as yet unspecified, is superposed upon the basic steady flow and write the dimensionless stream-function and temperature as

$$\left.\begin{aligned} \Psi(x,y,t) &= x^{3/4}\{f(\eta) + \epsilon\,\psi(x,\eta,t)\}, \\ \Theta(x,y,t) &= g(\eta) + \epsilon\,\theta(x,\eta,t), \end{aligned}\right\} \qquad (1)$$

where $\epsilon \ll 1$, $\eta = y/x^{\frac{1}{4}}$ and f, g satisfy ordinary differential equations whose properties are well established in the literature. In particular we note at this stage that $f'(\eta)$ has a maximum, at $\eta = \eta_0$ say, about which we may expand f, g as

$$\left.\begin{aligned} f &= a_0 + a_1(\eta-\eta_0) + a_3(\eta-\eta_0)^3 + O((\eta-\eta_0)^4), \\ g &= b_0 + b_1(\eta-\eta_0) + O((\eta-\eta_0)^2), \end{aligned}\right\} \qquad (2)$$

where the numerical values of the coefficients are readily calculated. Equations satisfied by ψ, θ defined in (1) are derived by substituting for Ψ, Θ in the time-dependent boundary-layer equations for a Boussinesq fluid, and neglecting terms $O(\epsilon^2)$. It is convenient to take the Laplace transform (parameter s) of the resulting equations, and further to introduce the variable $\xi = x^{\frac{1}{2}}$ Thus if a tilde denotes the transform of a variable we have, for unit Prandtl number,

$$\left.\begin{aligned} \frac{\partial^3\tilde\psi}{\partial\eta^3} + \frac{3}{4}f\frac{\partial^2\tilde\psi}{\partial\eta^2} - f'\frac{\partial\tilde\psi}{\partial\eta} + \frac{3}{4}f''\tilde\psi + \tilde\theta - \xi\left(s\frac{\partial\tilde\psi}{\partial\eta} - \frac{1}{2}f'\frac{\partial^2\tilde\psi}{\partial\eta\partial\xi} - \frac{1}{2}f''\frac{\partial\tilde\psi}{\partial\xi}\right) &= 0, \\ \frac{\partial^2\tilde\theta}{\partial\eta^2} + \frac{3}{4}f\frac{\partial\tilde\theta}{\partial\eta} + \frac{3}{4}g'\tilde\psi - \xi\left(s\tilde\theta + \frac{1}{2}f'\frac{\partial\tilde\theta}{\partial\xi} - \frac{1}{2}g'\frac{\partial\tilde\psi}{\partial\xi}\right) &= 0. \end{aligned}\right\} \qquad (3)$$

The boundary conditions for these equations will depend upon the particular nature of the superposed disturbance. For each of the examples which we consider in this paper it is possible to develop series solutions for $|s\xi| \ll 1$ in the form

$$\tilde\psi = \sum_{n=0}^{\infty}(s\xi)^n f_n(\eta), \quad \tilde\theta = \sum_{n=0}^{\infty}(s\xi)^n g_n(\eta).$$

In a similar manner, when $|s\xi| \gg 1$ asymptotic series solutions of (3) may be constructed, but since such an asymptotic solution cannot satisfy any conditions which may be imposed at $\xi = 0$ it must be completed by the addition of suitable eigensolutions. We now briefly consider these eigensolutions, for a more detailed discussion reference may be made to the work of Brown and Riley (1973, 1974). The structure of the eigensolutions is not, as may at first be expected, dependent upon local conditions far downstream but, as we have already asserted depends crucially upon conditions at the leading edge on account of the wavelike nature of the governing equations in terms of the time and streamwise co-ordinate. Thus the leading-edge effect associated with any disturbance which is superposed upon the basic flow will propagate through the boundary layer, in the streamwise direction, with a speed equal to the maximum speed of the basic steady flow which is, from (1) and (2), $dx/dt = a_1 x^{\frac{1}{2}}$ This corresponds, in terms of the variable ξ, to a uniform 'speed' $d\xi/dt = a_1/2$. Consequently, following the initiation of any disturbance, regardless of its nature, information about conditions at the leading edge will not reach the station ξ until a time $t_0 = 2\xi/a_1$ has elapsed. This leads us to assume that the eigensolutions of $\tilde{\psi}$, $\tilde{\theta}$ in (3) are dominated by the function which is the Laplace transform of zero for $t < t_0$ and unity for $t > t_0$, namely $\exp(-2\xi s/a_1)$. Accordingly we write

$$\tilde{\psi}(\xi,\eta,s) = \exp(-2s\xi/a_1)F(\xi,\eta) \ , \quad \tilde{\theta}(\xi,\eta,s) = \exp(-2s\xi/a_1)G(\xi,\eta), \tag{4}$$

and substitute in (3). The resulting equations for the eigensolutions F, G are to be solved subject to homogeneous boundary conditions. Since the disturbance is propagating from the leading edge with the maximum speed embedded within the boundary layer, the leading-edge effect is first manifested at $\eta = \eta_0$, and the eigensolutions are determined as $|s\xi| \to \infty$ in a region of thickness $O(|s\xi|^{-\frac{1}{4}})$ about $\eta = \eta_0$ Thus F and G are written as

$$\left.\begin{aligned} F(\xi,\eta) &\sim (s\xi)^{\beta}\exp(-\alpha(s\xi)^{\frac{1}{2}})\sum_{n=0}^{\infty} M_n(Y)(s\xi)^{-\frac{n}{4}}, \\ G(\xi,\eta) &\sim (s\xi)^{\beta+\frac{3}{4}}\exp(-\alpha(s\xi)^{\frac{1}{2}})\sum_{n=0}^{\infty} N_n(Y)(s\xi)^{-\frac{n}{4}}, \end{aligned}\right\} \tag{5}$$

where $Y = (\eta - \eta_0)(s\xi)^{\frac{1}{4}}$. Consideration of the first few terms of (5) gives $\alpha = 4{\cdot}845$ and $\beta = -2{\cdot}888$. The eigensolutions (5) satisfy neither the conditions at $\eta = 0$ nor those as $\eta \to \infty$ and corresponding inner and outer layers have to be introduced. Finally, we record the contribution from the leading eigensolution to the asymptotic form of the perturbation displacement thickness $\tilde{\psi}(\xi,\infty)$, skin friction coefficient $\partial^2\tilde{\psi}/\partial\eta^2|_{\eta=0}$ and heat transfer coefficient $\partial\tilde{\theta}/\partial\eta|_{\eta=0}$ for $|s\xi| \gg 1$, as

$$\left.\begin{aligned} \delta &\sim \exp\{-3.979\, s\xi - 4.845(s\xi)^{\frac{1}{2}} - 2.388\log(s\xi) + O(1)\}, \\ \tau &\sim \exp\{-3.979\, s\xi - 5.456(s\xi)^{\frac{1}{2}} - 2.138\log(s\xi) + O(1)\}, \\ \lambda &\sim \exp\{-3.979\, s\xi - 5.456(s\xi)^{\frac{1}{2}} - 1.638\log(s\xi) + O(1)\}. \end{aligned}\right\} \qquad (6)$$

NUMERICAL SOLUTION

Before we discuss particular examples we describe the numerical method of solution of (3), the criteria which it must fulfil, and the purpose for which the numerical solution has been exploited. As we remarked in section 1 there is, at this time, no satisfactory numerical solution of the prototype non-linear problem in which a fluid motion is induced by suddenly heating a semi-infinite vertical plate. An analytical description of certain key features of that problem has been given by Brown and Riley (1973). It has been argued that there is a close analogy between that problem and the problem under discussion, in particular in respect of the manner in which signals associated with the leading edge propagate along the boundary layer. Consequently if in the present problem it is possible to substantiate, by a combination of analytical and numerical methods, the arguments concerning the propagation of leading-edge effects which centre on the eigensolutions of section 2, then this should lead to a greater confidence in the ideas which have been advanced on the non-linear problem.

In the next section, for two particular problems, asymptotic series solutions for $|s\xi|\gg 1$ are derived. Since, as we have already noted the addition of a multiple of the eigensolutions (4) is necessary to complete them the difference between the numerical solution and the asymptotic series must be dominated by the leading eigensolution. It will be seen at once, from (6), that this difference is very small for $|s\xi|\gg 1$ and consequently a numerical method capable of achieving high accuracy is required. Furthermore since the disturbance to the steady state is ultimately confined to a very narrow region of thickness $O(|s\xi|^{-\frac{1}{2}})$ the numerical method must be sufficiently flexible to handle the situation in which the disturbance, initially spread over the whole boundary layer becomes confined to this thin region.

Satisfactory finite-difference methods to handle situations in which a boundary layer exhibits a multiple structure, particularly if the analytical details of this structure are available, have been devized. However we have chosen to use a hybrid method in which discretization of (3) is first carried out in the ξ-direction leaving a boundary-value problem, for a linear ordinary differential equation, in the η-direction to be solved at each ξ-wise step. This boundary-value problem is solved by representing the unknowns in the form of a finite Chebychev series. Thus we write the unknown quantities, at each step, in the form $\sum_{j=1}^{N} a_j T_{j-1}(z)$, where

T_k is the Chebychev polynomial of degree k, $z = (2\eta/\eta_\infty - 1)$ where η_∞ is the value of η chosen to represent the outer edge of the boundary layer and a_j $(j=1,\dots,N)$ denotes an unknown constant. These constants are determined by satisfying the appropriate boundary conditions (of which there are, say, M) exactly, and the differential equations exactly at the (N-M) 'selected points' $z_r = \cos\{(r-1)\pi/(N-M)\}$, $r = 1,\dots,N-M+1$. This choice of z_r guarantees (see Lanczos (1957)) a satisfactorily even distribution of the error over the interval $0 \leq z \leq 1$. The main advantages of this method are (i) that it is conceptually easy to use (ii) the Chebychev series converge very rapidly so that although they do not explicitly reflect the boundary-layer structure, they are sufficiently flexible to accommodate situations of the type under consideration (iii) a practical estimate of the error incurred is provided by the value of $|a_N|$. The main drawback of the method is that for the purpose of matrix inversion there is no particular structure in the matrix from which advantage may be taken. In the present case this is partly off-set by the fact that the differential equations under consideration are coupled.

With this method it was found, typically, that with $\eta_\infty = 13$ then to achieve solutions accurate to within $O(10^{-6})$ it was necessary to use not fewer than 30 polynomials in each series.

EXAMPLES

(i) $\Theta_{\eta=0} = 1 + \epsilon e^{i\omega t}$

This case, in which the plate is subjected to an oscillatory temperature, is the analogue of the oscillatory forced convection flow considered by Ackerberg and Phillips (1972), and Brown and Stewartson (1973). If we write

$$\psi(\xi,\eta,t) = e^{i\omega t}\,\tilde{\psi}(\xi,\eta)\,, \quad \theta(\xi,\eta,t) = e^{i\omega t}\,\tilde{\theta}(\xi,\eta), \tag{7}$$

(which is equivalent to taking a Fourier rather than Laplace transform) then $\tilde{\psi}$, $\tilde{\theta}$ satisfy (3) with $s = i\omega$. Series solutions for $\omega\xi \ll 1$ and the numerical solution for $\omega\xi = O(1)$ are obtained as described in sections 2 and 3 with boundary conditions

$$\tilde{\psi} = \frac{\partial\tilde{\psi}}{\partial\eta} = 0\,, \quad \tilde{\theta} = 1 \quad \text{on} \quad \eta = 0\,;\ \frac{\partial\tilde{\psi}}{\partial\eta}\,,\ \tilde{\theta} \to 0 \quad \text{as} \quad \eta \to \infty, \tag{8}$$

together with the condition at $\xi = 0$, in the original variables, $\partial\Psi/\partial y$, $\Theta = 0$, $x = 0$, $y > 0$.

For $\omega\xi \gg 1$ the dominant disturbance to the basic flow is confined to a layer, analogous to the Stokes shear-wave layer, of thickness $O((\omega\xi)^{-\frac{1}{2}})$. From the analysis which takes account of this we find, after some manipulation, the asymptotic

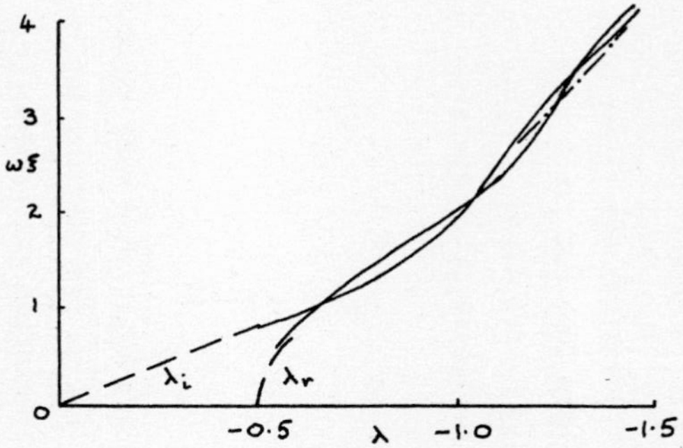

Fig. 1

The perturbation heat transfer $\lambda = \lambda_r + i\lambda_i$. Series solution ----, numerical solution ———, one term asymptotic solution -·-·-·- .

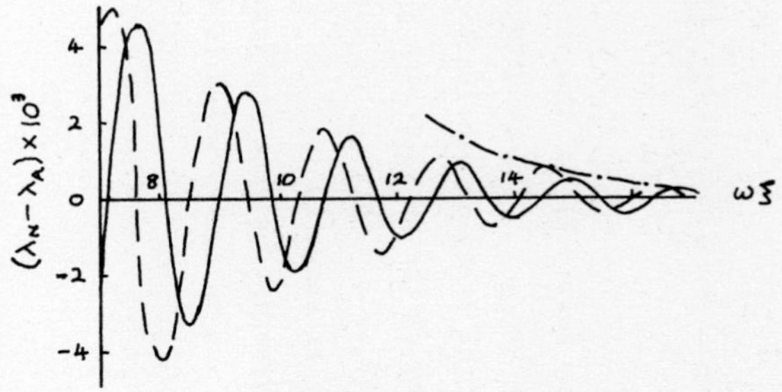

Fig. 2

The difference $\lambda_N - \lambda_A$ between the numerical and asymptotic solutions for the perturbation heat transfer. Real part ———, imaginary part - - -, amplitude of eigensolution (6) -·-·-·-.

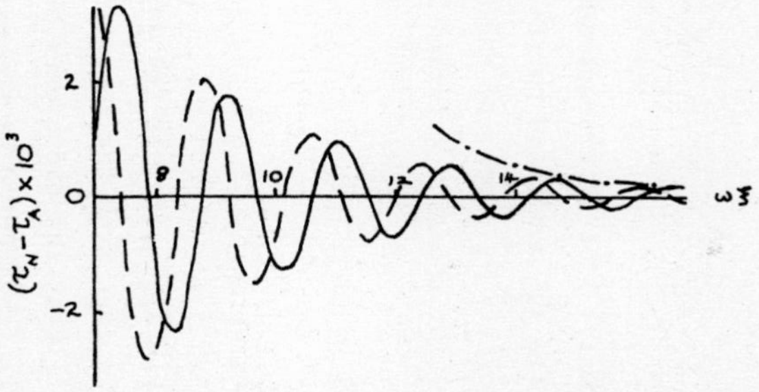

Fig. 3

The difference $\tau_N - \tau_A$ between the numerical and asymptotic solutions for the perturbation skin friction.

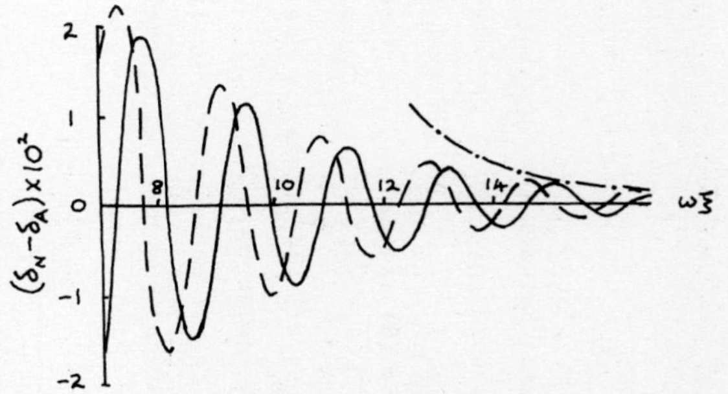

Fig. 4

The difference $\delta_N - \delta_A$ between the numerical and asymptotic solutions for the perturbation displacement thickness.

expansions for the disturbances to the displacement thickness, skin friction and heat transfer to be

$$\left.\begin{aligned} \delta_A &= (i\omega\xi)^{-\frac{3}{2}}\{0.5 - 0.5960(i\omega\xi)^{-\frac{3}{2}} + O((\omega\xi)^{-\frac{5}{2}})\}, \\ \tau_A &= (i\omega\xi)^{-\frac{1}{2}}\{0.5 - 0.05676(i\omega\xi)^{-\frac{3}{2}} + O((\omega\xi)^{-\frac{5}{2}})\}, \\ \lambda_A &= -(i\omega\xi)^{\frac{1}{2}}\{1 + 0.02838(i\omega\xi)^{-\frac{3}{2}} + 0.009398(i\omega\xi)^{-\frac{5}{2}} + O((\omega\xi)^{-3})\} \end{aligned}\right\} \quad (9)$$

The difference between the numerical solution (denoted by suffix N) and the expressions (9) should be represented by the eigensolutions (6) with $s = i\omega$. The results are shown in Figs 1 - 4. The differences $\lambda_N - \lambda_A$ etc. show the decaying oscillatory behaviour of (6) as $\omega\xi \to \infty$. In particular the frequency of the oscillations shown in Figs 2 - 4 is close to the predicted behaviour for $\omega\xi \gtrsim 7$. This may be due to the fact that all the eigensolutions (and only the first is shown in (6)) are dominated by the factor $\exp(-2i\omega\xi/a_1)$. By contrast the agreement between the decaying amplitudes is not satisfactory until $\omega\xi \approx 15$ (and then not entirely) which is at the limit of our numerical solutions.

(ii) $\Theta_{\eta=0} = 1 + \epsilon e^{\omega t}$

As a second example we choose a disturbance wall temperature which grows from zero, exponentially, from $t = -\infty$. To satisfy the conditions for linearization we require $\epsilon e^{\omega t} \ll 1$. As with example (i) we take advantage of the simple form of the wall temperature to write ψ, θ as

$$\psi(\xi,\eta,t) = e^{\omega t}\,\tilde{\psi}(\xi,\eta), \quad \theta(\xi,\eta,t) = e^{\omega t}\,\tilde{\theta}(\xi,\eta), \quad (10)$$

then upon substitution into the linearized equations $\tilde{\psi}$, $\tilde{\theta}$ satisfy (3) with $s = \omega$ which is real. The eigensolutions are again of the form (6), but since s is now real they are <u>all</u> dominated by the factor $\exp(-3.979\,\omega\xi)$ for $\omega\xi \gg 1$. This we hope to identify by comparing the numerical solution and asymptotic expansions. The latter, for the skin friction and heat transfer, are

$$\left.\begin{aligned} \tau_A &= (\omega\xi)^{-\frac{1}{2}}\{0.5 - 0.05676(\omega\xi)^{-\frac{3}{2}} + O((\omega\xi)^{-\frac{5}{2}})\}, \\ \lambda_A &= -(\omega\xi)^{\frac{1}{2}}\{1 + 0.02838(\omega\xi)^{-\frac{3}{2}} + O((\omega\xi)^{-\frac{5}{2}})\}. \end{aligned}\right\} \quad (11)$$

From Fig.5 we see that the asymptotic forms (11) are approached very rapidly indeed. In fact from the logarithmic plot of Fig. 6 all we can say with any certainty is that the difference between the numerical solution and asymptotic expansion is represented by the next term of the latter. No firm conclusions can

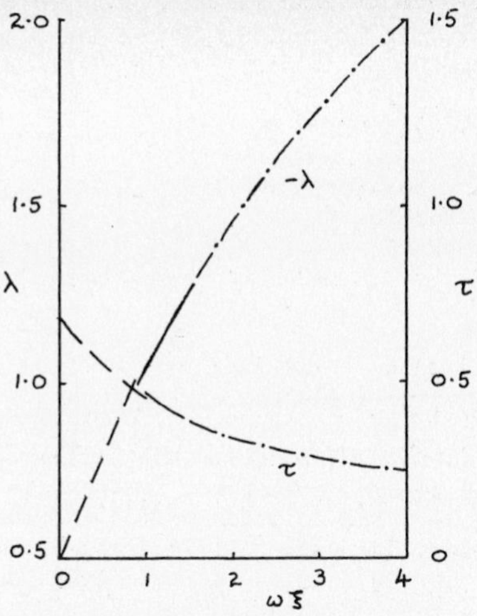

Fig. 5

The perturbation heat transfer λ and shear stress τ .

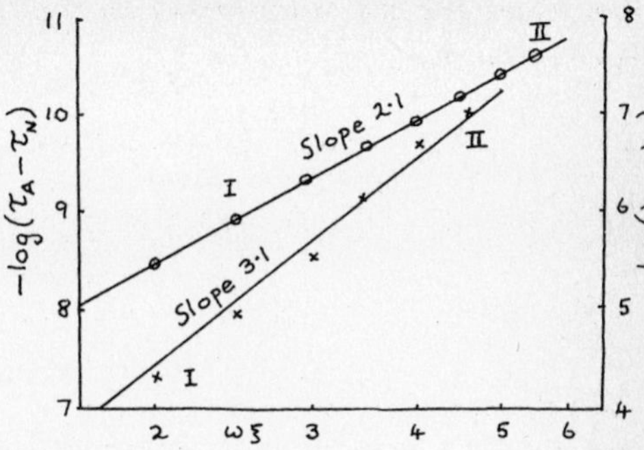

Fig. 6

Logarithmic plots of the difference between the numerical and asymptotic solutions for the skin friction and heat transfer compared with lines of slope - 3.1 and - 2.1 respectively. At I the actual differences are 2 parts in 1000 and at II 2 parts in 10,000.

be drawn from this observation, for whilst it cannot be denied that the results are consistent with eigensolutions which decay algebraically one might add that only by calculating a prohibitively large number of terms in the asymptotic expansion can we hope to isolate the very rapid exponential decay predicted for the eigensolutions.

In conclusion we state that we have established positive evidence in favour of the basic hypothesis that the eigensolutions are, in essence, a leading-edge phenomenon for this free-convection problem. None of our numerical results suggest otherwise. Although it is reasonable to suppose that the linear problem and the corresponding non-linear impulsive problem exhibit the same essential features, final confirmation of the conjectures concerning the latter must await further numerical work.

REFERENCES

Ackerberg, R. C. and Phillips, J. H. J. Fluid Mech. 51 (1972) 137.

Brown, S. N. and Riley, N. J. Fluid Mech. 59 (1973) 225.

Brown, S. N. and Riley, N. To appear in Z.A.M.P. (1974).

Brown, S. N. and Stewartson, K. Proc. Camb. phil. Soc. 73 (1973) 493.

Lanczos, C. Applied Analysis (Pitman 1957).

Stewartson, K. Q. Jl. Mech. appl. Math. 26 (1973) 143.

THE TRANSIENT DYNAMICS OF CHEMICALLY REACTIVE GASEOUS MIXTURES WITH TURBULENCE

W. C. Rivard, T. D. Butler, O. A. Farmer*
Los Alamos Scientific Laboratory
University of California
Los Alamos, New Mexico

INTRODUCTION

The RICE computer program has been developed to study the mixing dynamics and finite rate chemical reactions in multicomponent gaseous mixtures with particular application to continuous wave (cw) chemical lasers. RICE solves the full, time dependent, Navier-Stokes equations with coupled species transport, finite rate chemistry, and turbulence in two space dimensions. The solution method and computer program are described in detail by Rivard, Butler, and Farmer (1974, ref. 7). The mass and momentum equations are integrated by the ICE solution method developed by Harlow and Amsden (1971). Instead of solving a Poisson equation for the pressure field, a relaxation technique devised by Chorin (1966) and modified by Hirt and Cook (1972) is used. This procedure provides a more convenient treatment for the boundary conditions than was possible with the original ICE method. An internal energy equation is solved in place of a total energy equation to obtain a more accurate temperature field because of its importance in the species diffusion and chemical rate equations. The effects of turbulence are described by a turbulence model which uses a separate transport equation to determine the turbulent kinetic energy. Closure is obtained through flux approximations and an empirical determination of the turbulence scale after Launder, et.al. (1972).

The complete set of finite difference equations is made stable and more accurate by locally computing and canceling low order diffusional truncation errors at each time cycle as described by Rivard, et.al. (1973). Solutions obtained with this procedure are found to be sensitive to molecular transport properties in the Reynolds number range 0-2000. In this range, the velocity gradients, which are an important source of turbulence energy production, are realistically maintained. In spite of the added computations inherent in this stabilization procedure, a significant decrease in the total computational time per problem is often realized over methods that add artificial diffusion based on global considerations. This is a result of the more severe time step restrictions associated with the latter.

The RICE program has received extensive testing and has been used in a wide variety of applications. Results have been compared with experiments on the supersonic mixing of viscous jets [Rivard, Butler, and Farmer (1974), ref. 6] and with analytical solutions of incompressible flows. Successful applications have been made to cw chemical lasers, to studies of detonation and deflagration waves, and to studies of the effects of energy deposition on supersonic channel flows.

GOVERNING DIFFERENTIAL EQUATIONS

The following equations describe the mathematical model that is solved in RICE. The differential equations for mass, momentum, and specific internal energy for the mixture are

$$\frac{\partial \rho}{\partial t} + \frac{\partial \rho u_j}{\partial x_j} = \frac{\partial}{\partial x_j}\left(\sigma \frac{\partial \rho}{\partial x_j}\right) \quad , \tag{1}$$

*This work was performed under the joint auspices of the Air Force Weapons Laboratory (Project Order 74-032) and the United States Atomic Energy Commission.

$$\frac{\partial \rho u_i}{\partial t} + \frac{\partial \rho u_i u_j}{\partial x_j} = - \frac{\partial p}{\partial x_i} + \frac{\partial}{\partial x_j}\left(\pi_{ij} - R_{ij}\right) + \frac{\partial}{\partial x_j}\left[\sigma\left(\frac{\partial \rho u_i}{\partial x_j} + \frac{\partial \rho u_j}{\partial x_i} - 2\rho\, e_{ij}\right)\right] , \qquad (2)$$

$$\frac{\partial \rho I}{\partial t} + \frac{\partial \rho I u_j}{\partial x_j} = - p \frac{\partial u_j}{\partial x_j} + \pi_{ij} e_{ij} + \frac{2\mu\,\Delta q}{s^2} + \frac{\partial}{\partial x_j}\left(\sigma \frac{\partial \rho I}{\partial x_j} + \kappa \frac{\partial T}{\partial x_j}\right) + \frac{\partial H_j}{\partial x_j} + \dot{Q}_c \quad , \qquad (3)$$

where t and x_i are the time and position and ρ, u_j, and I are the density, velocity, and specific internal energy of the mixture. The pressure p is determined through the equation of state which is $p = (\gamma - 1)\rho I$ for a mixture of perfect gases where γ is the specific heat ratio and depends upon the mixture composition.

In Eq. (1) the correlation between the fluctuating mixture density and velocity, which describes the turbulent transport of mass, is modeled as a diffusion of ρ by the turbulent kinematic viscosity σ. The functional dependence of σ depends on the value of the turbulent Reynolds number, $Re_T = \rho s\sqrt{2q}/\mu$, where s is the length scale of the energy carrying eddies, q is the specific turbulent kinetic energy, and μ is the molecular shear viscosity. The length scale is determined from the mean velocity field while q is calculated through a separate transport equation described later. The value of σ is computed as $\sigma = .008\rho s^2 q/\mu$ for $Re_T \leqslant 5$ (low intensity turbulence) and as $\sigma = .02s\sqrt{2q}$ for $Re_T > 5$ (high intensity turbulence).

In Eq. (2), π_{ij} is the usual viscous stress tensor for a Newtonian fluid, and R_{ij} is the Reynolds stress modeled as

$$R_{ij} = 2\rho\left[\left(q + \sigma\, e_{kk}\right)\delta_{ij}/3 - \sigma\, e_{ij}\right] , \qquad (4)$$

where e_{ij} is the usual strain rate tensor. Turbulent shearing stresses exist when the off-diagonal components of e_{ij} are nonzero. The effects of compressibility on the turbulent transport of momentum are contained in the last, square bracket, term. This term describes the diffusion of momentum that results from the turbulent diffusion of mass and models the density-velocity correlations, the same as was done for the mass equation.

The decay of turbulent kinetic energy by viscosity produces internal energy through the positive definite term $2\mu\,\Delta q/s^2$ in Eq. (3) where $\Delta = 5$ for $Re_T \leqslant 5$ and $\Delta = Re_T$ for $Re_T > 5$. The transport of internal energy by turbulence is modeled similar to the transport of mass, i.e., as a diffusion of ρI by σ. This diffusion term is included in the parenthesis with the heat conduction term where κ is the molecular conductivity. The diffusion of thermal energy by interdiffusion of the species is described by the energy flux H_j while $\dot{Q}_c$ represents the heat released or absorbed by chemical reactions. These two quantities are defined later.

The specific turbulent kinetic energy, q, is determined from the transport equation

$$\frac{\partial \rho q}{\partial t} + \frac{\partial \rho q u_j}{\partial x_j} = - R_{ij} e_{ij} - (\sigma/\rho)\frac{\partial \rho}{\partial x_j}\frac{\partial p}{\partial x_j} + \frac{\partial}{\partial x_j}\left[\sigma \frac{\partial \rho q}{\partial x_j} + \mu \frac{\partial q}{\partial x_j}\right] - 2\mu\,\Delta q/s^2 \ . \qquad (5)$$

On the right side of this equation, the first term leads to creation of q from the mean velocity gradients (note that the effects of compressibility in this term may create or decay q), the second term describes the creation or decay from the effects of buoyancy or stratification, the third term accounts for the diffusion of q by the turbulent and molecular viscosities, and the fourth term models the decay of q by molecular viscosity.

The motion of the individual species is governed by the species transport equations

$$\frac{\partial \rho_\alpha}{\partial t} + \frac{\partial \rho_\alpha u_j}{\partial x_j} = \frac{\partial}{\partial x_j}\left[\rho D_\alpha \frac{\partial(\rho_\alpha/\rho)}{\partial x_j} + \sigma \frac{\partial \rho_\alpha}{\partial x_j}\right] + \left(\dot{\rho}_\alpha\right)_c \quad , \tag{6}$$

where subscript α refers to species α. The effects of molecular diffusion are modeled by Fick's law where D_α is the effective binary diffusion coefficient, and the energy flux due to interdiffusion of the species, H_j, appearing in Eq. (3) is

$$H_j = \sum_\alpha \rho h_\alpha D_\alpha \, \partial\left(\rho_\alpha/\rho\right)/\partial x_j \quad , \tag{7}$$

where h_α is the specific enthalpy. The turbulent transport of species α is modeled as the diffusion of ρ_α by the turbulent kinematic viscosity, σ. The rate of change of species density from chemical reactions $\left(\dot{\rho}_\alpha\right)_c$ is given by

$$\left(\dot{\rho}_\alpha\right)_c = M_\alpha \sum_r \left(b_{\alpha r} - a_{\alpha r}\right) \dot{\omega}_r \quad . \tag{8}$$

In this equation M_α is the molecular weight, $a_{\alpha r}$ and $b_{\alpha r}$ are the stoichiometric coefficients for reaction r, which is written as $\sum_\alpha a_{\alpha r} X_\alpha \rightleftharpoons \sum_\alpha b_{\alpha r} X_\alpha$ with X_α being the chemical symbol for species α, and $\dot{\omega}_r$ is the rate of reaction r given by the mass-action rate equation as

$$\dot{\omega}_r = K_{fr} \prod_\alpha \left(\rho_\alpha/M_\alpha\right)^{a_{\alpha r}} - K_{br} \prod_\alpha \left(\rho_\alpha/M_\alpha\right)^{b_{\alpha r}} \quad , \tag{9}$$

where K_{fr} and K_{br} are the forward and backward rate multipliers, respectively. The rate of heat release or absorption in Eq. (3) is

$$\dot{Q}_c = \sum_r \dot{\omega}_r Q_r \quad , \tag{10}$$

where Q_r is the heat of formation for reaction r.

SOLUTION METHOD

The finite difference approximations to the mass and momentum equations for the mixture are solved by the ICE implicit solution technique. Intermediate (tilde) values of the density and momenta, $\tilde{\rho}$ and $\widetilde{\rho u_j}$, are computed explicitly using quantities at time level n. The convective mass flux and pressure gradient are evaluated at intermediate time levels. The final solution values at time level (n+1) are obtained by adjusting the pressure and accordingly the density and momenta until $|D| < \varepsilon$ simultaneously in all computational cells where

$$D = \rho - \tilde{\rho} + \theta\,\delta t\left\{\left[(\rho u_1)_R - (\rho u_1)_L\right]/\delta x_1 + \left[(\rho u_2)_T - (\rho u_2)_B\right]/\delta x_2\right\} \quad . \tag{11}$$

The constant parameter θ determines the time level of the convective mass flux, $0 \leq \theta \leq 1$. The momenta components are located on the right and top boundaries of the computational cells. The right, left, top, and bottom boundaries are denoted by the subscripts R, L, T, and B, respectively. The pressure increment is computed from D as

$$\delta p = - \Omega D / \left[c^{-2} + 2\theta\phi\, \delta t^2 \left(\delta x_1^{-2} + \delta x_2^{-2} \right) \right] , \tag{12}$$

where Ω is a constant over-relaxation factor, c is the adiabatic sound speed, and ϕ determines the time level of the pressure gradient. The pressure, density, and momenta are updated from their previous values through the pressure increment as

$$\begin{aligned} & p = p + \delta p \quad , \qquad \rho = \rho + \delta p / c^2 \quad , \\ & \left(\rho u_1\right)_R = \left(\rho u_1\right)_R + \phi \delta t\, \delta p / \delta x_1 \quad , \qquad \left(\rho u_1\right)_L = \left(\rho u_1\right)_L - \phi \delta t\, \delta p / \delta x_1 \quad , \\ & \left(\rho u_2\right)_T = \left(\rho u_2\right)_T + \phi \delta t\, \delta p / \delta x_2 \quad , \qquad \left(\rho u_2\right)_B = \left(\rho u_2\right)_B - \phi \delta t\, \delta p / \delta x_2 \quad . \end{aligned} \tag{13}$$

When D, Eq. (11), is reduced in magnitude to less than ε for all computational cells, the final velocity is then determined from the final density and momenta by division.

Equation (3) is solved explicitly for the specific internal energy using the final density and velocity fields and omitting, until later in the time cycle, the effects of species diffusion and chemistry.

Similarly, the specific turbulent kinetic energy, q, is computed explicitly according to Eq. (5). The turbulence scale, s, and kinematic viscosity, σ, are then computed from the final velocity and q fields.

Species densities are computed from Eq. (6) in three stages that consider separately the effects of convection, diffusion, and chemistry, respectively. The first two stages are explicit calculations while the third is implicit. Local mass conservation is ensured in the first stage by computing the density of the Nth species as the difference between the mixture density and the sum of the first (N-1) species densities.

The values obtained from the first stage are used to compute the effects of molecular diffusion and are updated accordingly. Mass conservation is ensured during the second stage by requiring that the following constraint on the diffusion coefficients, which is evaluated by simple differences, be satisfied

$$\frac{\partial}{\partial x_j} \left[\rho \sum_\alpha D_\alpha \frac{\partial(\rho_\alpha/\rho)}{\partial x_j} \right] = 0 \ . \tag{14}$$

Equation (14) is enforced by limiting diffusion such that there is no net mass flux across any computational cell boundary. The energy flux due to the interdiffusion of species is computed from the known mass fluxes according to Eq. (7). The specific internal energy is then updated to include this effect.

The third stage updates the results of the second stage to their final values by including the effects of chemistry. Each reaction equation is solved according to the following procedure. The direction of reaction progress is determined from the algebraic sign of $\dot{\omega}_r$ in Eq. (9). The species that are being depleted in reaction r are then checked to find the one with minimum value of $\left|\rho_{\alpha}/\left[M_{\alpha}\left(b_{\alpha r} - a_{\alpha r}\right)\right]\right|$. This species is referred to as the "reference species" for reaction r. The density of the reference species at time level (n+1) is computed implicitly according to Eq. (9) after replacing $\dot{\omega}_r$ with $\dot{\rho}_{\beta}/[M_{\beta}(b_{\beta r} - a_{\beta r})]$ where β denotes the reference species. The time step for the calculation is chosen according to the ratio of the characteristic time for the reaction to the fluid dynamic time step and may be a submultiple of the latter. The implicit solution for the reference species insures that it, and the other species, will not be depleted by the reaction. The increment in the progress variable, $\delta\omega_r$, is computed from the change in density of the reference species. When all the reaction equations have been considered the final species densities are calculated from Eq. (8), the total rate of heat release is determined from Eq. (10), and the specific internal energy is updated to its final value. The final pressure is then computed from the equation of state.

The finite difference approximations used in RICE are all first order accurate in time and second order accurate in space for the convection terms except for the species transport equations which are second order accurate in both time and space. Without the addition of artificial diffusion these finite difference equations are numerically unstable. The mixture mass equation and the species transport equations are stabilized by locally computing and canceling their low order diffusional truncation errors. The momentum and energy equations have been successfully stabilized by two different procedures. In one case artificial diffusion is added through the molecular viscosity and thermal conductivity coefficients by specifying constant values for μ/ρ, λ/ρ, and κ/ρ based upon the most severe conditions expected in the computing mesh. The second procedure, which proves to be much better, is to locally compute and cancel their low order diffusional truncation errors similar to what is done for the mass and species transport equations. For a given number of computational cells, this procedure results in a significant increase in accuracy over the first and reduces the total computing time by about a factor of three in many cases by permitting a correspondingly larger time step.

NUMERICAL EXAMPLE

The complex fluid dynamics, species mixing, and chemical reactions that take place in the cavity of a continuous wave HF chemical laser provide a realistic example problem to illustrate the capabilities of the RICE program. A portion of the nozzle bank geometry and the computing region are shown in Fig. 1.

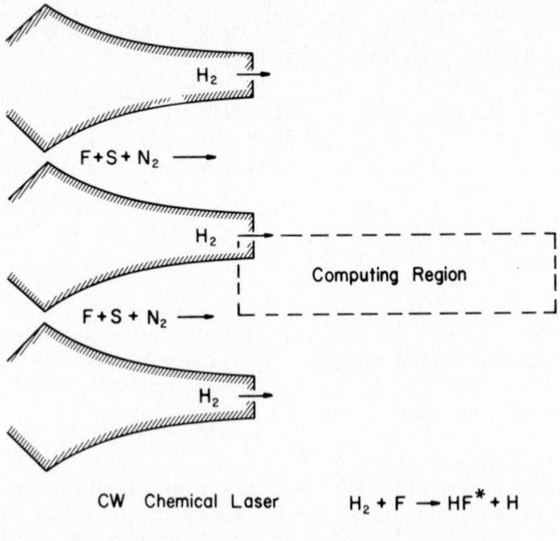

Figure 1. Nozzle geometry and computing region for a (cw) HF chemical laser

Fluorine atoms, produced by dissociation of SF_6, enter the lasing cavity at Mach 4.6 and a Reynolds number of 2000 based on centerline conditions. Hydrogen is injected into the cavity under high pressure through a narrow slit. The H_2 and F-atoms mix and react exothermally to produce vibrationally excited states of HF which serve as the lasing medium. For the purpose of this example, these excited states are denoted collectively as HF^*. The upper and lower boundaries of the computing region are reflective planes of symmetry. The left boundary has prescribed input while the right boundary provides continuous outflow.

Results obtained at steady state are shown in Fig. 2. This figure shows the velocity vector field and contours of the mixture density, the densities of F-atoms and HF^*, and the specific turbulent kinetic energy.† The velocity field shows strong transverse motions induced near the nozzle exit plane by the high pressure injection of hydrogen. An oblique shock is produced that compresses the fluorine stream as shown in the contours of mixture and F-atom densities. The velocity field also shows the diversion of the hydrogen stream after its initial collision with the fluorine stream and the resulting interaction with the upper symmetry boundary which produces a second oblique shock. The production of HF^* occurs along the diffuse interface of the reactant streams with some production occurring in the fluorine nozzle itself. A strong creation of turbulent kinetic energy occurs at the nozzle exit plane near the fluorine nozzle wall as a result of the combined axial and transverse velocity gradients. At this location, the turbulent kinematic viscosity σ and the molecular diffusion coefficient for H_2 are comparable while at about four nozzle radii downstream σ has decayed to a value an order of magnitude smaller than D_{H_2}.

REFERENCES

1. Chorin, A. J., AEC Research and Development Report, NYO-1480-61 (1966).

2. Harlow, F. H. and Amsden, A. A., J. Comp. Phys. 8, 197 (1971).

3. Hirt, C. W. and Cook, J. L., J. Comp. Phys. 10, 324 (1972).

4. Launder, B. E., Morse, A., Rodi, W. and Spalding, D. B., "The Prediction of Free Shear Flows - A Comparison of the Performance of Six Turbulence Models," Imperial College of Science and Technology Report, TM/TN/B/19 (1972).

5. Rivard, W. C., Butler, T. D., Farmer, O. A. and O'Rourke, P. J., "A Method for Increased Accuracy in Eulerian Fluid Dynamics Calculations," Los Alamos Scientific Laboratory Report, LA-5426-MS (1973).

6. Rivard, W. C., Butler, T. D. and Farmer, O. A., "Numerical Calculations of Viscous Supersonic Jet Mixing," Los Alamos Scientific Laboratory Report, (to be published in 1974).

7. Rivard, W. C., Butler, T. D. and Farmer, O. A., "RICE: A Computer Program for Multicomponent Chemically Reactive Flows at all Speeds," Los Alamos Scientific Laboratory Report, (to be published in 1974).

† The contour data given correspond to scaled, non-dimensional quantities.

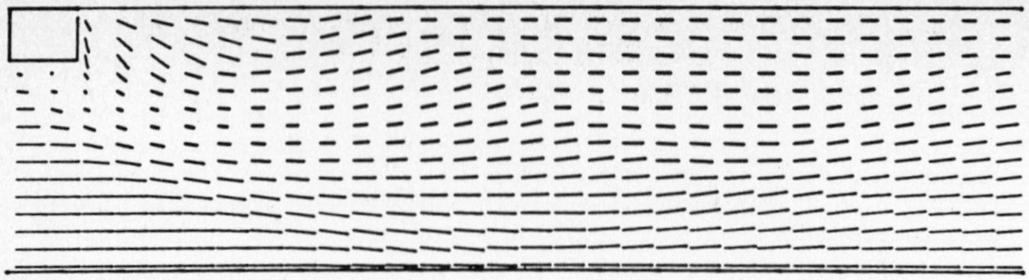

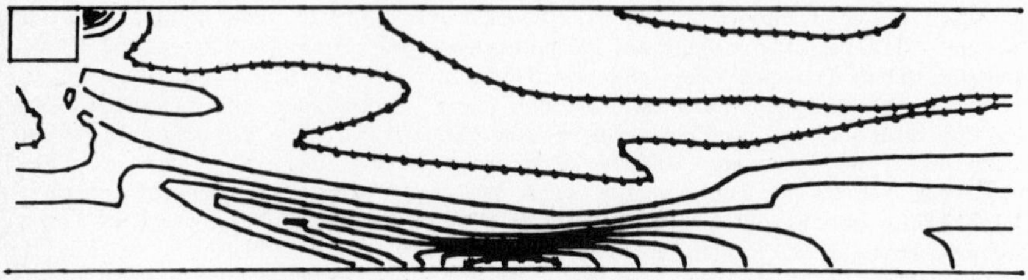

Density ρ, High (H) = 1.097, Low (L) = 0.151, Interval = 0.118

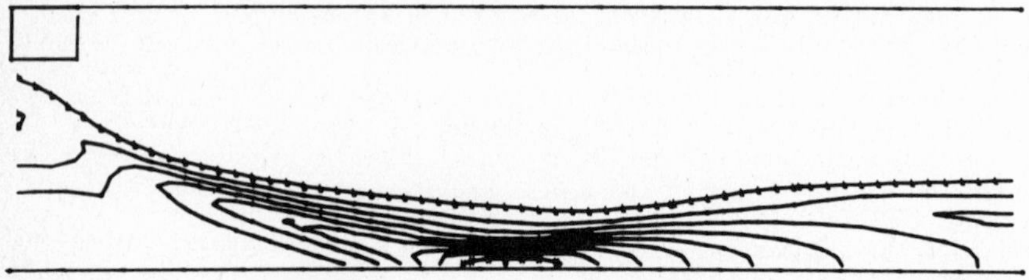

Density of F-atom, High (H) = 0.074, Low (L) = 0.008, Interval = 0.008

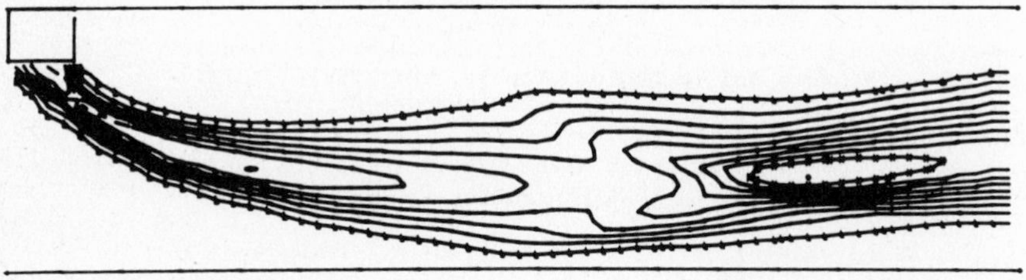

Density of HF^*, High (H) = 0.020, Low (L) = 0.002, Interval = 0.002

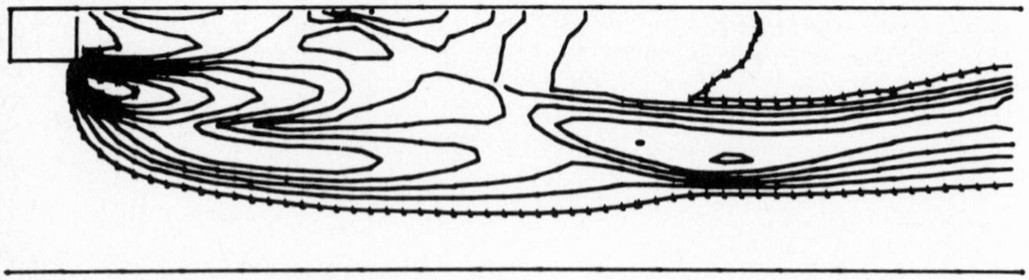

Specific turbulent kinetic energy q, High (H) = 0.025, Low (L) = 0.003, Interval = 0.003

Figure 2. Steady solution from RICE for the velocity field and contours of mixture density ρ, F-atom density, HF^* density, and specific turbulent kinetic energy q.

A GENERALIZED HYPERBOLIC MARCHING TECHNIQUE FOR THREE-DIMENSIONAL SUPERSONIC FLOW WITH SHOCKS

Arthur W. Rizzi, Andrew Klavins,* and Robert W. MacCormack

Computational Fluid Dynamics Branch
Ames Research Center, NASA
Moffett Field, California, 94035

INTRODUCTION

Numerous finite-difference procedures have been developed in recent years to compute supersonic flow fields about relatively simple bodies. These techniques are based on either Eulerian or Lagrangian systems or their combination and take advantage of the hyperbolic character of the governing equations to integrate initial Cauchy data in a coordinate direction along which the local velocity component is everywhere supersonic. Common to all these existing techniques (e.g., Babenko et al. [1964], Thomas et al. [1972], Kutler et al. [1973], and Marconi and Salas [1973]) is the requirement that the initial data lie on a plane normal to the marching direction (usually the body axis) and that this plane advance downstream undistorted. This condition simplifies the coordinate geometry and difference equations, but ignores the orientation of the initial data's cones of influence.† For some flows, e.g., those with large incidence angles, the angles between streamlines of the initial data and the marching direction can be so great that the velocity component in that direction is actually subsonic. In such cases, existing finite-difference methods are inappropriate and a more general procedure is needed that admits initial conditions situated on any arbitrary surface and integrates them forward along a general curvilinear coordinate that nearly conforms to the local streamline, thus distorting and deforming the initial data surface as it advances. This paper introduces such a generalized, nonorthogonal coordinate system and develops finite-difference operators that solve the three-dimensional, steady, inviscid conservation equations of fluid flow referenced to this general frame. The paper also presents a new method of alining the difference mesh to the bow shock wave. This technique eliminates both the necessity of differencing the free-stream flow properties as well as the spurious fluctuations that usually arise in the conservative variables when they are differenced across the discontinuity.

MATHEMATICAL FORMULATION

The nonorthogonal curvilinear system with contravariant coordinates x^i and covariant base vectors $\vec{g}_i$ $(i = 1,2,3)$ is introduced where x^1 and x^2 lie in the initial data surface and x^3 is in the approximate direction of the streamlines‡ (Fig. 1). A reciprocal set of directions is associated with this system and are called contravariant field vectors $\vec{g}^m \equiv \text{grad}\, x^m$ $(m = 1,2,3)$ defined as the gradients of the curvilinear coordinates and related to the base vectors by $\vec{g}^m \cdot \vec{g}_i = \delta^m_i$. Between these coordinates and a rectangular Cartesian system z_m with unit base vectors $\vec{a}_m$ exists the functional relationship

$$x^i = x^i(z_1, z_2, z_3) \qquad i = 1,2,3 \tag{1}$$

The steady, inviscid equations of fluid dynamics written in divergence form with respect to the curvilinear system are

$$\left.\begin{aligned} \text{div}\, \rho\vec{v} &= (1/\sqrt{g})(\partial/\partial x^i)(\sqrt{g}\rho u^i) = 0 \\ \text{div}\, T &= (\vec{a}_m/\sqrt{g})(\partial/\partial x^i)\left\{\sqrt{g}[\rho u^i w_m + p(\partial x^i/\partial z_m)]\right\} = 0 \end{aligned}\right\} \text{ repeated indices are summed} \tag{2}$$

---

*ASEE Summer Faculty Fellow 1972-73.

†While the method of characteristics does, in principle, account for the proper cones of influence, in practical applications of the method to three-dimensional flow, the coordinate mesh has usually been alined with the body axis, rather than with the local flow direction in order to simplify the geometry (see the comprehensive survey of this technique by Chushkin [1968]).

‡To minimize confusion, exponents are not used on the coordinates; superscripts denote only the contravariant components.

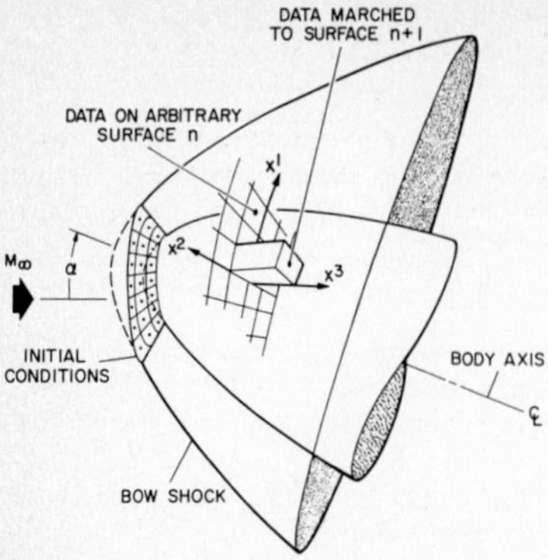

Fig. 1. Generalized coordinates and flow-field discretization.

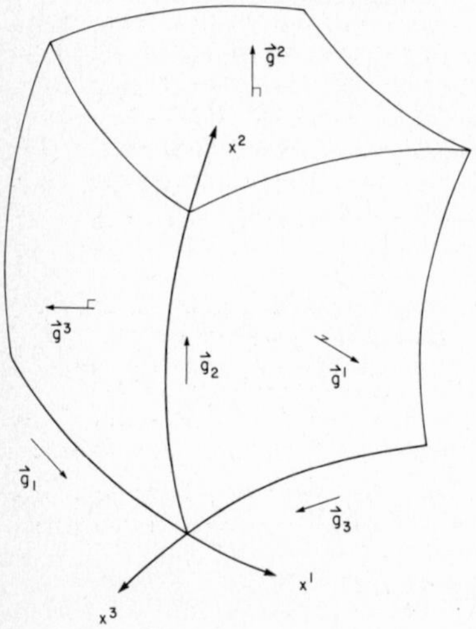

Fig. 2. Formation of typical volume cell by the surfaces x^i = constant.

where T is a second-order tensor, $\sqrt{g}$ is the Jacobian $\partial(z_1,z_2,z_3)/\partial(x^1,x^2,x^3)$, and the flow properties are velocity $\vec{v} = u^i\vec{g}_i = u_\ell\vec{g}^\ell = w_m\vec{a}_m$, density ρ, enthalpy h, and pressure p.§ Equations (2) express the conservation of mass and momentum, and the set is completed by an equation of state $h = h(p,\rho)$ and the steady energy equation $h + (1/2)v^2 = h_{tot}$ = constant. A more physical interpretation of the contravariant coordinates and covariant directions can be achieved by using Green's transformation to cast Eqs. (2) into integral form

$$\left.\begin{aligned} \oiint \rho\vec{v}\cdot\hat{\eta}\,ds &= \sum_m \iint_{S^m} \rho\vec{v}\cdot\vec{g}^m\,ds = 0 \\ \oiint T\cdot\hat{\eta}\,ds &= \sum_m \iint_{S^m} T\cdot\vec{g}^m\,ds = 0 \end{aligned}\right\} \quad (3)$$

where $\vec{S}^m = S\vec{g}^m$ is one of the six surfaces (with normal $\vec{g}^m$) of the hexahedron formed by the coordinate surfaces x^m = constant as illustrated in Fig. 2. Equations (2) and (3) are mathematically equivalent, but each has its particular advantage. The numerical analysis of a method's accuracy and stability is easier to execute with the partial differential form (2), whereas the actual computations are more conveniently carried out in the integral form (3) because use of Eqs. (2) requires calculating either analytically or numerically the Jacobian $\sqrt{g}$ and the term $\partial x^i/\partial z_m$ at each cell, which for practical body shapes may be time consuming and inaccurate. The integral form in effect replaces explicit use of these functions with the products of vector quantities that can be determined numerically at each mesh cell very quickly by simple geometrical expressions, and therein lies its advantage. Although some computational fluid dynamicists have used integral forms in solving flow problems, no one we are aware of has brought out the equivalence of the integral and curvilinear tensor differential forms, probably because of the classical bias toward orthogonal coordinates. However, we emphasize that these two formulations are identical, and consequently the accuracy and stability criteria of a chosen difference scheme when applied to one are exactly the same criteria when applied to the other. The differential form is very useful for establishing the physical and numerical domains of dependence which determines the condition for numerical stability. It is also vital to the analysis of the accuracy of a given numerical procedure.

---

§Note that expressing the tensor T as $T = T^{im}\vec{g}_i\vec{a}_m$, the sum of tensor products of both bases $\vec{g}_i$ and $\vec{a}_m$, results in derivatives of only Cartesian unit vectors. So doing leaves Eqs. (2) in strong conservation form although requiring use of mixed velocity components u^i and w_m (see Vinokur [1974] for further discussion of this point).

NUMERICAL PROCEDURE

Difference Operators

A two-step, dimensionally split difference method of second-order accuracy that is analogous to the familiar MacCormack predictor-corrector scheme and used for numerically solving Eqs. (3) on each computational cell similar to the one displayed in Fig. 2 is

$$\left.\begin{aligned} \tilde{T}^{n+1/2}_{j,k} \cdot \vec{S}^{n+1} &= -T^{n}_{j,k} \cdot \vec{S}^{n} - T^{n}_{j,k} \cdot \vec{S}_{k+1} - T^{n}_{j,k-1} \cdot \vec{S}_{k} \\ T^{n+1/2}_{j,k} \cdot \vec{S}^{n+1} &= (1/2)(-T^{n}_{j,k} \cdot \vec{S}^{n} + \tilde{T}^{n+1/2}_{j,k} \cdot \vec{S}^{n+1} - \tilde{T}^{n+1/2}_{j,k+1} \cdot \vec{S}_{k+1} - \tilde{T}^{n+1/2}_{j,k} \cdot \vec{S}_{k}) \end{aligned}\right\} \quad (4a)$$

$$\left.\begin{aligned} \tilde{T}^{n+1}_{j,k} \cdot \vec{S}^{n+1} &= -T^{n+1/2}_{j,k} \cdot \vec{S}^{n} - T^{n+1/2}_{j,k} \cdot \vec{S}_{j+1} - T^{n+1/2}_{j-1,k} \cdot \vec{S}_{j} \\ T^{n+1}_{j,k} \cdot \vec{S}^{n+1} &= (1/2)(-T^{n+1/2}_{j,k} \cdot \vec{S}^{n} + \tilde{T}^{n+1}_{j,k} \cdot \vec{S}^{n+1} - \tilde{T}^{n+1}_{j+1,k} \cdot \vec{S}_{j+1} - \tilde{T}^{n+1}_{j,k} \cdot \vec{S}_{j}) \end{aligned}\right\} \quad (4b)$$

where now the affixal notation no longer refers to the contravariant and covariant directions but rather to the geometrical location in the difference mesh. The subscripts j and k denote the discrete location of the cell j,k along the axes x^1 and x^2, while $\vec{S}_j$ and $\vec{S}_k$ are the areas of the sides facing these directions. The superscripts denote the position $x^3 = \sum_n \Delta x^3(n)$ to which the solution has been advanced during each cycle of the operators, and $\vec{S}^n$ is the area facing x^3. This set of equations can be written more conveniently in operator notation. Let $L_k(\Delta x^3)$ denote the operation performed by the set of Eqs. (4a) in advancing the solution from $T^n_{j,k}$ to $T^{n+1/2}_{j,k}$, i.e., $T^{n+1/2}_{j,k} = L_k(\Delta x^3)T^n_{j,k}$ and let $L_j(\Delta x^3)$ be similarly defined by Eqs. (4b). Stability conditions can be determined analytically for each operator by a locally linear analysis of Eqs. (2). For L_k

$$\Delta x^3_k \leq \min_{j,k} \left\{ \frac{\Delta x^2}{(\sqrt{g^{22}/g^{33}})[\nu^2\nu^3 - c^2 \cos\sigma + c\sqrt{q^2 - c^2 \sin^2\sigma}]/(\nu^3\nu^3 - c^2)} \right\} \quad (5)$$

where $c = c(\rho,p)$ is the local speed of sound, $\nu^i = \vec{v} \cdot \vec{g}^i/\sqrt{g^{ii}} = u^i/\sqrt{g^{ii}}$ is the component of v along the $\vec{g}^i$ direction, while $q = (\nu^2\nu^2 + \nu^3\nu^3 - 2\nu^2\nu^3 \cos\sigma)^{1/2}$ is the component of $\vec{v}$ lying in the plane defined by $\vec{g}^2$ and $\vec{g}^3$ and $\cos\sigma = \vec{g}^2 \cdot \vec{g}^3/\sqrt{g^{22}g^{33}}$ and $g^{ii} = (\partial x^i/\partial z_m)(\partial x^i/\partial z_m)$ (summed on m only). An analogous relation determining Δx^3_j can also be written for the L_j operator. The symmetric sequence

$$L_k(\Delta x^3)L_j(\Delta x^3)L_j(\Delta x^3)L_k(\Delta x^3) \quad (6)$$

of these operators will then advance the numerical solution from $T^n_{j,k}$ to $T^{n+2}_{j,k}$ with uniformly second-order accuracy in the space variables. This sequence is stable if the necessary (CFL) condition

$$\Delta x^3 \leq \min(\Delta x^3_j, \Delta x^3_k) \quad (7)$$

is satisfied. MacCormack and Paullay [1972] and Rizzi and Inouye [1973] have shown how split operators analogous to these allow larger integration steps than unsplit ones and reduce the required computation time.

Boundary Conditions and Shock Treatment

The mesh is constrained by segmenting the initial data surface into small quadrilaterals in a way that the innermost row (x^2 = constant) lies on the body and the outermost row coincides with the shock surface. This arrangement simplifies specification of the conditions at the outer and inner edges of the overall mesh, referred to as entrance (shock wave) and streamline (body) boundaries. Along the entrance boundary the flow variables are held fixed at their supersonic free-stream values, while across cell faces coincident with a streamline boundary, such as an impervious body, no transport is allowed. The only variable actually specified at such a cell face is the pressure, which can be related to the flow properties in the adjacent interior mesh cells by the component of the momentum equation normal to a streamline.

At the outer edge of the mesh the pressure just downstream of the bow wave specifies the inclination of the shock at that cell. The surface with this inclination is determined from the steady shock relations, and it becomes the outer edge of the mesh for the next step n+1. Thus $\vec{g}^2$ at the outermost row of cells is identical to the outward normal of the shock. The difference operators then conserve mass and momentum across the discontinuity. But this is precisely the condition used to derive the shock relations, so that in effect applying the difference operators across the shock reduces numerically to the analytic jump conditions. This procedure, which we term mesh alining, simplifies the shock-fitting logic and also avoids differencing in the freestream. Furthermore, it averts local errors and inconsistencies in the influx to the outermost cells and eliminates the oscillatory nature of the flow properties near the shock, which is a common shortcoming with conventional shock-capturing techniques using fixed meshes.

The method permits the mesh to be nonequispaced. It constructs an arbitrarily oriented surface (x^3 = constant) at the n+1 step, constrained only by the CFL condition [Eq. (7)] on Δx^3, and segments this into an array of mesh cells from which the difference operators can advance the solution forward. This new surface can be completely general in shape; and for the cases presented here we have chosen it to be circular conical, defined at each step by the location of its apex A, vertex angle θ, and axis orientation angle β, and illustrated in Fig. 3. As the solution proceeds downstream, A advances while θ increases monotonically toward 90° and β approaches zero so that at the final step the surface degenerates into a plane normal to the body axis.

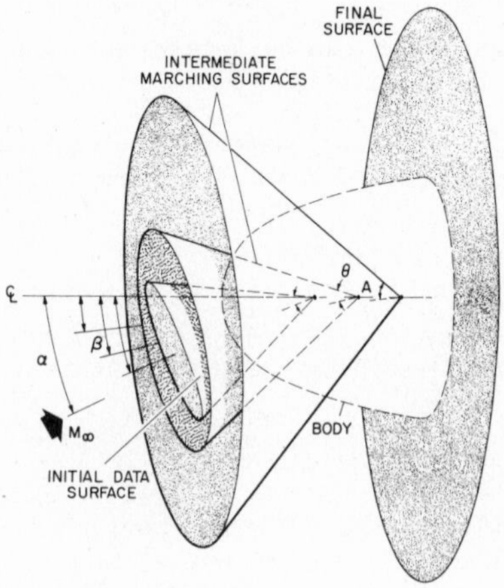

Fig. 3. Illustration of position, orientation, and vertex angle of conical data surfaces for successive steps.

NUMERICAL RESULTS

Computed results are shown in Figs. 4 and 5. They compare well in Fig. 4 with values from experiment and also the method of characteristics for a blunted cone traveling in helium at M_∞ = 14.9 and 20° incidence. The sectional views indicate the presence of a (captured) crossflow shock caused by the turning of the supersonic flow at the leeward plane of symmetry. Figure 5 displays the shock shape for a smooth three-dimensional body, described by a series of third-degree polynomials in plan and profile views and ellipses in cross section, traveling in air at M_∞ = 21.7 and 30° incidence. The embedded shock, identified by an abrupt jump in the pressure, appears at 2.5 R_b and grows steadily downstream. A plot of the pressure through the cross-section shock layer at 5 R_b illustrates the sharp pressure rise at the bow shock and the smooth monotonic variation behind it that is obtained by the meshing alining procedure.

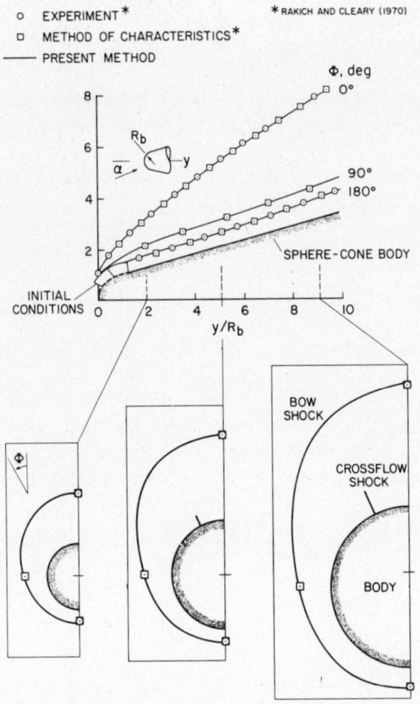

Fig. 4. Bow shock shape for 15° sphere-cone; $\alpha = 20°$, $M_\infty = 14.9$, $\gamma = 5/3$.

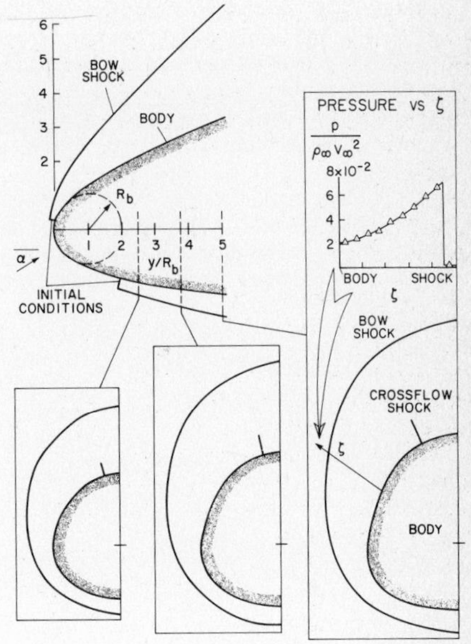

Fig. 5. Bow shock shape and pressure distribution for a three-dimensional body; $\alpha = 30°$, $M_\infty = 21.7$, $p = 7/5$.

CONCLUDING REMARKS

The method presented demonstrates that nonorthogonal curvilinear coordinates can be useful for computing practical fluid problems and in fact are the underlying basis of the integral form of the governing equations. Generalized coordinates allow the solution to advance in an arbitrary direction and at the same time simplify application of the body boundary conditions. Furthermore, this procedure lends itself to a new scheme of alining the mesh with the bow shock, which is simple to implement, accurate, and consistent with the interior flow. For the results presented, the arbitrary surface on which the solution was advanced was conical, although for these two particular cases with a different orientation of the initial conditions a simple planar surface would have sufficed. However, a planar surface would not suffice for flows with sharply curving streamlines, and a closer correspondence between the marching direction and the local streamline would be necessary. One practical example of this is a supersonic stream flowing smoothly past the fuselage of a wing-body configuration and then turning sharply at the leading edge of the wing. Such a flow, however, could be computed by using a less restrictive surface. Another very interesting endeavor would be to use a surface more general than conical, undoubtedly piecewise fitted, and attempt to advance each individual mesh cell downstream at its own local maximum CFL condition.

REFERENCES

Babenko, K. I., Voskresenskii, G. P., Lyubimov, A. N., and Rusanov, V. V. Prostranstvennoye obtekaniye gladkikh tel ideal'nym gazom. Nauka, Moscow (1964), Engl. Trans. Three-Dimensional Flow of Ideal Gas Past Smooth Bodies. NASA TT F-380 (1966).

Chushkin, P. I. Numerical Method of Characteristics for Three-Dimensional Supersonic Flows, in Progress in Aeronautical Sciences (ed. D. Küchemann), vol. 9, Pergamon Press, Oxford (1968).

Kutler, P., Reinhardt, W. A., and Warming, R. F. Multishocked, Three-Dimensional Supersonic Flowfields with Real Gas Effects. AIAA J., vol. 11, no. 5, 657-664 (1973).

MacCormack, R. W., and Paullay, A. J. Computational Efficiency Achieved by Time Splitting of Finite-Difference Operators. AIAA Paper No. 72-154 (1972).

Marconi, F., and Salas, M. Computation of Three-Dimensional Flows About Aircraft Configurations. Computers & Fluids, vol. 1, 185-195 (1973).

Rakich, J. V., and Cleary, J. W. Theoretical and Experimental Study of Supersonic Steady Flow Around Inclined Bodies of Revolution. AIAA J., vol. 8, no. 3, 511-518 (1970).

Rizzi, A. W., and Inouye, M. Time-Split Finite-Volume Method for Three-Dimensional Blunt-Body Flow. AIAA J., vol. 11, no. 11, 1478-1485 (1973).

Thomas, P. D., Vinokur, M., Bastianon, R., and Conti, R. J. Numerical Solution for Three-Dimensional Inviscid Supersonic Flow. AIAA J., vol. 10, no. 7, 887-894 (1972).

Vinokur, M. Conservation Equations of Gasdynamics in Curvilinear Coordinate Systems. J. Comp. Phys., vol. 14, no. 2, 105-125 (1974).

THE SPLIT NOS AND BID METHODS FOR THE STEADY-STATE NAVIER-STOKES EQUATIONS

by
Patrick J. Roache*
Aerodynamics Research Department
Sandia Laboratories
Albuquerque, New Mexico 87116

Introduction

In Ref. 1, we presented two methods, the LAD and NOS methods, for solving the steady-state Navier-Stokes equations in finite-difference forms. These methods are non-conventional, being neither time-dependent nor even time-like in their iterations. They are based on recent advances in solving linear equations by direct (non-iterative) methods. Some improvement over time-dependent methods was demonstrated. In the present paper (and in the forthcoming Refs. 2 and 3), we present two new methods which are further improvements over the LAD and NOS methods, and over any time-dependent methods.

The vorticity transport equation is

$$\zeta_t = - \mathrm{Re}\nabla \cdot \vec{V}\zeta + \nabla^2\zeta \tag{1}$$

where ζ is vorticity, $\vec{V}$ is velocity, and Re is Reynolds number. This is solved iteratively with the Poisson equation for stream function ψ,

$$\nabla^2\psi = \zeta \tag{2}$$

and the boundary conditions which couple ζ on the boundary with internal values of ψ. The NOS method was defined as

$$\nabla^2\zeta^k - \mathrm{Re}\nabla \cdot (\vec{V}^{k-1}\zeta^k) = 0 \tag{3}$$

The usual second-order centered differences are used throughout. The nonlinear velocities are lagged in the iteration, and the boundary values of ζ are under-relaxed by a factor r, in the usual way. This method requires a direct method for second-order linear equations which is general, i.e., it allows first-order terms (Ref. 4). The LAD method was defined by

$$\nabla^2\zeta^k = \mathrm{Re}\nabla \cdot (\vec{V}^{k-1}\zeta^{k-1}) \tag{4}$$

This method requires only the more generally available fast Poisson solvers. The NOS method converges much faster than LAD for flow-through problems typified by the developing channel-flow problem, but it unfortunately requires the re-initialization of the linear solver at each iteration, which is time-consuming.

---

* Now Staff Scientist, Science Applications, Inc., Albuquerque, New Mexico 87108.

Split NOS Method

The Split NOS method combines some virtues of both the LAD and NOS methods. Defining the initial guess at the velocity field as $\vec{V}^0$, we write the total velocity V as $\vec{V} = \vec{V}^0 + \vec{V}'$. Then the Split NOS method is

$$\nabla^2 \zeta^k - \text{Re}\nabla \cdot (\vec{V}^0 \zeta^k) = \text{Re}\nabla \cdot (\vec{V}'^{k-1} \zeta^{k-1}) \qquad (5)$$

The linear operator need not be re-initialized at each iterative step, yet it has some of the advection information. The performance is equal to the NOS method at low Re, and is actually an improvement at high Re, as shown in Figure 1. The problem of developing channel flow was computed in as few as 11 iterations, each of which requires computer time comparable to a single time-step using an explicit method.

For this Split NOS method, the linear solver must be re-initialized only when Re is changed. It is also possible to improve the guess on $\vec{V}^0$ by re-initializing the linear solver, once or periodically; if done at each iteration, the Split NOS reverts to the simple NOS.

BID Method

For recirculating flows as typified by the driven cavity problem, the convergence rates of the LAD, NOS and Split NOS are approximately the same; they are better than the FTCS method (Refs. 1, 2), representative of time-dependent methods, but are not as good as anticipated. Numerical experiments (Refs. 1, 2) demonstrated that the lagging boundary values of vorticity were responsible for this less-than-anticipated performance. In Refs. 1 and 2, we suggested that the biharmonic-driver method, or BID method, might overcome this difficulty. Combining the vorticity transport and Poisson equations gives

$$\frac{\partial}{\partial t}(\nabla^2 \psi) = - \text{Re}\nabla \cdot (\vec{V}\nabla^2 \psi) + \nabla^2(\nabla^2 \psi) \qquad (6)$$

The BID method is then defined as

$$\nabla^4 \psi^k = \text{Re}\nabla \cdot (\vec{V}^{k-1} \nabla^2 \psi^{k-1}) \qquad (7)$$

Figure 2-a is a schematic of the usual second-order 13-point biharmonic operator for $\Delta x = \Delta y$ at interior points; Figure 2-b shows the modification required near a corner. Near the moving "lid" of the cavity, an additional term containing the lid speed appears only in the right member of equation (7).

Since the steady-state boundary conditions on ψ are known and need not be iterated (or under-relaxed), the BID method was expected to converge rapidly. Using direct biharmonic solvers (Refs. 5, 6), we have now verified this expectation for the driven cavity problem. The results are shown in Figure 3. Unequivocal convergence is obtained in 8 iterations at Re = 20, and 6 iterations at Re = 10, for mesh sizes from $\Delta = 1/10$ to $1/100$. Note that the steady-state solution for Re = 0 is attained at the first iteration (although a second iteration is required to mechanically verify the convergence criterion), demonstrating clearly that the method is not at all time-like.

Although the BID, LAD and NOS methods are all second-order accurate, the answers are not identical because the BID method is based on a fourth-order partial differential equation while the others use two second-order equations. Comparisons (Ref. 3) indicate that the two-equation approach (LAD, NOS) is more accurate for the same mesh size. However, the BID method does not require the determination of an under-relaxation factor r, and its iterative convergence rate is entirely insensitive to mesh size; for recirculating flow problems, it appears to be preferable. For flow-through problems at moderate to high Re, the Split NOS method is clearly preferable to the NOS method.

Extensions and Comparisons

The LAD and BID methods, which contain no advection information in the linear operator, both fail to converge for high Re flow-through problems (channel flow), while the NOS and Split NOS converge rapidly for this problem. (The NOS methods are analogous to an analytical solution by perturbation methods using a large Re expansion, while the LAD and BID methods are analogous to a small Re expansion.) This obviously suggests the consideration of methods based on a fourth-order operator which includes advection terms.

The fourth-order method analogous to the (simple) NOS method would be the FOD (fourth-order driver) method

$$\nabla^4\psi^{k+1} - Re\nabla\cdot(\vec{V}^k\nabla^2\psi^{k+1}) = 0 \tag{8}$$

Similarly, a method analogous to Split NOS would be the Split FOD method

$$\nabla^4\psi^{k+1} - Re\nabla\cdot(\vec{V}^0\nabla^2\psi^{k+1}) = Re\nabla\cdot(\vec{V}'^k\nabla^2\psi^k) \tag{9}$$

By analogy with the results for the LAD and NOS methods, we infer that these methods would probably converge in fewer iterations than BID, and would probably accomplish the high Re iterative convergence for flow-through problems. For the FOD method, equation (8), each iteration would require a new initialization of the linear solver, so that this method would certainly be more time-consuming than the BID method. However, the Split FOD method, equation (9), would not require initialization at each iteration but only for Re changes, and would almost surely be an excellent method for flow-through problems. These FOD methods require the use of a general block-5 linear solver. The Buzbee-Dorr biharmonic solver (Ref. 6) is not adaptable to the general block-5 problem, but the Bauer-Reiss solver (Ref. 5) is. The details of treating inflow and outflow boundaries using the biharmonic approach may be found in Ref. 3.

The Split NOS approach appears to be adaptable to 3D and to the primitive (u-v-P) equations, and also to compressible flows. Originally (Refs. 1, 2) we were pessimistic about using this approach for compressible flow because of fundamental difficulties with solving a steady-state equation for mass density ρ without using a time-like approach. Rather than attempt the solution for ρ directly, we are now of the opinion that the compressible flow problem can be solved with this type of steady-state approach, by solving for the pressure rather than the density. As in the incompressible flow problem in primitive variables, a Poisson equation for pressure would be formulated by higher differentiation of the momentum equations. The continuity equation would be used to simplify the terms, but the source term would still be more complex than the incompressible term. Solutions for subsonic compressible viscous flows are expected without difficulty; for supersonic flows, the linear operator in the pressure equation may require modification to account for the limited domain of influence in the inviscid limit. We are presently unsure of the resolution of this problem for supersonic flow.

Morihara and Cheng (Ref. 7) have presented a method for solving the steady-state Navier-Stokes equations which is not time-like. Like the present methods, theirs is based on solving a sequence of linear problems. As with the NOS method, each iteration gives the solution of a steady-state linearized problem. They use a form of the Navier-Stokes equations in which pressure has been eliminated and the velocity terms appear up to third-order partial derivatives. The iterative convergence is aided by a quasilinearization treatment of the advective terms. Direct comparisons of the iteration convergence with the present methods are difficult, because Morihara and Cheng used a coordinate transformation to study a developing channel flow problem; they also used a different convergence criterion in different variables, and sometimes used initial conditions from coarse-mesh solutions. By "stacking" Re cases so that lower Re solutions were used as initial conditions, they obtained convergence in 3 or 4 iterations. Using BID on stacked Re runs for the driven cavity problem, we obtained our convergence in 6 or 7 iterations.

It would appear that the method of Morihara and Cheng may give even faster iterative convergence than the present BID method, but the advantage would probably not persist in computer time for large problems. Like the NOS method or the proposed FOD method, their method requires the re-initialization of a Gaussian elimination routine at each iteration. (The use of their equations involving third-order partial derivatives also increases the computing time and the algebraic complexity.) For large problems, this substantially increases the computer time, and also is likely to limit the mesh size of the problem due to round-off error considerations.

The real advantage of the Morihara-Cheng method over BID would appear to be in the use of the velocity variables, which may give faster truncation-error convergence. Our experiments (Refs. 1, 2) with the NOS and Split NOS methods would imply that the excellent iteration-convergence results of the Morihara-Cheng method are due as much to the use of the velocity variables with known steady-state boundary conditions as to the use of quasilinearization. The obvious method to try would be the combination of the Morihara-Cheng method with the basic linearization idea of the Split NOS method. This method would use velocity variables, giving known steady-state boundary conditions to speed iteration convergence and possibly better truncation-error convergence, and yet the linear algebra routine would be initiated only once for each Re case, thus offsetting the increase in number of iterations by a decrease in computer time.

The basic idea in all these methods for the steady-state Navier-Stokes equations is to use direct (non-iterative) linear solvers at each iteration for the nonlinear equations; the same idea has now been used by Martin and Lomax in a series of papers (e.g., Ref. 8) to solve inviscid subsonic and transonic problems, with great success.

References

1. Roache, P. J., "Finite-Difference Methods for the Steady-State Navier-Stokes Equations," Sandia Laboratories, Albuquerque, N. M., SC-RR-72 0419, December 1972. See also, Proc. Third International Conference on Numerical Methods in Fluid Dynamics, Vol. I, Paris, France, July 1972, H. Cabannes and R. Temam, Eds., Springer-Verlag, Berlin, 1973.

2. Roache, P. J., "The LAD, NOS and Split NOS Methods for the Steady-State Navier-Stokes Equations," to appear in Computers and Fluids, 1974.

3. Roache, P. J. and Ellis, M. A., "The BID Method for the Steady-State Navier-Stokes Equations," submitted for publication to Computers and Fluids.

4. Roache, P. J., "A Direct Method for the Discretized Poisson Equation," SC-RR-70-579, Sandia Laboratories, Albuquerque, N. M., February 1971. See also, Proc. Second International Conference on Numerical Methods in Fluid Dynamics, Berkeley, Calif., September 1970, M. Holt, Ed., in Lecture Notes in Physics Series, Vol. 8, Springer-Verlag, Berlin, 1970.

5. Bauer, L. and Reiss, E. L., "Block Five Diagonal Matrices and the Fast Numerical Solution of the Biharmonic Equation," Mathematics of Computation, Vol. 26, No. 118, April 1972, pp. 311-326.

6. Buzbee, B. L. and Dorr, F. W., "The Direct Solution of the Biharmonic Equation on Rectangular Regions and the Poisson Equation on Irregular Regions," LA-UR-73-636, Los Alamos Scientific Laboratories, N. M., 1973. Also, submitted for publication to SIAM J. on Numerical Analysis.

7. Morihara, H. and Cheng. R. T., "Numerical Solution of the Viscous Flow in the Entrance Region of Parallel Plates," J. Computational Physics, Vol. 11, 1973, pp. 550-572.

8. Martin, E. D. and Lomax, H., "Rapid Finite-Difference Computation of Subsonic and Transonic Aerodynamic Flows," AIAA Paper No. 74-11, presented at AIAA 12th Aerospace Sciences Meeting, January 30-February 1, 1974, Washington, D. C.

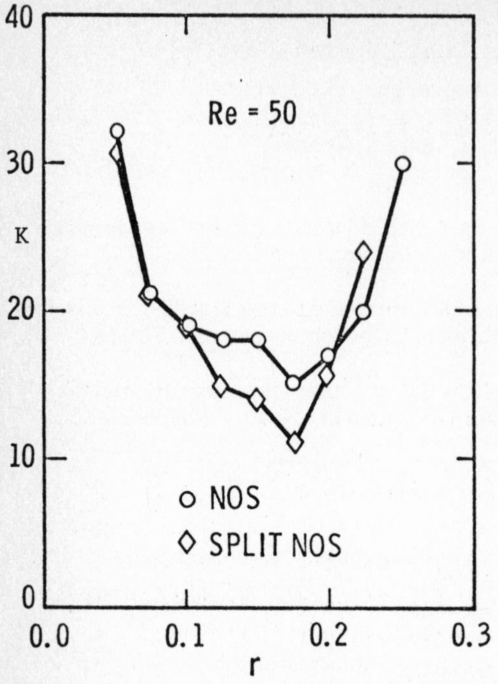

Figure 1. Convergence Rate for Split NOS Method. Channel flow problem, $\Delta y=1/10$, $\Delta x=1$, Re=50, Woods' second-order equation for wall vorticity, ζ. K is the number of iterations to convergence. The abscissa r is the under-relaxation factor for wall ζ.

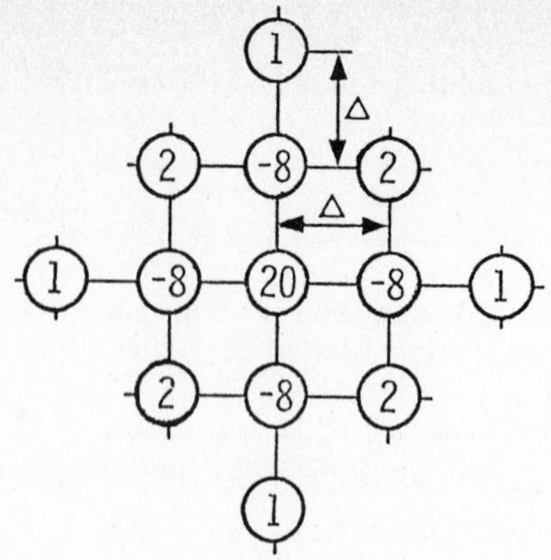

a. operator for interior points

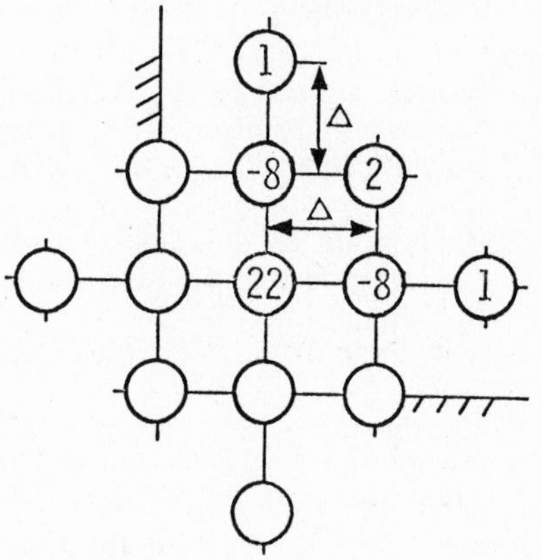

b. operator adjacent to the lower left corner

Figure 2. Finite Difference Biharmonic Operators for BID Method with $\Delta x = \Delta y$.

Re	K	Δ
0	2(1)	1/10-1/100
10	6	"
20	8	"
50	14-18	"

Figure 3. Convergence Rate for BID Method. Driven cavity problem. K is the number of iterations to convergence. No underrelaxation is required for the BID method.

SOME PROPERTIES OF THE AXISYMMETRIC GAS FLOW ABOUT THE POWER-SHAPE BODIES

V.V. Rusanov

Institute of Applied Mathematics,
Moscow, U S S R

I. Introduction

There are plenty of papers treating the flow about power- shape bodies [I]. The reason is that in the Newtonian theory [2] such bodies have a number of optimal properties. Thus, if the equation of body generatrix is $r = f(z)$, then among the bodies of revolution with infinite aspect ratio the power-shape body $r = z^n$ has the minimum wave drag at $n = n^* = 0.75$.

The Newtonian theory, as well as its modifications [3] , is rather a rough approximation to the real flow. More accurate approximation is given by the hypersonic theory, based on the analogy with the non-steady flow around the cylindrical piston expanding with the velocity $r = t^n$. The approximation obtained by this theory is sufficiently good for $M_\infty \to \infty$ and in this case $n^* \approx 0.7$.

But for the finite M_∞ the hypersonic theory gives the parameters of the flow which differ considerably from the real ones. The determination of n^* in this case is an important problem having both theoretical and practical significance.

Bearing that in mind, the accurate computations have been made of the flow about power-shape bodies placed in the supersonic stream. The results obtained give the possibility to evaluate the difference between the accurate and approximate solution as well as to compute the function n^* more exactly.

All the computations have been made by the finite-difference method described in 4 .

2. The Notations

$r = f(z) = z^n$, $0 < n < 1$ - the equation of the body generatrix.

$r_b(z)$ and $r_w(z)$ - the radii of the body and the shock wave at some co-ordinate z .

$\kappa = 1.4$ - the ratio of specific heats.

$p_b(z)$ and $p_w(z)$ - the pressure on the body and after the shock wave.

$C_x(z)=4(\rho_\infty U_\infty^2 z^{2n})^{-1}\int_0^z(p_b-p_\infty)f(z)f'(z)\,dz$ - the wave drag coefficient, where p_∞ is the pressure of the undisturbed stream. It is convenient to use the aspect ratio $\lambda=z/r=z^{1-n}$ instead of co-ordinate z for the comparison between the flows about different bodies.

3. The Flow about the Nose Part of the Body

In fig. I the shape of the shock wave, sonic line and characteristics are plotted in the meridional plane for $n=0.75$, $M_\infty=10$. Such pattern is typical for the values of $M_\infty \geq 3$ and $0.5<n<1$. If n increases the sonic point on the body surface moves from its apex and sonic line approaches more closely the body surface. In accordance with the terminology of paper [5] the transonic region has type III.

As n or M_∞ decreases the sonic point moves to the body apex and the type of the transonic region changes to II. The typical pattern of the sonic line and characteristics in these cases are represented in fig. 2 for $n=0.125$, $M_\infty=20$. For all $0<n<0.5$ the sonic point on the body surface is situated near the point of the maximal curvature of the generatrix, but there is some displacement in the direction of the axis, which increases with the M_∞ increasing.

4. The Shape of the Shock Wave

The shape of the shock wave may be described by the ratio r_b/r_w as function of λ. In the hypersonic theory this function is constant and r_b/r_w depends on the n only. In fig. 3 the ratio r_b/r_w is plotted as function of $\ln\lambda$ for several values of n and M_∞. As $\ln\lambda\to\infty$ the ratio r_b/r_w for $M_\infty=\infty$ approximates the value given by the hypersonic theory. The curves for $M_\infty\neq\infty$ have essentially different shapes. In these cases the ratio r_b/r_w is near to its hypersonic theory value only for large M_∞ and not very large λ.

It is known that if $n\leq 0.5$ the semisimilar solution of the hypersonic equations does not exist. Because of that we have only the numerical solution to obtain the flow about body with $n\leq 0.5$. The computations show that for $n<0.5$ at $M_\infty=\infty$ the shape of the shock wave is asymptotically represented by the formula $r_w\cong C_0\sqrt{z}$ where C_0 is the function of n.

For $n=0.5$, i.e. for the paraboloid, the formula contains the logarithmic term: $r_w\cong(C_0+C_1\ln z)\sqrt{z}$. So for the $n\leq 0.5$ and $M_\infty=\infty$ $\lim(r_b/r_w)=0$ as $\lambda\to\infty$.

5. The Pressure Distribution on the Body and the Wave Drag Coefficient

In fig. 4 the ratio ρ_b/ρ_w is plotted as a function of $\ln\lambda$ for the same n and M_∞ as in fig. 3. As well as r_b/r_w the ratio ρ_b/ρ_w is constant in the hypersonic theory and depends on n only.

If $M_\infty=\infty$ and $\lambda\to\infty$ then the ratio ρ_b/ρ_w approaches the value given by the hypersonic theory.

For $M_\infty\neq\infty$ $\rho_b/\rho_w\to 1$ as $\lambda\to\infty$. The shape of the curves ρ_b/ρ_w depends weakly on n and much more depends on M_∞. The ratio ρ_b/ρ_w is near to its hypersonic theory value only for large M_∞ and not very large λ.

One of the important properties of the flow about body is the wave drag coefficient C_x. In the hypersonic theory the value $C_x\lambda^2=const$ and depends on n only. In fig. 5 the values of $C_x\lambda^2$ are plotted as functions of $\ln\lambda$ for some n and M_∞. The values of C_x have been computed using the pressure distribution along the body surface obtained in numerical solution*).

As before, the difference from the hypersonic theory is rather significant especially at small M_∞. On the other hand the diagrams confirm the well-known fact, that the deviation from the hypersonic theory is the least at large M_∞ and at not very large λ.

However, at these values of M_∞ and λ the value of deviation is sufficient to make the analysis of the accurate numerical results appropriate for practical purposes.

The value of $n^*(\lambda,M_\infty)$ is especially interesting, i.e. the value of n which gives the minimum $C_x(n,\lambda,M_\infty)$ at fixed λ and M_∞. In fig. 6 the value of n^* is plotted as function of $\ln\lambda$ for different M_∞. One can see that the optimal value n^* may differ considerably from both value 0.75 of the Newtonian theory and from value $\approx$0.70 of the hypersonic theory.

---

*) The detailed tables of pressure distribution for $0.5\leq n\leq 0.75$ and $2\leq M_\infty\leq\infty$ are contained in [6].

REFERENCES

I. Guiraud, J.P., a.o. Bluntness effects in hypersonic small disturbance theory. In: "Basic developments in fluid dynamics". V. I. Holt (ed.). N.Y.-London, (I965).

2. Hayes, W.D., Probstein. Hypersonic flow theory. N.Y.-London, Ac. Press, (I966).

3. Miele, A. Theory of optimum aerodynamic shapes. N.Y.-London, Ac. Press, (I965).

4. Rusanov, V.V., Liubimov, A.N. The flows about blunt bodies, Moscow, "Nauka", (I970) (in Russian).

5. Rusanov, V.V. Some properties of the transonic gas flow in the vicinity of the sonic surface. Preprint N I9, Inst.Appl.Math., (I973) (in Russian).

6. Rusanov, V.V., Nazhestkina, E.N. The wave drag of the power-shape bodies of revolution (axisymmetrical flow). Preprint N 33, Inst.Appl.Math., (I972) (in Russian).

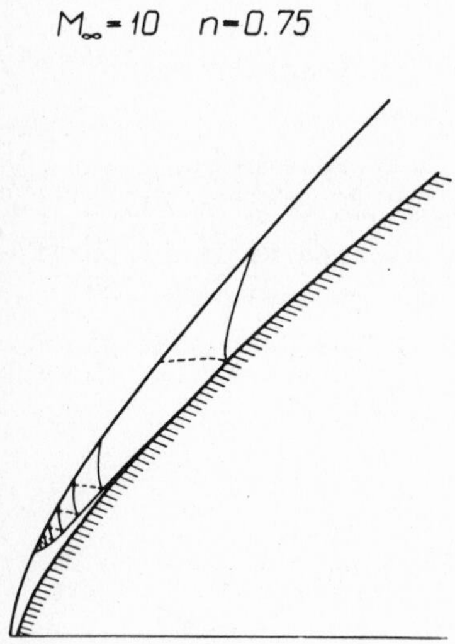

Fig. 1

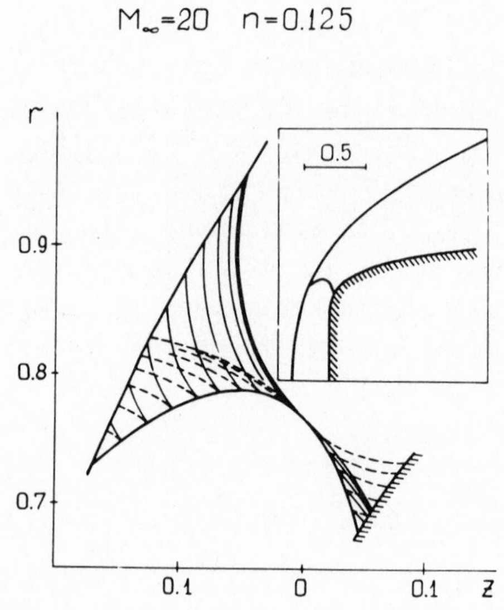

Fig. 2

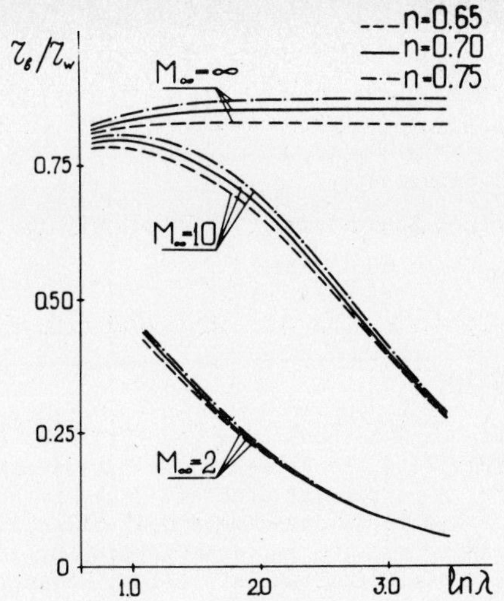

Fig. 3

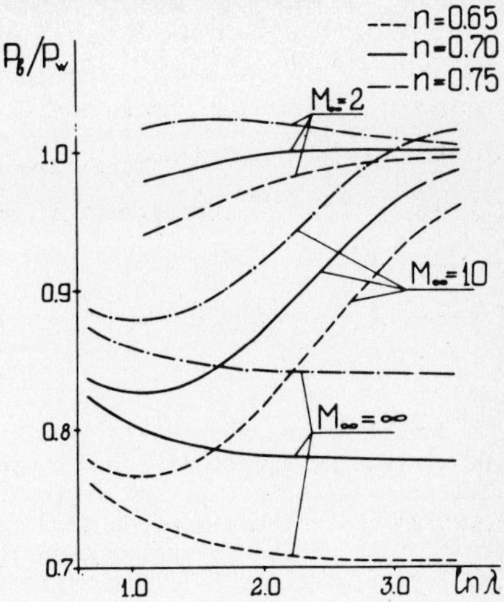

Fig. 4

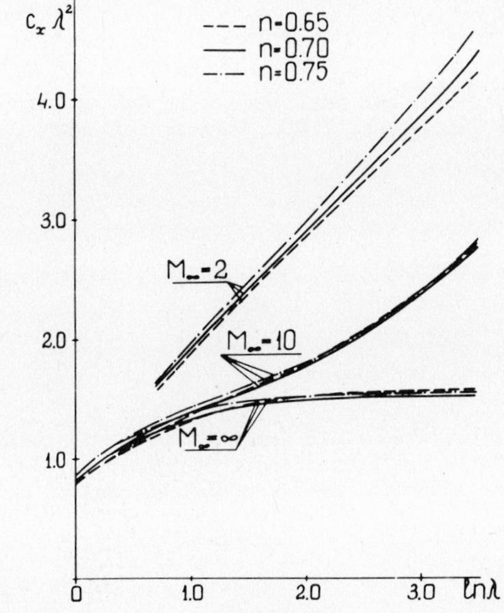

Fig. 5

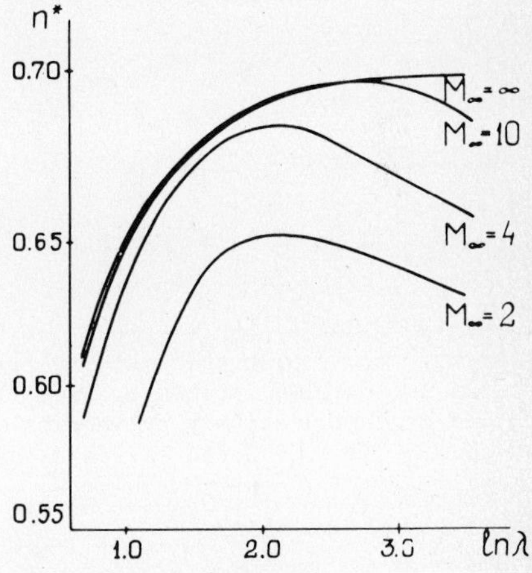

Fig. 6

A STUDY OF DIFFERENCE METHODS FOR RADIALLY SYMMETRIC SHOCKED FLOWS

Nobuyuki Satofuka

Kyoto Technical University
Department of Mechanical Engineering
Matsugasaki, Sakyo-ku, Kyoto 606, JAPAN

1. INTRODUCTION

In recent years, a number of finite difference methods have been proposed for solving systems of equations for a compressible flow. As a result, it has become a difficult task to select an optimum method for a particular problem. Most of the attempts to evaluate these methods were conducted only in one-dimensional plane shock problems for which the governing equations can be written in conservation form[Emery (1968), Taylor et al(1972)]. For cylindrical or spherical shock problem, however, the equations cannot be put into conservation form. To the author's knowledge, any attempt to evaluate these methods has not yet been performed for this case.

In this paper, the second order two step difference methods of Richtmyer(1963), Burstein(1967) and MacCormack(1969) are tested for the imploding shock problem previously solved by Payne(1957) and several attempts by the author to extend these standard schemes to the case of cylindrical or spherical symmetry are presented.

2. BASIC EQUATIONS AND TEST PROBLEM

2.1 The Equations for Radially Symmetric Flow. The Eulerian equations for the time-dependent one-dimensional flow of an inviscid, non heat conducting gas with constant specific heats ratio can be written in a quasi-conservation form as follows:

$$W_t + F_r + Z = 0 \tag{1}$$

where

$$W = \begin{pmatrix} \rho \\ m \\ E \end{pmatrix}, \quad F = \begin{pmatrix} m \\ m^2/\rho + p \\ (E+p)m/\rho \end{pmatrix}, \quad Z = (\alpha-1)\begin{pmatrix} m/r \\ m^2/\rho r \\ (E+p)m/\rho r \end{pmatrix} \tag{2}$$

in which ρ, m, E, p, t and r are respectively the fluid density, momentum, total energy per unit volume, pressure, time and the space coordinate of symmetry. Thus when α=2,3 we have respectively cylindrical or spherical symmetry. In the case of an ideal gas, specific energy E can be stated as

$$E = \frac{p}{\gamma - 1} + \frac{m^2}{2\rho} \tag{3}$$

where γ is the ratio of the specific heats. Non-dimensionalizing the system (2) by the initial values of the flow variables in the inner region, the resulting non-dimensional system is identical to the system (2) exept the pressure terms which are to be multiplied by a factor of $1/\gamma$. Hereafter these non-dimensional equations will be used.

2.2 The Converging Shock Problem.

The converging cylindrical shock problem previously solved by Payne(1957) was selected as a test problem. It constitutes a stringent test for the finite difference methods, since a shock, a contact surface and a rarefaction wave must all be represented.

Initially, a cylindrical diaphragm of radius r_0 separates the two uniform regions of a gas at rest as in a shock tube with outer pressure and density being higher than inner one. After the diaphragm rupture at t=0, a shock wave is created and travels into the low pressure region followed by a contact surface, while an expansion fan travels into the high pressure region. An illustrative sketch of the flow pattern is shown in Fig. 1.

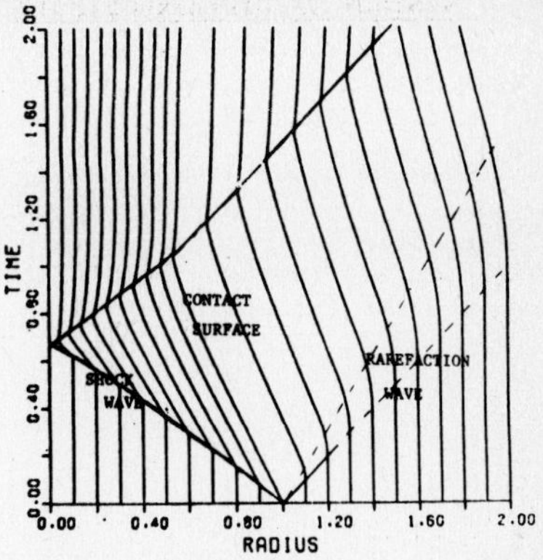

Fig. 1 Schematic of flow pattern

3. NUMERICAL METHODS AND RESULTS

3.1 Basic Finite Difference Methods.

In the case of $\alpha=1$ or plane symmetry, there exist a number of well established difference methods integrating the equation (1). Among these methods, the two step schemes of Richtmyer(1963), Burstein(1967) and MacCormack(1969) are most extensively applied. In these schemes, however, for $\alpha=2,3$ method of differencing the undifferenciated term Z can be uncertain. Therefore, five difference methods listed in Table 1 were applied to the test problem with the diaphragm pressure ratio $p^*=4$ and 100. The results for all five schemes were essentially the same. It should be noted that in contrast to the plane shock problem, the scheme 1 produces less overshoot and oscillations behind the shock than the scheme 2 or 3.

TABLE 1 Basic difference methods

Scheme	1st step	Differencing methods for equation (1)
1	Richtmyer	$W_i^{n+1}=W_i^n - \lambda(F_{i+1/2}^{n+1/2} - F_{i-1/2}^{n+1/2}) - \Delta t Z_i^{n+1/2}$
1A	"	$W_i^{n+1}=W_i^n - \lambda(F_{i+1/2}^{n+1/2} - F_{i-1/2}^{n+1/2}) - \Delta t(Z_{i+1/2}^{n+1/2} + Z_{i-1/2}^{n+1/2})/2$
2	Burstein	$W_i^{n+1}=W_i^n - \lambda[(F_{i+1}^n - F_{i-1}^n)/2 + \tilde{F}_{i+1/2}^{n+1} - \tilde{F}_{i-1/2}^{n+1}] - \Delta t(Z_i^n + \tilde{Z}_i^{n+1})/2$
2A	"	$W_i^{n+1}=W_i^n-\lambda[(F_{i+1}^n-F_{i-1}^n)/2+\tilde{F}_{i+1/2}^{n+1}-\tilde{F}_{i-1/2}^{n+1}]-\Delta t[Z_i^n+(\tilde{Z}_{i+1/2}^{n+1}+\tilde{Z}_{i-1/2}^{n+1})/2]/2$
3	MacCormack	$W_i^{n+1}=[W_i^n + \tilde{W}_i^{n+1} - \lambda(\tilde{F}_i^{n+1} - \tilde{F}_{i-1}^{n+1}) - \Delta t\tilde{Z}_i^{n+1}]/2$

* $\lambda=\Delta t/\Delta x$

3.2 Coordinate Transformation. In such a problem as converging shock wave, resolution of the flow fields near the center is of primary importance. Then it is convenient to transform the radial coordinate r to a uniform grid in x so that its image in r is most dense in the neighborhood of the center. The relation between x and r is defined by

$$x = \frac{(1 + H/R)r}{1 + r} \qquad (0 \leq r \leq R) \tag{4}$$

where H is a parameter controlling the non-uniformity of mesh spacing. The transformed equations and the components of W, F and Z are

$$W_t + F_x + Z = 0,$$

$$W = \begin{pmatrix} \rho \\ m \\ E \end{pmatrix}, \quad F = \begin{pmatrix} ym \\ y(m^2/\rho + p) \\ y(E+p)m/\rho \end{pmatrix}, \quad Z = \begin{pmatrix} [(\alpha-1)/r + \eta]m \\ (\alpha-1)m^2/\rho r + \eta(m^2/\rho + p) \\ [(\alpha-1)/r + \eta](E + p) \end{pmatrix} \tag{5}$$

where $y = dx/dr$ and $\eta = -dy/dx$. The transformed systems (5) were solved using scheme 1A.

The results of the density profiles at non-dimensional time t=0.6 for the case of p*=4 calculated with three different values of H(H=0.5, 1.0 and 2.0) are illustrated in Fig. 2. Changes of the mesh spacing with H can be clearly seen in the figure. The profile obtained with H=0.5 shows large overshoot and oscillations at the shock, while those with H=1.0 and 2.0 exhibit much smaller overshoot. Except in the neighborhood of the shock, all three values of H give practically the same profile.

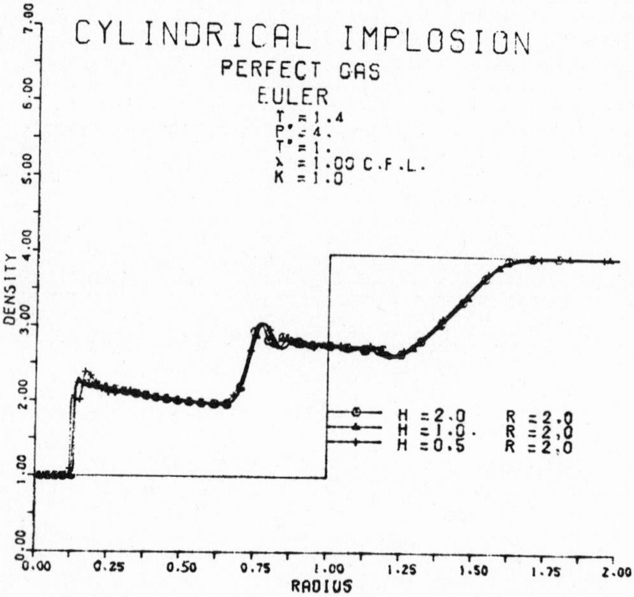

Fig. 2 Density profiles for p*=4 calculated using non-uniform mesh

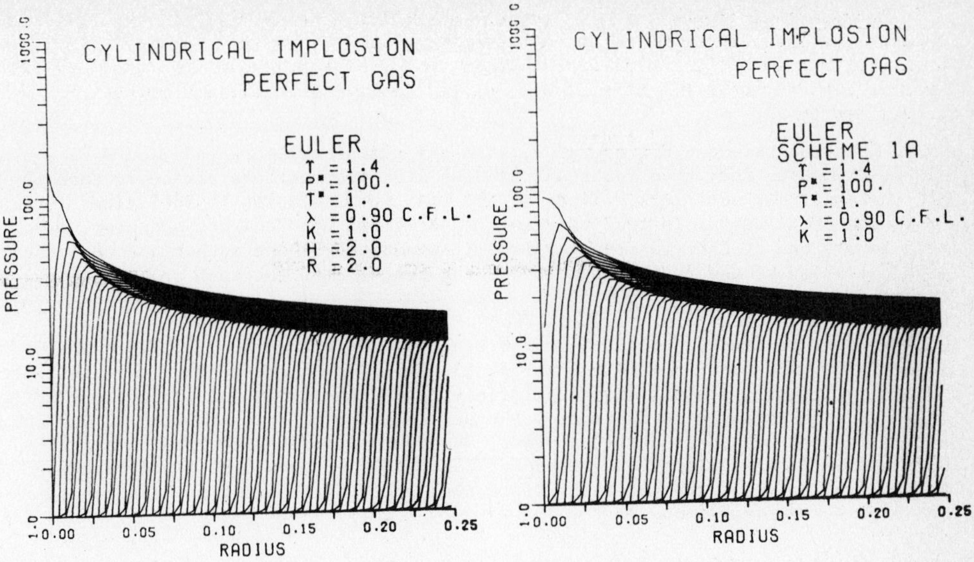

(a) Non-uniform mesh, H=2.0 (b) Uniform mesh

Fig. 3 Comparison of pressure profiles in the neighborhood of the center

For p*=100, pressure profiles in the neighborhood of the center are presented in Fig. 3 in which frame (a) is of non-uniform mesh with H=2.0 while frame (b) shows those of uniform mesh. The cases of calculations for p*=100 are summarized in Table 2. It can be clearly seen from Table 2 that considerable saving in computational time can be achieved by using the non-uniform mesh spacing.

TABLE 2 Cases of calculation for p*=100

Case	H	Δx	Δr		No.of grid	CPUT
			r≃0	r≃0.25		
1	2.0	0.0025	0.0025	0.0032	89	38.8
2	2.0	0.004	0.004	0.0051	56	17.3
3	1.0	0.004	0.0027	0.0042	76	22.7
4	0.5	0.004	0.0016	0.0036	105	31.5
5	1.0	0.005	0.0033	0.0052	61	16.2
6	0.5	0.005	0.002	0.0045	84	22.8
7	uniform mesh		0.005		50	29.1

3.3 Four Step Iterative Method. Abarbanel and Goldberg(1972) have proposed an iterative method which uses the original Lax-Wendroff method(1964) successively. Their scheme can be written as

$$W_i^{n+1,s+1} = W_i^n + [(1-\theta)L(W_i^n) + \theta L(W_i^{n+1,s})] \tag{6}$$

where L is the Lax-Wendroff operator, s is the number of iterations and θ is a real number of $0 \leq \theta \leq 1$. They have reported that the iteration is more effective than the additional explicit artificial viscosity and that one iteration is sufficient to suppress the oscillation. Instead of using the original Lax-Wendroff operator, an extension to use one of the two step versions is attempted by the author and the results of application of the four step method using scheme 1 to the test problem appear to be promising.

In Fig. 4, density profiles at t=0.6 calculated with θ=0, 0.2, 0.5 and 1.0 are illustrated. when θ=0, the method is identical to the scheme 1, then the profiles show significant overshoot and oscillations behind the shock and the contact surface, which is a characteristic feature of the second order methods. For θ=0.2, 0.5 and 1.0, since the method is of first order, somewhat diffused shock profiles are obtained as is the case of the method of Lax(1954). It should be noted that much better resolution of the contact surface than other first order methods can be achieved by the iteration.

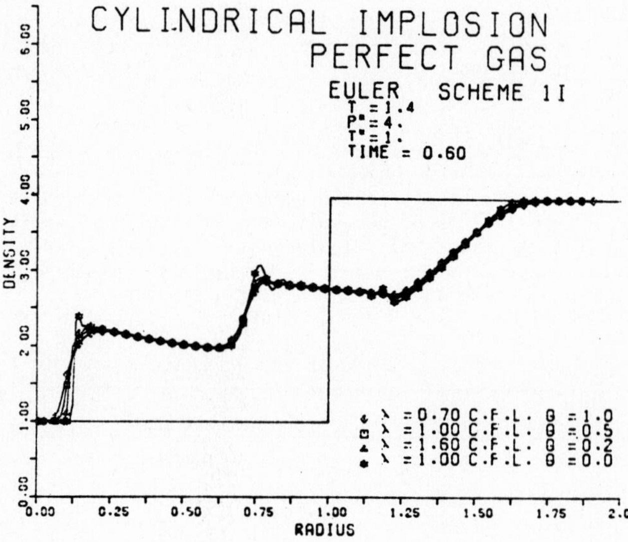

Fig. 4 Density profiles calculated by four step iterative method

3.4 Time-Splitting Cartesian Method. To remove the singularity introduced at the center along with the polar coordinate, Lapidus(1971) has proposed the Cartesian method which employs the equations in rectangular coordinates. As is well known, however, the stability condition for the two step versions of Lax-Wendroff method for two dimensions is not optimal. These severe stability restrictions can be alleviated by using the concept of time-splitting.

First order time-splitting method can be written as

$$W_{i,j}^{n+1} = L_x L_y W_{i,j}^n \tag{7}$$

where L_x, L_y are one-dimensional Lax-Wendroff operator in x and y directions, respectively. Following Gourlay and Morris(1970), one of the second order time-splitting methods of Strang(1968) is written as

$$W_{i,j}^{n+1} = L_{y/2} L_x L_{y/2} W_{i,j}^{n} \qquad (8)$$

where $L_{y/2}$ is the Lax-Wendroff operator in y direction with a half span in time.

Density profiles obtained by incorporating the scheme (7) and (8) into the Cartesian method are presented in Fig. 5. It is interesting that practically the same profiles can be obtained by first and second order time-splitting techniques. Therefore considerable reduction in computational time may be expected by using the scheme (7).

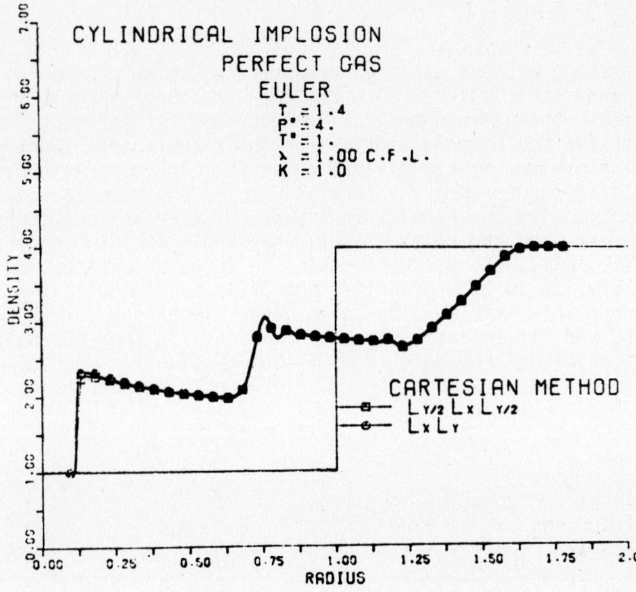

Fig. 5 Density profiles by time-splitting Cartesian Methods

REFERENCES

[1] Abarbanel, S. and Goldberg, M. J. Comp. Phys., 10, 1 (1972)
[2] Emery, A. F. J. Comp. Phys., 2, 306 (1968)
[3] Gourlay, A. R. and Morris J. Ll. J. Comp. Phys., 5, 229 (1970)
[4] Lapidus, A. J. Comp. Phys., 8, 106 (1971)
[5] Lax, P. D. Comm. Pure Apll. Math., 7, 159 (1954)
[6] Lax, P. D. and Wendroff, B. Comm. Pure Apll. Math., 17, 381 (1964)
[7] MacCormack, R. W. AIAA Paper 69-354 (1969)
[8] Payne, R. B. J. of Fluid Mechanics, 2, 185 (1957)
[9] Richtmyer, R. D. NCAR Technical Note 63-2 (1963)
[10] Rubin, E. L. and Burstein, S. Z. J. Comp. Phys., 2, 178 (1967)
[11] Strang, W. G. SIAM J. Numer.Anal., 5, 506 (1968)
[12] Taylor, T. D., Ndefo, E. and Masson, B. S. J. Comp. Phys., 9, 99 (1972)

SOME RESULTS USING RELAXATION METHODS FOR TWO- AND THREEDIMENSIONAL TRANSONIC FLOWS *

W. Schmidt, S. Rohlfs, R. Vanino

Dornier GmbH Aerodynamics Department
D-799 Friedrichshafen, Germany

1. INTRODUCTION

In the past relaxation methods has been demonstrated to be a powerful numerical tool for obtaining steady-state solutions to the two- and threedimensional transonic potential equations. The basic numerical procedure, first introduced by Murman and Cole (8) accounts for the mixed elliptic-hyperbolic character of the governing equations by using a mixed finite-difference scheme. The general procedure is to employ centered differences when the flow is locally subsonic and one-sided differences when it is locally supersonic. In a paper at the previous conference Bailey and Ballhaus (1) extended the mixed elliptic-hyperbolic relaxation method to the inviscid transonic small disturbance equation in three dimensions. In this paper we use an approach similar to that of Bailey and Ballhaus. In particular, we consider transonic flow over thin lifting wings with sweep, taper and curved leading edges and about non-lifting and lifting wing-body combinations. To show the possibly large influence of viscosity in transonic flows a twodimensional comparison of wind tunnel and relaxation method results using the displacement-thickness concept for viscous effects is done.

2. BASIC EQUATIONS AND BOUNDARY CONDITIONS

Expanding the potential in Krupp's (4) twodimensional form in three dimensions as

$$\Phi\,(x,y,z) = U_\infty\,\{x + \frac{\delta^{2/3}}{M_\infty}\,\phi\,(x,y,z) + \ldots\} \tag{1}$$

the small disturbance transonic perturbation potential equation can be written

$$[(1-M_\infty{}^2) - (\kappa+1)\,\delta^{2/3}\,M_\infty\,\phi_x]\,\phi_{xx} + \phi_{yy} + \phi_{zz} = 0 \tag{2}$$

Not fully consistent with the expansion procedure but according to Krupp (4) in better agreement with the exact values the pressure coefficient is chosen as

$$cp = -2\,\frac{\delta^{2/3}}{M_\infty^{3/4}}\,\phi_x \tag{3}$$

* This work was partially supported by the Ministry of Defense of the Federal Republic of Germany under ZTL contract T/R 720/R 7600/32 008

where M_∞ is the free-stream Mach number and δ is the thickness to chord ratio of the wing root section. In small disturbance theory the flow tangency conditions at the surface are with $\vec{n}$ = surface normal vector

$$n_x \cdot \frac{M_\infty^{1/2}}{\delta^{2/3}} + n_y \phi_y + n_z \phi_z = 0 \tag{4}$$

and for wings linearised and applied to the wing mean plane (z = 0)

$$\phi_z (x,y,0) \Big|_{u,l} = \frac{M_\infty^{1/2}}{\delta^{2/3}} \frac{\partial}{\partial x} f (x,y) \Big|_{u,l} \tag{5}$$

where $\frac{\partial f}{\partial x}$ are the slopes of the upper and lower surfaces and include angle of attack, camber and thickness. Like in twodimensional flow on the right sides of Eq. 4,5 a factor $M_\infty^{1/4}$ is neglected.

For lifting wings and wing-body combinations the Kutta condition is applied on the wing, thus forcing the flow to leave all subsonic trailing edges smoothly. In the small disturbance theory the Kutta condition is satisfied by requiring that the pressure (ϕ_X) be continous across the trailing edge. The vortex sheet is assumed to be straight and lie in the wing mean plane z = 0 with the conditions that ϕ_X and ϕ_Z be continous and ϕ be discontinous through it. This jump in potential is independent of x at any span station $y = y_0$ and is equal to the circulation about the wing section defined by

$$\gamma (y_0) = - \oint d\phi (x,y_0,z) = \Delta\phi (y_0) \; ; \; \phi_x, \phi_z \text{ continous} \tag{6}$$

The outer flow boundary conditions some distance away for a lifting wing are that the perturbation velocities, ϕ_y and ϕ_z, do not vanish due to the wing lift and thickness. Basing on Klunker (3) one approximate analytic expression for the whole far field is used.

$$\phi_{Lift} = \frac{z}{4\pi} \left(1+\frac{x}{R}\right) \int_{-b/2}^{+b/2} \frac{\gamma (\eta)}{(y-\eta)^2 + z^2} \, d\eta$$

$$\phi_{Thick} = \frac{M_\infty^{3/4}}{\delta^{2/3}} \frac{x}{4\pi R^3} \int_{S_W} d (\zeta,\eta) \, dS \tag{7}$$

where $R^2 = x^2 + (1-M_\infty^2) (y^2+z^2)$, $d (\xi,\eta)$ = local thickness at the wing station (ξ,η), S_W = wing surface, b = wing span.

As conditions at the downstream boundary, i.e. Trefftz plane, for wings the twodimensional Laplace equation is used. Comparisons with computations using conditions found by relaxing the Laplace equation with boundary condition, Eq. (6), along with the rest of the flow field, show no difference. For wing-body combinations the latter method is used.

Sometimes a design-method is very helpful, in which for aerofoil, resp. wing, wing-body combination design the pressure coefficient c_p (and hence ϕ_X) is known over all or part of the aerofoil, wing, wing-body surface and it is required to determine the shape which produces it. The core of such a design procedure is the treatment of the boundary conditions. In a similar manner like Langley (5) and Steger and Klineberg (10) we are using the vorticity equation

$$\frac{\partial}{\partial_y} \ (\phi_x) = \phi_{xy} = \phi_{yx} \ ; \int \phi_{yx} \, dx = \phi_y = \text{direct b.c.} \tag{8}$$

in order to get a boundary condition in ϕ_z, resp. ϕ_y instead of ϕ_x, the numerical details of which will be shown later.

3. NUMERICAL PROCEDURE

The basic feature of the numerical method is to solve the transonic pertubation equation in a rectangular grid box with variable mesh size. Like Bailey and Ballhaus (1) we account for the mixed elliptic-hyperbolic nature of the equation by central differences for the streamwise derivatives when the coefficient of ϕ_{xx} is positive and backward differences when the coefficient is negative. The y- and z-derivatives are replaced everywhere by the usual centered formula except at the centerline y = 0, where symmetry condition is used. The resulting set of nonlinear algebraic equations are solved iteratively by a line-relaxation algorithm.

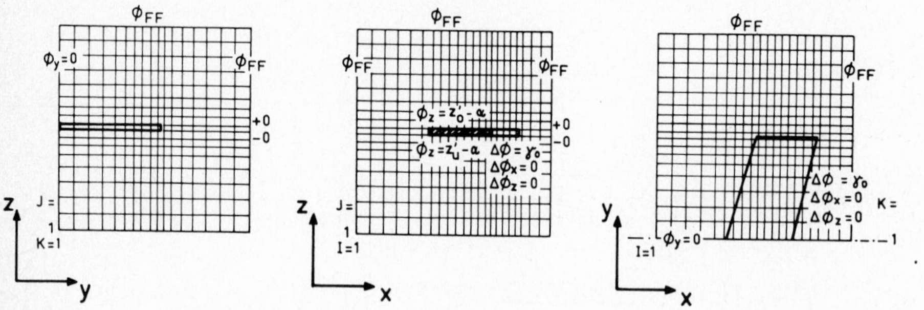

THREEDIMENSIONAL GRID ARRANGEMENT

Figure 1

Fig. 1 shows the grid arrangement in the planes x = const., y = const., z = const. In the z = const.-plane care must be taken of the wing planform. Good results are reached using a variable mesh size that is defined by the leading edge being always midway between two adjacent x = const., resp. y = const. lines in the plane z = 0.

The wing boundary conditions are incorporated in the same way as in twodimensional flow by Krupp (4) in the first plane above, resp. below the wing plane z = 0. The resulting difference equation is

$$\phi_{zz_{i,J+1,k}} = \frac{2}{t(t+2s)} \{\phi_{i,J+2,k} - \phi_{i,J+1,k}\} - \frac{2}{(t+2s)} \phi_{z_{i,J,k}} \Big|_u \tag{9}$$

where J is the z = 0 plane-number in the grid and s = mesh size between plane J and J + 1, $t = s\sqrt{3}$ is the mesh size between J + 1 and J + 2. $\phi_{z i,J,k}$ is the left side of boundary condition Eq. (5). The value of $\phi_{i,J,k}$ on the wing mean plane itself is found by linear extrapolation.

For the design problem the numerical formulation of the boundary condition according to Eq. (8) is, using for better convergence two relaxation parameters:

$$\phi_{xz_{i,J,k}}^{\nu+1/2} = -\frac{2s+t}{s(s+t)} \phi_{x_{i,J,k}}^{\nu+1/2} + \frac{s+t}{s \cdot t} \phi_{x_{i,J+1,k}}^{\nu} - \frac{s}{t(s+t)} \phi_{x_{i,J+2,k}}^{\nu} \tag{10}$$

where

$$\phi_{x_{i,J,k}}^{\nu+1/2} = -\omega_1 \cdot \frac{1}{2} Cp_{i,k} + (1-\omega_1)\, \phi_{x_{i,J,k}}^{\nu} \tag{11}$$

or Eq. 10 integrated

$$\phi_{z_{i,J,k}}^{\nu+1/2} = \int_{x_{LE}}^{x_i} \phi_{xz_{i,J,k}}^{\nu+1/2}\, dx + \phi_{z_{i_{LE},J,k}} \tag{12}$$

and the final updated value

$$\phi_{z_{i,J,k}}^{\nu+1} = \omega_2\, \phi_{z_{i,J,k}}^{\nu+1/2} + (1-\omega_2)\, \phi_{z_{i,J,k}}^{\nu} \tag{13}$$

The main difference in incorporating the boundary condition for bodies is the direct calculation of the potential $\phi_{i,J,k}$ on, resp. inside the boundary and the use of body boundary equation Eq. (4) as difference equation instead ot the potential equation Eq. (2).

$$\phi_{i,J,k}^{\nu+1/2} = \{ \frac{M_\infty^{1/2}}{\delta^{2/3}} n_x + \frac{n_y}{\Delta_y} \phi_{i,J,k+1}^{\nu} + \frac{n_z}{\Delta_z} \phi_{i,J+1,k}^{\nu} \} / (\frac{n_y}{\Delta_y} + \frac{n_z}{\Delta_z}) \tag{14}$$

with the relaxation

$$\phi_{i,J,k}^{\nu+1} = \omega \phi_{i,J,k}^{\nu+1/2} + (1-\omega)\, \phi_{i,J,k}^{\nu} \tag{15}$$

This $\phi_{i,J,k}$ is used as well for the computation of ϕ_{zz} for the first field points J + 1 of the line (i,k) as for ϕ_{yy} for the point J of the line (i, k+1).

4. RESULTS

The first set of calculations to be discussed is for the swept and tapered RAE wing "C" having a 5.4 % thick RAE 101 aerofoil section.

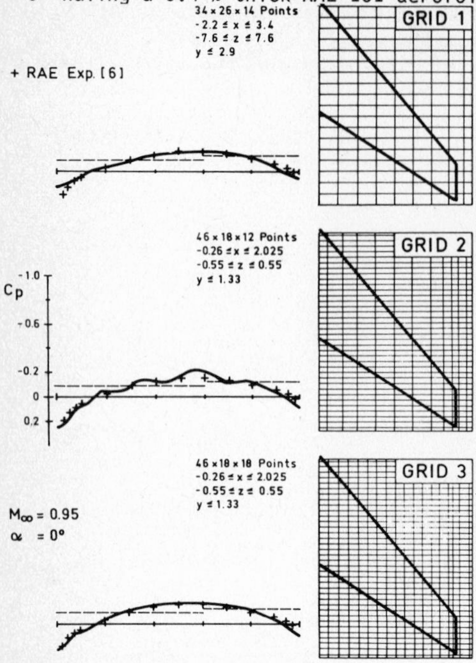

INFLUENCE OF GRID-ARRANGEMENT ON THE PRESSURE IN THE ROOT SECTION

SWEPT AND TAPERED RAE WING "C" - $\Lambda = 3.6$, $\lambda = 1/3$, $\varphi_{25} = 47.6°$

Figure 2

Fig. 2 shows the influence of mesh arrangement and size on the computed results compared with Lock's (6) RAE experimental data. Even grid 1 which is primarily for the first iterations gives after 200 iterations good results. Grid 2 has for the leading edge sweep a bad Δ_y/Δ_x-ratio which leads to wavy results. Grid 3 is the same in x- and z-direction but correct in y-direction. The results are again smooth but near the leading edge in better agreement than for grid 1.

In Fig. 3 for the same wing the calculated pressure distributions and isobars for angle of attack 0° and 3° are shown at $M_\infty = 0.95$. The results for zero angle of attack are obtained after 40 iterations with grid 2 ($\hat{=}$ 7 min CPU IBM 370/155), the 3° case after 200 iterations with grid 3 ($\hat{=}$ 20 min CPU). This includes angle of attack steps 1.5 and 2.25^o.

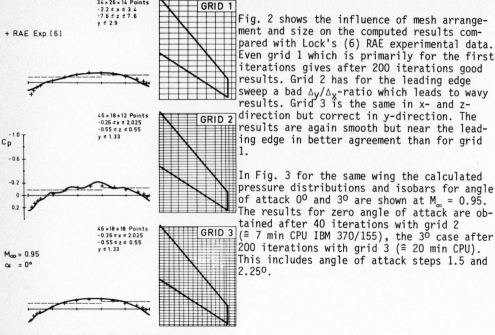

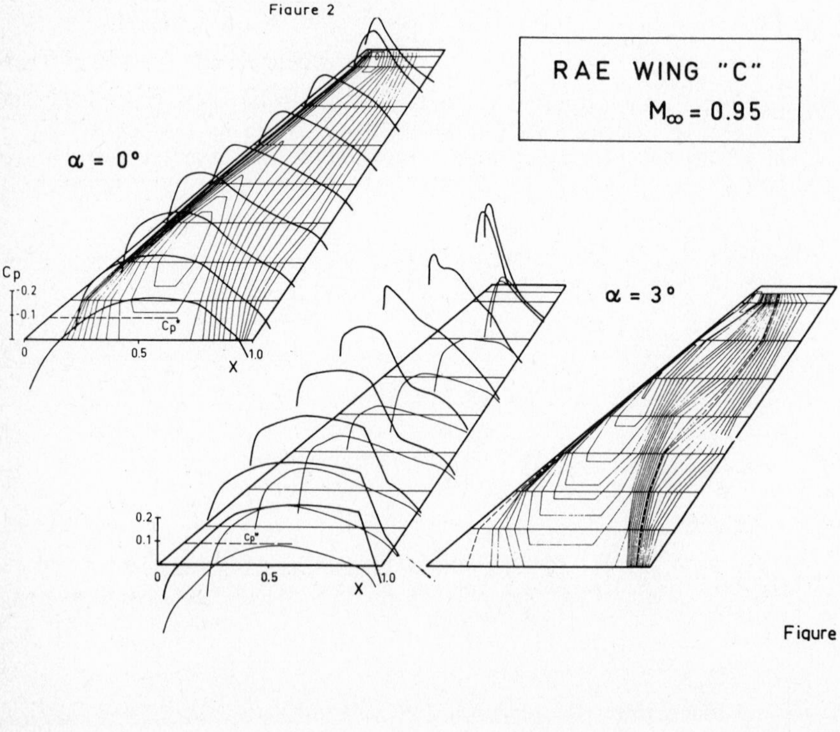

Figure 3

Comparisons with Monnerie's (7) experiments on ONERA wing "M6" having a 9.8 % thick ONERA "D" aerofoil are shown in Fig. 4 and 5.

Fig. 4 shows the converged solutions for the three grids. Only a grid as fine as grid 3 produces numerical results which are reasonable good. The pressure distributions and shock positions in Fig. 5 are calculated with grid 3. The lower surface pressure agrees very good, while for the upper surface with forward and rear shock differences increase in spanwise direction.

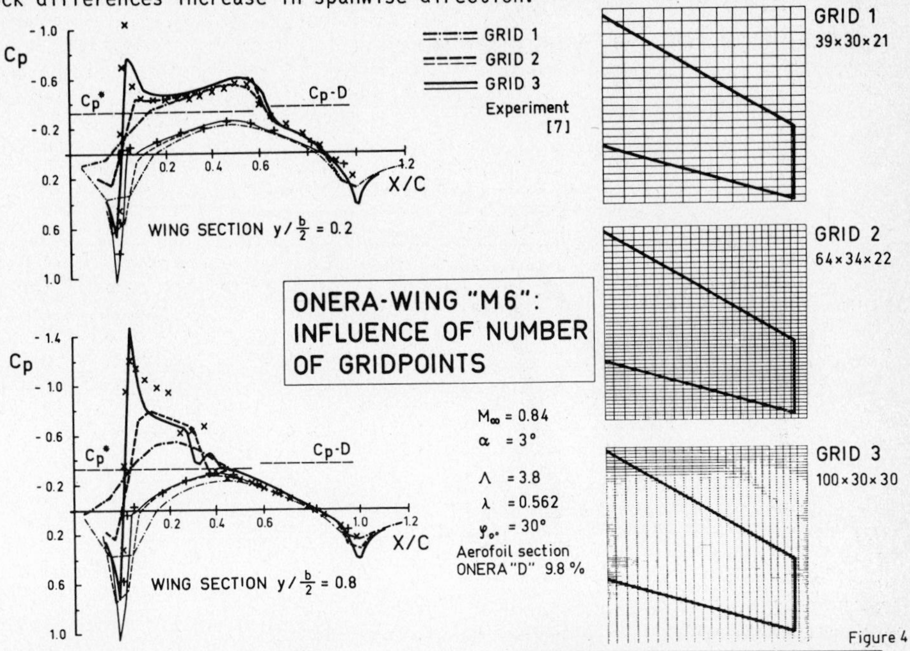

Figure 4

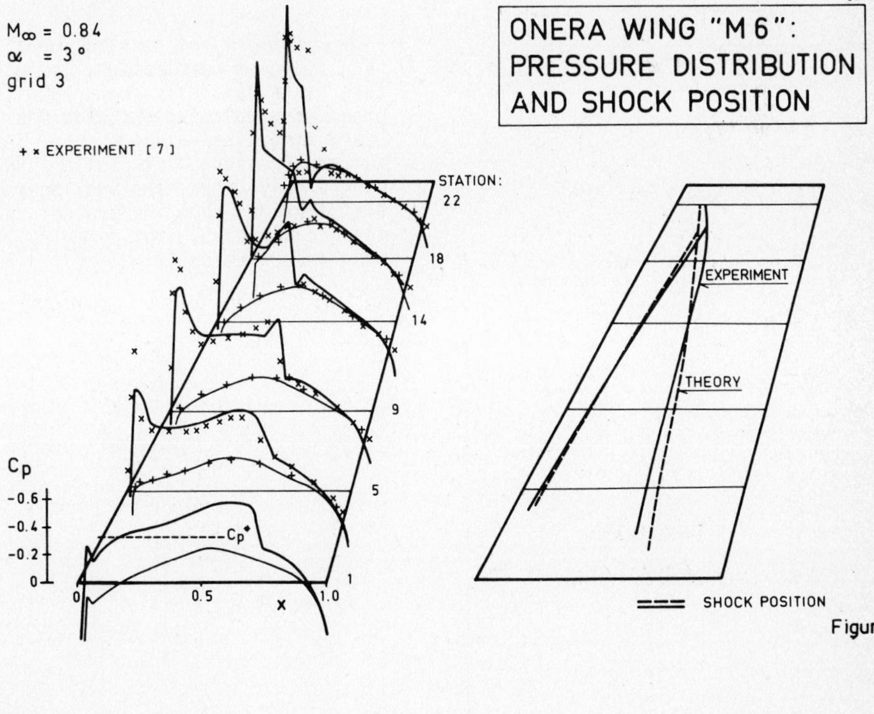

Figure 5

Fig. 6 shows the calculated wing presure distribution and the isobars for a wing with a curved leading edge. This wing was designed with a supercritical aerofoil section and tested as a wing body model at FFA, details of which will be presented by Gustavsson and Vanino (2).

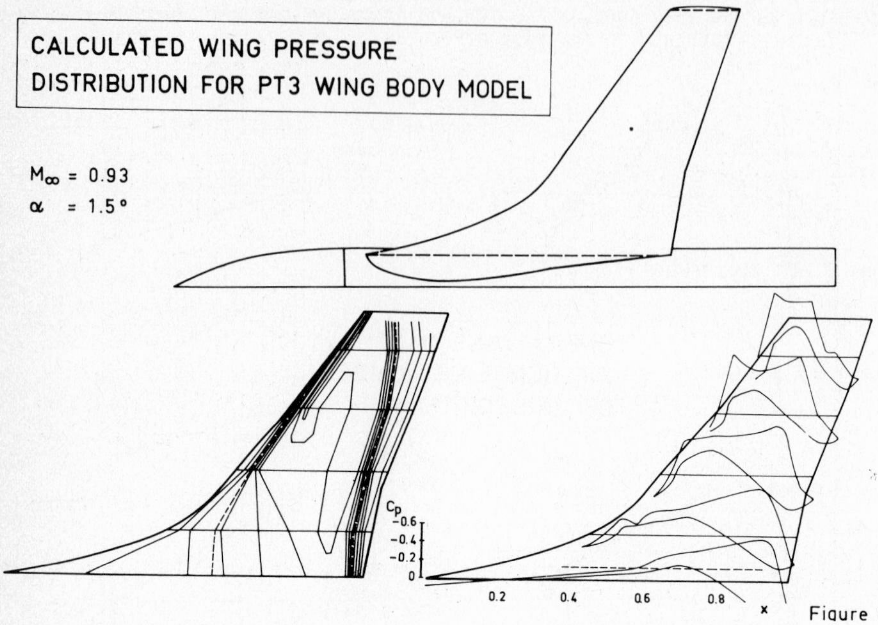

Figure 6

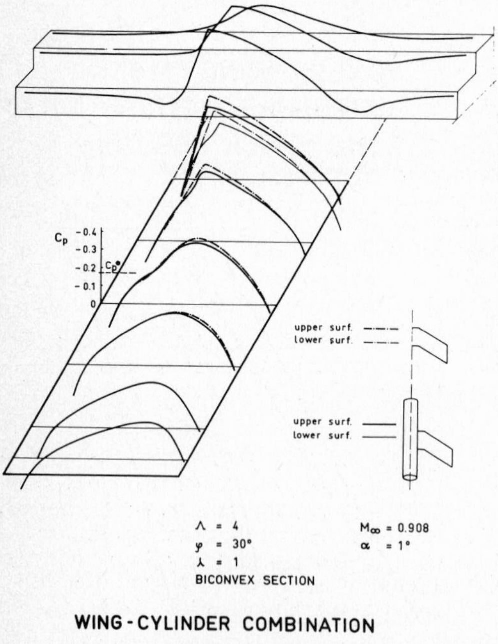

Fig. 7 shows first results for a lifting wing-cylinder combination. The wing pressure distribution is compared with exposed wing calculations. The most inboard wing station shows upper and lower surface pressure distributions the others only upper. The body pressure distributions show the typical interference increase, from symmetry line to wing-intersection.

Figure 7

5. DISCUSSION

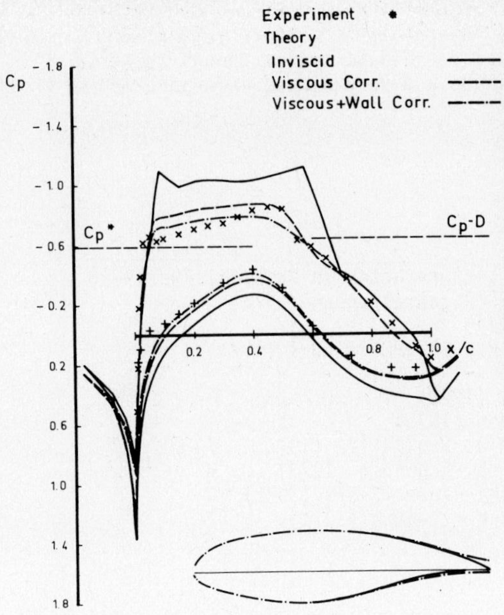

INFLUENCE OF VISCOSITY AND WIND TUNNEL CORRECTIONS

Figure 8

From these and further calculations (9) it can be concluded that the shown method may be applied to transonic flow problems for wings even with large sweep and taper. To improve the results - as shown in Fig. 8 for two-dimensional flow by Zimmer and Stanewski (11) - boundary layer displacement effects must be considered in supercritical flow.

For the calculation of yawed wings the small disturbance transonic perturbation potential equation has to be solved in the yawed coordinate system $\bar{x}$, $\bar{y}$, $\bar{z}$. Transforming the exact transonic potential equation

$$x = \bar{x}\cos\Theta + \bar{y}\sin\Theta, \qquad y = -\bar{x}\sin\Theta + \bar{y}\cos\Theta \quad (16)$$

and expanding the potential as

$$\Phi\,(x,y,z) = \bar{U}_\infty\,\{\bar{x} + \bar{y}\tan\Theta + \phi\,(\bar{x},\bar{y})\},$$

$$\bar{U}_\infty = U_\infty \cos\Theta \quad (17)$$

the small disturbance equation can be written

$$\begin{aligned}\{(1-\bar{M}_\infty)^2-\bar{M}_\infty^{\,2}[(\kappa+1)\phi_{\bar{x}} + (\kappa-1)\phi_{\bar{y}}\tan\Theta]\}\,\phi_{\bar{x}\bar{x}} \\ + \{(1-\bar{M}_\infty^{\,2}\tan^2\Theta)-\bar{M}_\infty^{\,2}\,[(\kappa-1)\phi_{\bar{x}} + (\kappa+1)\phi_{\bar{y}}\tan\Theta]\}\,\phi_{\bar{y}\bar{y}} \\ + \phi_{\bar{z}\bar{z}} - 2\,\bar{M}_\infty^{\,2}\tan\Theta\,\phi_{\bar{x}\bar{y}} = 0\end{aligned} \quad (18)$$

There are additional nonlinear terms compared with Eq. (2) that must be used. For infinite yawed wings with y-derivatives of the perturbation potential equal zero Eq. (18) becomes identically the twodimensional small disturbance transonic potential equation.

The first results for lifting wing-body-combinations incorporating an infinite long cylinder as body are very promising. Area-ruled bodies as well as non-circular cross-sections can be treated. The next extension of the present method will be the calculation of combinations with semi-infinite bodies, though problems might arise from the use of the small disturbance equation Eq. (2) and boundary condition Eq. (4) in the body nose region. For the future calculation of arbitrarily shaped wing-body-combinations a method is in development incorporating finite volume technique and relaxation method. The basic idea of this method is to divide the flow field between the body and a far field boundary in arbitrarily shaped finite volumes for which the governing equations are solved in integral from using a relaxation method. The main advantages are an orthogonal coordinate system though the grid arrangement can be optimized for the body shape and very simple incorporation of arbitrary body-boundaries.

6. REFERENCES

[1] Bailey, F.R. and Ballhaus, W.F., Lecture Notes in Physics, 19 (1972)
[2] Gustavsson, S.A.L., Vanino, R., DGLR Jahrestagung (1974)
[3] Klunker, E.B., NASA TN-D 6530 (1971)
[4] Krupp, J.A., The Boeing Company, Rep. D 180-12958-1 (1971)
[5] Langley, M.J., ARA Memo 143 (1973)
[6] Lock, R.C., Private Communication (1973)
[7] Monnerie, B., Euromech 40, Schweden (1973)
[8] Murman, E.M. and Cole, J.D., AIAA-J, 9, 114-121 (1971)
[9] Rohlfs, S., Vanino, R., Euromech 40, Schweden (1973)
[10] Steger, J.L., Klineberg, J.M., AIAA Paper 72-679 (1972)
[11] Zimmer, H., Stanewsky, E., DGLR Jahrestagung (1974)

RECENT RESULTS FROM THE GISS MODEL OF THE GLOBAL ATMOSPHERE

Richard C. J. Somerville[1]

Institute for Space Studies
Goddard Space Flight Center, NASA
New York, New York 10025, USA

INTRODUCTION

Large numerical atmospheric circulation models are in increasingly widespread use both for operational weather forecasting and for meteorological research. The results presented here are from a model developed at the Goddard Institute for Space Studies (GISS) and described in detail by Somerville et al. (1974). This model is representative of a class of models, recently surveyed by the Global Atmospheric Research Program (1974), designed to simulate the time-dependent, three-dimensional, large-scale dynamics of the earth's atmosphere.

MODEL STRUCTURE

The fundamental equations are the equation of motion, the equation of continuity, the equation of state, the first law of thermodynamics, the hydrostatic equation and a conservation equation for water vapor:

$$\frac{d\underline{V}}{dt} + f\,\underline{k} \times \underline{V} + \nabla_\sigma \Phi + \sigma \alpha \nabla \pi = \underline{F} \qquad (1)$$

$$\frac{\partial \pi}{\partial t} + \nabla_\sigma \cdot (\pi \underline{V}) + \frac{\partial}{\partial \sigma}(\pi \dot{\sigma}) = 0 \qquad (2)$$

$$p\,\alpha = RT \qquad (3)$$

$$\frac{d\theta}{dt} = \frac{1}{c_p}\frac{\theta}{T} Q \qquad (4)$$

$$\frac{\partial \Phi}{\partial p} = -\alpha \qquad (5)$$

$$\frac{dq}{dt} = -C + E \qquad (6)$$

Here

$\underline{V}$ = horizontal velocity

t = time

f = Coriolis parameter

$\underline{k}$ = vertical unit vector

---

[1]Present address: National Center for Atmospheric Research, Boulder, Colorado 80303, USA.

σ = $(p - p_t)/(p_s - p_t)$, the vertical coordinate

p = pressure

p_t = pressure at top of model atmosphere = const

p_s = pressure at bottom of model atmosphere

α = specific volume

π = $p_s - p_t$

$\underline{F}$ = horizontal frictional force

R = gas constant

T = temperature

θ = potential temperature

c_p = specific heat at constant pressure

Q = heating rate per unit mass

Φ = geopotential

q = water vapor mixing ratio

C = rate of condensation

E = rate of evaporation

The domain is global. Finite-difference resolution is currently nine levels in the vertical with a horizontal grid of four degrees of latitude by five degrees of longitude and a time step of five minutes. The numerical method is due to Arakawa (1972). Simulation of one day of real time requires about one hour of IBM 360/95 computer time.

The physical complexity of the model lies largely in the source terms ($\underline{F}$, Q, C, and E) in the above equations. The heating rate Q, for example, includes contributions from condensation, convection, diffusion and radiation. These processes, in the real atmosphere, typically occur on space scales too small to be explicitly resolved on the model grid. They must, therefore, be treated parametrically.

Precipitation and cloudiness occur in the model due to both grid-scale supersaturation and parameterized sub-grid-scale convection. Model clouds, generated by supersaturation or convective activity, have radiative properties which depend on their altitude and on the process which created them. Solar radiation is treated by a parameterization based on detailed radiative transfer calculations, including multiple scattering, performed outside the model. Terrestrial radiative heating rates are determined via a highly non-grey calculation of fluxes and flux divergences in twice the number of model layers. Inputs to the radiation routines include the model generated cloud and water vapor fields. Land temperature is determined from heat flux calculations. Surface fluxes and sub-grid-scale processes are treated parametrically using drag and eddy coefficients. At the lower boundary, realistic distributions of continents, topography, sea-surface temperatures, soil moisture, albedo, and ice and snow cover are specified. The interactions of the principal physical processes in the model are represented schematically in Figure 1.

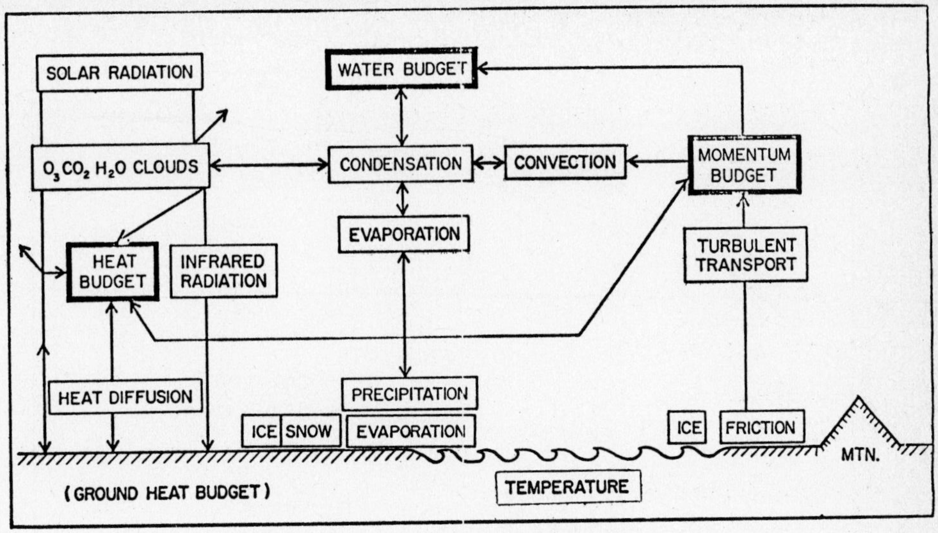

Figure 1. Interactions of physical processes in the GISS model.

ATMOSPHERIC PREDICTABILITY

Historically, large-scale numerical atmospheric models were first used to produce weather forecasts for a few days by integrating initial value problems from initial data supplied by conventional, synoptic meteorological observations. Today, such models comprise the core of the process by which operational forecasts are routinely produced in many countries. A second major application of large-scale models has been the simulation of the general circulation by constructing statistics from extended integrations, which may be compared with observed climatological data. In test integrations (Somerville et al., 1974), the GISS model has shown good short-range forecasting skill and has produced realistic simulations of climatology, including diabatic heating, hydrology, mean and eddy transports, and energetics.

The quality of short-range forecasts, however, degrades rapidly after two or three days. This degradation is due, of course, to both imperfect initial data and model deficiencies. An estimate of the importance of errors in the initial state can be obtained by performing two model integrations which differ only in initial condition. The first integration is regarded as the true evolution of the atmosphere. The second is regarded as a forecast, and the difference in initial states is regarded as an error due to an imperfect observing system. The error, or difference between forecast and true evolution, increases with time and eventually reaches an asymptotic value characteristic of the difference between two randomly chosen states of the model atmosphere. The error typically requires a few weeks to reach this asymptotic value, after which time the forecast is useless. This type of predictability study has been carried out with most major atmospheric models, and results of two such studies with the GISS model are shown in Figure 2.

In the first experiment (dashed line), the initial error was confined to the temperature field and consisted of a random error of mean zero and rms amplitude 0.5 deg K at all grid points. In the second experiment (solid line), the error was more nearly realistic, consisting of random errors of 3 m sec^{-1} in both components of horizontal wind and 1 deg K in temperature at all grid points, and 3 mb in pressure at all surface grid points. In both experiments, the errors decrease in the first day due to internal adjustment in the model, and then increase with similar growth rates and asymptote after perhaps three weeks. Such experiments indicate promise for extended-range forecasting, but demonstrate that observational errors alone ultimately limit predictability.

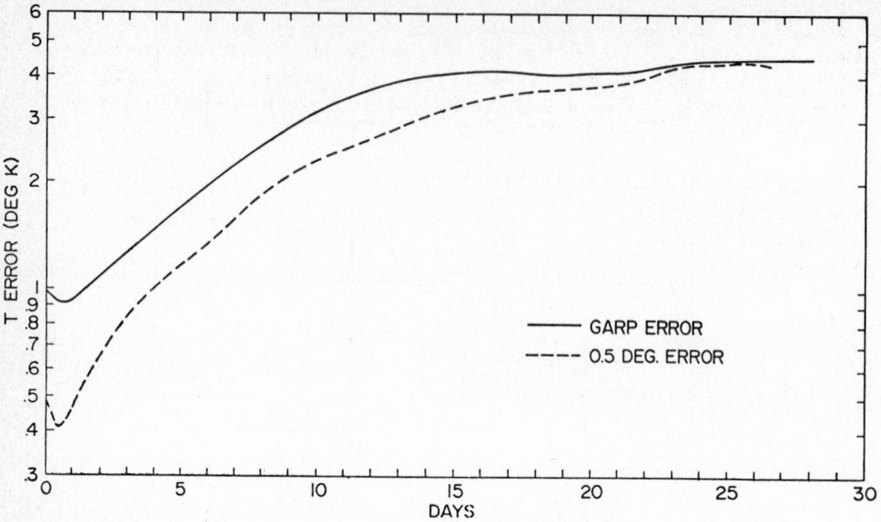

Figure 2. Mean rms temperature error as a function of forecast duration in two predictability studies (see text).

Actual forecasts, verified against the real atmosphere rather than against a simulated atmosphere, obviously degrade more quickly, due both to larger observational errors and to model deficiencies. Figure 3 shows the horizontally averaged temperature error as a function of height and time in actual forecasts with the GISS model. The results shown are the mean of six two-week forecasts begun from and verified against conventional meteorological data over most of the Northern Hemisphere in December and January.

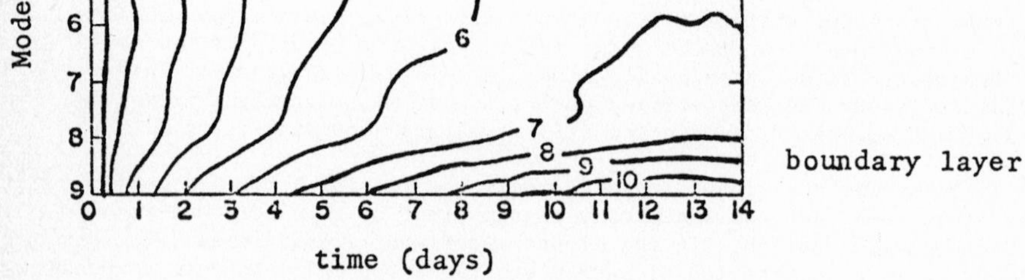

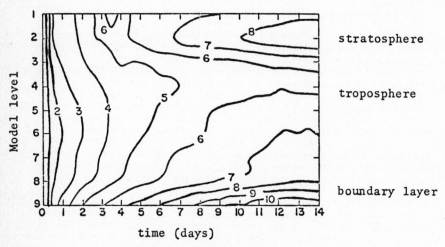

Figure 3. Mean rms temperature error (deg K) in six two-week forecasts (see text).

In this figure, model levels correspond roughly to pressure in hundreds of mb. Note the rapid growth of error at all levels in the first day and the subsequent large errors, originating near the tropopause (level 2) and in the boundary layer (level 9), and propagating into the interior of the atmosphere. Similar results have been obtained by Miyakoda et al. (1972), using another large model. Figure 3 may perhaps be interpreted as an indication of three principal sources of error: observational inadequacies and truncation error (particularly in the first few days) and imperfect model physics (particularly in the boundary layer and tropopause). Progress in weather forecasting thus depends both on better observations and on numerical and physical improvements in models.

ACKNOWLEDGMENT

Large-scale atmospheric modeling is necessarily a team effort, and the above preliminary results were obtained in collaboration with L. M. Druyan, W. J. Quirk, and P. H. Stone, with the assistance of many scientists and computer personnel at GISS.

REFERENCES

Arakawa, A., 1972: Design of the UCLA atmospheric general circulation model. Tech. Rept. No. 7, Dept. of Meteorology, University of California at Los Angeles.

Global Atmospheric Research Program, 1974: Modelling for the First GARP Global Experiment. GARP Publications Series No. 14. (Available from Secretariat, World Meteorological Organization, Case postale No. 5, CH-1211, Geneva 20, Switzerland.)

Miyakoda, K., G. D. Hembree, R. F. Strickler, and I. Shulman, 1972: Cumulative results of extended forecast experiments: I. Model performance for winter cases. Monthly Weather Review, 100, 836 - 855.

Somerville, R. C. J., P. H. Stone, M. Halem, J. E. Hansen, J. S. Hogan, L. M. Druyan, G. Russell, A. A. Lacis, W. J. Quirk, and J. Tenenbaum, 1974: The GISS model of the global atmosphere. Journal of the Atmospheric Sciences, 31, 84 - 117.

RADIATING FLOWS WITH INTENSIVE EVAPORATION OVER BLUNT BODIES

Stulov, V.P., Mirsky, V. N.
Computing Center, Academy of Sciences,
Moscow, U.S.S.R.

The problem of radiating flow over the front surface of a body with intensive evaporation is solved. A strong radiative heat transfer in aerodynamics takes place in the conditions when gas is rather dense; thus the gasdynamical model of flow is adopted. Gas moves in two layers: the shock layer and the vapour layer, divided by a contact surface. Radiative heat flux is absorbed partly by the vapour layer; the rest of it reaches the bodie's surface and causes evaporation.

1. Formulation of the problem and physical model of gas. Fig.1 shows the scheme of flow, which is governed by usual equations of radiative gasdynamics. The Rankine - Hugoniot relations are satisfied at the shock wave S. It is assumed that freestream does not radiate, thus $I^+_{\nu s} = 0$ where + denotes direction towards the body. The conditions at the contact surface are

$$p_1 = p_2 , \quad v_{n1} = v_{n2} = 0 \tag{I}$$

$$I^+_{\nu c} = I_{\nu 1} , \quad I^-_{\nu c} = I_{\nu 2}$$

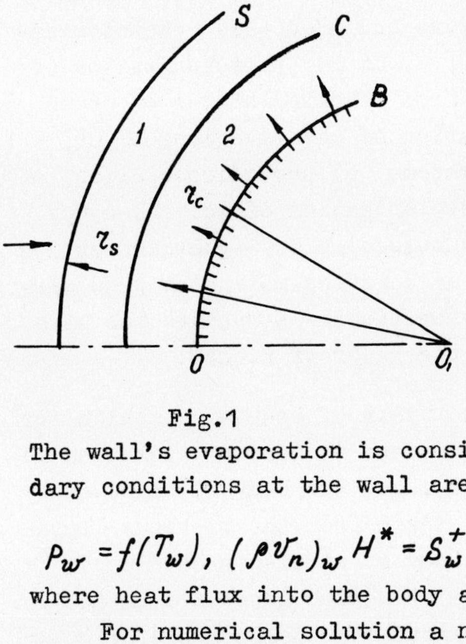

Fig.1

The wall's evaporation is considered to be equilibrium. The boundary conditions at the wall are

$$p_w = f(T_w), \ (\rho v_n)_w H^* = S^+_w - \varepsilon_w \sigma T^4_w , \ I_{\nu w} = \varepsilon_w B_{\nu w} \tag{2}$$

where heat flux into the body and reflection are neglected.

For numerical solution a new physical model is developed for radiating equilibrium gas mixture consisting of C, O, N, H and electrons. Such a model is correct for planetary atmospheres and for vapours of thermal protection materials consisting of carbon with different resins. The total absorption coefficient of mixture is expressed as follows

$$\varkappa_\nu = \sum_{i=1}^{N} \sigma_{\nu i} n_i = \rho \sum_{i=1}^{N} \varkappa_{\nu i}(T) x_i(\rho, T) \tag{3}$$

While calculating $\varkappa_\nu$ it is necessary to take into account not only the main components but also some components of small concentration. An approximate method of calculation of equilibrium composition $x_i(\rho, T)$ was developed. The main idea of the method is the allocation of component with maximum concentration and the simplification of equations of active masses' law and equations of material balance. As a result one gets explicite analytical expressions for $x_i(\rho, T)$

Absorption coefficient in individual optical transitions have been calculated by means of analytical approximations as functions of temperature and frequency. Contribution of individual spectral lines was neglected. Series of lines converging to ionization thresholds have been taken into account approximately by displacing the thresholds. In the considered equilibrium mixture of gases the following transitions were included : I) photoionization from ground states of atoms O, N, C, 2) photoionization of atomic hydrogen, 3) photoionization from excited states of atoms O,N,C and ions N^+, 4) photoionization of negative ions N^-,O^-, 5) absorption in molecular band systems, 6) photodissociation of O_2, 7) photoionization of NO, 8) photoionization of molecules O_2, N_2, C_2, H_2, CO, CN, 9) absorption by molecules C_3 (photoionization and band system λ 4050Å). While calculating molecular band systems smoothing of rotational and vibrational structure was made according to the model developed in the paper of French.

2. Numerical Method. The numerical solution of boundary problem for nonlinear integro - differential equations is done by iteration method. In each iteration the flowfield about the body is calculated with a given distribution of div S taken from the previous iteration. We use for this problem the method of straights developed in the paper of Stulov for a perfect gas flow with intensive injection. Unknown functions are evaluated by Lagrange's polynomials of polar angle

$$f = \sum_{j=0}^{N} \bar{f}_j \, \theta^{2j}, \qquad v = \sum_{j=0}^{N} \bar{v}_j \, \theta^{2j+1} \tag{4}$$

Here v is the tangential component of velocity. Coefficients are expressed through quantities of unknown functions on rays $\theta_i = const$ .Calculation of derivatives with respect to θ with the help of equations (4) and substitution into the initial system of equations yields the approximate system of ordinary differential equations. This transformation is done in the both layers; the complete boundary problem is solved by successive solution of the boundary problems for layers I and 2.

To calculate the contribution of radiation with given distribution of thermodynamical parameters the well known approximation of locally one - dimensional slab is used. It is conveniently to use exponential approximation for integro- exponential function

$$E_3(\tau) \cong a \exp(-b\tau) \qquad (5)$$

Here τ is an optical coordinate in the layer. Radiative heat flux in an elementary spectral interval $[\nu_e, \nu_{e+1}]$ in an arbitrary point of the layer consisting of two parts divided by $x = x_c$,is expressed as follows

$$\int_{\nu_e}^{\nu_{e+1}} S_\nu^{(1)} d\nu = A^{(1)} - \exp[-b(\tau_c - \tau)] G_c^{(1)}, \qquad 0 \leq \tau \leq \tau_c \qquad (6)$$

$$\int_{\nu_e}^{\nu_{e+1}} S_\nu^{(2)} d\nu = A^{(2)} + \exp[-b(\tau - \tau_c)] G_c^{(2)} - \exp[-b(\tau_w - \tau)] G_w^{(2)}, \qquad (7)$$

$$\tau_c \leq \tau \leq \tau_w$$

Here $\tau = 0$ is the shock wave, $\tau = \tau_w$ is the surface of the body. Terms $G_i^{(j)}$ are the contributions of external media to radiation of the j-th layer

$$G_c^{(1)} = A^{(2)}(\tau_c) + \exp[-b(\tau_w - \tau_c)] G_w^{(2)},$$

$$(8)$$

$$G_c^{(2)} = A^{(1)}(\tau_c), \qquad G_w^{(2)} = a \varepsilon_w \int_{\nu_e}^{\nu_{e+1}} B_{\nu w} d\nu$$

Finite difference expressions for radiation of layer A depend on the value of optical thickness

$$A = a\sum_{k=n_a}^{n} \tilde{B}_{ek}\, b\Delta\tau_k - a\sum_{k=n}^{n_b} \tilde{B}_{ek}\, b\Delta\tau_k , \quad \tau_b - \tau_a \ll 1 \qquad (9)$$

$$A = a\sum_{k=n_a}^{n_b} \tilde{B}_{ek}\,[\exp(-b|\tau-\tau_{k+1}|) - \exp(-b|\tau-\tau_k|)], \qquad (10)$$

$$\tau_b - \tau_a \sim 1$$

$$A = a\int_{\nu_e}^{\nu_{e+1}} B_\nu\, d\nu\,[\exp(-b(\tau_b-\tau)) - \exp(-b(\tau-\tau_a))], \qquad (11)$$

$$\tau_b - \tau_a \gg 1$$

$$\tilde{B}_{ek} = \frac{1}{2}\left(\int_{\nu_e}^{\nu_{e+1}} B_{\nu k}\, d\nu + \int_{\nu_e}^{\nu_{e+1}} B_{\nu k+1}\, d\nu\right)$$

The total radiation heat flux is the sum of equation (6),(7) over the whole frequency range.

3. Influence of radiation on flowfield in the two layers. The calculation of hypersonic air flow about a sphere was conducted. Let's consider the main features of flow with v_∞ = 16 km/sec, p_∞ = = 10^{-3} atm for sphere of radius R = 2 m; the thermal protection of the body consists of carbon phenolic with specific heat of sublimation H^* = 3675 + 601 exp (- 0.983 p_w) cal/ g. Under these conditions the thickness of the vapour layer is of the same order as the thickness of the shock layer (ε_1 = 0.0353, ε_2 = 0.0267). The temperature profiles show that radiation of the shock layer causes its considerable cooling. This radiation is absorbed partly by the vapour layer and the rest of it causes evaporation of the bodie's surface. Absorption and heating of the vapour layer is realized mainly near the contact surface, i.e. in the external part of the layer. The fraction of the absorbed energy depends significantly on spectral composition of the coming radiation. In Fig. 2, the monochromatic one - side heat flux (in kwt/ cm^2) is presented

$$\tilde{S}_\nu = \frac{1}{\Delta\nu_e}\int_{\nu_e}^{\nu_{e+1}} S_\nu\, d\nu \qquad (12)$$

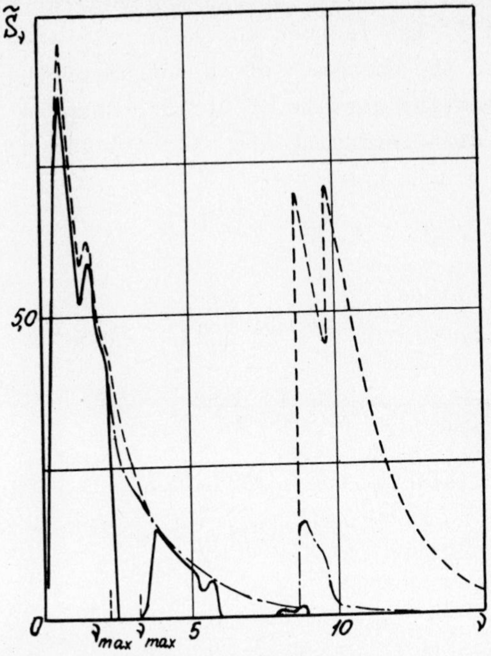

Fig.2

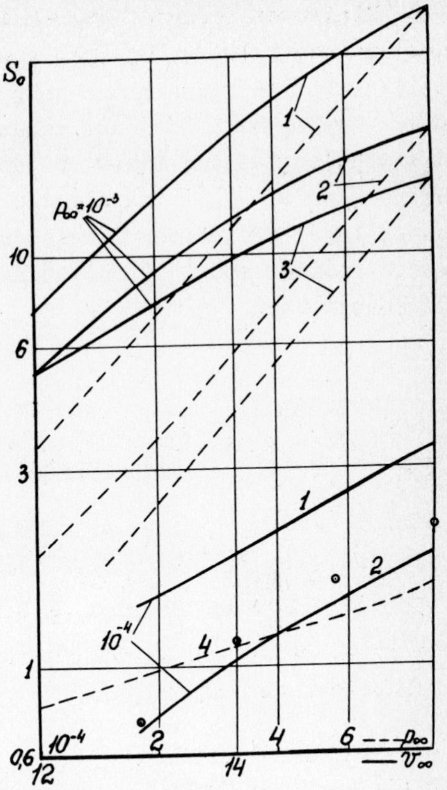

Fig.3

for three sections along axis of the flow; the dashed line shows the spectrum of layer 1 radiation at the shock wave, i.e.the radiation into freestream, "dash - point" line represents radiation of the shock layer at line C into the vapour layer, solid line shows the spectrum of radiation coming into the bodie's surface. It is seen that the vapour layer absorbs radiation in the vacuum ultraviolet region, in band systems of C_3 $\lambda 4050\mathring{A}$ and in CO 4th positive band system.

4. <u>Dependence of radiation on the freestream conditions</u>. In Fig.3 the total radiative heat flux is presented in characteristic points at the flow axis. Curves 1,2,3,4 shows S_s^-, S^+_c, S_w^+ and S_w^-, respectively, where subscripts s, c, w denote conditions at the shock wave, the contact line and the bodie's surface. The solid lines show functions of v_∞ (the lower scale), dashed lines show

functions of ρ_∞ (the upper scale). The fraction of radiation absorbed by the vapour layer slightly depends on ρ_∞ because the absorption takes place only in a thin external zone of the vapour layer which does not depend on the whole thickness. However distinction between lines 2 and 3 is reduced monotonically with decreasing of V_∞ mainly due to the change of spectrum of the shock layer radiation. Circles show the results of Stulov, Shapiro where another model of radiating air (including the spectral lines) have been used.

References

French, E.P. AIAA Journal, 2, No. 12, 2209 - 2211 (1964)

Stulov, V.P. Izv. AN SSSR Mech. Zhidk. i Gasa, No.2, 89 - 97, (1972)

Stulov, V.P. and Shapiro, E.G. Izv. AN SSSR Mech.Zhidk. i Gasa, No, 1, 154 - 160 (1970)

STEADY VISCOUS INCOMPRESSIBLE FLOW AROUND A GIVEN CONTOUR [*)]

Andrzej M. STYCZEK

1. Formulation of the physical problem

The aim of this paper is to present a method for determining the plane steady viscous and incompressible flow past an obstacle placed in the infinite stream. The velocity distribution far from the obstacle is subject only to the principle of the conservation of mass. In order to solve the problem we can "truncate" the infinite region of flow and replace it by an annulus, whose exterior boundary lies sufficiently far from the contour of the obstacle.

In this paper we confine our attention to the flows inside such annular regions, with given velocity distributions on their boundaries.

2. Formulation of the mathematical problem

The annular region of flow will be denoted by Ω and bounded by an interior curve C and an exterior curve C_∞, both curves possessing continuous tangents.

The physical problem previously formulated may now be recasted in terms of the stream function ψ as the following boundary value problem:

$$\text{(1)} \qquad \left.\psi\right|_{\partial\Omega} = \psi_0(P), \quad \left.\frac{d\psi}{dn}\right|_{\partial\Omega} = \psi_1(P) \; ; \qquad P \subset \partial\Omega = C + C_\infty$$

posed on the curves C and C_∞, for the nonlinear elliptic partial differential Navier-Stokes equation:

$$\text{(2)} \qquad \Delta^2 \psi = R \cdot J(\Delta\psi, \psi)$$

of the fourth order.

---

*) This is part of the Ph.D.-Thesis, presented by the author at the Warsaw University of Technology, under direction of Professor W.J. Prosnak.

In this equation Δ^2 denotes the biharmonic operator, J - the usual Jacobian, and R - the Reynolds number.

3. Method of solution

The boundary value problem (1) for the equation (2) was solved by the method of successive approximations. The functions $\hat{\varphi}_k$ (k=1,2,3,...), which enter the sequence of the successive approximations satisfy the condition (1) at the boundary of Ω and the equations:

$$(3) \qquad (g + \Delta^2)\hat{\varphi}_{k+1} = g\hat{\varphi}_k + R\cdot J(\Delta\hat{\varphi}_k, \hat{\varphi}_k)$$

$$g = \text{const} > 0;$$

in the interior of the region.
In these equations g denotes a positive constant.

We shall prove that the sequence $\{\hat{\varphi}_k\}$ thus constructed, converges in the mean square sense to the solution of the nonlinear boundary value problem (1)+(2), provided that the initial function $\hat{\varphi}_0$ has been chosen sufficiently "small".

The proof is based on the following properties:

a) The operator $A = g + \Delta^2$ is a Friedrichs′ operator and the inverse of its Friedrichs′ extension $\tilde{A}^{-1}$ transforms the space L^2 into the Sobolev space $W_2^{(4)}$, according to a theorem on the differentiation of the weak solutions (see e.g.[1])

b) The Ehrling (see e.g.[2]) inequality is satisfied:

$$\|u\|^2_{W_2^{(1)}} \leq \text{const}\cdot g^{-2(1-\frac{1}{4})}\left[g^2\|u\|^2_{L^2} + \|u\|^2_{W_2^{(4)}}\right]$$

c) The following norms are equivalent:

$$\|u\|^2_{L^2} + \sum_{i+j=4}\left\|\frac{\partial^4 u}{\partial x^i \partial y^j}\right\|^2_{L^2} \stackrel{df}{=} \|u\|^2_{W_2^{(4)}} \sim$$

$$\sim |u|^2_{*} \stackrel{df}{=} \|Au\|^2_{L^2} \qquad u \in C_0^{(4)} \text{ - dense in } W_p^{(1)}$$

In course of the proof we introduce the difference

$$\hat{R}_{k+1} = \hat{\varphi}_{k+1} - \hat{\varphi}_k$$

and take into account the inequality

$$2ab \leq a^2 + b^2$$

as well as the identities

$$(\hat{R}_k, J(\Delta\hat{\psi}_{k-1}, \hat{R}_k)) = (\Delta\hat{\psi}_{k-1}, J(\hat{R}_k, \hat{R}_k)) = 0$$

$$(\hat{R}_k, J(\Delta\hat{R}_k, \hat{\psi}_k)) = (\hat{\psi}_k, J(\Delta\hat{R}_k, \hat{R}_k)).$$

We obtain successively:

$$\text{(4)} \qquad \| R_{k+1} \|^2_{\varkappa} = \| P \|^2_{L^2}$$

$$P = g\hat{R}_k + R\left[J(\Delta\hat{R}_k, \hat{\psi}_k) + J(\Delta\hat{\psi}_{k-1}, \hat{R}_k)\right]$$

$$\| P \|^2_{L^2} \leqslant g^2 \| \hat{R}_k \|^2_{L^2} + 2R^2\Big[\| J(\Delta\hat{R}_k, \hat{\psi}_k) \|^2_{L^2} +$$

$$+ \| J(\Delta\hat{\psi}_{k-1}, \hat{R}_k) \|^2_{L^2}\Big] + 2gR \| \hat{\psi}_k \|_{L^2} \| J(\Delta\hat{R}_k, \hat{R}_k) \|_{L^2} .$$

By virtue of the Sobolev Theorem (see e.g. [3]) and the Hölder inequality we show that

$$\| J(\Delta a, b) \|^2_{L^2} \leqslant \text{const} \cdot \| D^1 b \|^2_{L^{(2p)}} \cdot \| D^3 b \|^2_{L^{(2q)}} \leqslant$$

$$\leqslant \text{const} \| b \|^2_{W_2^{(l_1)}} \cdot \| a \|^2_{W_2^{(l_2)}}$$

where

$$a, b \in C_0^4 , \quad l_1 \geqslant 2 - \frac{1}{p} , \quad l_2 \geqslant 4 - \frac{1}{q} , \quad \frac{1}{p} + \frac{1}{q} = 1.$$

The Ehrling inequality (b) and the inequalities of Friedrichs-Poincare $\| u \|^2_{L^2} \leqslant (g^2 + 2g\lambda_1^2 + \lambda_2^2)^{-1} \| u \|^2_{\varkappa}$, $\lambda_1, \lambda_2 > 0$, lead to the following estimate

$$\text{(5)} \qquad \| P \|^2_{L^2} \leqslant \Big\{ \frac{g^2}{g^2 + 2g\lambda_1^2 + \lambda_2^2} + \frac{2R^2_{\varkappa}}{g\sqrt{g}} \left(\| \hat{\psi}_k \|^2_{\varkappa} + \| \hat{\psi}_{k-1} \|^2_{\varkappa} \right) +$$

$$\frac{2R_{\varkappa}}{\sqrt{g\sqrt{g}}} \frac{g}{\sqrt{g^2 + 2g\lambda_1^2 + \lambda_2^2}} \| \hat{\psi}_k \|_{\varkappa} \Big\} \| \hat{R}_k \|^2_{\varkappa} , \quad R_{\varkappa} = R \cdot \text{const}.$$

Taking (4) into account we obtain:

$$\text{(6)} \qquad \| \hat{R}_{k+1} \|^2_{\varkappa} \leqslant \{ \cdots \} \| \hat{R}_k \|^2_{\varkappa} ,$$

Thus, the transformation $\hat{R}_k \longrightarrow \hat{R}_{k+1}$ is contractive for $k = k_1$ if

$$\|\hat{\psi}_{k_1-1}\|_{\varkappa}, \quad \|\hat{\psi}_{k_1}\|_{\varkappa} \leq Y$$

(7)

$$Y = \frac{g^{13/4}}{2R_{\varkappa}\sqrt{g^2+2g\lambda_1^2+\lambda_2^2}}\left[\sqrt{1+4\beta\frac{2g\lambda_1^2+\lambda_2^2}{g^2}}-1\right],$$

for each $0<\beta<1$.

If each entry of the sequence $\hat{\psi}_k$ fulfills the condition

$$\|\hat{\psi}_k\|_{\varkappa} \leq Y,$$

then the transformation

$$\hat{R}_k \longrightarrow \hat{R}_{k+1}$$

is contractive for each k.

In order to satisfy the above condition it is enough to choose the initial function $\hat{\psi}_0$ subject to the constraint:

$$\text{(8)} \quad \|\hat{\psi}_0\|_{\varkappa} \leq \left(1-\sqrt{\frac{g^2+(2g\lambda_1^2+\lambda_2^2)\beta}{g^2+2g\lambda_1^2+\lambda_2^2}}\right)\frac{g^{13/4}}{2R_{\varkappa}\sqrt{g^2+2g\lambda_1^2+\lambda_2^2}} \cdot$$

$$\cdot\left(\sqrt{1+4\beta\frac{2g\lambda_1^2+\lambda_2^2}{g^2}}-1\right).$$

Then $\|\hat{R}_k\|_{\varkappa} \longrightarrow 0$ and there exists the limit function $\hat{\psi}_\infty \in W_2^{(4)}$.

In order to show that $\hat{\psi}_\infty$ represents the solution of the boundary value problem (1)+(2) in the mean square sense, we notice that

$$\|\Delta^2\hat{\psi}_\infty - RJ(\Delta\hat{\psi}_\infty,\hat{\psi}_\infty)\|^2_{L^2} = \|A\hat{\psi}_\infty - g\hat{\psi}_\infty - RJ(\Delta\hat{\psi}_\infty,\hat{\psi}_\infty)\|^2_{L^2} =$$

$$= \|A\hat{\psi}_\infty - A\hat{\psi}_N + g\hat{\psi}_{N-1} - g\hat{\psi}_\infty - RJ(\Delta\hat{\psi}_\infty,\hat{\psi}_\infty) + RJ(\Delta\hat{\psi}_{N-1},\hat{\psi}_{N-1})$$

$$+ A\hat{\psi}_N - g\hat{\psi}_{N-1} - RJ(\Delta\hat{\psi}_{N-1},\hat{\psi}_{N-1})\|^2_{L^2} \leq$$

$$\leq \Big\{\|\hat{\psi}_\infty - \hat{\psi}_N\|_{\varkappa} + \frac{g}{\sqrt{g^2+2g\lambda_1^2+\lambda_2^2}}\|\hat{\psi}_\infty - \hat{\psi}_{N-1}\|_{\varkappa} +$$

$$+ 2R_{\varkappa}g^{-3/4}\|\hat{\psi}_\infty - \hat{\psi}_{N-1}\|_{\varkappa}\Big\}^2 \longrightarrow 0.$$

The construction of the sequence $\hat{\psi}_k$ requires solving the sequence of the Dirichlet boundary value problems (1) for the linear equations

(9) $$(g + \Delta^2)\hat{\Psi}_k = f$$

with the known right-hand sides.
By using the variational method (see e.g.[3]) we minimize the Ritz functional

(10) $$l(\hat{\Psi}_k) = \int_\Omega \left[g\hat{\Psi}_k^2 + (\Delta\hat{\Psi}_k)^2 - 2f\hat{\Psi}_k \right] d\Omega = \min.$$

4. Algorithm

Suitable change of the independent space variables transforms the annular region Ω into the regular annulus bounded by two concentric discs. Then, by developing the sought functions $\hat{\Psi}_k$ into the trigonometric series and by using the quadrature formulae with respect to the radius we obtain the following difference analogue of (10):

(11) $$\frac{l_s}{2\pi} = \sum_{\substack{(i)\\ \text{mesh points}}} r_i \, H(\Psi_{oi}, \Psi_{1i}, \ldots, \bar{\Psi}_{1i} \ldots \bar{\Psi}_{Ni}) ,$$

where Ψ_{ki} denotes the k-th Fourier amplitude at the i-th point of the mesh.
Having performed the appropriate differentiations we finally arrive at the equations:

(12) $$A_i^2 \vec{Z}_{i+2} + A_i^1 \vec{Z}_{i+1} + A_i^o \vec{Z}_i + A_i^{-1} \vec{Z}_{i-1} + A_i^{-2} \vec{Z}_{i-2} = \vec{F}_i$$

for the unknown vectors $\vec{Z}_i$, defined as follows:

$$\vec{Z}_i = \text{Re}\{\Psi_{oi}, \text{Re}\,\Psi_{1i}, \ldots \text{Im}\,\Psi_{1i}, \text{Im}\,\Psi_{2i}, \ldots\} .$$

The algebraic system (12) may be solved by the method of factorization, applicable to the systems of the general form:

$$\vec{Z}_{i+2} = \mathcal{L}_{i+2} \vec{Z}_{i+1} + \beta_{i+2} \vec{Z}_i + \vec{\jmath}_i .$$

5. Numerical Results

The above algorithm served as a basis for some numerical computations performed on a small automatic computor ODRA 1204, whose speed was less then 10000 additions per second. These technical difficulties posed strong restrictions on the range of the Reynolds number as well as on the shape of the region, which was taken to be the regular annulus bounded by the two concentric discs. The results are represented by the figures 1, 2, 3, 4 and 5, whose interpretation is self-evident.

6. Acknowledgements

Thanks are due to Professor Włodzimierz Prosnak my scientific advisor, and to my colleague - Zenon Nowak Ph.D., whose help and criticism have much contributed to the present form and the contents of this paper.

7. Bibliography

1 Bers, L., John, F., Schechter, M.: "Partial Diff. Equations", Interscience, 1964 N.York, London, Sydney

2 Maurin, K.: "Methods of Hilbert spaces" PWN, 1965, Warszawa

3 Михлин, С.Е.: "Вариационные методы в математмческой физике" Наука, 1970, Москва.

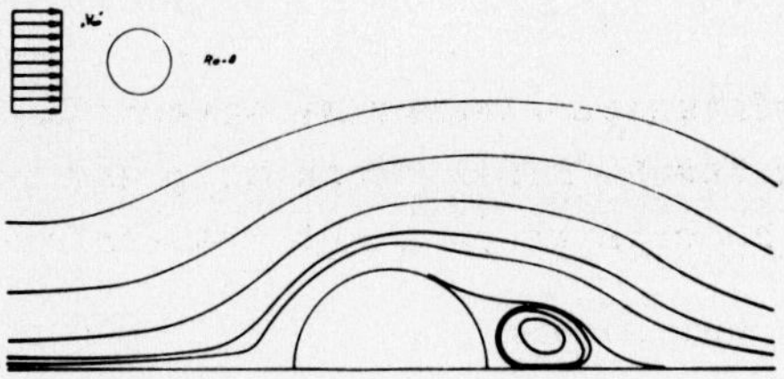

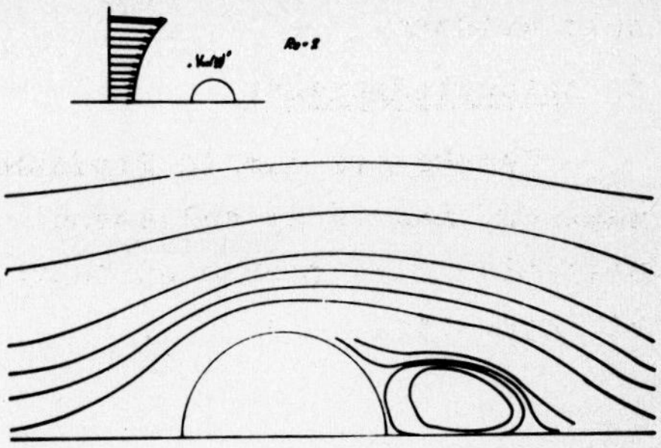

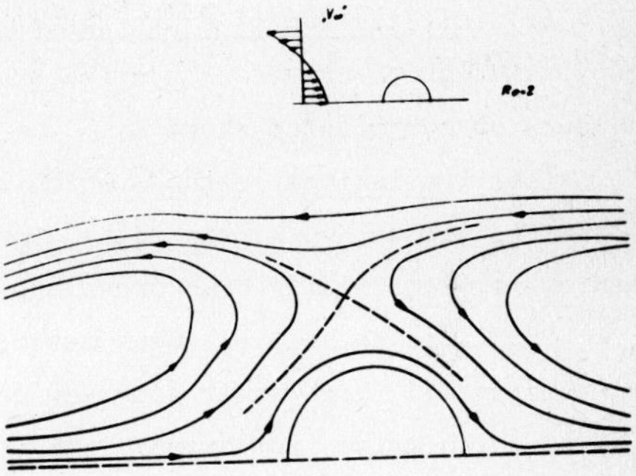

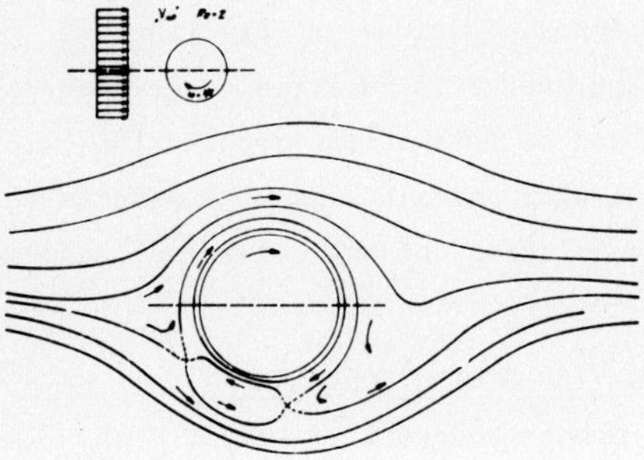

SOLUTION OF THE NAVIER-STOKES EQUATIONS BY FINITE ELEMENT METHODS

Roger TEMAM

University of Paris South, Orsay, France, and
University of California, Berkeley, U.S.A.

François THOMASSET

IRIA - 78 Rocquencourt, France

INTRODUCTION

The finite element method is a very powerful method for the solution of boundary value problems, introduced by civil engineers. The method is based on two ideas:

(i) utilization of a non-rectangular discretization grid which allows a better approximation of curved boundaries and is particularly appropriate in the case of domains of complicated shape.

(ii) variational methods which are directly related to variational principles (like the virtual work principle), and which give a very natural framework for implementing systematically high order approximation methods.

For fluid mechanics, these methods are also promising and suitable in similar circumstances: high order accuracy or the treatment of domains with complicated shapes which arise, for example, in oceanography, or magnetohydrodynamics and plasma physics.[1]

Our purpose here is to describe a set of finite element methods which are now available for the computation of viscous incompressible flows. We will recall rapidly the variational formulation of Navier-Stokes equations and describe the methods and their accuracy. For more details the reader is referred to References [4] and [6] for the analysis of these methods, and to forthcoming reports [8] which will describe their practical implementation and will contain a general program valid for all geometries (including an automatic triangulation of the domain) cf. also [1] and [7].

1. VARIATIONAL FORMULATION OF THE STOKES AND NAVIER-STOKES EQUATIONS

We assume that the fluid fills a domain Ω whose boundary Γ is at rest. The nondimensional Navier-Stokes equations can be written as

$$(1)_1 \qquad -\frac{1}{Re}\Delta u + (u\cdot\nabla)u + \nabla p = f \quad \text{in } \Omega$$

$$(1)_2 \qquad \nabla u = 0 \quad \text{in } \Omega$$

$$(1)_3 \qquad u = 0 \quad \text{in } \Gamma$$

where u and p are the fluid velocity and pressure, and f represents forces per unit volume.

---

[1] See the papers of J. P. Boujot, and C. K. Chu, H. C. Lui, this volume.

Let us denote by V the set of divergence free vector functions vanishing on Γ. Then $u \in V$ and if v is any other element of V we can multiply $(1)_1$ by v, integrate in Ω and integrate by parts using Green's formula. The pressure disappears, and we obtain

$$(2) \qquad \frac{1}{Re}((u,v)) + b(u,u,v) = (f,v) \qquad ,$$

where

$$((u,v)) = \sum_{i,j=1}^{n} \int_\Omega \frac{\partial u_i}{\partial x_j}\frac{\partial v_i}{\partial x_j}\,dx \quad , \quad (f,v) = \int_\Omega fv\,dx \quad ,^1$$

$$b(u,u,v) = \sum_{i,j=1}^{n} \int_\Omega u_i \frac{\partial u_j}{\partial x_i} v_j\,dx \qquad .$$

The variational form of the problem (1) is:

<u>To find $u \in V$ satisfying (2) for each $v \in V$.</u>

It can be proved that if u satisfies (2) then there exists a function p such that $(1)_1$ holds and problems (1) and (2) are therefore equivalent.

<u>Remarks</u>

(i) The nonlinear term $(u\cdot\nabla)u$ can be written as

$$\frac{1}{2}(u\cdot\nabla)u + \sum_{i=1}^{n} D_i(u_i u)$$

and therefore the term b in (2) can be replaced as well by

$$\tilde{b}(u,u,v) = \frac{1}{2}\int_\Omega u_i \left(\frac{\partial u_j}{\partial x_i} v_j - u_j \frac{\partial v_j}{\partial x_i}\right) dx \qquad ,$$

with the property that $\tilde{b}(u,u,u) = 0$, even if $\nabla u \neq 0$.

(ii) For the Stokes problem we just drop the b term.

(iii) If Γ is not at rest, and has a given velocity $\phi(x)$ the same formulation is applicable. Let Φ denote an extension of ϕ inside Ω as a divergence free vector function. Then we have <u>to find</u> $u \in \Phi + V$ <u>such that</u> (2) <u>holds for each</u> $v \in V$. Non-homogeneous problems can be handled without any additional difficulties.

2. FINITE ELEMENT METHODS

2.a. Variational Approximation

Once the problem is written in its variational form (2), the principle of variational approximation is the following one:

---

[1] n = the dimension of space, = 2 or 3, $x = (x_1,x_2)$ or (x_1,x_2,x_3), $dx = dx_1dx_2$ or $dx_1dx_2dx_3$.

Let V_h be a finite dimensional subspace of V, which must be specified. We replace the problem (2) by the problem:

Find $u_h \in V_h$ such that

(3) $$\frac{1}{Re}((u_h,v)) + b(u_h,u_h,v) = (f,v) \quad , \quad \forall \; v \in V_h$$

This amounts to solving a linear system in the Stokes problem case or a system of quadratic equations otherwise.

The finite element methods allow us to construct very simple spaces V_h such that the matrix of the linear system associated with (3) is sparse, almost as sparse as for finite difference methods.

Let us now describe some spaces V_h.

2.b. Triangulation of Ω

We cover Ω as completely as possible by a family of triangles which have no common points or have a common vertex or a whole common side.

Starting with a given coarse triangulation of Ω, one can refine it several times either by adding a new vertex inside a triangle T and dividing T into 3 new triangles, or by dividing T into 4 similar triangles.

Special care is necessary for the treatment of boundary vertices. All these procedures and the ordering of the nodes are parts of standard programs of automatic triangulation.

The finite element method is convergent if the maximum $\rho(h)$ of the diameters ρ_T of the triangles goes to zero, while the angles θ do not become too large or too small:

(4) $$0 < \theta_o \leq \theta \leq \theta_1 < \pi \quad ;$$

$\rho(h)$ plays the role of the mesh in finite differences. Condition (4) is important when one constructs a triangulation of Ω.

2.c. Description of the space V_h

The space V_h is a subspace V, or approximates such a subspace.

Linear nonconforming elements.

The nodal values are the values of u_h at the mid points of the sides of the triangles. The functions u_h are linear in each triangle T and take arbitrary values at the mid points ($= 0$ or ϕ at boundary mid points). Moreover, $\operatorname{div} u_h = 0$ in each triangle.

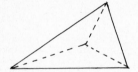

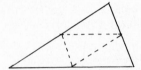

The error for the L^2 norm is of the order of $\rho(h)^2$.

Finite elements of degree two.

The nodal values are the values of u_h at the vertices and mid points of the sides. The functions u_h are polynomials of degree 2 in each triangle T and take arbitrary values at each node (0 or ϕ at boundary nodes). Moreover, $\int_T \operatorname{div} u_h \, dx = 0$ in each triangle T.

The error, for the L^2 norm is of order $\rho(h)^2$, or $\rho(h)^3$ with a slight modification of the space V_h using the so-called bulb function; see [2] and [6].

Finite elements of degree three.

Available in the three dimensional case.

Finite elements of degree four (n = 2).

There are 21 nodal values on a given triangle. The function u_h is a polynomial of degree four on each triangle. The error is of order $\rho(h)^5$ (L^2 norm).

2.d. Other problems.

(i) The solution of the discrete problems (4) is not obvious. Several appropriate iterative procedures are available. For further details, see [6].

(ii) The evolution problems can be reduced to a sequence of problems of stationary type by a standard discretization in the time variable. Each stationary type problem can be solved by finite element methods.

2.e. Figures.

Figures 1 and 2 show the automatic triangulation of the domain Ω limited by two non-concentric circles. Figures 3 and 4 show the stream lines of the flow in a cavity at Re = 100; see [8].

REFERENCES

[1] J. P. Boujot, "Le chauffage ohmique d'un plasma, etude numerique;" this volume.

[2] Crouzeix and P. A. Raviart, to appear.

[3] M. Fortin, Thesis, University of Paris, 1972.

[4] M. Fortin, Proceedings of the 3rd International Conference on Numerical Methods in Fluid Mechanics.

[5] R. Temam, Navier-Stokes Equation, Lecture Notes No. 9, Mathematics Department University of Maryland, 1973.

[6] R. Temam, Navier-Stokes Equations, Theory and Numerical Analysis, North Holland, a book to appear in 1975.

[7] F. Thomasset, Thèse de 3ème cycle, University of Paris South, 1973.

[8] F. Thomasset, Report, IRIA-Laboria, Rocquencourt, France, to appear.

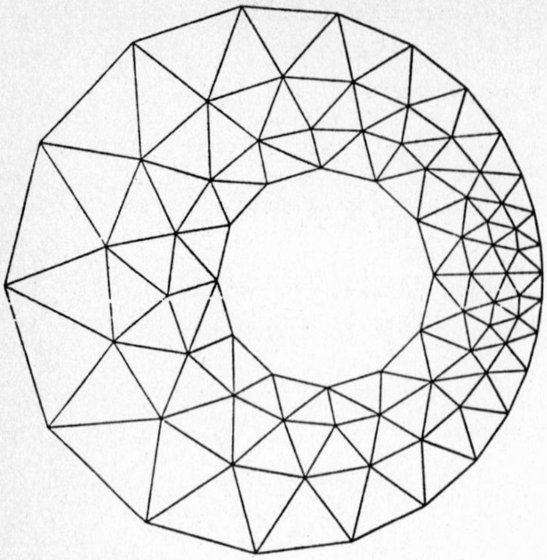

Fig. 1.

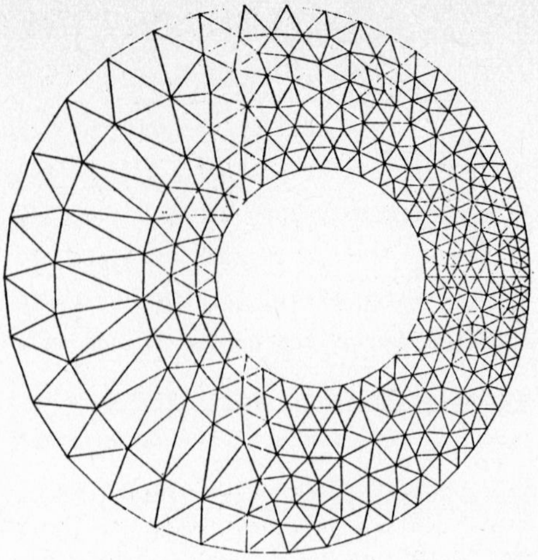

Fig. 2.

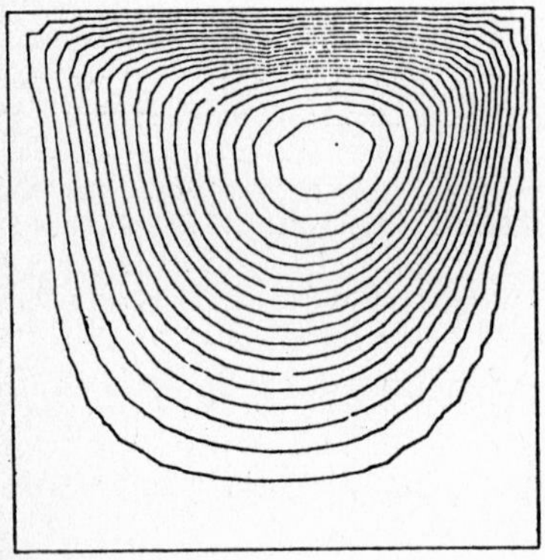

Fig. 3.

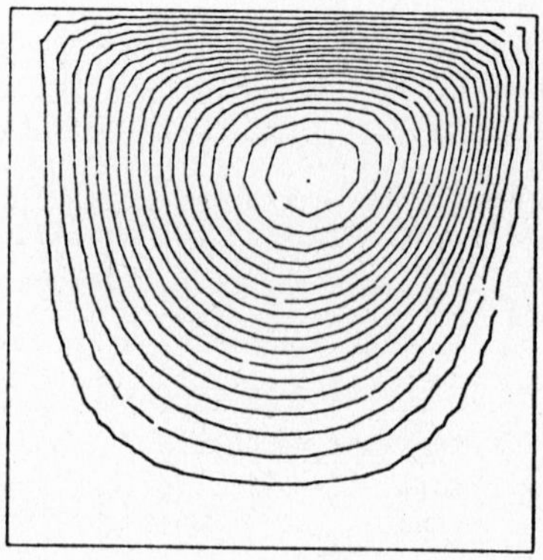

Fig. 4.

CALCULATIONS ON THE HYPERSONIC RAYLEIGH PROBLEM

B.W. Thompson
University of Melbourne, Australia

The Rayleigh problem is a conceptual experiment in which a fluid is set in motion entirely by viscous entrainment with a moving solid boundary. Thus no part of the field responds to an inviscid analysis. The compressible case is one of very great difficulty if the boundary speed is large, because mathematically it becomes a "free boundary" problem with singularities in the solution in the neighbourhood of both the free and solid boundary which necessarily inhibit all efforts to impose boundary conditions upon them. Transformations to put the problem into a numerically tractable form are discussed.

1. INTRODUCTION

By far the greatest number of extant successful solutions of compressible flow problems have been based on the inviscid model, at least over the major portion of the flow space. The reasons for this are almost obvious: coefficients of viscosity and heat conduction in gases are of the same order of magnitude and so neglect of one requires neglect of the other. This done, the equations of motion each lose their highest order term and so drop by a unit of order, and the energy equation reduces to a single term which integrates directly into an algebraic relation between pressure and density on each streamline. The resulting measure of simplification of the mathematical task can hardly be exaggerated.

Physically the inviscid model reduces the flow to the opposition of inertial and elastic forces, and this indicates that there will be certain cases in which no meaningful information will be extracted from its use. Chief of these must be those situations in which viscous entrainment by solid boundaries is a major factor determining the distribution of energy throughout the flow space. Here we are forced to accept the difficulties of the fully viscous model and just make the best we can of them.

Aside from the fact that mathematically the viscous model differs from the inviscid one not only by involving differential equations of higher order, but also in the degree of interaction subsisting between them, not a great deal has been catalogued of the fine structure of solutions of the viscous equations and the typical difficulties arising in special subregions of the flow. Because these difficulties can be very great, it is a useful and certainly non-trivial research exercise to choose for analysis an example with very rudimentary external geometry, with the object of exciting specific effects outside the scope of an inviscid model whilst retaining the possibility that a measure of theoretical analysis can accompany its numerical outworking, so that a solution is obtained which can be sorted into individual terms algebraically, whose properties can be related back to the physical parameters and numerical feasibility of the problem posed. We may then hope to use this experience as a reference standard for the behaviour of solutions of less transparent cases.

Such an example is afforded by the hypersonic Rayleigh problem. Here an infinite half-space of compressible gas is bounded by an initially motionless flat plate which is both inflexible and thermally insulating. Initially the gas is in total equilibrium, and at time $t = 0$ the plate starts to move in its own plane, entraining the gas, and eventually acquiring high speed. The disturbance thus created in the gas moves outwards from the plate as a wave whose front moves at roughly the local sound speed, and leaves a two-dimensional motion and temperature readjustment behind it. Ahead the gas is undisturbed.

In seeking to further particularize the example, our guiding principle is that laid down above: to concentrate on effects of viscous rather than purely elasto-inertial origin. If the plate moves finitely from rest there will be a transition from subsonic to supersonic flow which will be a mathematical complication, and arises from the latter cause. It can be avoided by supposing that the motion of the plate is impulsively started, and is for all times a uniform large constant velocity V. This is the form of the Rayleigh problem to be discussed below.

2. MATHEMATICAL EQUATIONS

The problem posed above has been treated analytically by Stewartson (1955) for t large, where the following equations will be found derived. Part of the aim of the present investigation is to integrate the problem long enough in time from t = 0 to see Stewartson's solution taking shape. If x is a coordinate in the direction of V and y is normal to the plate, and u, v are components of the gas velocity in the x, y directions, then with $\frac{D}{Dt} = \frac{\partial}{\partial t} + v\frac{\partial}{\partial y}$ we have

$$\text{(i)}\quad \frac{D\rho}{Dt} + \rho\frac{\partial v}{\partial y} = 0 \qquad\qquad \text{(ii)}\quad p = \rho RT$$

$$\text{(iii)}\quad \rho\frac{Du}{Dt} = \frac{\partial}{\partial y}\mu\frac{\partial u}{\partial y} \qquad\qquad \text{(iv)}\quad \rho\frac{Dv}{Dt} = -\frac{\partial p}{\partial y} + \frac{4}{3}\frac{\partial}{\partial y}\mu\frac{\partial v}{\partial y}$$

$$\text{(v)}\quad \rho C_p\frac{DT}{Dt} - \frac{Dp}{Dt} = \frac{\partial}{\partial y}\frac{\mu}{\sigma}C_p\frac{\partial T}{\partial y} + \mu\left\{\left(\frac{\partial u}{\partial y}\right)^2 + \frac{4}{3}\left(\frac{\partial v}{\partial y}\right)^2\right\}. \tag{2.1}$$

Here μ is the local viscosity, ρ the density, p the thermodynamic pressure, T the temperature and R the gas constant. C_p and σ are the specific heat at constant pressure and Prandtl number respectively. Later we shall assume $\sigma \approx 1$, and as usual denote the ratio of specific heat by γ. We shall also adopt a model of linear dependence of μ on T, so that if μ_0, T_0 are the values in the undisturbed gas we have $\mu = \mu_0 T/T_0$ in general.

To provide a far space boundary for equations (2.1) we shall denote the front of the disturbance at time t by Y(t). The boundary conditions are then

(i) at $y = Y(t)$: $u = v = 0$, $T = T_0$, $p = p_0$, $\rho = \rho_0$ (all constant); $Y(0) = 0$.

(ii) at $y = 0$ $(t > 0)$: $u = V$, $v = 0$, $\partial T/\partial y = 0$ (2.2)

The well-known substitutions of Vu, Vv, $\rho_0\rho$, ρ_0V^2p, V^2T/R and $\mu_0 t/\rho_0 a_0^2$ where $a_0^2 = \gamma p_0/\rho_0$, replace u, v, ρ, p, T, t by variables which are generally O(1) in the flow space; and it is also convenient to replace y by a scaled stream function Ψ, defined by replacing y by $\mu_o V a_o^{-2} y$, Y by $\mu_o V a_o^{-2} Y$ and then setting

$$\psi = \begin{cases} y & \text{if } y \geq Y(t) \\ \int_0^y \rho\, dy & \text{if } 0 \leq y < Y(t) \end{cases} \tag{2.3}$$

as basic space variable normal to the plate. These manoeuvres replace (2.1) by

$$\text{(i)}\quad \frac{\partial\rho}{\partial t} + \rho^2\frac{\partial v}{\partial\psi} = 0 \qquad\qquad \text{(ii)}\quad p = \rho T$$

$$\text{(iii)}\quad \frac{\partial u}{\partial t} = \gamma\frac{\partial}{\partial\psi}p\frac{\partial u}{\partial\psi} \qquad\qquad \text{(iv)}\quad \frac{\partial v}{\partial t} + \frac{\partial p}{\partial\psi} = \frac{4}{3}\gamma\frac{\partial}{\partial\psi}p\frac{\partial v}{\partial\psi}$$

$$\text{(v)}\quad \frac{\partial T}{\partial t} - \frac{\gamma - 1}{\gamma\rho}\frac{p}{t} = \frac{\gamma}{\sigma}\frac{\partial}{\partial\psi}p\frac{\partial T}{\partial\psi} + (\gamma - 1)\,p\left\{\left(\frac{\partial u}{\partial\psi}\right)^2 + \frac{4}{3}\left(\frac{\partial v}{\partial\psi}\right)^2\right\} \tag{2.4}$$

with associated boundary conditions ($M_0 = V/a_0$)

$$\text{(i)}\quad \text{at } y = Y(t):\quad \psi = y,\ u = v = 0,\ \rho = 1,\ p = T = 1/\gamma M_0^2;\quad Y(0) = 0$$

$$\text{(ii)}\quad \text{at } \psi = 0,\ t > 0:\quad u = 1,\ v = 0,\ \partial T/\partial\psi = 0. \tag{2.5}$$

Condition (2.5i) brings us to the first mathematical difficulty of the problem. If V is large then p and T almost vanish at $\psi = Y(t)$ - i.e. at the wave front. Now if p vanishes, the solutions u, v, T of (2.4iii-v) must all go singular. Thus in all cases we must expect u, v, T to have lines of singularity so close to $\psi = Y(t)$ as to seriously interfere with numerical analysis. To make progress mathematically, it seems that the best course is to treat the problem as being a perturbation of the basic case $1/M_0^2 = 0$, and begin by solving the latter case as furnishing the leading order terms of the associated expansions. For this, we replace (2.5i) by

$$\text{at } y = Y(t):\quad \psi = y,\ u = v = 0,\ \rho = 1,\ p = T = 0;\quad Y(0) = 0 \tag{2.5i'}$$

When we attend to the earlier effect on numerical analysis of starting the plate impulsively we find it has caused all initial-time characteristics in the (t, ψ) plane to pass through the origin, and the starting data is all concentrated in this one singular point. There is, however, a device which "spreads out" our initial data in a more conventional (and numerically feasible) layout. This is to take a similarity transformation by putting $\psi = \eta t^{\frac{1}{2}}$ and replacing v by $vt^{\frac{1}{2}}$. Then if we introduce the operator $\mathcal{L} = t\dfrac{\partial}{\partial t} - \frac{1}{2}\eta\dfrac{\partial}{\partial\eta}$ and set $m = \gamma/\sigma$ and $\rho^{-1} = R$ (this symbol is <u>not</u> the gas constant now, of course) in (2.4) we arrive at

$$\text{(i)}\quad \mathcal{L}R = t\frac{\partial V}{\partial\eta} \qquad\qquad \text{(ii)}\quad T = pR$$

$$\text{(iii)}\quad \mathcal{L}u = \gamma\frac{\partial}{\partial\eta}p\frac{\partial u}{\partial\eta} \qquad\qquad \text{(iv)}\quad \mathcal{L}v + \tfrac{1}{2}v + \frac{\partial p}{\partial\eta} = \frac{4}{3}\gamma\frac{\partial}{\partial\eta}p\frac{\partial v}{\partial\eta}$$

$$\text{(v)}\quad \frac{R}{\gamma}\mathcal{L}p + tp\frac{\partial v}{\partial\eta} = m\frac{\partial}{\partial\eta}p\frac{\partial T}{\partial\eta} + \frac{\gamma - 1}{\gamma}p\left\{\left(\frac{\partial u}{\partial\eta}\right)^2 + \frac{4}{3}t\left(\frac{\partial v}{\partial\eta}\right)^2\right\}. \tag{2.6}$$

For boundary conditions, we set $t^{-\frac{1}{2}}Y(t) = \eta_0(t)$ say, and obtain

$$\text{(i)}\quad \text{For } \eta \geqslant \eta_0(t)\quad u = v = 0,\ R = 1,\ p = T = 0,$$

$$\text{(ii)}\quad \text{For } \eta = 0\ (t > 0):\quad u = 1,\ v = 0,\ \partial T/\partial\eta = 0. \tag{2.7}$$

To see what sort of boundary (2.7i) places on the (t, η) plane it is necessary to know the value of $\eta_0(0)$ and this is not prescribed as yet.

3. THE STARTING DATA ON $0 \leqslant \eta \leqslant \eta_0(0)$

If we let $t \to 0$ in equations (2.6), (2.7) we get

$$\text{(i)}\quad \frac{dR}{d\eta} = 0,\ \text{ so } R = 1,\ \text{ so} \qquad\qquad \text{(ii)}\quad p = T$$

$$\text{(iii)}\quad \gamma\frac{d}{d\eta}p\frac{du}{d\eta} = -\tfrac{1}{2}\eta\frac{du}{d\eta} \qquad\qquad \text{(iv)}\quad \frac{4}{3}\gamma\frac{d}{d\eta}p\frac{dv}{d\eta} = \frac{dp}{d\eta} - \tfrac{1}{2}\eta\frac{dv}{d\eta} + \tfrac{1}{2}v$$

$$\text{(v)}\quad m\frac{d}{d\eta}p\frac{dp}{d\eta} = -\frac{1}{2\gamma}\eta\frac{dp}{d\eta} - (\gamma - 1)p\left(\frac{du}{d\eta}\right)^2 \tag{3.1}$$

with end conditions

$\eta = 0$ (plate): $u = 1,\ p' = 0,\ v = 0.$

$\eta = \eta_0(0)$ (wave front): $u = p = v = 0.$ (3.2)

This can be converted into an initial-value problem with just one initial value missing by writing $p = p_0P$, $u = 1 + p_0^{\frac{1}{2}} U$, $\eta = p_0^{\frac{1}{2}} X$ where $p_0 = p(0)$ in equations (3.1 iii, v). Then $P(0) = 1$, $P'(0) = 0$, $U(0) = 0$ and p_0 cancels out completely. Earlier workers assumed that η_0 could be found by "shooting" the resulting equations from $X = 0$ with test values of $U'(0)$ until an X emerged, say X_0, with $P = 0$. Then $U(X_0) = -1/p_0^{\frac{1}{2}}$ giving p_0, and $\eta_0(0) = p_0^{\frac{1}{2}}X_0$. If this is tried one arrives at the situation depicted in Figure 1:

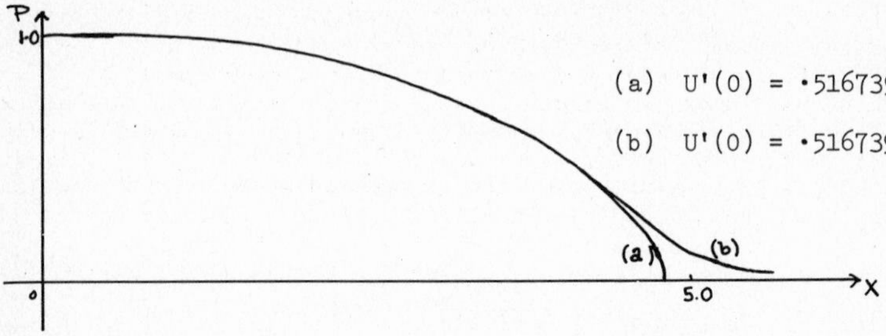

Fig 1. Result of Shooting P from X = 0 to a Hoped-For End Point

Neither curve is acceptable because p' should be bounded at $\eta_0(0)$.

This happens because the condition $p = 0$ at $\eta = \eta_0(0)$ makes $p(\eta)$ singular at this point in general, by (3.1 v). We should be shooting in the other direction, which is feasible if $|\eta_0(0)| < \infty$ and $|p'(\eta_0(0))| < \infty$. The latter is indeed necessary on physical grounds to deny infinite temperature gradient at the wave front, and is found to imply the former. It thus becomes possible to normalize the space variable and set $\eta = \eta_0(0)\,(1 - z)$ so that $0 \leqslant z \leqslant 1$. Writing $e_0 = \eta_0(0)$ and substituting $e_0\, u$, $e_0\, v$, $e_0^2\, p$, $e_0^2\, T$ for u, v, p, T respectively in (3.1, 2) we obtain

$$\text{(i)}\quad \frac{d}{dz}\, p\, \frac{du}{dz} = \frac{1}{2\gamma}\,(1 - z)\frac{du}{dz} \qquad \text{(ii)}\quad \frac{d}{dz}\, p\, \frac{dp}{dz} = \frac{1}{2\gamma m}(1 - z)\frac{dp}{dz} + \frac{1-\gamma}{\gamma m}\, p\left(\frac{du}{dz}\right)^2$$

$$\text{(iii)}\quad \frac{d}{dz}\, p\, \frac{dv}{dz} = -\frac{3}{8\gamma}\left\{2\,\frac{dp}{dz} - v - (1 - z)\,\frac{dv}{dz}\right\} \tag{3.3}$$

subject to (i) $u = p = v = 0$ at $z = 0$ (wave front)

(ii) $u = 1/e_0$ $v = 0$, $p' = 0$ at $z = 1$ (plate) (3.4)

These equations can be solved in series near $z = 0$ to finally yield us some of the analytical information we seek concerning the behaviour of the gas near the wave front. The result is that to leading orders

$$\text{(i)}\quad p = z[D + dz + \dots\ ] \qquad \text{(ii)}\quad u = z^m[A + az + \dots\ ]$$

$$\text{(iii)}\quad v = z[C + cz + Bz^{\alpha-1} + bz^{\alpha} + \dots\ ] \qquad (\alpha = \tfrac{3}{4}\, m) \tag{3.5}$$

where A, B, ... a, b, ... are constants which can all be expressed as known functions of A and B. α and m are not integers but lie roughly between 1 and 5/3.

4. NUMERICAL SOLUTION OF EQUATIONS (3.3) AND (3.4)

The form of (3.5) indicates that if we introduce variables $Y_1 \dots Y_6$ given by

$$p = Dz(1 + Y_1) \qquad pp' = D^2z(1 + Y_2) \qquad u = Az^m(1 + Y_3)$$

$$pu' = mADz^m(1 + Y_4) \qquad v = Cz(1 + Y_5) + z^\alpha(B + bz), \qquad pv' = CDz(1 + Y_6) + \alpha BDz^\alpha[1+z(1-\tfrac{3}{4}\alpha)] \tag{4.1}$$

then $Y_1 \dots Y_6$ will be sufficiently well-behaved near 0 to yield accurate determinations by a standard step-by-step numerical algorithm. If we select test values of A we can interpolate them to satisfy the end condition $p'(1) = 0$, and $\eta_0(0)$ emerges immediately as $1/u$ (1). For $\sigma = 1$, $\gamma = 5/3$ we thus find $\eta_0(0) = 1{\cdot}70627$, $A = 0{\cdot}95818$. Once A is found, B can be similarly treated to satisfy $v(1) = 0$ in (3.3 iii). For $\sigma = 1$, $\gamma = 5/3$ we find $B = -2{\cdot}237$. Graphs of p, u, v and p' at t = 0 are plotted in Figure 2.

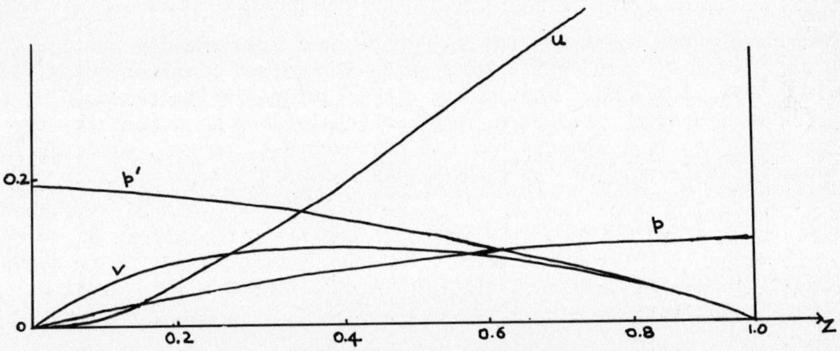

Fig 2. Graphs of Key Variables at t = 0.

5. THE SOLUTION FOR t > 0

By writing $z = 1 - \eta/\eta_0(t)$ in equations (2.6, 7) and scaling u, v, p, T as in §3 with $e_0 = \eta_0(t)$ we can get a system of partial differential equations similar to (3.3) in that series solutions near z = 0 are again possible. R needs no further scaling. The series are

$$\text{(i)}\quad p = z[D + dz + tEz^\alpha + \dots], \qquad \text{(ii)}\quad u = z^m[A + az + tHz^\alpha + \dots]$$

$$\text{(iii)}\quad v = z[C + cz + Bz^{\alpha-1} + bz^\alpha + \dots], \qquad \text{(iv)}\quad T = z[D + hz + tGz^\alpha + \dots]$$

$$\text{(v)}\quad R = 1 - tz[K + kz + Nz^{\alpha-1} + nz^\alpha + \dots]$$

where A, B, ... a, b, ... are independent of z and finite when t = 0. Similar substitutions to (4.1), e.g. $p = Dz[1 + Y_1 + tEz^\alpha]$ again give variables with tractable numerical behaviour, except in the case of the density.

Putting $R = 1 - tw$ and $w = Kz(1 + Y_7) + Nz^\alpha + nz^{\alpha+1}$, we find that Y_7 is ill-behaved at the endpoint z = 1. This effect is essentially generated by the starting value of w in the limit $t \to 0$. This latter satisfies the equation

$$\tfrac{1}{2}(1 - z)\frac{dw}{dz} + w = \frac{dv}{dz} \tag{5.1}$$

with $w = 0$, $w' = K$ at $z = 0$ and necessarily $w = v'(1)$ at $z = 1$.

Apparently we have enough initial data to integrate (5.1) out to $z = 1$, but of course the condition $w(1) = v'(1)$ overspecifies the problem unless a fortunate accident supervenes. To see that this does happen, we put $1 - z = y$ in (5.1) and let $v = v(y) = v_1 + \frac{1}{6} v_3 y^3 + \ldots$ where v_1, v_3 are constants and we note by (3.3 iii) that $d^2v/dz^2 = 0$ at $z = 1$, so no term in y^2 is needed. This yields the solution near $y = 0$:

$$w = v_1 - v_3 y^2 \log y - 2y^2 \int_0^y \tau^{-3} \{v'(\tau) - v_1 - \tfrac{1}{2} v_3 \tau^2\} d\tau + Sy^2 \tag{5.2}$$

in which $v_3 \neq 0$ but S is arbitrary, the condition $w = v_1 = dv/dz$ at $z = 1$ being obviously met. It follows that we can integrate (5.1) from $z = 0$ to a convenient intermediate value and then use S to match this solution with (5.2) at that point. For $\gamma = 5/3$, $\sigma = 1$ matching at $z = \cdot 5$ yields $S \approx - 0{\cdot}529$.

Unfortunately the equations for $t > 0$ are much more heavily coupled than in the limit $t \to 0$, and A, B and $\eta_0(t)$ have to be determined simultaneously for each t. Added to this is the fact that w and v are coupled in the analogue of (5.1). Very complicated iteration procedures are thus called for in advancing the solution in time, and full success has not been achieved as yet, although the fact that the only variable not yet tamed is Y_7 may indicate that success is not far off.

REFERENCE

Stewartson, K., Proc. Camb. Phil. Soc., 51, 1, 202-19 (1955).

STUDIES OF FREE BUOYANT AND SHEAR FLOWS BY THE VORTEX-IN-CELL METHOD

J.A.L. Thomson and J.C.S. Meng
Physical Dynamics, Inc.
P.O. Box 1069
Berkeley, California 94701

I. INTRODUCTION

In making a decision to choose an ideal method for modelling and engineering system analysis, one often sets certain requirements: It must be fast in terms of computational efficiency. It must yield an accurate and reliable solution, and it must not diverge under strain conditions, for example, when there is steep gradient, high spin rotation - for vortex development, sharp interface - for density discontinuities, etc. This would provide a useful check against existing codes. It must be illustrative and demonstrative of physical phenomenon so that it is capable of direct use in system analysis. It must be flexible and versatile in treating different problems. It must be easy to implement on new problems; in other words, it is usable and does not require massive numerical method machinery. Along these guidelines we have developed the VIC method to study the hydrodynamics of the following problems:

1. Gravitational Interface Instabilities
   (a) Rayleigh-Taylor Instability

   - natural atmosphere thermals
   - smoke stack plumes

   (b) Saffman-Taylor Instability for Two Fluids in a Porous Medium
   - oil water interface in sand
   - low density plasmas for a perturbed ionosphere
2. Aircraft Wakes in Terminal Operation
   - transport of the vortex wake in a turbulent wind shear field
3. Gravity Current
   - for atmospheric and oceanic fronts
   - for stabilization and spreading of thermals

Basically these are low speed flows of moderate Reynolds number with shear or density continuities, the equations are elliptic and flow inevitably will develop small scale motion to cause mesh problems, that is, steep gradient will arise to form sharp interface, and high spin will take place to form vortex; these are the characteristics of the incompressible shear buoyant flows. One can even treat the interface as diffusive, but due to the nature of the flow, it will evolve into a sharp interface. In fact, it is advantageous to treat the problem from the point of view of the sharp interface, because in doing this we deal with the interface only so that it reduces a 2D plus time calculation to a 1D plus time calculation where 1D is the dimension along the interface. One can discretize the interface and calculate the interactions among the elements; in this case they are discrete vortices. It is a precise simulation of a real problem and it is a limiting model for the sharp interface; however, our experiments show that it gives wrong fine spatial structures, plus the fact that the total number of operations is proportional to N^2, where N is the number of discrete vortices on the interface. This will become costly as N increases.

To be free of erroneous fine structure, we would like to construct a method capable of sub-grid spatial resolution. Neither a pure Eulerian nor a pure Lagrangian method can be satisfactory. The former has poor resolution since the sharp interface does not fall right on the grid, the latter produces fine structure errors. A mixed Eulerian-Lagrangian seems to be the only alternative. One can discretize the interface but calculate the flow field on a fixed grid. The velocity field is calculated from the stream function which produces no sharp edges, therefore no fine structure, while the density or vorticity field comes from the Lagrangian elements so that discontinuities can be allowed to develop. Due to the ellipticity of the problem, the variable of the entire flow domain is required to be calculated; this coincides with the basic theme of economy of the FFT, that is, the Fourier component of the entire domain will be transformed at once. This is a perfect match to make it possible to solve the stream function-vorticity equation in a very efficient way.

In summary, the numerical method combines the concepts of the particle-in-cell (PIC) method, Green's function and the fast Fourier transform (FFT) methods and may be dubbed the vortex-in-cell (VIC) method. The vortices are treated as the marker particles on an Eulerian grid with the velocity field solved from the updated vorticity distribution. The velocity field can either be calculated from the distribution of the vortices by Green's function formalism or by the FFT. The computation time required for Green's function formalism is proportional to the square of the numbers of the vortices N^2, while that of the FFT is fixed to $M_xM_y \ln (M_xM_y)$ where M_x, M_y are the number of mesh points in the x and y directions. Therefore, the FFT formalism can deal with a problem of a very large number of vortices without increasing the computation time. The basic economy obtained by the VIC method is not only due to the efficient numerical aspects of the scheme but also is due to the basic idea that the flow is dominated by the rotational motion generated by the discretized vortices; it is wasteful to compute the passive portion of the flow which stays passive and irrotational.

II. FORMULATION OF THE PROBLEM: FOR INVISCID FLUIDS

For inviscid fluids the vorticity equation takes the form

$$\frac{d\vec{\zeta}}{dt} = -\nabla \frac{1}{\rho} \times \nabla p + \vec{g} \tag{1}$$

and to first order in the density difference (since $p = \rho_o \vec{g} + O\left(\frac{\Delta\rho}{\rho}\right)$),

$$\frac{d\vec{\zeta}}{dt} = -\frac{\nabla\rho}{\rho_o} \times \vec{g} = \frac{g}{\rho_o}\frac{\partial\rho}{\partial x}\vec{n}_z \quad . \tag{2}$$

For two-dimensional motion in the (x,y) plane, the vorticity is effectively a scalar (i.e., has only a z component). Of particular interest is the case of two uniform immiscible fluids of slightly differing density. In this case vorticity is generated only at the interface between the two fluids, the remainder of the flow remains irrotational.

The total circulation of a given (i^{th}) fluid element $\vec{\Gamma}_i = \int \vec{\zeta}_i \, dx'dy'$ is determined by

$$\frac{d\vec{\Gamma}_i}{dt} = \frac{g(\rho_+ - \rho_-)}{\rho_o} \Delta y_i \, \vec{n}_z \tag{3}$$

where Δy_i is the length or height of the fluid element in the vertical direction, and $\vec{n}_z$ is the unit vector perpendicular to the plane of motion.

A convenient numerical analysis of the evolution of the fluid motion can be obtained by dividing up the interface into a number of discrete fluid elements and approximating the circulation of each element as being concentrated into a line vortex having circulation Γ_i. The quantity Δy_i is then to be interpreted as the separation between adjacent vortices. The evaluation of the fluid motion then reduces to the problem of following the motion of the individual discrete vortices. The velocity of the ith vortex is a summation over contributions from all other vortices:

$$\frac{d\vec{r}_i}{dt} = \vec{u}_i = \sum_{j \neq i}^{N} \frac{\vec{\Gamma}_i}{2\pi} \times \frac{(\vec{r}_i - \vec{r}_j)}{|\vec{r}_i - \vec{r}_j|^2} \tag{4}$$

This equation of motion, plus the relation determining the circulation growth rate (Eq. 3) in which Δy_i is replaced by $\frac{1}{2}(y_{i+1}-y_{i-1})$, yields a direct deterministic procedure for following the motion; this is termed Green's function formalism.

Instead of the Green's function form for the velocity field, the velocity may be expressed in terms of a stream function $\vec{\Psi}$:

$$\vec{u} = \nabla \times \vec{\Psi} \tag{5}$$

defined such that $\nabla \cdot \vec{\Psi} = 0$ (the vanishing of $\nabla \cdot \vec{\Psi}$ is automatically fulfilled in two-dimensional motion). The stream function satisfies a Poisson equation with the vorticity as the source function.

$$\nabla^2 \Psi = \zeta \quad . \tag{6}$$

In the present analysis we will be concerned primarily with fluids where the fractional variation of the density and viscosities are small ($\Delta\rho/\rho$, $\Delta\eta/\eta$, $<< 1$). The case of large density difference across the interface with the considerations of surface tension effect were discussed in Thomson and Meng (1973).

When there is no stratification, Eq. (4) is reduced to dimensionless form by introducing the following characteristic dimensions:

length R:	initial cylinder radius, or wavelength
time T:	$\dfrac{2\pi R^2}{\Gamma_o}$
Circulation Γ_o:	total circulation assumed to be distributed on the cylinder or on the interface.

When there is density difference, Eqs. (3) and (4) may be reduced to dimensionless form by introducing the following characteristic dimensions:

length:	R = the initial cylinder radius
time:	$T = \left[g(\Delta\rho/\rho)/2\pi R\right]^{-\frac{1}{2}}$
circulation:	$\Gamma_o = \left[2\pi g(\Delta\rho/\rho)R^3\right]^{\frac{1}{2}}$

In terms of the dimensionless distances, $\xi = x/R$, $\eta = y/R$, the dimensionless time $\tau = t/T$, and dimensionless circulation $\gamma = \Gamma/\Gamma_o$, the equations of motion become

$$\frac{d\gamma_i}{d\tau} = \Delta\eta_i \tag{7}$$

$$\frac{d\xi_i}{d\tau} = -\sum_{j\neq i} \gamma_j(\eta_j-\eta_i)/\left[(\xi_j-\xi_i)^2 + (\eta_j-\eta_i)^2\right] \tag{8}$$

$$\frac{d\eta_i}{d\tau} = \sum_{j\neq i} \gamma_j(\xi_j-\xi_i)/\left[(\xi_j-\xi_i)^2 + (\eta_j-\eta_i)^2\right] \tag{9}$$

where

$$\Delta\eta_i = \tfrac{1}{2}(\eta_{i+1}-\eta_{i-1}) \quad .$$

<u>Porous Medium or Viscous Flows</u>

When the viscosity forces dominate the inertial forces, and for the flow of a fluid between two parallel plates, the flow is locally Poisuelle-like and the viscous term is dominated by the curvature of the velocity profile in the direction normal to the plates, so that $\nabla^2 \vec{u} \simeq -8\vec{u}/d^2$ where d is the plate separation and $\vec{u}$ is the centerline velocity. We then have

$$\vec{u} = -\frac{d^2}{8\eta\rho}\nabla p + \frac{d^2}{8\eta}\vec{g} \quad . \tag{10}$$

Taking the curl of Eq. (10) we obtain for the vorticity

$$\vec{\zeta} = \frac{d^2}{8}\left[-\nabla\left(\frac{1}{\rho\eta}\right)\times\nabla p + \nabla\left(\frac{1}{\eta}\right)\times\vec{g}\right] \quad . \tag{11}$$

To first order in the density and viscosity variations ∇p may be replaced by ∇p_o where $\nabla p_o = -8\eta_o\rho_o/d^2\,\vec{U}_o + \rho_o\vec{g}$. In other words

$$\vec{\zeta} = -\frac{\nabla(\eta\rho)}{\eta_o\rho_o}\times\vec{U}_o + \frac{\nabla\rho}{\rho_o}\times\left(\frac{d^2 g}{8\eta}\right) \tag{12}$$

where $\vec{U}_o$ and $\vec{g}$ are in the same direction.

For two uniform fluids separated by a sharp interface, the total circulation of a given (i^{th}) fluid element is determined by

$$\Gamma_i = \left[\left(\frac{\mu_+-\mu_-}{\mu_o}\right)U_o + \left(\frac{\rho_+-\rho_-}{\rho_o}\right)\frac{d^2 g}{8\eta}\right]\Delta y_i \quad . \tag{13}$$

III. SOME FEATURES OF THE VIC METHOD

Spreading

One characteristic length in the VIC method, that is, the cell size for the FFT, not only sets the limit on the accuracy of the solution to $\frac{1}{\text{cell size}}$ but also defines the highest gradient resolvable. When the identifying agent, for example, the mass or circulation carried by the particle, is spread over a disc instead of the infinitesimal areas covered by a point, one additional dimension is introduced, and that will enhance the resolution of the scheme and help to alleviate the high gradient associated with the particle representation. Consider the following three distributions: the first one is a step function with narrow width and its Fourier transform will produce noise at very high frequency; in the second distribution, less noise is produced at high frequency as a result of spreading the quantity smoothly; in the last case, which is the best representation one could have, no noise is generated at high frequency. Using the last distribution one can have a scheme free of noise inherent to many of the PIC methods. We have adopted here a distribution similar to the last case; the figure illustrates the relationship between the mesh, identified by indices I and J, and the particle located at $[X(M), Y(M)]$. Let us spread the quantity to a disc of a dimension of one cell size,

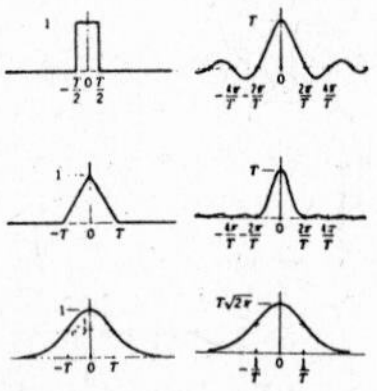

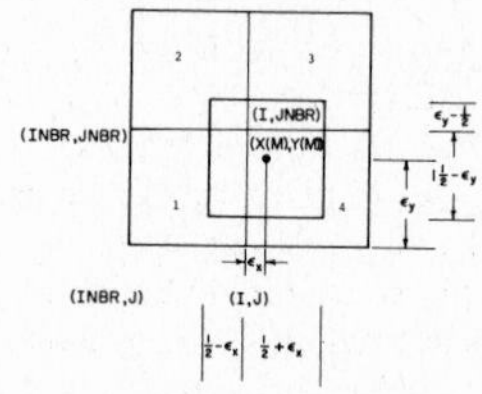

which is the minimum size of the disc if one is sure that there is approximately one particle per cell throughout the computation, and further assume a distribution according to $1/\{1+[[x-X(M)]/\Delta x]^2\}\{1+[[y-Y(M)]/\Delta y]^2\}$ where Δx, Δy are the half widths and should be equal to the separation distance between two vortices. The weighting factors can be obtained by integrating the above relation over the overlapping area in each cell. The spreading of the physical quantities, such as the mass or vorticity or heat, according to the above relation was found effective in removing irregular motion of the vortices. The effect is similar to that due to the finite core radius in the Green's function formalism, basically because it introduces a mesh-independent length scale over which one can distribute the quantity arbitrarily. This greatly enhanced the capability of the VIC method to deal with flows with high gradient.

Damping

Most of the physical problems treated by the VIC method are basically unstable, that is, amplification of small errors always exists. It is therefore desirable to eliminate as many sources of noise generation by the numerical procedure as possible to obtain a realistically smooth solution. One of the sources is due to the finite mode Fourier transform. The error is known as aliasing errors and was discussed in detail by Cooley and Tukey (1965). Basically, the mesh size sets the highest wavenumber representable by the Fourier series; in other words, any higher wavenumber components are assumed negligible. This is also to say that any strong variation over a mesh size must be removed artificially, and this can be achieved conveniently through the damping procedure in the Fourier space. One commonly applied damping is the function

$$e^{-\beta(k_x^4+k_y^4)} \quad .$$

We have found $\beta = 2$ to be sufficient in very unstable flows.

Repacking the Vortex Procedure

One problem which is always associated with the PIC method is the continual adding or removing particles from the calculation. Depending upon the physical nature of the problem, particles may be congested to yield unrealistically high gradients of flow variables or the number of particles in a region of interest is so low that no realistic representation of flow is possible. From the point of view of maintaining a uniform accuracy, a repacking procedure which can add or delete particles as necessary must be applied. It is also desirable from the economy point of view, since higher than a limiting particle number density will not improve accuracy, but lower than a limiting particle density will reduce accuracy. Without being committed to using a large number of particles throughout the computation, there is no alternative but to adopt a repacking procedure. Since the particles do carry physical variables - mass, vorticity, heat - the repacking procedure must be based upon the laws of conservation. For the VIC method, these are vorticity and centroid of the vorticity.

One parameter in the calculation serves as a convenient indicator to apply the repacking procedure, that is, the separation distance between any neighboring vortices. When a preset separation is exceeded or reduced, vortices are added or deleted and the vorticity is redistributed among the added and the original vortices to satisfy the conservation laws. For example, assume that between the ith and $i+1^{th}$ vortices a vortex is added. The circulation of the new vortex is assumed to be $1/3\ (\Gamma_i+\Gamma_{i+1})$ and the circulation on the original particles is reduced to $2/3\ \Gamma_i$ and $2/3\ \Gamma_{i+1}$, accordingly. The new vortex position is determined by a linear relation which keeps the centroid of the vorticity unchanged.

Implication of Periodic Boundary Conditions

The efficiency of the FFT is accompanied with a necessary restriction which states that the boundary condition must be periodic. The domain dimensions are automatically implied to be periodic lengths so that one must realize that images do exist at one periodic length apart even though they are not visible in the domain treated. It also implies that when a vortex is swept out of the flow domain at one side, another vortex must be introduced to the flow domain at a distance of one periodic length away from the original position. One should always make certain that no significant influence of the images is exerted upon the flow; this in general will require the region of interest to be centered in the domain and to readjust the domain whenever the flow becomes close to the boundaries.

IV. NUMERICAL COMPUTATIONS WITH THE DISCRETE VORTEX MODEL

Vortex consideration furnishes a powerful ally in attacking many of the complex problems of nonlinear rotational flows. We shall, in this context, establish a working and efficient numerical basis for such an approach by emphasizing the manner in which flow motion is generated by the vorticity and how the subsequent evolution develops.

IV.1 Rise of a Buoyant Cylinder (Thermal)

Scorer (1958) suggested that the behavior of plumes of smoke, when they have been bent over by a crosswind and become nearly horizontal, can conveniently be discussed in terms of a line source of buoyancy. Turner (1959) made a study of this in a water channel and observed that the plumes bent over in this way tend to split sideways into two concentrated regions with a clear space between them. He found that the flow in planes perpendicular to the axis of the plume is very like that in a vortex pair, with a region of fast rise in the center and slower regions on each side.

Equations (7) to (9) have been used to calculate the time dependent motion in two dimensions following the release of an initially uniform circular cylinder of light weight fluid in a homogeneous heavier fluid. Since the density gradients in this example are limited to the (deforming) surface of the cylinder, the motion may be followed by following the history of the vortex sheet which comprises the cylinder boundary (see Fig. 1.a). The results of the calculation are shown in Figs. 1.a to 1.d. The cylinder boundary at time zero was divided uniformly into N points (61 points on the half circle). The velocity of each point was calculated at successive time increments according to Eqs. (8) and (9). The circulation of each point was calculated from Eq. (7) as a function of time, the initial values being taken equal to zero (i.e., no initial motion).

The initial motion of the cylinder appears to be simply an upward displacement without sensible distortion. By the time the net displacement is of the order of ½ the initial cylinder radius, the beginning of vortex development is evident (Fig. 1.b). The vortex appears well developed by the time the buoyant region has risen about one diameter (Fig. 1.c). By this time most of the vorticity is concentrated in the vortex region. The rate of change of the total circulation of this region is obtained by summing Eq. (7) over the entire vortex sheet

$$\frac{d}{d\tau}\left(\sum_i \gamma_i\right) = \delta\eta \qquad \text{or} \qquad d\Gamma/dt = g\,\frac{\Delta\rho}{\rho}\,\delta y$$

where $\delta\eta$ is the thickness of the cap on the axis of symmetry. Thus, the value of the vortex circulation grows during the vortex development but saturates when the vortex has fully developed. The subsequent motion (after vortex formation) of the buoyant and entrained material has been discussed by others (particularly by J.S. Turner (1959) and also by T. Fohl (1967)). It is important to point out here that the flow entrainment is by the vortex interaction process, not by turbulence; the basic motion is an inviscid laminar but rotational flow, not a turbulent motion as many believed.

Although the present calculation was carried out for a cylindrical configuration, essentially similar results are anticipated for spherical buoyant bubbles.

When the cylinder (see Fig. 1.d) has risen more than one diameter the set of point vortices form an irregular distribution within a finite cloud. The original vortex sheet is now so convoluted as to be impossible to follow. Although the numerical model cannot be a good model of the small scale structure at such times, it is interesting to note that the large scale motion agrees reasonably well with theoretical expectations. This may be seen as follows.

Turner (1959) has shown that the circulation of each vortex approaches a constant value after vortex formation. This may be seen from Kelvin's theorem which states that around any closed circuit C, $d\Gamma/dt = \oint_C 1/\rho\, \nabla p \cdot d\vec{s}$. After vortex formation, the density along a path threading the center of the vortex is essentially constant and equal to the ambient value and $d\Gamma/dt \to 0$. When Γ is constant, the rise velocity varies inversely as the separation of the vortex pair $V \sim 1/R$. The upward momentum increases at a constant rate $d(MV)/dt = F_B$ where F_B is the (constant) buoyant force. Since M is proportional to T^2 in two dimensions and RV is constant, the separation R increases linearly with time $R \sim t$. Since the rise velocity of the vortex pair varies as Γ/R, the net rise distance y increases logarithmically (in two dimensions) with time $y \sim \ell n\, t$. In Fig. 2 we show that the time dependence of the width and height of the rising vortex pair agree reasonably well with the expected dependence. (For Fig. 2, see the inserts on Fig. 1.)

In three dimensions the expansion rate will have a different time dependence. Since here the mass varies as R^3, the momentum equation reduces to $dR^2/dt \sim$ constant after torus formation (when $RV \sim$ constant). Here $R \sim t^{\frac{1}{2}}$ and $dz/dt \sim t^{-\frac{1}{2}}$. Thus the radius of the torus increases linearly with height ($R \sim z$).

When the fast Fourier transform (FFT) is applied to solve the stream function, a vortex system of much larger numbers of particles can be employed costing essentially the same amount of computation time. For example, in this case the Green's function approach using 61 points took .246 seconds per time step. The stream function approach using 591 points took .426 seconds per step, while the same approach using 41 points cost .31 seconds per step. One drawback is, however, that the finite mode Fourier transform does produce aliasing errors. Figures 1.e through 1.h show the result obtained from the stream function using 200 points through the same period as that in Figs. 1.a to 1.d.

IV.2 Saffman-Taylor Instability

Long, narrow convecting cells, that is, the "salt fingers", are commonly observed when hot salty water is poured over cold fresh water. A very similar phenomenon occurs at the interface of two superposed viscous fluids when they are forced by gravity and an imposed pressure gradient through a porous medium. The practical examples, in addition to those already mentioned in the introduction, are oil-water interface in sand or in shale and fresh air-smoke interface in a peat moss or a granular coal bed fire. Saffman and Taylor (1958) studied the finger-like structure in a Hele-Shaw cell and found that the ratio of the width of the finger to the spacing of the fingers is almost always equal to ½.

The general equation (Eq. 12) is composed of two diffusive gradients. A flow system in this context is usually called the doubly diffusive convection. One typical example is hot, salty water overlying cold, fresh water.

Equation (13) determines the circulation at the discrete vortex (x_i, y_i). From that the velocity field is calculated from Eq. (4). To reduce these equations into dimensionless forms, we should notice that the flow is characterized by two quantities: the acceleration $g\Delta\rho/\rho$ and the time k/η, where k is the permeability and η is the kinematic viscosity. The time scale is derived as the time that it takes the viscosity to diffuse across the void area in a porous medium which is represented by k. From these two variables, we can get the following characteristic dimensions:

$$\text{length } R: \ \frac{1}{2\pi}\frac{g\Delta\rho}{\rho}\left(\frac{k}{\eta}\right)^2, \quad \text{time } T: \ \frac{k}{\eta}, \quad \text{circulation } \Gamma_o: \ \frac{1}{2\pi}\left(\frac{g\Delta\rho}{\rho}\right)^2\left(\frac{k}{\eta}\right)^3 .$$

Equation (13) becomes

$$\gamma_i = \left(\frac{\Delta\mu}{\mu}\,\widetilde{U}_o + 1\right)\Delta\eta_i \tag{14}$$

where $\widetilde{U}_o$ is the dimensionless variable $U_o/R\,T$ and $\eta_i = y_i/R$, $\gamma_i = \Gamma_i/\Gamma_o$. Eq. (4) is reduced to Eqs. (8) and (9).

In order to determine the maximum allowable time step Δt, it is necessary to find the order of magnitude of the terminal velocity. We shall attempt to estimate this quantity by two means.

First, consider that the finger is replaced by a sphere of fluid accelerated under the effective gravitational force $g\Delta\rho/\rho$ and decelerated by the viscous force $\mu\nabla^2 u \sim \mu\, U_{terminal}/k$ times the surface area of the sphere. Therefore, $U_t \cong 4/3\; g\Delta\rho/\rho\; k/\eta$. In terms of the characteristic length and time, we have $\widetilde{U}_t = 8\pi/3$.

Second, if we assume $U_o \equiv 0$ for simplicity, the vorticity generated is that due to the terms $k/\eta\; \nabla\rho/\rho_o \times \vec{g}$ only. From Fig. 3.a, the vorticity is maximum where $\partial\rho/\partial x$ is the maximum. The resultant motion will be to lift up the center and push down the external edges. At some later time, Fig. 3.b shows the growing finger where the vorticity of the opposite sign also appears, but the resultant motion is a further acceleration in the same trend as in Fig. 3.a. Assuming the final stage of the finger structure is that depicted in Fig. 3.e, we can estimate the velocity at the center top due to the vortices distributed on the now vertical interface which is of length ℓ and width h,

$$u_t \sim \frac{2}{\pi} \int_0^{\ell/2} \frac{\Gamma dy}{\sqrt{h^2+y^2}} \sim \frac{2k}{\eta\pi} \frac{g\Delta\rho}{\rho} \ell n \left(\frac{\ell}{h}\right)$$

if $h << \ell/2$ or $\tilde{u}_t \cong 9.2$, when $\ell \cong 10h$.

From these two estimations, one can say that Δt should be approximately $1/\tilde{u}_t \cong .1$ and, in fact, we set $\Delta t \cong .01$ to be sure of a stable time integration.

The initial disturbance corresponds to that shown in Fig. 3.a. The interface is perturbed by a gaussian displacement at the center. At $\tau = .22$, Fig. 3.b shows that the center has risen while the edges of the gaussian displacement are depressed. Fig. 3.c shows that as the finger grows the spacing between the vortices on the top becomes large, so that unless a method by which vortices can be added to this region is implemented, one will not obtain good resolution in this region. Notice that in Fig. 3.d at $\tau = .68$, one vortex has been added into the center region. This re-packing procedure and its aspects of economy was explained in Section III.

The sharp edges appearing in Fig. 3.d are a result of two vortices rotating around each other. Figure 3.e shows the final result at $\tau = .98$, where the final number of vortices is N = 146. Figures 3.f through 3.j are the counterparts of Figs. 3.a to 3.e, but are solved by the FFT scheme; the finger structure reveals more realistic configurations than the Green's function solution.

IV.3 <u>Transport of Aircraft Trailing Vortices</u>

By wing theory, the lift or the wing loading is linearly proportional to the circulation about the wing cross section, and it is well known that the wing load [so is the circulation S(x)] can be approximated by the elliptic curve

$$\frac{S(x)}{S_o} = \sqrt{1 - \left(\frac{x}{R}\right)^2}$$

where S_o is the maximum circulation at x = 0 and R is the wing span.

The vortex sheet is divided into N strips along the direction of flight, each segment of which contains a circulation $\Gamma_i(x)$ given by

$$\frac{\Gamma(x_i)}{S_o} = \left[S(x_i) - S(x_{i+1})\right]/\Delta x \quad . \tag{15}$$

The wind profile near the ground is known to exhibit a logarithmic dependence upon the elevation y, $U = u_\tau/\kappa \, \ell n \, (y/y_o) + \text{constant}$, where u_τ is the friction velocity and is usually given by the relation $u_\tau \cong 1/30 \, U_{\text{at height of 1 km}}$, the κ is the Kármán constant and is equal to .42 for most applications. The y_o is the roughness parameter.

From the above equation the vertical wind shear can be obtained by taking the derivative with respect to y which yields $\Gamma(y) = u_\tau/\kappa y$. This circulation is assigned to each mesh point as a discrete vortex.

The limitation of a purely FFT scheme for studying the present problem is its small time step which is essentially determined by the angular velocity of a pair of vortices with the shortest separation distance. The separation distance between two trailing vortices is in the order of Wing Span/Number of Vortices on the Sheet, which is about 1.2 meters for the Boeing 747 case, so that the time step will be $(1.2)^2/\Gamma_o \cong .014$ seconds. To cover the time elapse of interest of 60 seconds, it would require 4000 steps or about 30 minutes cpu time. It is therefore desirable to obtain a numerical scheme which can handle the same number of vortices without being limited to the small integration time step.

One important physical phenomenon gives a hint in solving this dilemma; that is, the vortices on the vortex sheet are coagulated into two well defined tip vortices within only a few seconds after the plane flies by. So for all practical purposes, one can regard the trailing vortex sheet as two tip vortices.

The velocity contribution from the tip vortices is obtained by the Green's function scheme which gives accurate results. The contribution from the wind shear vortices and the vortices on the trailing vortex sheet, excluding those in the tip vortex core, is still calculated by the FFT scheme. Since there are only two tip vortices, the total number of additional operations using Green's function scheme to calculate the velocity will be $\sim 4\cdot(N-2)$ where N is the total number of vortices, including the tip vortices. The total number of operations will be $4\cdot(N-2) + M_x M_y/4 \ \ell n\ (M_x M_y/4)$; one therefore has a numerical method which possesses the precision of the Green's function scheme and the economy of the FFT scheme. The reason that the Green's function scheme will not be limited by the small time step is because one can calculate the angular displacements directly due to the tip vortices instead of calculating the linear velocity. In other words, one can calculate accurately the displacement of a point, which rotates many times about a center, by multiplying the angular velocity to the time step while an integration of the linear velocity must be done in a much smaller time step in order to stay on the same orbit.

Here we will apply a methodology which includes the turbulently diffusing vortex core established by Owen (1970) and the scheme mentioned above. In other words, at $t = 0$, the initial trailing vortex sheet is given by Eq. (15), with a vortex core defined by $R = 2/\sigma\sqrt{\nu t}$ and $\sigma = 1/\Lambda\ (\nu/\Gamma_o)^{\frac{1}{4}}$ and $\Lambda \cong 1$. All the vortices in the core will be absorbed by the tip vortices with the new location of the tip vortices at the centroid of the original distribution of vortices inside the core. As the core expands, more trailing vortices are incorporated into the tip vortices and the tip vortices locations updated to the centroid of the original vortex distribution. The velocities due to the tip vortices and the wind shear are then superposed.

A study of the trailing vortex shed from a Boeing 747 aircraft, at a height of 120 meters above the runway, using a 32×32 grid was carried out. The trailing vortex was represented by 25 discrete vortices over half of the wing span, each assigned a circulation value according to Eq. (15). The wind shear vorticity is distributed over the flow domain on a 17×32 mesh, and the images are obtained by the symmetry condition in the vertical direction. The buoyant engine exhaust is also represented by 25 vortices; the temperature difference with respect to the ambient air is assumed to be 10°K. Figure 4.a shows a Boeing 747 trailing vortex and its buoyant exhaust in a 17×32 grid. Figure 4.b shows the rolling up of the vortex sheet after 7 seconds, no skewness is observed at this time. Figure 4.c shows the overall picture of the vortex system at t = 17 seconds, the wind shear vortices near the ground where the vorticity is maximum are swept up and mutual induction between those wind shear vortices and the tip vortices may be expected to emerge. Figure 4.d shows the skewed configuration at t = 7 seconds. The trailing vortices are transported nearly 170 meters to the left from the original position, and the position of the wind shear vortices delineate clearly the wind profile. Figure 4.e shows the locations of the vortices at t = 37 seconds; the downwind vortex is observed to rise as a result of the interaction with the wind shear. The trajectories of the vortices are marked in Fig. 4.f.

IV.4 Gravity Current

We now turn to a phenomenon which is commonly treated as a hydraulic jump problem rather than from the point of view of the vortex interactions. It corresponds to the intrustion of a heavier fluid (a front or "nose") into a fluid of lighter density. Examples of this flow are found in the atmosphere; in a weather front (say, a sea breeze), in front of a gravity current which is usually termed "Sudanese haboob", at river-sea junction; at the intrusion of salt water under fresh water when a lock gate is opened.

In hydraulics this is called the lock exchange problem. Many experiments have been made in this area. A summary can be found in Turner (1973). Benjamin (1968) showed that the front must have a shape of head behind which there is a turbulent region and an abrupt drop to a layer of uniform depth. Kármán (1940) showed that the shape of the nose or head at the front is 60° to the horizontal.

Figure 4.a shows the initial geometry and its velocity vector plot. A circular cylinder of fluid of intermediate density is formed by, for example, mixing of fluids on a stably stratified density discontinuity. The lines show the location of the vortices: 171 points altogether, distributed non-uniformly over the first quadrant, with the higher number density near the thermocline and fewer on the top. This is necessary to ensure good resolution of the nose geometry. The vorticity is initially zero. Then Eqs. (7) through (9) are applied to advance the calculation. Due to the lower fluid density over the cylinder and higher underneath, the buoyant force will flatten the cylinder; if there is no vorticity generated, the circular cylinder will simply be flattened into a thin layer. Due to the vorticity generated by the buoyance, there forms an advancing nose which is called the gravity current or weather front in meteorology. Notice that the maximum velocity at $\tau = 0$ is small. As it develops, Fig. 5.b shows at $\tau = .6$ the flattening wake and its velocity distribution. The velocity has grown to 1.638 in terms of the variables defined by I.2. Figure 5.c shows the well defined nose shape at $\tau = 1.6$. The nose has a slope of nearly 60° half included angle as predicted by Kármán (1940). The nose advancing velocity is bounded by $1/\sqrt{2} \leq v/\sqrt{g\Delta\rho/\rho \, H} \leq \sqrt{2}$ if the ratio of the intruded layer depth H to the overlying layer depth d is bounded by $\frac{1}{2} \geq H/d \geq 0$.

At $\tau = 1.6$, one can estimate from the above relations in terms of the presently-defined non-dimensional variables that the maximum velocity (the velocity at the nose) is approximately $\sqrt{2\pi}$, which agrees with what is shown in the velocity vector plots. At $\tau = 2.2$, the solution is shown in Fig. 5.d.

REFERENCES

Benjamin, T.B., "Gravity Currents and Related Phenomena", J. Fluid Mech. 31, 2, 209-248, 1968.

Cooley, J.W. and Tukey, J.W., "An Algorithm for the Machine Calculation of Complex Fourier Series", Math. of Comput., 19, 297-301, 1965.

Fohl, T., "Optimization of Flow for Forcing Stack Wastes to High Altitudes", J. Air Poll. Control Assoc. 17, 730-733, 1967.

Karman, Th. von, "The Engineer Grapples with Non-linear Problems", American Mathematical Society, Bulletin 46, 615-683, 1940.

Owen, P.R., "The Decay of a Turbulent Trailing Vortex", Aeronautical Quarterly, 21, 69-78, 1970.

Saffman, P.G. and Taylor, G.I., "The Penetration of a Fluid into a Porous Medium or Hele-Shaw Cell Containing a More Viscous Liquid", Roy. Soc. of London, A. 245, 312-329, 1958.

Scorer, R.S., Natural Aerodynamics, Pergamon Press, London, 1958.

Thomson, J.A. and Meng, J.C.S., "Development of Vortex Structures in Buoyant and Shear Flows", Physical Dynamics Report PD-73-034, RADC-TR-74-117.

Turner, J.S., "A Comparison between Buoyant Vortex Rings and Vortex Pairs", J. Fluid Mech. 7, Part 3, 419-432, 1959.

Turner, J.S., Buoyancy Effects in Fluids, Cambridge University Press, 1973.

Woods, J.D., " On Richardson's Numbers as a Criteria for Laminar-Turbulent-Laminar Transition in the Ocean and Atmosphere", Radio Sci. 4, 1289, 1969.

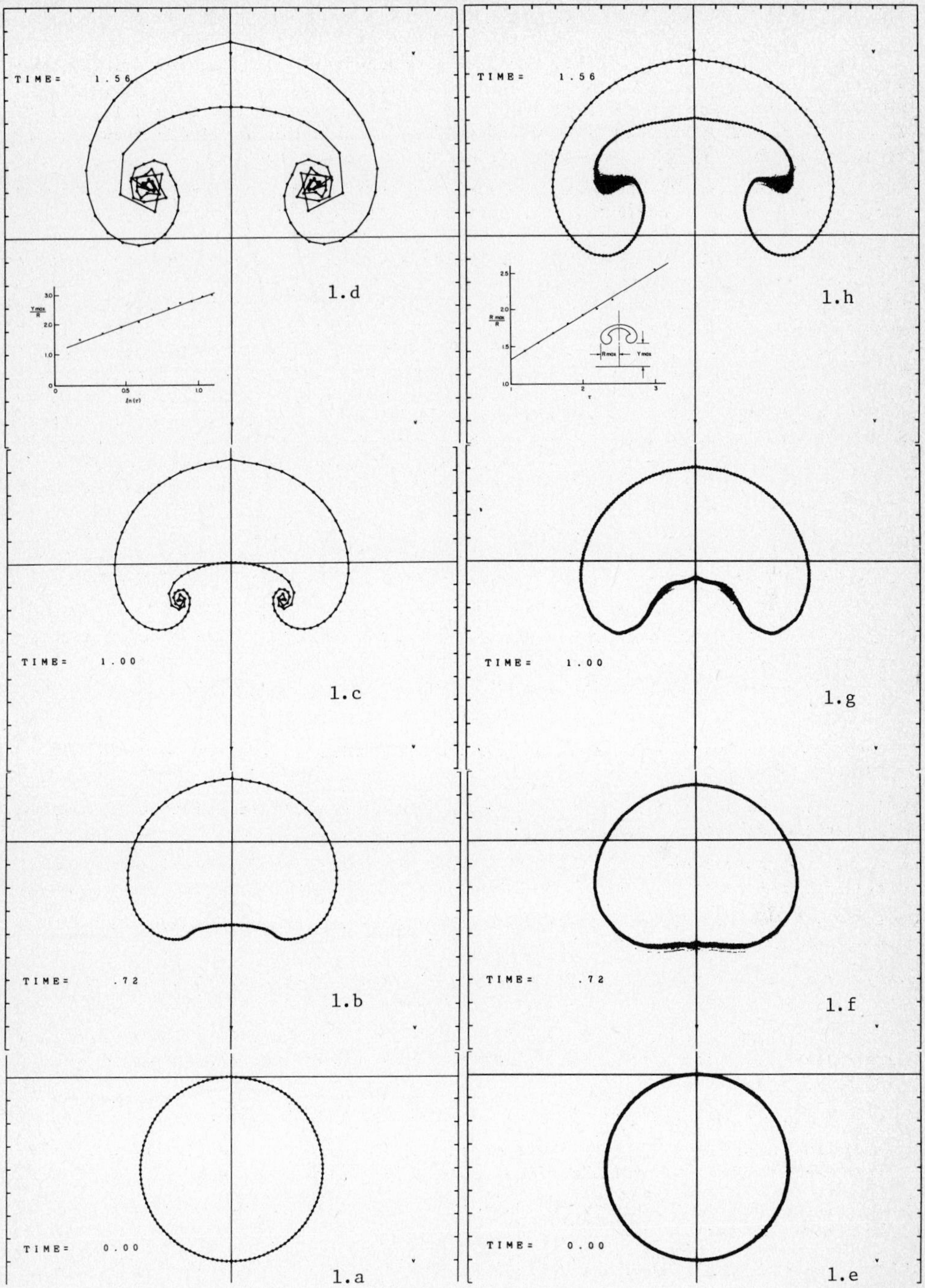

XBL 746-1023

Figure 1.

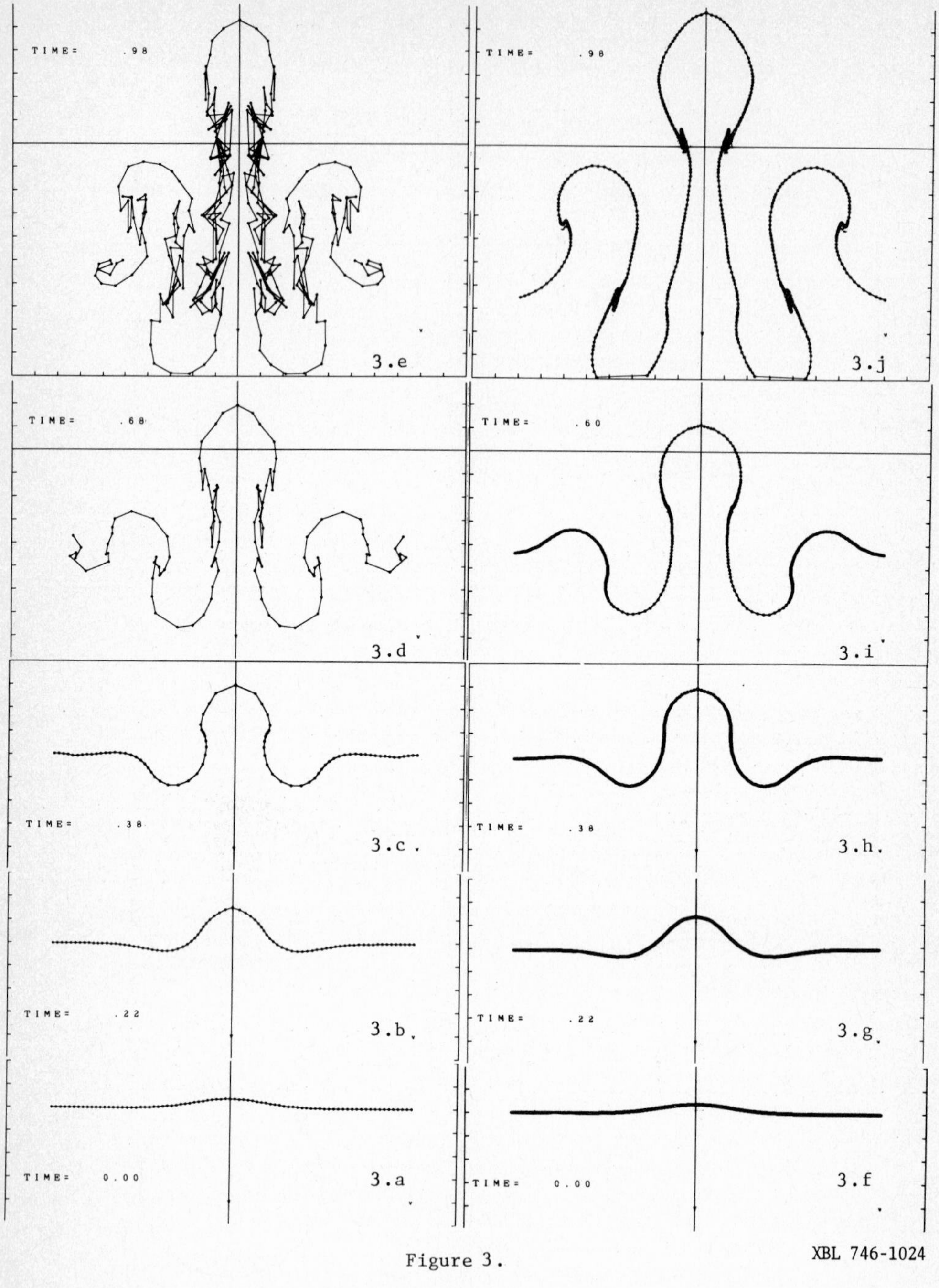

Figure 3.

XBL 746-1024

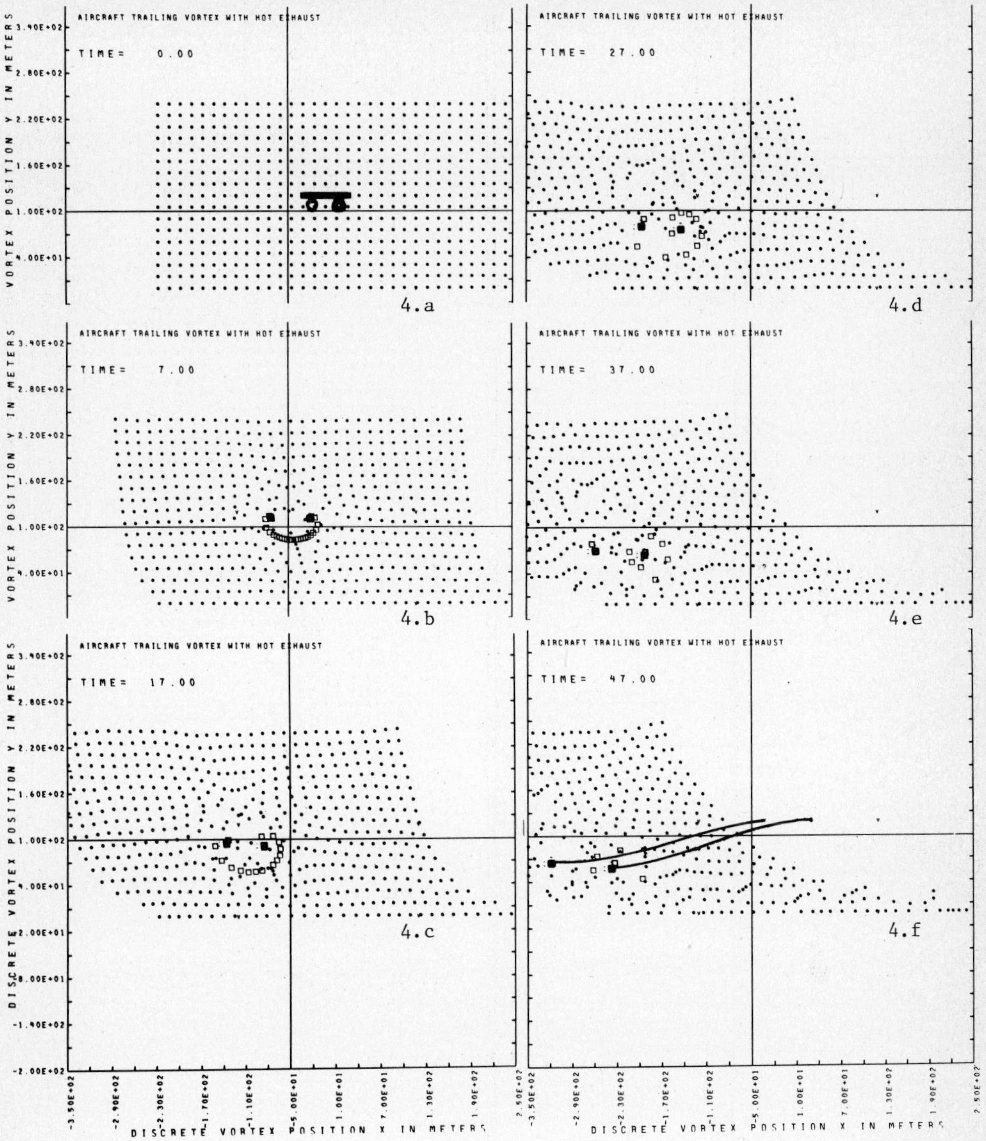

XBL 746-1025

Figure 4.

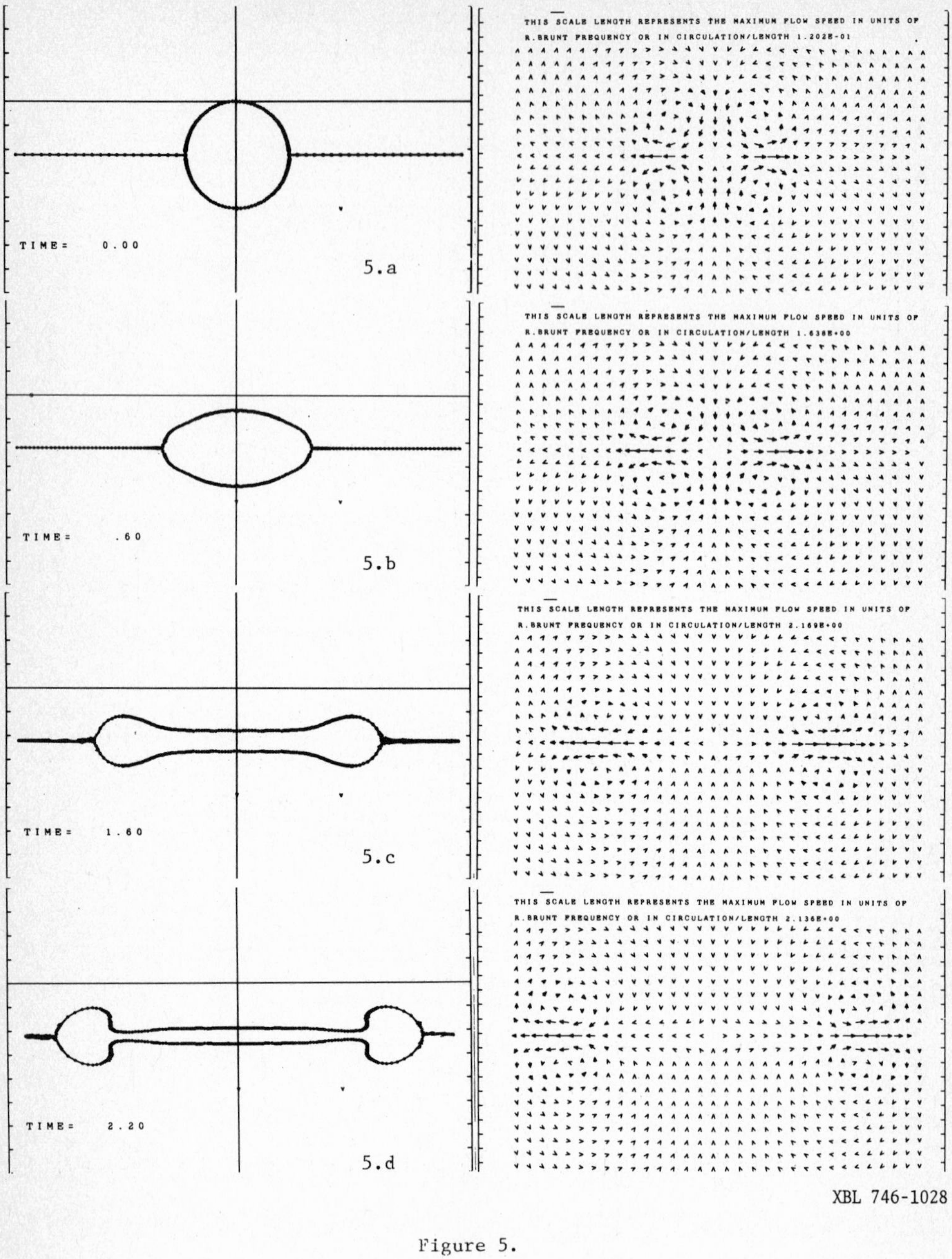

XBL 746-1028

Figure 5.

TIME - DEPENDENT ONE - DIMENSIONAL LIQUID - BUBBLE FLOW

P. van Beek

Delft University of Technology
Delft, the Netherlands

ABSTRACT

In this report a numerical method is proposed to integrate a one-dimensional time-dependent model for the flow of a liquid containing gas bubbles. A linear analysis of this method is carried out. First appropriate boundary conditions are given. Then a correct approximation at the boundaries is derived, necessary to solve the Lax-Wendroff scheme uniquely for all mesh points.
The method is applied to calculate the propagation of discontinuities in a liquid-bubble mixture.

MATHEMATICAL MODEL

Under certain circumstances a flowing liquid-bubble mixture can be described by conservation of mass and momentum written down for liquidphase and gasphase separately. The one-dimensional dimensionless equations read as follows:

a) for the gasphase

$$\frac{\partial}{\partial t}(\alpha\rho) + \frac{\partial}{\partial x}(\alpha\rho u_g) = 0$$

$$\frac{\partial}{\partial t}(\alpha\rho u_g) + \frac{\partial}{\partial x}(\alpha\rho u_g^2) = -R$$

b) for the liquidphase

$$\frac{\partial}{\partial t}(1-\alpha) + \frac{\partial}{\partial x}((1-\alpha)u_\ell) = 0$$

$$\frac{\partial}{\partial t}((1-\alpha)u_\ell) + \frac{\partial}{\partial x}((1-\alpha)u_\ell^2) + \frac{\partial}{\partial x}((1-\alpha)p) = R + \tau_w$$

α is the volume fraction occupied by the gasphase, p is the pressure and ρ the mass density of the gasphase relative to the mass density of the liquidphase. Furthermore u_g and u_ℓ denote gas- and liquidvelocity respectively. The term R describes the resistance of the gas bubbles due to the velocity difference $u_g - u_\ell$. Friction between liquidphase and wall (of a tube e.g.) is represented by τ_w.

For a complete list of physical assumptions see van Beek ([1]).
The system of equations has the conservation law property

$$(1)\quad \frac{\partial u}{\partial t} + \frac{\partial}{\partial x} f(u) = b(u)$$

with

$$u = \begin{bmatrix} 1-\alpha \\ \alpha\rho \\ m_\ell \\ m_g \end{bmatrix} \qquad f(u) = \begin{bmatrix} m_\ell \\ m_g \\ \frac{m_\ell^2}{1-\alpha} + (1-\alpha)p \\ \frac{m_g^2}{\alpha\rho} + \alpha p \end{bmatrix} \qquad b(u) = \begin{bmatrix} 0 \\ 0 \\ R - \tau_w \\ -R \end{bmatrix}$$

Here m_ℓ and m_g are defined as $(1-\alpha)u_\ell$ and $\alpha\rho u_g$ respectively.
The Jacobian $A(u) = \frac{\partial f}{\partial u}$ of the system is :

$$A(u) = \begin{bmatrix} 0 & 0 & 1 & 0 \\ 0 & 0 & 0 & 1 \\ -u_\ell^2 + c_g^2 \frac{\rho}{\alpha} & c_g^2(1/\alpha - 1) & 2u_\ell & 0 \\ 0 & c_g^2 - u_g^2 & 0 & 2u_g \end{bmatrix}$$

Note : $c_g = \sqrt{\frac{p}{\rho}}$ is the isothermal speed of sound in pure gas.
The eigenvalues of A(u) are : $\lambda_1 = u_g + c_g$, $\lambda_2 = u_g - c_g$, $\lambda_3 = u_\ell + c_g\sqrt{\frac{\rho}{\alpha}}$, $\lambda_4 = u_\ell - c_g\sqrt{\frac{\rho}{\alpha}}$.
These eigenvalues are real and different, so we call system (1) hyperbolic. It is assumed that the flow does not become critical, so that $|u_g| < c_g$ and $|u_\ell| < c_g\sqrt{\frac{\rho}{\alpha}}$
Thus λ_1 and λ_3 are positive, λ_2 and λ_4 are negative.

BOUNDARY CONDITIONS

Appropriate boundary conditions can be derived for a system of hyperbolic differential equations in characteristic form. Consider

$$(2)\quad \frac{\partial v}{\partial t} + \Lambda(x,t)\frac{\partial v}{\partial x} + B(x,t)\cdot v = g(x,t)$$

with

$$\Lambda = \begin{bmatrix} \Lambda_1 & 0 \\ 0 & \Lambda_2 \end{bmatrix}$$

where Λ_1(rxr) and Λ_2((n-r) x (n-r)) are a positive and a negative diagonalmatrix respectively.

Partition v in a corresponding way :

$$v = \begin{bmatrix} v^{I} \\ \hdashline v^{II} \end{bmatrix} \begin{matrix} \updownarrow r \\ \updownarrow n-r \end{matrix}$$

It can be proved (see Thomée ([5])) that (2) leads to a well posed problem in the region $0 \leq x \leq 1$, $t \geq 0$ if the initial value $v(x,0)$ is given on $0 \leq x \leq 1$ and if at the boundaries $x = 0$ and $x = 1$ relations are given of the form :

$$v^{I}(0,t) = S_0(t)\, v^{II}(0,t) + c_0(t) \tag{3}$$

$$v^{II}(1,t) = S_1(t)\, v^{I}(1,t) + c_1(t) \tag{4}$$

where $S_0(t)$ and $S_1(t)$ are $r\text{x}(n-r)$ and $(n-r)\text{x}r$ matrices respectively. To derive appropriate boundary conditions for the two phase problem, (1) has to be transformed to quasi-linear characteristic form. This can be done by the transformation $du = T(u).dv$, where $T(u)$ is the matrix of eigenvectors of $A(u)$. The resulting nonlinear system then has to be linearised, in order to get a system of form (2).

If at $x = 0$ r relations between the components of $u(0,t)$ are prescribed, so that

$$L_0(t)\, u(0,t) = a_0(t) \qquad (L_0(t) = r \times n) \tag{5}$$

then the linearised transformed relation

$$L_0(t)\, T(U)\, v(0,t) = a_0(t) \tag{6}$$

has to be considered.

If relation (6) can be put into form (3), and if the same can be done for the boundary at $x = 1$ then the linearised system is well posed.

<u>Example</u>

(7) Prescribe $\begin{cases} \text{at } x = 0 : (1-\alpha)(0,t) \text{ and } \alpha\rho(0,t) \\ \text{at } x = 1 : m_{\ell}(1,t) \text{ and } m_g(1,t) \end{cases}$

$L_0(t)$ in relation (5) now becomes :

$$\begin{bmatrix} 1 & 0 & 0 & 0 \\ 0 & 1 & 0 & 0 \end{bmatrix}$$

Calculating the eigenvectors of $A(u)$ it follows :

$$T(u) = \begin{bmatrix} 1 & 1 & 1 & 1 \\ 0 & A & 0 & B \\ \lambda_3 & \lambda_1 & \lambda_4 & \lambda_2 \\ 0 & \lambda_1 A & 0 & \lambda_2 B \end{bmatrix}$$

with $A = \dfrac{\lambda_3\lambda_4 - \lambda_1(2u_\ell - \lambda_1)}{c_g^2(1/\alpha - 1)}$ and $B = \dfrac{\lambda_3\lambda_4 - \lambda_2(2u_\ell - \lambda_2)}{c_g^2(1/\alpha - 1)}$

Relation (6) becomes :

$$\begin{bmatrix} 1 & 1 \\ 0 & A \end{bmatrix} v^{I}(0,t) + \begin{bmatrix} 1 & 1 \\ 0 & B \end{bmatrix} v^{II}(0,t) = a_0(t)$$

This leads to a relation of form (3) if and only if $A \neq 0$. It can be easily shown that $A \approx 0$ if $\alpha = \rho$ (notice to this end that $u_g << c_g$ and $u_\ell << c_g$). Because $\rho = O(10^{-3})$ it is safe to require that $0.01 < \alpha \leq 1$ in order to have $A \neq 0$. Prescribing $m_\ell(1,t)$ and $m_g(1,t)$ at $x = 1$ leads to a well posed problem if $\lambda_2\lambda_4 B \neq 0$, the latter inequality appearing by a similar analysis as was made for the boundary at $x = 0$. On assumption $\lambda_2 \neq 0$ and $\lambda_4 \neq 0$. As for A, it appears that $B \approx 0$ if $\alpha = \rho$.
Conclusion :

if $0.01 < \alpha \leq 1$ then the linearised system (1), with initial conditions and boundary conditions (7) is well posed.

Note : if at $x = 0$ or at $x = 1$ $\alpha\rho$ and m_g are prescribed as functions of t, then it is not possible to obtain a boundary condition of form (3) or (4). This is proved in van Beek ([1]).

DISCRETISATION OF THE CONTINUOUS PROBLEM

For the approximation u_i^n of the solution $u(i\Delta x, n\Delta t)$ at the interior points the two-step version of the Lax-Wendroff scheme, as introduced by Richtmeyer, is chosen. This choice is based on the well known favourable properties of this conservative difference scheme for compressible flow problems, where discontinuities may develop. The difference scheme has to be adapted to the non-homogeneous system (1), which is straightforward (see e.g. Gourlay and Morris ([2])). The result is :

(8a) $$u_{i+1/2}^{n+1/2} = 1/2\left(u_{i+1}^n + u_i^n\right) - 1/2\,\frac{\Delta t}{\Delta x}\left(f_{i+1}^n - f_i^n\right) + \frac{\Delta t}{4}\left(b_{i+1}^n + b_i^n\right)$$

(8b) $$u_i^{n+1} = u_i^n - \frac{\Delta t}{\Delta x}\left(f_{i+1/2}^{n+1/2} - f_{i-1/2}^{n+1/2}\right) + \frac{\Delta t}{2}\left(b_{i+1/2}^{n+1/2} + b_{i-1/2}^{n+1/2}\right)$$

Note : $f_i^n = f(u_i^n)$ and $b_i^n = b(u_i^n)$.
Let the mesh points on $0 \leq x \leq 1$ be numbered from 0 to L ($L\Delta x = 1$). If the approximation u_i^n is known for $i = 0(1)L$ at some time $n\Delta t$, then scheme (8a,b) defines u_i^{n+1} for $i = 1(1)L-1$.
For the two phase problem two components of u_0^{n+1} and two components of u_L^{n+1} are known (e.g. by (7)). The other two components of u_0^{n+1} and u_L^{n+1} have to be determined by an additional boundary approximation. This has to be done with care in order to avoid instabilities at the boundaries (see e.g. Kreiss ([3])). For a system in characteristic form it seems natural to determine the components of $v_0^{II\,n+1}$ and similarly those of $v_L^{I\,n+1}$ by extrapolation from the interior points, because

$v^{II}(x,t)$ is propagated towards $x = 0$ and $v^{I}(x,t)$ towards $x = 1$.
It is proved by Kreiss et al.([3] and [4]) that the following types of boundary approximations give rise to a stable difference method for an initial-boundary value problem :

$$(9)\quad \text{at } x = 0: \quad D_+^j \, v_0^{II\,n+1} = 0 \qquad (j = 1,2,3,\cdots)$$

$$\text{at } x = 1: \quad D_-^j \, v_L^{I\,n+1} = 0 \qquad (j = 1,2,3,\cdots)$$

$$(10)\quad \text{at } x = 0: \quad v_0^{II\,n+1} = v_0^{II\,n} - \frac{\Delta t}{\Delta x}\Lambda_2 D_+ v_0^{II\,n}$$

$$\text{at } x = 1: \quad v_L^{I\,n+1} = v_L^{I\,n} - \frac{\Delta t}{\Delta x}\Lambda_1 D_- v_L^{I\,n}$$

Note : $D_+ v_i^n = v_{i+1}^n - v_i^n$ and $D_- v_i^n = v_i^n - v_{i-1}^n$

To derive an appropriate boundary approximation for system (1) relations (9) and (10) have to be transformed to the original variable u. To this end $T^{-1}(u)$ has to be calculated, for locally $v = T^{-1}(u).u$ holds.

$$T^{-1}(u) = \frac{1}{2c_g\sqrt{\beta/\alpha}}\begin{bmatrix} -\lambda_4 & (\lambda_3 - 2u_g)/P & 1 & 1/P \\ 0 & -(\lambda_2\sqrt{\beta/\alpha})/A & 0 & \sqrt{\beta/\alpha}/A \\ \lambda_3 & (2u_g - \lambda_4)/Q & -1 & -1/Q \\ 0 & (\lambda_1\sqrt{\beta/\alpha})/B & 0 & -\sqrt{\beta/\alpha}/B \end{bmatrix}$$

where $P = \dfrac{\lambda_1\lambda_2 + \lambda_3(\lambda_3 - 2u_g)}{c_g^2(1/\alpha - 1)}$ and $Q = \dfrac{\lambda_1\lambda_2 + \lambda_4(\lambda_4 - 2u_g)}{c_g^2(1/\alpha - 1)}$

For an example of a boundary treatment see van Beek [1].

SHOCKS IN LIQUID - BUBBLE MIXTURES

The relevant variables behind and in front of the shock (indicated by the lower indices 1 and 0 respectively) can be predicted by the Hugoniot relations derived from system (1). It can be proved (see van Beek ([1])) that these Hugoniot relations possess only two different solutions under most physical circumstances, viz.

1) solution 1, with a shockspeed much smaller than c_g and which satisfies the Hugoniot relations for the gasphase trivially, so $\alpha_0 p_0 = \alpha_1 p_1$ and $u_{g0} = u_{g1}$
2) solution 2, with a shockspeed of the same order of magnitude as c_g. It appears for this solution that $\alpha_0 \simeq \alpha_1$.

If a discontinuity is introduced such that $\alpha_0 p_0 = \alpha_1 p_1$ then it is propagated accor-

ding to solution 1, thus with low velocity.
If a discontinuity is introduced such that $\alpha_0 p_0 \neq \alpha_1 p_1$, while there is a fairly large difference between α_0 and α_1 then the states on both sides of the discontinuity can never satisfy the Hugoniot relations, because neither $\alpha_0 p_0 = \alpha_1 p_1$ nor $\alpha_1 \simeq \alpha_0$ holds.
Therefore two shocks are generated.
The propagation of both kinds of discontinuities can be simulated by the finite difference scheme as described in the foregoing section. It appears that artificial viscosity has to be added in order to avoid violent oscillations. Then in both cases the calculated solutions satisfy the Hugoniot relations very well. If a discontinuity of the second kind is introduced indeed two shocks are generated. For an extensive discussion of the results see van Beek ([1]).

REFERENCES

[1] Beek, P. van, One-dimensional liquid-bubble flow, Report N.A.-9, Mathematical Institute of the Delft University of Technology, to appear.

[2] Gourlay, A.R., and Morris, J.Ll., Finite difference methods for nonlinear hyperbolic systems II, Math. Comp., vol. 22, 1968.

[3] Kreiss, H.-O., Boundary conditions for difference approximation of hyperbolic differential equations, Agard Lecture Series no. 64, 1973.

[4] Gustafsson, B., Kreiss, H.-O., Sundström, A., Stability theory for mixed initial-boundary value problems II, Math. Comp., vol. 26, 1972.

[5] Thomée, V., Estimates of the Friedrichs-Lewy type for mixed problems in the theory of linear hyperbolic differential equations in two independent variables, Math. Scand., vol. 5, 1957.

DRAG OF A FINITE FLAT PLATE

A.E.P. Veldman and A.I. van de Vooren

Mathematical Institute, University of Groningen, The Netherlands

1. INTRODUCTION

It has been shown by Stewartson [1] and Messiter [2] that for large values of the Reynolds number Re the frictional drag at one side of a finite flat plate of length ℓ is given by

$$\frac{D}{\frac{1}{2}\rho U_\infty^2 \ell} = \frac{1.328}{Re^{1/2}} + \frac{c}{Re^{7/8}} + O(Re^{-1}) \qquad (1.1)$$

The second term, which has been overlooked for a long time, is due to increased shear stress in a region $O(Re^{-3/8})$ ahead of the trailing edge.

It is the purpose of the present note to give results for the coefficient c, the increased shear stress, the pressure, the wake velocity and the displacement thickness, all at a distance $O(Re^{-3/8})$ from the trailing edge.

2. MATHEMATICAL FORMULATION OF THE PROBLEM

First approximations of viscous flow along a flat plate and in the wake are given by the Blasius and Goldstein [3] solutions. These are solutions of the boundary layer equations without pressure term. Near the trailing edge, however, these solutions are not valid as is most easily seen from the Goldstein solution. This solution implies a pressure gradient of $O(Re^{-1/2})$ with a coefficient behaving like $x^{-2/3}$ for $x \to 0$, $x = 0$ being the trailing edge. Hence, the assumption that the pressure term is smaller than the other terms in the boundary layer equation - these are $O(1)$ - is no longer valid for x sufficiently small.

Independently, Stewartson [1] and Messiter [2] showed that the Blasius and Goldstein solutions no longer hold for $|x| \leq O(Re^{-3/8})$. By evaluating carefully the order of magnitude of all terms in the Navier-Stokes equations they came to the conclusion that there exists in y-direction, normal to the plate, a triple-deck structure. In the three decks different approximations to the N.S. equations hold. The main deck is the middle deck where $y = O(Re^{-1/2})$ and where the flow is inviscid (constant vorticity along streamlines). Since this flow does not satisfy the no-slip condition at the plate, there is a viscous sublayer (lower deck) with $y = O(Re^{-5/8})$. The upper deck is required for smoothing out the disturbances of the main deck, which are larger than elsewhere in the boundary layer. In the upper deck, where $y = O(Re^{-3/8})$, exists potential flow.

In accordance with the method of matched inner and outer expansions we introduce new variables - dependent and independent - which together with their derivatives are all of order 1 for $Re \to \infty$.

For the *middle deck* we have

$$x^* = Re^{3/8} x, \quad \tilde{y} = Re^{1/2} y. \qquad (2.1)$$

together with the following asymptotic expansions for stream function ψ and pressure p.

$$\left.\begin{aligned} \psi(x,y;Re) &= Re^{-1/2} f(\tilde{y}) + Re^{-5/8} \tilde{\psi}_1(x^*,\tilde{y}) + \ldots \quad && Re \to \infty \\ p(x,y;Re) &= \qquad\qquad\quad Re^{-1/4} p_1(x^*,\tilde{y}) + \ldots && x^*,\tilde{y} \text{ fixed} \end{aligned}\right\} \qquad (2.2)$$

Here, $f(\tilde{y})$ denotes the Blasius solution satisfying

$$2f''' + ff'' = 0$$

with the boundary conditions $f(0) = f'(0) = 0$, $f'(\infty) = 1$.

With u and v denoting horizontal and vertical velocities, we have the following orders of magnitude

$$u = O(1), \quad v = O(Re^{-1/4}), \quad p = O(Re^{-1/4})$$

By evaluating the order of magnitude of all terms in the N.S. equations it is found that terms $O(Re^{1/4})$ are the most important ones. The first equation becomes

$$f'\tilde{\psi}_{1_{x^*\tilde{y}}} - f''\tilde{\psi}_{1_{x^*}} = 0$$

where subscripts denote partial derivatives.

The solution of this equation is

$$\tilde{\psi}_1(x^*,\tilde{y}) = f'(\tilde{y})G(x^*), \tag{2.3}$$

where $G(x^*)$ is unknown at this moment but has the property that $G(-\infty) = 0$, since for $x^* \to -\infty$ the solution should match the Blasius solution.

The second equation leads in the usual way to the result that p_1 will be independent of $\tilde{y}$ and hence we shall write $p_1(x^*)$ instead of $p_1(x^*,\tilde{y})$.

We now consider the *lower deck*, where

$$x^* = Re^{3/8}x, \quad y^* = Re^{5/8}y \tag{2.4}$$

$$\left.\begin{aligned} \psi(x,y;Re) &= Re^{-3/4}\psi_1^*(x^*,y^*) + \ldots \qquad Re \to \infty \\ p(x,y;Re) &= Re^{-1/4}p_1(x^*) + \ldots \qquad x^*,y^* \text{ fixed} \end{aligned}\right\} \tag{2.5}$$

Orders of magnitude are

$$u = O(Re^{-1/8}), \quad v = O(Re^{-3/8}), \quad p = O(Re^{-1/4})$$

In the new variables the first N.S. equation becomes

$$\psi^*_{1_{y^*}}\psi^*_{1_{x^*y^*}} - \psi^*_{1_{x^*}}\psi^*_{1_{y^*y^*}} = -p_{1_{x^*}} + \psi^*_{1_{y^*y^*y^*}} \tag{2.6}$$

Elimination of p_1 by differentiation to y^* yields

$$\psi^*_{1_{y^*}}\psi^*_{1_{x^*y^*y^*}} - \psi^*_{1_{x^*}}\psi^*_{1_{y^*y^*y^*}} = \psi^*_{1_{y^*y^*y^*y^*}} \tag{2.7}$$

The boundary conditions are

at the plate $x^* < 0, \quad y^* = 0 : \quad \psi_1^* = 0, \quad \psi^*_{1_{y^*}} = 0$

in the wake $x^* > 0, \quad y^* = 0 : \quad \psi_1^* = 0, \quad \psi^*_{1_{y^*y^*}} = 0$

Other boundary conditions follow from the matching condition with the inviscid solution. This says that the coordinate expansion for $\tilde{y} \to 0$ of the asymptotic expansion of the middle-deck solution should correspond to the coordinate expansion for $y^* \to \infty$ of the asymptotic expansion of the lower-deck solution. The first one is, using (2.2) and (2.3)

$$\tilde{y} \to 0 \quad \psi(x,y;Re) = Re^{-1/2}\cdot\tfrac{1}{2}a_1\tilde{y}^2 + Re^{-5/8}a_1\tilde{y}\,G(x^*) + \ldots$$

where $a_1 = f''(0) = 0.33206$, the Blasius constant. The second expansion then must be

$$y^* \to \infty \quad \psi(x,y;Re) = Re^{-3/4}\{\tfrac{1}{2}a_1 y^{*2} + a_1 y^* G(x^*) + \ldots\}$$

Therefore we may write, using (2.5)

$$y^* \to \infty \qquad \psi_1^*(x^*,y^*) = \tfrac{1}{2}a_1 y^{*2} + a_1 y^* G(x^*) + a_1 H(x^*), \tag{2.8}$$

where the last term comes from the first omitted term in (2.2).

For known $G(x^*)$ and $H(x^*)$, eq. (2.8) corresponds to boundary conditions for ψ_1^*, $\psi_{1_{y^*}}^*$ and $\psi_{1_{y^*y^*}}^*$ at $y^* \to \infty$. Together with the conditions at $y^* = 0$, we would have five boundary conditions for the fourth-order parabolic differential equation (2.7). This means that there exists a relation between $G(x^*)$ and $H(x^*)$.

The pressure being independent of y^* can be obtained by substitution of eq. (2.8) into (2.6), followed by integration to x^*. The result is

$$p_1(x^*) = a_1^2\{H(x^*) - \tfrac{1}{2}G^2(x^*)\} \tag{2.9}$$

In order to obtain a second relation between the functions $G(x^*)$ and $H(x^*)$, we have to consider the potential flow in the *upper deck*. The variables in the upper deck are

$$x^* = Re^{3/8}x, \quad \hat{y} = Re^{3/8}y \tag{2.10}$$

The asymptotic expansions are

$$\left.\begin{aligned} \psi(x,y;Re) &= Re^{-3/8}\hat{y} - Re^{-1/2}\beta + Re^{-5/8}\hat{\psi}_1(x^*,\hat{y}) + \dots && Re \to \infty \\ p(x,y;Re) &= \qquad\qquad Re^{-1/4}\hat{p}_1(x^*,\hat{y}) + \dots && x^*,\hat{y} \text{ fixed} \end{aligned}\right\} \tag{2.11}$$

The first two terms of the expansion for ψ follow from substitution in (2.1) of the coordinate expansion of $f(\tilde{y}) = \tilde{y}-\beta$ if $\tilde{y} \to \infty$. Matching with the middle deck solution yields

$$\hat{\psi}_1(x^*,0) = G(x^*) \text{ and } \hat{p}_1(x^*,0) = p_1(x^*) \tag{2.12}$$

Orders of magnitude in the upper deck are

$$u = O(1), \quad v = O(Re^{-1/4}), \quad p = O(Re^{-1/4})$$

At the outer edge of the upper deck ($r^* = \sqrt{x^{*2}+\hat{y}^2} > O(Re^{-3/8})$) v and p have decreased to $O(Re^{-1/2})$, the usual order of magnitude outside the boundary layer.

In the upper deck the viscous terms are small compared to inertia and pressure terms and since the oncoming stream ($y > O(Re^{-1/2})$) has no vorticity, we have potential flow in the upper deck and $\hat{\psi}_1$ is a harmonic function. The linearized form of Bernoulli's law, $u = 1-p$, holds and hence

$$\hat{\psi}_{1_{\hat{y}}}(x^*,0) = -p_1(x^*) \tag{2.13}$$

The boundary conditions for the harmonic function $\hat{\psi}_1$ then are

$$\hat{y} = 0 \qquad \hat{\psi}_1(x^*,0) = G(x^*) \text{ and } \hat{\psi}_{1_{\hat{y}}}(x^*,0) = a_1^2\{\tfrac{1}{2}G^2(x^*)-H(x^*)\} \tag{2.14}$$

This, apparently, is too much and hence there must exist a second relation between $G(x^*)$ and $H(x^*)$. These two functions are completely determined by the two relations and with them, the whole problem is determined.

3. TRANSFORMATION TO BOUNDED VARIABLES

A difficulty for the numerical evaluation is that all variables - dependent and independent - are unbounded. The independent variables are kept bounded by transformation to finite regions, while the dependent variables are kept finite by subtraction of the terms which become infinite. These terms are determined from the asymptotic behaviour of the dependent variables for large values of the independent variables.

A study of the asymptotic behaviour was performed by Stewartson [1] and Messiter [2]. The latter, however, has overlooked the appearance of eigenfunctions in the viscous as well as in the potential region. For details in the derivation of the asymptotics we refer to [1] and [2].

In the viscous region the behaviour of the streamfunction for $x^* \to \infty$ is given by

$$\psi_1^*(x^*,y^*) = x^{*2/3} f_0(\eta) + b_1 x^{*-1/3}\{2f_0(\eta) - \eta f_0'(\eta)\} + O(x^{*-2/3}). \tag{3.1}$$

where $\eta = y^*/x^{*1/3}$ and $f_0(\eta)$ is the Goldstein solution [3], which satisfies

$$f_0''' + \tfrac{2}{3} f_0 f_0'' - \tfrac{1}{3} f_0'^2 = 0 \text{ with } f_0(0) = f_0''(0) = 0,\ f_0''(\infty) = a_1 \tag{3.2}$$

For $\eta \to \infty$ we have $f_0(\eta) = \frac{1}{2} a_1 (\eta + A_1)^2$ + exp. small terms (3.3)

where $A_1 = 1.2881$.

The term of order $x^{*-1/3}$ in (3.1) is an eigenfunction. It is multiplied by an unknown constant b_1.

For $x^* \to -\infty$ the behaviour of the streamfunction in the viscous region is

$$\psi_1^*(x^*,y^*) = \tfrac{1}{2} a_1 x^{*2/3} \eta^2 + O(x^{*-2/3}) \tag{3.4}$$

From eqs. (2.8), (3.1) and (3.4) we derive the behaviour of G and H for $|x^*| \to \infty$.

$$\left.\begin{aligned} x^* \to \infty \quad & G(x^*) = A_1 x^{*1/3} + b_1 A_1 x^{*-2/3} + O(x^{*-1}) \\ & H(x^*) = \tfrac{1}{2} A_1^2 x^{*2/3} + b_1 A_1^2 x^{*-1/3} + O(x^{*-2/3}) \\ x^* \to -\infty \quad & G(x^*) = O(x^{*-1}), \quad H(x^*) = O(x^{*-2/3}) \end{aligned}\right\} \tag{3.5}$$

The expansions for G will be used as boundary conditions for $\hat{\psi}_1$ in the potential region, see eq. (2.14). Using polar coordinates we obtain for $r^* \to \infty$

$$\hat{\psi}_1(r^*,\theta) = r^{*1/3} A_1 \{-3^{-1/2} \sin(\theta/3) + \cos(\theta/3)\} + O(r^{*-2/3}) \tag{3.6}$$

From eq. (2.13) the pressure then follows as

$$\left.\begin{aligned} x^* \to \infty \quad & p_1(x^*) = 3^{-3/2} A_1 x^{*-2/3} + O(x^{*-5/3}) \\ x^* \to -\infty \quad & p_1(x^*) = -2.3^{-3/2} A_1 |x^*|^{-2/3} + O(x^{*-5/3}) \end{aligned}\right\} \tag{3.7}$$

In the *potential region* we now introduce the dependent variable

$$\psi_p = \hat{\psi}_1 - r^{*1/3} A_1 \{-3^{-1/2} \sin(\theta/3) + \cos(\theta/3)\} \tag{3.8}$$

which remains finite everywhere. Corresponding terms in G and H are also subtracted

$$\left.\begin{aligned} x^* \geq 0 \quad & G^+(x^*) = G(x^*) - A_1 x^{*1/3}, \quad H^+(x^*) = H(x^*) - \tfrac{1}{2} A_1^2 x^{*2/3} \\ x^* \leq 0 \quad & G^-(x^*) = G(x^*), \quad H^-(x^*) = H(x^*) \end{aligned}\right\} \tag{3.9}$$

The independent variable r^* has been transformed to a finite variable σ by means of

$$\sigma(r^*) = r^{*1/3} (Z + r^{*1/3})^{-1} \tag{3.10}$$

The constant Z was determined empirically during the computations. It was found that Z = 3 gave a good distribution of the meshpoints over the whole interval $r^* \in [0,\infty)$. The harmonic function ψ_p satisfies

$$\sigma^2_{r*}(\psi_p)_{\sigma\sigma} + (\sigma_{r*r*} + \frac{1}{r*}\sigma_{r*})(\psi_p)_\sigma + \frac{1}{r*^2}(\psi_p)_{\theta\theta} = 0 \tag{3.11}$$

with the conditions following from (2.14)

$$\left.\begin{array}{ll} \psi_p(\sigma,0) = G^+(x*) , & \psi_p(\sigma,\pi) = G^-(x*) \\ \psi_p(1,\theta) = 0 , & \psi_p(0,\theta) = G^+(0) = G^-(0) \end{array}\right\} \tag{3.12}$$

$$\left.\begin{array}{ll} \theta = 0 & (\psi_p)_\theta = -a_1^2 r*\{H^+ - \tfrac{1}{2}G^{+2} - G^+A_1 r*^{1/3}\} + 3^{-3/2}A_1 r*^{1/3} \\ \theta = \pi & (\psi_p)_\theta = a_1^2 r*\{H^- - \tfrac{1}{2}G^{-2}\} + 2.3^{-3/2}A_1 r*^{1/3} \end{array}\right\} \tag{3.13}$$

In the *viscous region* we replace eq. (2.7) by

$$\psi^*_{1_{y*y*}} = K^*_1, \quad K^*_{1_{y*y*}} = \psi^*_{1_{y*}} K^*_{1_{x*}} - \psi^*_{1_{x*}} K^*_{1_{y*}} \tag{3.14}$$

For $x* \leq 0$ we introduce as new dependent variables

$$\psi_v = \psi^*_1 - \tfrac{1}{2}a_1 y*^2 - a_1 y* G^-(x*), \quad K = K^*_1 - a_1 \tag{3.15}$$

which keeps ψ_v finite for $y* \to \infty$ and for $x* \to \infty$, see eqs. (2.8) and (3.5).

The independent variables are kept finite by aid of the transformations

$$\sigma(x*) = |x*|^{1/3}(Z + |x*|^{1/3})^{-1}, \quad \mu(x*,y*) = y*\{Z(\alpha+|x*|^{1/3})+y*\}^{-1} \tag{3.16}$$

The first formula has been taken equal to (3.10) because of the convenience for matching. The latter formula is the combination of $\tau = y*/Z(\alpha+|x*|^{1/3})$ and $\mu = \tau/(1+\tau)$, where the purpose of the first is to introduce a variable behaving like η for large $|x*|$ without degenerating for $x* \to 0$. The constant α was taken equal to 7.

The first equation (3.14) becomes $(\psi_v)_{y*y*} = K$ and is used as an equation for ψ_v. Using boundary conditions $\psi_v = 0$ at $y* = 0$ and $(\psi_v)_{y*} \to 0$ as $y* \to \infty$, we can write

$$\psi_v(\sigma,\mu) = - \int_0^\mu (\mu_{y*})^{-1}\{\int_{\bar{\mu}}^1 (\mu_{y*})^{-1} K(\sigma,\bar{\bar{\mu}})d\bar{\bar{\mu}}\}d\bar{\mu} \tag{3.17}$$

The second equation (3.14) is used as an equation for K

$$\mu^2_{y*}K_{\mu\mu} + \mu_{y*y*}K_\mu = \psi^*_{1_{y*}}(K_\sigma \sigma_{x*} + K_\mu \mu_{x*}) - \psi^*_{1_{x*}} K_\mu \mu_{y*} \tag{3.18}$$

The ψ^*_1-derivatives are regarded as coefficients and therefore have not been written in the new variable ψ_v. The boundary conditions are

$$K(\sigma,1) = 0, \quad \int_0^1 (\mu_{y*})^{-1} K(\sigma,\mu)d\mu = a_1 G^-(\sigma) \tag{3.19}$$

The second condition follows from $(\psi_v)_{y*} = -a_1 G^-$ for $y* = 0$.

The *wake region* ($x* \geq 0$) is divided into a lower and an upper region as was also done by Smith [4]. Independent variables in the *lower region* are $x*$ and η. The $x*$-variable is transformed to σ as in (3.16). The lower region lies between $\eta = 0$ and $\eta = \alpha$. Dependent variables are defined by

$$\left.\begin{array}{l} \psi^*_1 = x*^{2/3}\psi_w = x*^{2/3}\{f_0(\eta) + \bar{\psi}_w(\sigma,\eta)\} \\ K^*_1 = K_w = f''_0(\eta) + \bar{K}_w(\sigma,\eta) \end{array}\right\} \tag{3.20}$$

The point $x* = 0$, $y* = 0$ is a singular point of the problem. The value of K^*_1 changes from, say $K_1^* = a_2$, for $x^* = 0^-$ to $K_1^* = 0$ for $x = 0^+$. Introduction of the η-coordinate removes this singularity. For $x^* = 0^+$, ψ_w satisfies

$$(\psi_w)_{\eta\eta\eta\eta} + \frac{2}{3}\psi_w(\psi_w)_{\eta\eta\eta} = 0, \quad \psi_w(0) = (\psi_w)_{\eta\eta}(0) = 0, \quad (\psi_w)_{\eta\eta}(\infty) = a_2 \tag{3.21}$$

while the fourth boundary condition is that for $\eta \to \infty$, ψ_w should have no term linear in η (see [1]).

The value of α is chosen such that the conditions at ∞ in both (3.2) and (3.21) could be applied at $\eta = \alpha$. This implies that the exponential terms for $\eta \to \infty$ in $\psi_w(0,\eta)$ and $f_0(\eta)$ are less than 10^{-8} for $\eta \geq \alpha$. The value $\alpha = 7$ fulfills this condition.

For a good distribution of the meshpoints we take equidistant steps in the variable t, defined by

$$\eta = t + (\alpha-1)t^3, \qquad t \in [0,1] \tag{3.22}$$

Except for $\sigma = 0$ the variables $\overline{\psi}_w$ and $\overline{K}_w$ are used. The differential equations follow by substitution of (3.20), (3.22) and the first equation (3.16) into eqs. (3.14).

The transformations of the independent variables in the *upper region* are such that they are smoothly connected to all adjacent regions. We take for x^* again (3.16) and y^* is transformed by

$$\mu = y^*/(\alpha Z+y^*) \tag{3.23}$$

New dependent variables are defined by

$$\psi_v = \psi_1^* - \tfrac{1}{2}a_1 y^{*2} - a_1 y^* G^+(x^*) - a_1 A_1 y^* x^{*1/3} - \tfrac{1}{2}A_1^2 x^{*2/3}, \quad K = K_1^* - a_1 \tag{3.24}$$

The equations for ψ_v and K again follow from substitution into (3.14). For the equations in the *wake region* we have as boundary conditions

$$t = 0: \; \overline{\psi}_w = \overline{K}_w = 0; \quad \mu = 1: \; (\psi_v)_{y^*} = K = 0 \tag{3.25}$$

Furthermore, at the line $\eta = \alpha$, smooth transition must occur from the lower to the upper region. The value, found for $\psi_v(\sigma,1)$ is equal to $a_1 H^+(\sigma)$.

4. THE ITERATION PROCESS

The equations have been solved by means of finite differences. A grid was used, equally spaced in the σ, μ, t and θ variables. When in the σ-interval [0,1] N points were chosen, the μ- and t-intervals were covered with 2N points each and the θ-interval with N points.

Line iteration has been used along curves σ = constant, starting with $\sigma = 0$ and marching in the direction of increasing σ. In the viscous region the equations have been discretized by the usual Crank-Nicholson method for parabolic equations, whereas in the potential region central discretization formulas have been used. For $x^* > 0$ new values at the line σ are obtained by aid of previous values at the line $\sigma - h$ ($h = 1/N$, steplength), but for $x^* < 0$ previous values at the line $\sigma + h$ must be used in order to keep the method stable.

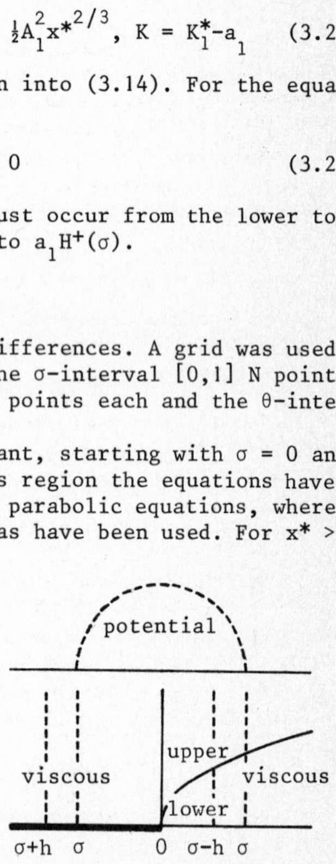

In each iteration step new values were calculated for ψ_p, ψ_v, $\overline{\psi}_w$, G^+, G^- and K. The complete system of equations was split into 3 parts: potential region, viscous layer $x^* > 0$ and viscous layer $x^* < 0$. An iteration step for a line σ = constant proceeds in the following way.

1°. Split the solution of eq. (3.18) with boundary conditions (3.19) into 2 parts, K_1 and K_2, so that the total solution can be written as

$$K = K_1 + G^- K_2 \qquad (4.1)$$

By integrating twice in μ-direction we can write $\psi_v = \psi_{v_1} + G^- \psi_{v_2}$ so that $H^-(\sigma) = \{\psi_{v_1}(\sigma,1) + G^- \psi_{v_2}(\sigma,1)\}a_1^{-1}$ is analytically expressed in G^- with numerically calculated coefficients. In shorter notation

$$H^-(\sigma) = H_1^-(\sigma) + G^-(\sigma)H_2^-(\sigma) \qquad (4.2)$$

2°. Solve the K-equations in the wake region. Two boundary conditions for K are known. Integrate twice to obtain ψ_v and $\overline{\psi}_w$.
As special value of ψ_v at $y^* = \infty$ we have

$$a_1 H^+(\sigma) = \psi_v(\sigma,1) = -\int_0^\infty \{\int_y^\infty K\, d\overline{y}\}dy \qquad (4.3)$$

The integral has to be calculated by integration through lower and upper regions, leading to a complicated form when written out in t- and μ-variables.

3°. In the potential region eq. (3.11) can be discretized for $\theta = \pi i/N$, $i = 1,2,\dots,N-1$. For the N+1 unknown values of ψ_p two equations must be added. These will be (3.13) combined with (3.12), (4.2) and (4.3). This is explained for the second of eqs. (3.13). We replace H^- by (4.2) and G^- in (4.2) by $\psi_p(\sigma,\pi)$ according to (3.12). The non-linear term G^{-2} in (3.13) is taken from the previous iteration step. We obtain the following equation

$$(3 - 2a_1^2 kr^* H_2^-)\psi_p(\sigma,\pi) - 4\psi_p(\sigma,\pi-k) + \psi_p(\sigma,\pi-2k) = 2a_1^2 kr^*\{H_1^- - \tfrac{1}{2}G^{-2}\} + 4.3^{-3/2} kA_1 r^{*1/3}$$

where $k = \pi/N$. An analogous formula is derived for the first equation (3.13). Then ψ_p can be solved and from (3.12) new values for G^+ and G^- are obtained, leading also to new values of the streamfunction.

For $\sigma = 0$ a slightly different procedure is followed. From the solution of (3.18) with (3.19), using values at $\sigma = h$ in the region $x^* < 0$, we obtain new values for K_1 and K_2 in (4.1). A new value for G^- is derived from neighbouring values of G^+ and G^-. The value of K(0,0) in (4.1) then leads, by aid of (3.15), to a new value of $K_1^*(0,0) = a_2$ with which eq. (3.21) is solved.

Starting values for the iteration process are $a_2 = a_1$ and $\psi_v = \psi_p = \overline{\psi}_w = 0$.

5. RESULTS

The final calculations have been performed with three different grids, N = 10, 20 and 40. The computation time on a CDC 6600 for these grids is about 1/2, 3 and 30 minutes, respectively.

Fig. 1 contains results for the value of $K(x^*,0)$ defined by eq. (3.15). It shows that the shear stress along the plate increases in the $x = O(Re^{-3/8})$ region by a factor $\{K(0,0) + a_1\}/a_1 = 1.352$. Moreover, fig. 1 presents results for the horizontal velocity, $Re^{-1/8}u(x^*,0)$, on the wake centerline and for the pressure, $Re^{-1/4}p_1(x^*)$, in the boundary layer ($y \leq O(Re^{-1/2})$). The vertical velocity at the outer edge of the boundary layer equals $-Re^{-1/4}G'(x^*)$. The decrease in displacement thickness due to the $Re^{-3/8}$-region is given by $Re^{-5/8}G(x^*)$, which is small compared to the displacement thickness in the Blasius and Goldstein regions where it is $O(Re^{-1/2})$. For $x \to 0^+$ but $x > O(Re^{-3/8})$, the decrease in displacement thickness is $A_1 x^{1/3} Re^{-1/2}$.

The coefficient c in the drag formula (1.1) is obtained by integration of the shear stress. Its value appears to be 2.651 ± .003. This is in excellent agreement with the value 2.644 obtained recently by Melnik [5], while it differs somewhat more from the value 2.69 given by Jobe [6]. It is very remarkable that (1.1), with neglect of the order term agrees well, down to Re = 1, with experiments performed by Janour [7] in the range Re = 12 to 2300 and with numerical calculations by Dennis [8] for Re = 1 to 200. The difference for Re = 1 in the total drag is about 7%.

For a more complete description of the investigation the reader is referred to the first author's thesis [9].

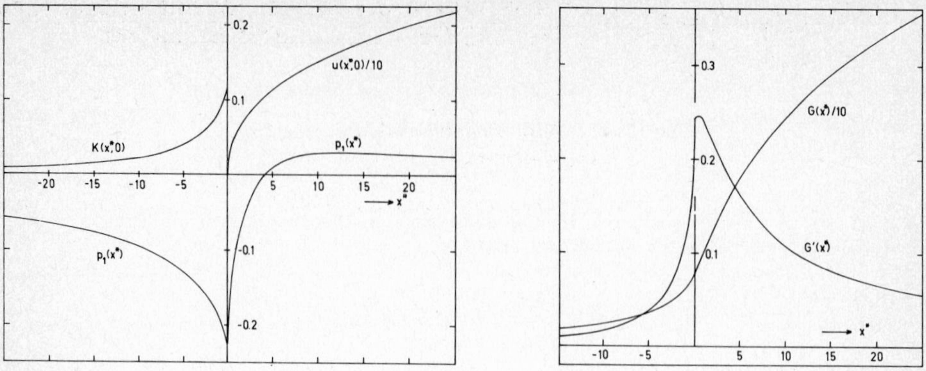

Fig. 1. Results for pressure, velocity, shear stress and displacement thickness.

Acknowledgement. The first author wishes to acknowledge the partial support by the Netherlands organization for the advancement of pure research (Z.W.O.).

REFERENCES

1. K. Stewartson: "On the flow near the trailing edge of a flat plate", Mathematika, 16, 106-121 (1969).
2. A.F. Messiter: "Boundary layer flow near the trailing edge of a flat plate", Siam J. Appl. Math. 18, 241-257 (1970).
3. S. Goldstein: "Concerning some solutions of the boundary layer equations in hydrodynamics", Proc. Camb. Phil. Soc. 26, 1-30 (1930).
4. F.T. Smith: "Boundary layer flow near a discontinuity in wall conditions", J. Inst. Maths. Applics. 13, 127-145 (1974).
5. R.E. Melnik: Private communication.
6. C.E. Jobe: "The numerical solution of the asymptotic equations of trailing edge flow", Thesis. Ohio State University. (1973).
7. Z. Janour: "Resistence of a plate in parallel flow at low Reynolds numbers", NACA TM 1316 (1951).
8. S.C.R. Dennis: Private communication.
9. A.E.P. Veldman: Thesis to be published. Math. Institute, University of Groningen.

NUMERICAL SOLUTION OF THE UNSTEADY NAVIER-STOKES EQUATIONS FOR THE OSCILLATORY FLOW OVER A CONCAVE BODY

by

George F. Widhopf and Keith J. Victoria
The Aerospace Corporation
P.O. Box 92957
Los Angeles, California 90009

INTRODUCTION

In the past few decades, the hypersonic flow over concave and spiked geometric shapes has been investigated experimentally by many investigators (e.g., Mair [1952], Bogdonoff and Vas [1959],[1962], Crawford [1959], Maull [1960], Wood [1961], Centolanzi [1963] and Holden [1966]) in order to determine the general flow characteristics and specifically to determine the attendant effects encountered for these shapes. In the course of these experimental investigations, a family of significantly different flow fields was observed, depending upon the specific geometric characteristics of the individual shape. The resultant general trends, for a limited variation of shapes, were correlated and described initially by Maull [1960], Wood [1961] and later by Holden [1966].

In brief, the observed flow fields included a class of steady attached flows as well as a spectrum of steady separated flows, where the flow reattached itself either on the afterbody face or at the shoulder. However, for a distinct class of shapes an oscillatory flow was observed which in general was characterized by a separated region that periodically grew and then collapsed, resulting in an oscillation of the entire shock layer. In most of these flows, embedded shock waves, resulting in strong viscous-inviscid interactions, were observed which are dominant precursors for these subsequent experimentally observed flow phenomena. Similar flow fields have been observed for space shuttle configurations (Goldman and Obremski [1973]) where local oscillatory flow fields have also been obtained under certain conditions.

It was determined by Maull [1960], Wood [1961] and Holden [1966] that the geometry of the body was the dominant controlling factor in the determination of which type of flow would eventually result. Specifically, the degree of compression encountered in the concavity was the dominant controlling parameter, together with the radius of the afterbody shoulder and the length of the slender forward section in relation to the afterbody diameter. When the primary oscillatory frequency measured in these studies and that of Kubota [1973] was nondimensionalized with respect to the freestream velocity and the afterbody diameter to form the Strouhal number, it was observed that the Reynolds number is not an important factor. Indeed, the results of these measurements, which spanned a two orders of magnitude Reynolds number range, show a minor effect on the Strouhal number and exhibit a slow variation of Strouhal number with freestream Mach number.

It should be remarked that the primary experimental diagnostic tool used in most of these investigations was photographic in nature. Some investigations included surface pressure and heat transfer measurements, the most detailed of which is the work of Holden [1966]. However, flow field measurements which can be used to determine the detailed fluid mechanics of the varied flow environments are not presently available.

Considering the complicated nature of the observed flow fields, which involve strong viscous-inviscid interactions, embedded shocks, separated regions and, in some cases, transient variations, the numerical simulation of any of these flow fields is a formidable task. In view of the observation that the nondimensional oscillatory frequency does not show a strong variation with Reynolds number, a low Reynolds number numerical simulation may provide an understanding of the important flow mechanisms and structure involved in these various cases.

The investigation described in this paper is an initial step undertaken to study the flow over these classes of shapes. In view of the complicated nature of these flow fields, the numerical solution of the unsteady Navier-Stokes equations was undertaken to calculate these various flow environments. In the course of this investigation, an interesting flow field has been calculated which is oscillatory in nature, but is not separated. This particular type of flow field has not been observed experimentally. A general description of the resultant flow structure is described in this paper, together with the flow mechanisms which are hypothesized as the phenomena which cause and control the oscillatory nature of the flow field.

The specific geometric shape which was considered is schematically described in Figure 1. The basic configuration, for which the flow field structure will be discussed in more detail, is indicated by the designation I. The unsteady Navier-Stokes equations were solved using an accurate, time-dependent numerical finite difference approximation (leap-frog/Dufort-Frankel) previously described by Victoria and Widhopf [1973]. In order to compute the solution in a computational domain with rapidly changing mesh spacing, it was necessary to utilize a Crocco [1965] difference approximation for the continuity equation in order to stabilize the density calculation in the freestream. Extensive testing of this approximation by varying the weighting factor, Γ, so that the Crocco differencing approaches second order accuracy, showed that this approximation did not affect the results. The equations were solved in the body oriented curvilinear coordinate system indicated in Figure 1. The entire flow structure was computed at each instant in time since the equations were solved subject to boundary conditions at the wall and in the freestream. Thus, the shock jump conditions were computed as an integral part of the solution and shocks were not treated as discontinuities. The freestream conditions at which the computations were performed included a Mach number of 10 and a Reynolds number based on freestream properties and the initial nose radius, R_n, of 333.

DISCUSSION OF RESULTS

The flow about the concave shape, designated I, was determined to be oscillatory in nature exhibiting a periodic temporal oscillation which initially becomes apparent downstream of the forebody pressure minimum. The flow is not separated and the nondimensional frequency of the oscillation is approximately a factor of 36 higher than the nondimensional frequency experimentally observed for the separated oscillatory flows previously described. The oscillation is characterized by a high frequency movement of the shock layer which results in local changes in surface pressure of approximately a factor of two near the shock inflection point. Typical surface pressure distributions are shown in Figure 2 at various times during the oscillatory period. Referral to the scale in Figure 1 provides an indication of the relative location of any point on the pressure distribution with respect to the shock inflection point. The steep pressure rise as the shock inflection is reached is depicted as well as the local maxima and minima which result from multiple reflections of the diffuse third leg of the lambda shock from the sonic line and the diffuse slipstream which originates from the lambda shock triple point. Thus, the streamwise resolution, which is indicated on the scale of Figure 2, by the body station index, J, allows for the determination of this complex structure. The progression of the pressure pulse which results from the oscillation of the shock layer is apparent by comparing the surface pressure distribution at various times in the oscillatory period, as depicted in Figure 2. The transient variations of the surface pressure at various selected streamwise locations are shown in Figure 3. Here it is seen that a well defined period (approximately 226 times the time increment used to advance the solution in time) is defined at each spatial location.

Initially the wave form is harmonic but becomes increasingly nonlinear as the shock wave inflection point is approached. Two periods have been depicted to exhibit the repetition of the oscillation. It should be remarked that the calculation has been carried out for approximately ten complete periods with no apparent damping. The spatial amplitude variation during a single period is seen to increase as the flow progresses downstream, reaching approximately a factor of two variation from maximum to minimum magnitude near the shock wave inflection point. Shown in Figure 4 is the envelope of the surface pressure maxima and minima encountered during one period of the oscillation, together with the time average over a single period. Two perspective views of a composite of Figures 2 and 3 are shown in Figures 5 and 6. These figures give a perspective view of the amplification in space and time of the surface pressure during the oscillation. The first view shows the time variation of the surface pressure distribution and depicts the progression of the pressure pulse toward the shock inflection point. The second view gives a better perspective of the temporal and spatial variation downstream of the triple point. Both views include 220 time cycles of the periodic oscillation.

It is interesting to observe the velocity profiles (Figure 7) at various stations at the times the surface pressure is at the maximum and minimum at a given streamwise location. Note the movement of the shock layer together with the appearance of an inflection point in the velocity profile at the time which corresponds to the pressure minimum at that station. Observe that the inflection point is no longer apparent for the profile which corresponds to the pressure maximum. In the region of favorable pressure gradient, the velocity profiles fill out and the inflection point near the wall disappears altogether. From hydrodynamical considerations, this might indicate an instability in the flow. Thus, any existing perturbation is amplified, and the amplification becomes increasingly nonlinear as it progresses downstream toward the shock triple point. The amplification is due to the converging nature of the forward shock wave and the compression geometry of the concavity. Although a detailed comparison with results of linear stability analyses is not possible due to the vast differences in this flow and those considered in most stability analyses, an estimation of the most unstable frequency as defined by Mack [1969] at this freestream Mach number is in agreement with respect to order of magnitude.

The waveforms of Figure 3 were analyzed using discrete complex Fourier series to determine the specific frequency content and streamwise amplification of the wave components. The essential results are plotted in Figure 8 which shows the spatial amplification of the absolute value of the complex amplitude, A(n), associated with each of the first five frequencies plus ten and twenty. The amplification at the fundamental frequency (n=1), which was taken to be 226 times the time step used to advance the solution in time, is shown along with the amplification at the first four harmonics (n = 2 to 5). The DC level of pressure (n=0) undergoes approximately an order of magnitude compression during which the fundamental frequency and its first few harmonics undergo approximately two orders of magnitude amplification at sequentially delayed spatial locations and at increased rates. Once the adverse pressure is removed, the amplitudes rapidly decrease. There is significant nonlinear interaction between the frequencies in a narrow region near the pressure maximum, both upstream and downstream of that location.

A perspective view of the variation of the temperature profile downstream of the stagnation point at one point in time is shown in Figure 9. Here the shock layer growth is apparent, together with the recompression and interaction pattern resulting from the embedded shock wave. The diffuse slipstream which originates from the lambda shock triple point is very apparent, as well as the shock layer growth on the afterbody.

When the flow field is initially established, the initial flow pulse is not aware of the presence of the afterbody. Thus, there must be some upstream communication through the subsonic layer near the wall, which will result in an adjustment of the flow pattern near the shock wave triple point, to account for the presence of the afterbody. A conjecture is that this triple point structure is not a steady flow pattern and the flow pattern may eventually adjust itself in a cyclic manner in order to accommodate this flow pattern near the inflection point. In this regard, an alteration of the afterbody geometry would allow for an evaluation of the effect of afterbody on the flow field. Thus, flow fields were calculated over the configurations designated II and III, which are shown in Figure 1. A typical distribution of the thickness of the shock layer, as defined by the point where the temperature reaches the freestream value, is shown in Figure 1 for each of the geometric shapes considered. As can be seen, the freestream location is not appreciably altered except downstream of the shock wave triple point. However, the shock layer thickness near the lambda shock and the embedded shock wave strength are appreciably different. Shown in Figures 10 and 11 are the transient surface pressure variations at locations at or upstream of the alternation in the afterbody shapes. Here the variation of the surface pressure which was calculated for configuration I has been superimposed on these results by shifting the respective time scales. It is immediately evident that the period of the oscillation has not been altered by these geometric changes. However, there is some alteration in the magnitude of the surface pressure at each location depicted. The major changes are near the inflection point since the pressure gradient is proportional to the actual compression encountered for each shape and subsequent relief through expansion. Thus, the afterbody geometry does not appreciably alter the nature of the oscillation, indicating that the flow structure in the vicinity of the lambda shock is not a dominant controlling parameter as to the determination of the nature of the oscillation with regard to its period and characteristics far upstream.

As a further test, another shape was investigated whose downstream body shape was chosen to be approximately parallel to the shock wave and thus would not produce a converging shock layer. This afterbody was tangent to the forebody at station J=32 with a slope of 19.2 degrees. This flow simulation was started from the final result from a previous oscillatory solution and, indeed, converged to a steady state. Thus, the presence of the compression and "channeling" effect of bow shock and body surface is necessary for the oscillation to occur and/or amplify and an oscillatory flow is not calculated for a geometry that should be steady.

Thus, the implication which still is apparent is that the flow near the tip is hydrodynamically unstable, the fact being implied by the presence of an inflection point in the streamwise velocity profile. Thus, any disturbance is initially linearly amplified and then in an increasingly nonlinear fashion as the flow approaches the shock wave inflection point.

In view of the above, it is interesting to determine the nature of the inviscid flow field solution for the flow over the identical shape. Shown in Figure 4 is the result of an inviscid time-dependent calculation (Reddall [1973]) which did not oscillate but, rather, converged to a steady state. The time averaged Navier-Stokes result and the inviscid result are in relatively good agreement except in the region of the lambda shock where the inviscid result is significantly lower. The velocity profiles calculated for the inviscid case do not exhibit inflection points. This may indicate the essential viscous nature of the problem, that is in order to calculate the instability the no-slip condition at the wall is important.

The fact that the period of the oscillation is approximately 226 times the time step increment used to advance the calculation in time, is a strong indication that the oscillation is not of numerical origin, since the most unstable frequencies, for the leap-frog finite difference approximation used in these calculations, are orders of magnitude larger (Kentzer [1972]). It can be shown that any numerical instability is proportional to the time increment utilized, so that if the time step is changed the frequency of any numerical oscillation should also change correspondingly for the same initial conditions. Shown in Figure 3 is the result of a calculation performed with the original time step reduced by a factor of the square root of two. Here the results have been shifted with respect to the time scales in order to overlay the results for comparison purposes. As can be seen from these comparisons, there is no change in the period and only minor variations in the amplitude at stations near the lambda shock. Thus, this provides additional evidence that the oscillation is a physical one.

Shown in Figure 12 are streamwise distributions of the surface heat transfer rate. Specifically, the maximum and minimum values are shown, together with the time average over an oscillatory period. As can be seen, there is a large variation in the heat transfer during a period, especially near the lambda shock. Specific transient variations at selected locations are shown in Figure 13. As can be seen, the heat transfer temporal variation is initially harmonic and becomes increasingly nonlinear as the lambda shock is approached. There is a slight phase shift with respect to the corresponding surface pressure variations of approximately a lag of 10 cycles.

CONCLUSIONS

An interesting oscillatory flow field over a concave shape, which has not been observed experimentally, has been calculated by numerically solving the unsteady Navier-Stokes equations. The unique characteristic of this flow is that it is not separated and yet is oscillatory in nature. All previously observed oscillatory flow fields contain separated regions. The general characteristics of the flow pattern have been described, together with the hypothesized flow mechanism which produces the oscillation. The flow field appears to be hydrodynamically unstable which results in an oscillatory variation of the flow field downstream of the forebody pressure minimum. Various changes in the afterbody shape do not alter the period of the oscillation and results in minor changes in the magnitude of the pressure oscillation. A number of numerical experiments have been performed and described, the results of which indicate that the oscillation is of physical origin. Further work is continuing to understand these flow fields and the underlying governing flow mechanisms. Experimental verification of this flow phenomenon is presently being pursued by Holden [1974] and a more detailed description of the flow will be forthcoming pending the results of this experimental study.

REFERENCES

Bogdonoff, S. M. and Vas, I. E., "Preliminary Investigations of Spiked Bodies at Hypersonic Speeds," Journal of Aerospace Sciences, Vol. 26, No. 2, pp. 65-74, February 1959.

Bogdonoff, S. M. and Vas, I. E., "Some Experiments on Hypersonic Separated Flows," ARS Journal, pp. 1564-1572, October 1962.

Centolanzi, F. J., "Heat Transfer to Blunt Conical Bodies Having Cavities to Promote Separation," NASA Technical Note D-1975, NASA Ames Research Center, Moffett Field, California, July 1963.

Crawford, D. H., "Investigation of the Flow Over a Spiked-Nose Hemisphere-Cylinder at a Mach Number of 6.8," NASA TN D-118, NASA Langley Research Center, Langley Field, Virginia, December 1959.

Crocco, L., "A Suggestion for the Numerical Solution of the Steady Navier-Stokes Equations," AIAA Journal, Vol. 3, No. 10, pp. 1824-1832, 1965.

Goldman, R. L. and Obremski, H. J., "Experimental Investigation of Hypersonic Buzz on a High Cross-Range Shuttle Configuration," AIAA Paper No. 73-157, AIAA 11th Aerospace Sciences Meeting, Washington, D.C., January 1973.

Holden, M. S., "Experimental Studies of Separated Flows at Hypersonic Speeds, Part I: Separated Flows Over Axisymmetric Spiked Bodies," AIAA Journal, Vol. 4, No. 4, pp. 591-599, April 1966.

Holden, M. S., Private Communication, 1974.

Kentzer, C. P., "Group Velocity and Propagation of Numerical Errors," AIAA Paper No. 72-153, Presented at the AIAA 10th Aerospace Sciences Meeting, San Diego, California, January 1972.

Kubota, T., Private Communication, 1973.

Mack, L. M., "Boundary-Layer Stability Theory," Notes Prepared for the AIAA Professional Study Series, High-Speed Boundary-Layer Stability and Transition, San Francisco, California, June 14-15, 1969.

Mair, W. A., "Experiments on Separation of Boundary Layers on Probes in Front of Blunt-Nosed Spiked Bodies in a Supersonic Air Stream," Philos. Mag., Vol. 43, No. 342, pp. 695-716, July 1952.

Maull, D. J., "Hypersonic Flow Over Axially Symmetric Spiked Bodies," Journal of Fluid Mechanics, Vol. 8, Part 4, pp. 584-592, 1960.

Reddall, W. F., III, Private Communication, 1973.

Victoria, K. J. and Widhopf, G. F., "Numerical Solution of the Unsteady Navier-Stokes Equations in Curvilinear Coordinates: The Hypersonic Blunt Body Merged Layer Problem," Lecture Notes in Physics, 19, Proceedings of the Third International Conference on Numerical Methods in Fluid Mechanics, Vol. II, Problems in Fluid Mechanics, July 1973, Springer-Verlag.

Wood, C. J., "Hypersonic Flow Over Spiked Cones," Journal of Fluid Mechanics, Vol. 12, Part 4, pp. 614-624, 1961.

ACKNOWLEDGMENTS

The authors would like to acknowledge the constructive criticism and suggestions given by Professor Toshi Kubota of the California Institute of Technology and Dr. Thomas D. Taylor of The Aerospace Corporation throughout the course of this study. Special acknowledgment is due to Dr. William S. Helliwell of The Aerospace Corporation who made the major modifications to the computer code which were necessary to carry out these types of calculations. Thanks are also due to Mr. James B. Carey of The Aerospace Corporation for his work in obtaining some of the numerical results and displays.

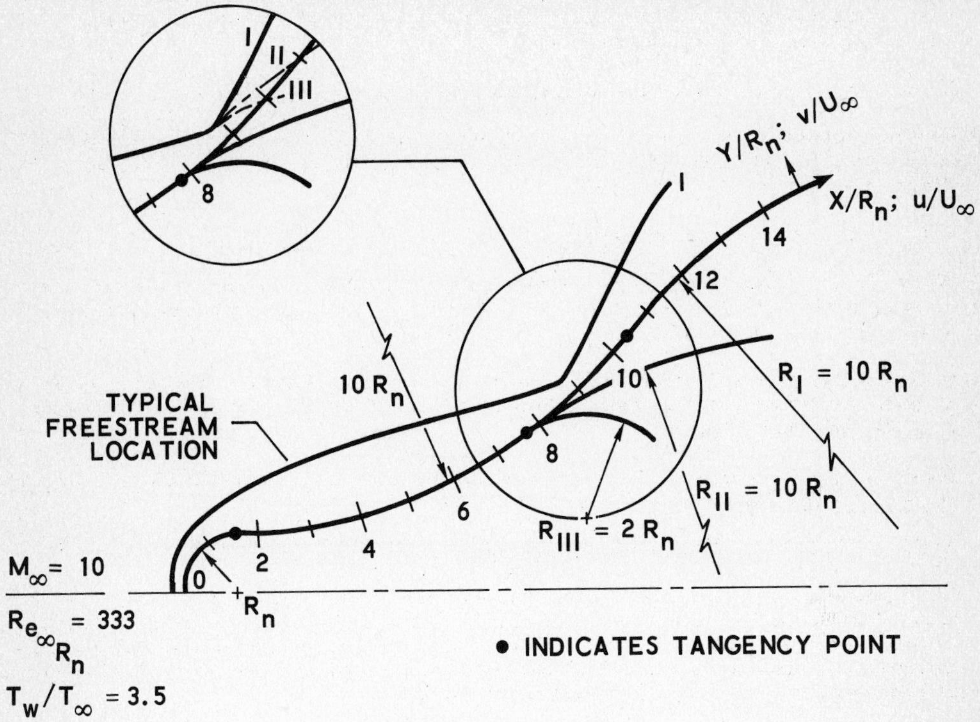

Fig. 1. Schematic of Axisymmetric Body Configurations and Freestream Locations

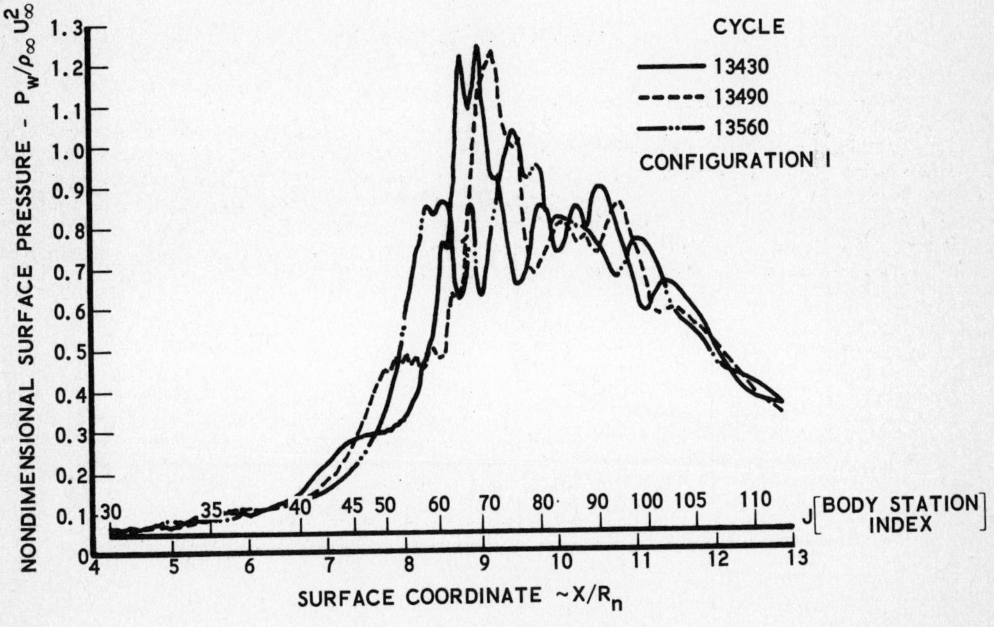

Fig. 2. Surface Pressure Distributions at Various Times During the Oscillation

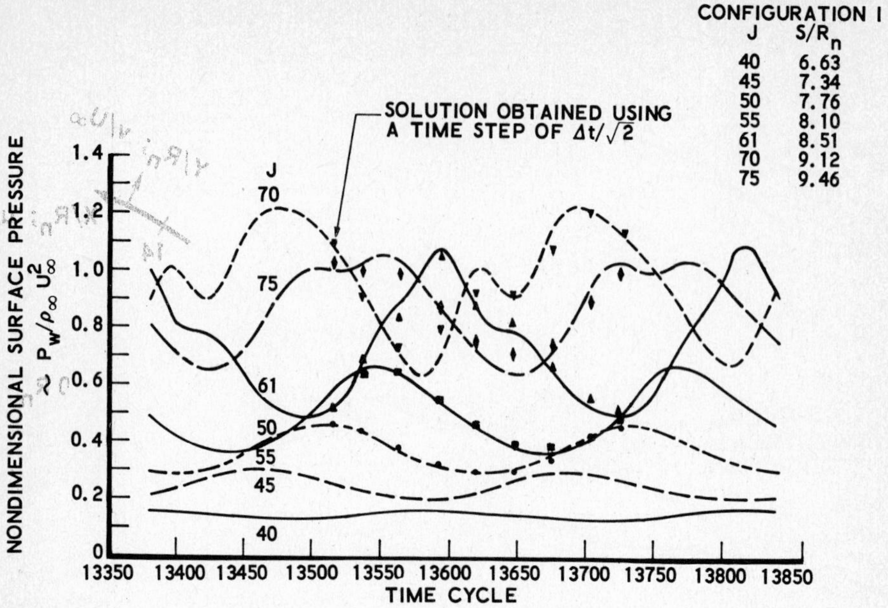

Fig. 3. Temporal Variation of Surface Pressure at Various Stations

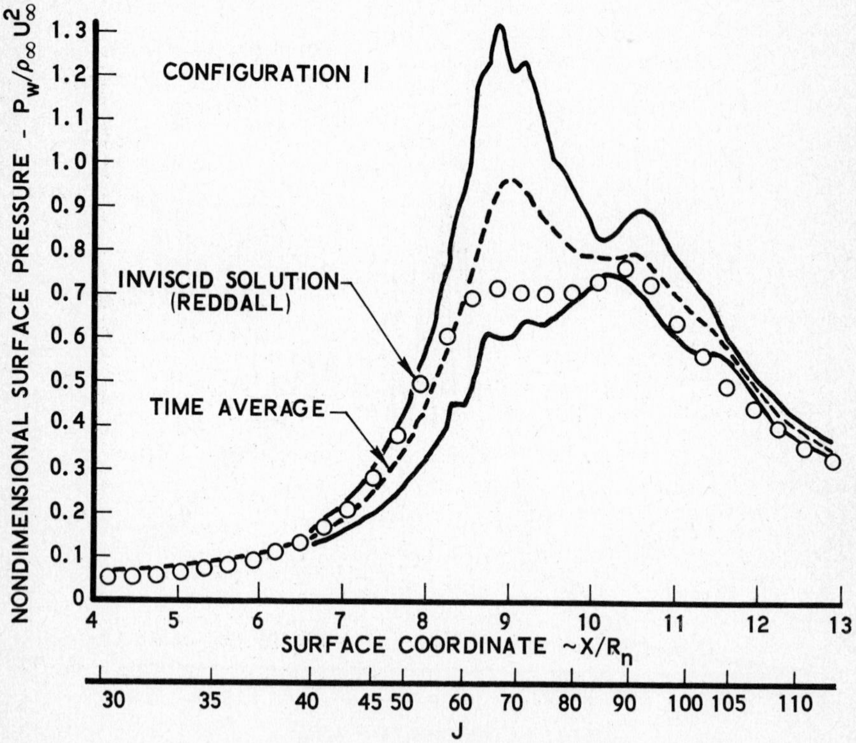

Fig. 4. Envelope of Surface Pressure Maxima and Minima

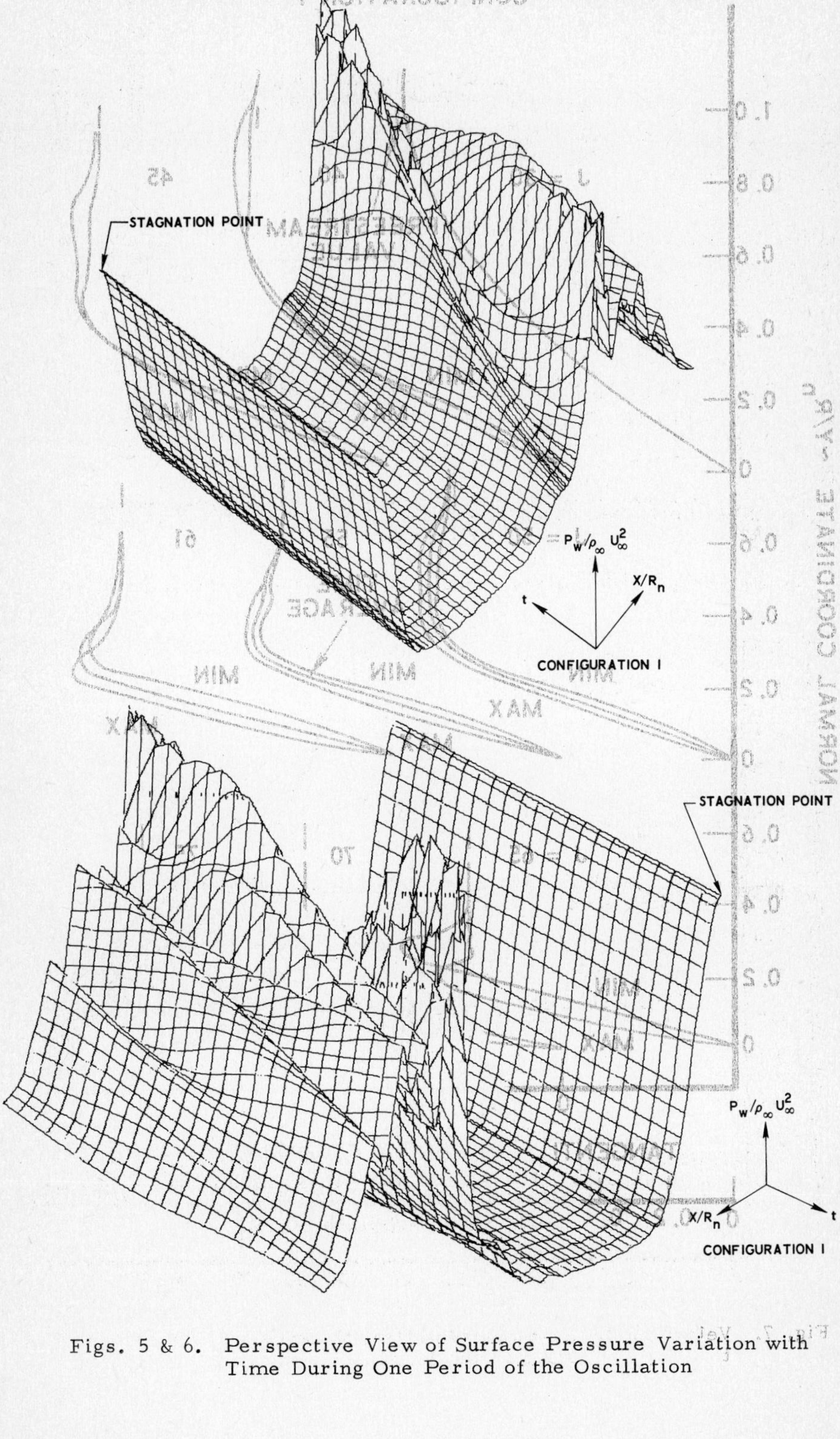

Figs. 5 & 6. Perspective View of Surface Pressure Variation with Time During One Period of the Oscillation

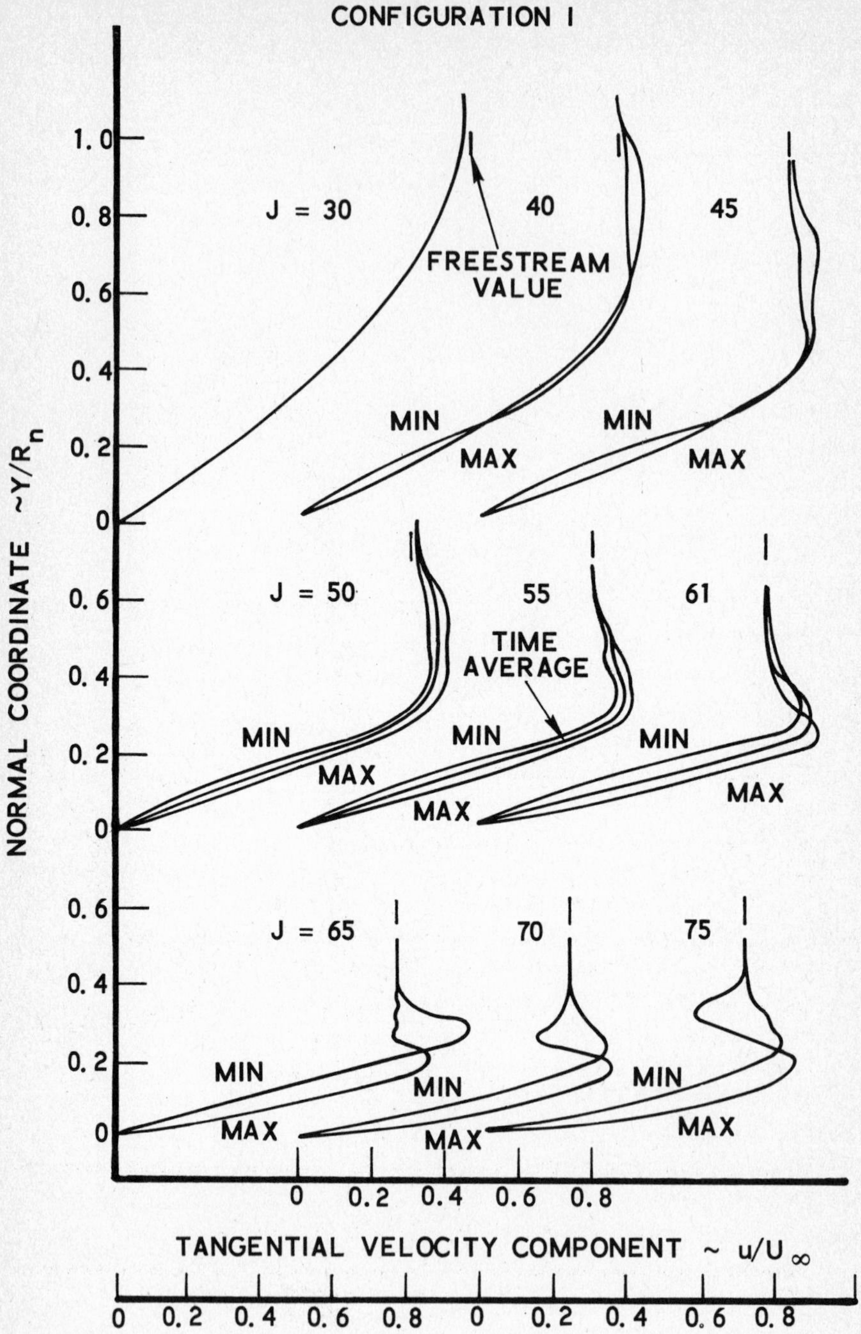

Fig. 7. Velocity Profiles at Various Stations at Times Corresponding to a Surface Pressure Maximum and Minimum

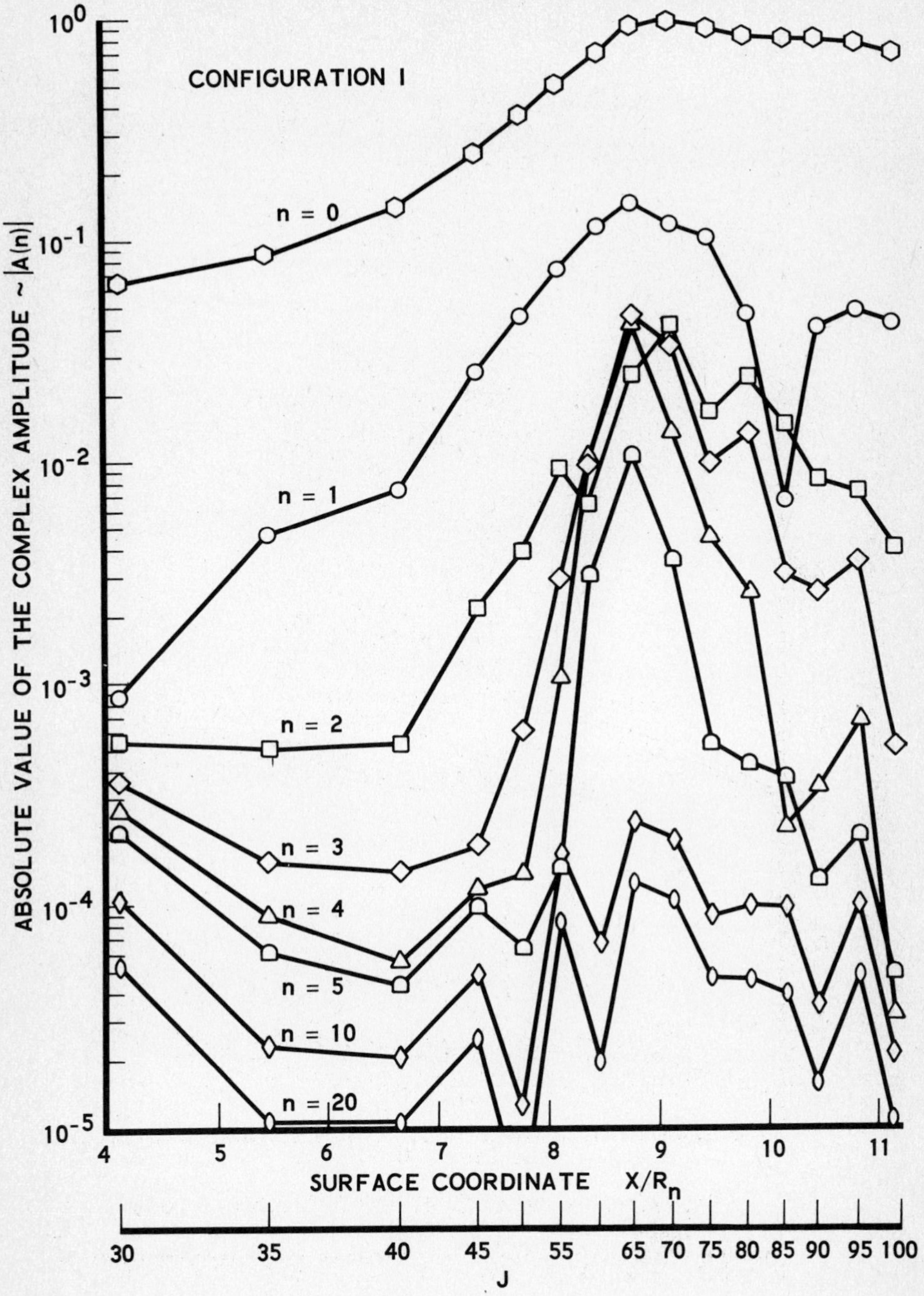

Fig. 8. Spatial Amplification at Various Frequencies

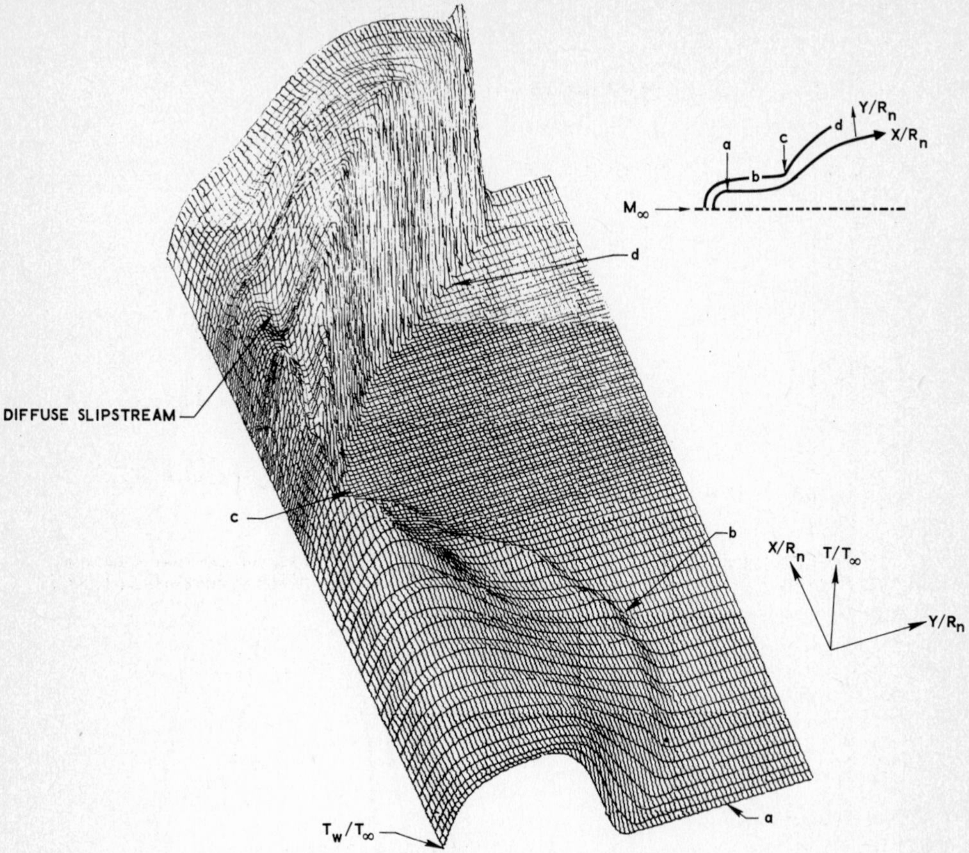

Fig. 9. Perspective View of Shock Layer Temperature Distribution for Configuration I

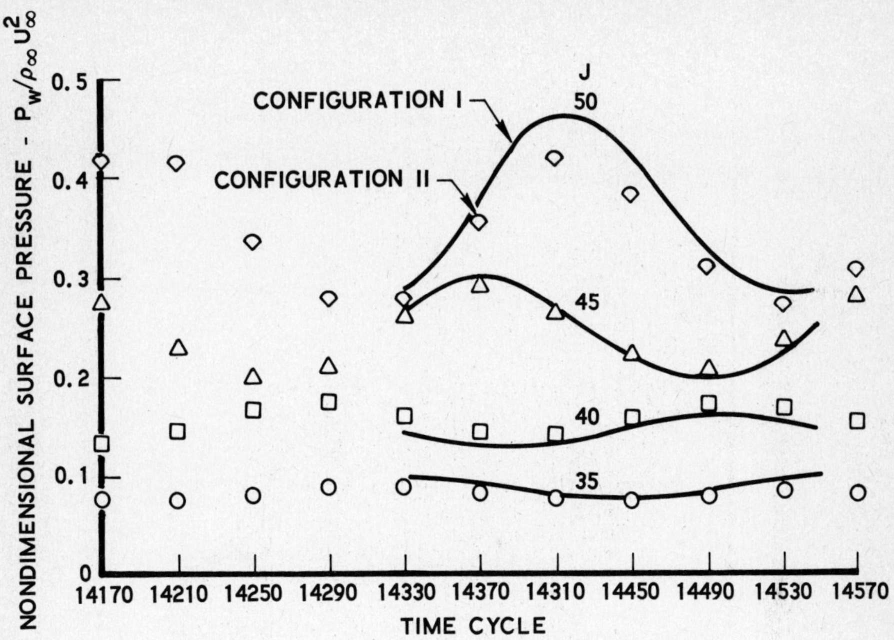

Fig. 10. Comparison of Temporal Variations of Surface Pressure for Configurations I and II

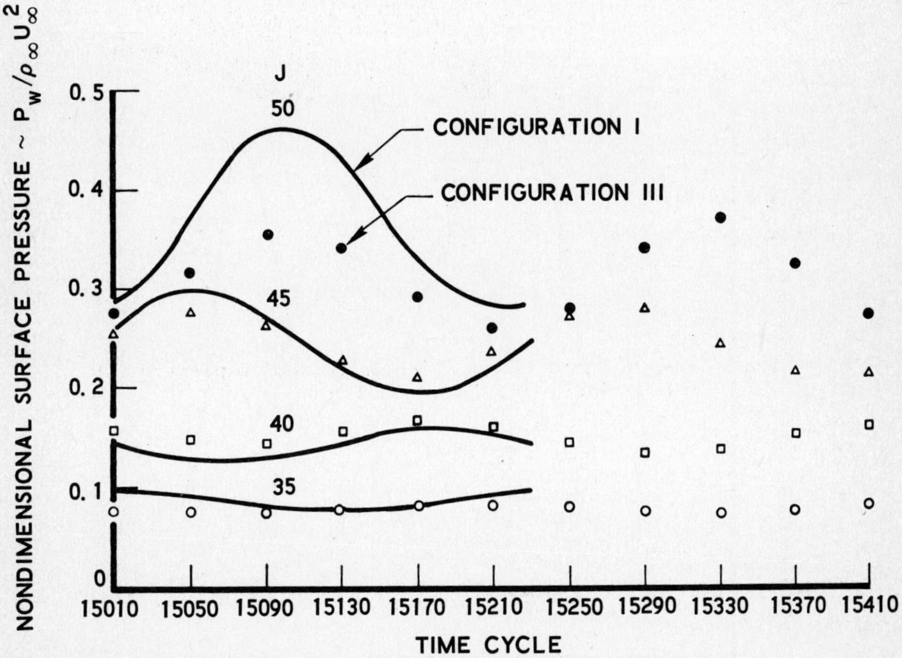

Fig. 11. Comparison of Temporal Variations of Surface Pressure for Configurations I and III

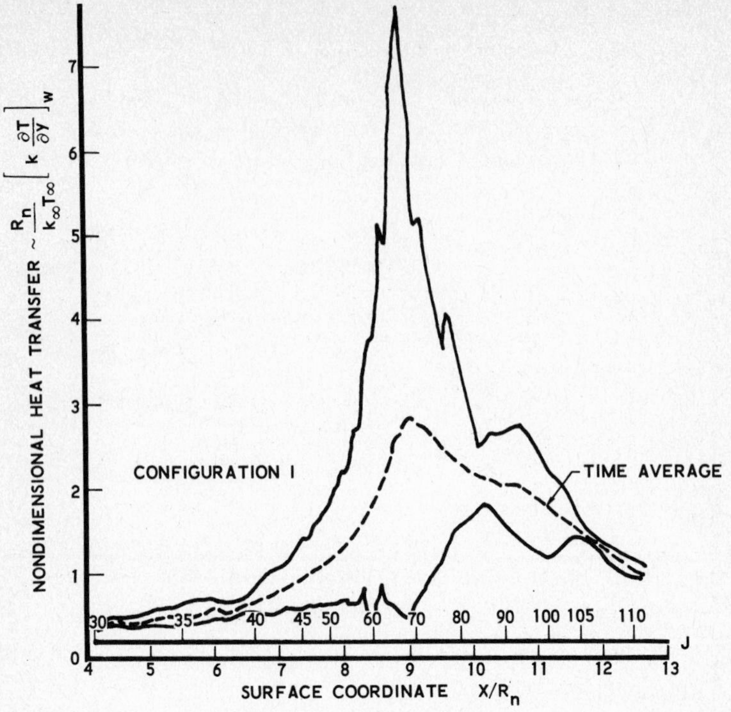

Fig. 12. Envelope of Surface Heat Transfer Maxima and Minima

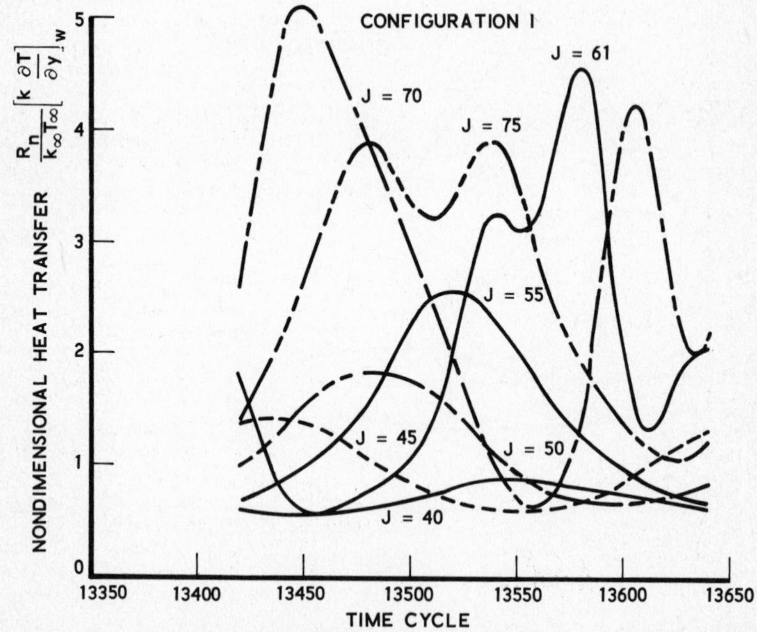

Fig. 13. Temporal Variation of Surface Heat Transfer at Various Stations

A REVERSE FLOW COMPUTATION IN THE THEORY OF SELF-INDUCED SEPARATION

P.G.Williams

Department of Mathematics, University College London

1. INTRODUCTION

1.1 Background. The aim here is to extend into the reverse flow region an earlier computation in the asymptotic theory of self-induced separation (Stewartson and Williams, 1969). This theory is concerned with the classical problem, represented schematically in Fig.1, of supersonic flow past a wall with separation of the wall boundary layer upstream of some disturbing feature, such as a ramp or an incident shock. Fig.1 also shows, outlined by broken lines, the region surrounding the separation point to which the theory applies. As indicated in the diagram, its scale in the x direction is $O(\epsilon^3)$, where ϵ is the small parameter defined by

$$\epsilon = (\mathrm{Re})^{-\frac{1}{8}}, \quad \mathrm{Re} = U^*_\infty x^*_S / \nu^*_\infty \qquad (1)$$

(* characterizes physical quantities). On this length scale, first approximations for the pressure p* and skin friction τ^* can be obtained by solving a self-contained boundary value problem in an inner layer of thickness $O(\epsilon^5)$ adjacent to the wall. It is the solution of this problem that we wish to continue into the reverse flow region.

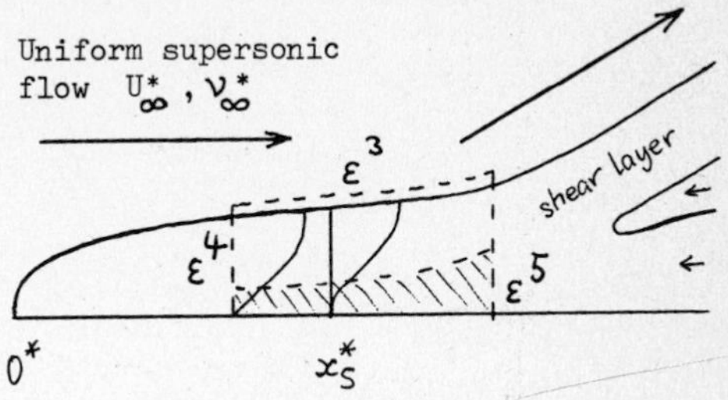

Figure 1. Inner layer (shaded)

1.2 The inner layer problem. Inner layer variables x,y,u,p can be introduced such that the governing equation is precisely the incompressible boundary layer equation

$$u_{yy} + u_y \int_0^y u_x \, dy - uu_x = p_x , \qquad (2)$$

where the pressure p(x) is a function of x only. This fuction is unknown, but there is an additional condition on it in the peculiar outer boundary condition

$$u \rightarrow y - A(x) \quad \text{as} \quad y \rightarrow \infty, \qquad (3)$$

$$\text{where} \qquad A(x) = \int_{-\infty}^{x} p \, dx. \qquad (4)$$

As initial condition we have

$$u \rightarrow y \quad \text{as} \quad x \rightarrow -\infty. \qquad (5)$$

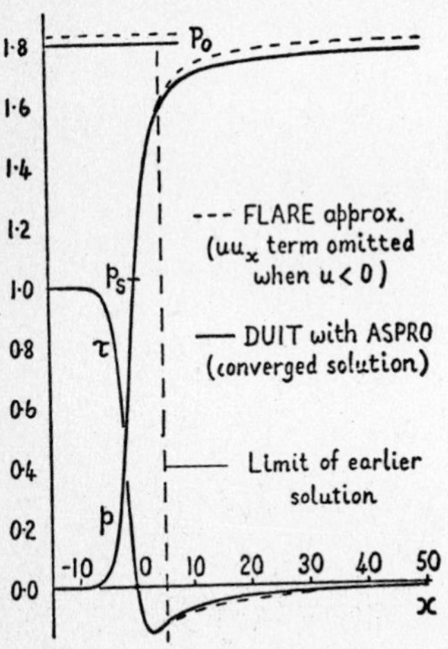

Figure 2. Results for p and τ.

In Stewartson and Williams (1969) these equations were integrated numerically from a small positive pressure kick upstream, through separation, to a point indicated in Fig.2 where the rapid expansion of the inner layer caused difficulties because of a fixed outer boundary. This occurred before instabilities due to the reverse flow became noticeable. Later computations which allowed the outer boundary to move out clearly showed the growth of an instability whose origin was traceable to the separation point. To continue the solution downstream it is therefore necessary to impose an asymptotic boundary condition of some kind as $x \to \infty$. One would hope at least that p would tend to a constant, say P_0 (the 'plateau pressure'?), and this seemed possible from the earlier upstream results - in fact it was guessed that p might tend to something like $P_0 \approx 1.8$. With the guidance of later downstream results obtained on the basis of the FLARE approximation below, an appropriate downstream structure was derived (Stewartson and Williams, 1973). The expansions involved in the asymptotic profile (ASPRO for brevity) are described briefly in section 2. Their incorporation in a downstream-upstream iterative scheme (DUIT) is discussed in section 3.

1.3 The FLARE approximation. A stable, but approximate, downstream solution can be obtained without such a downstream condition by using an idea first put forward by Flugge-Lotz and Reyhner (1968), namely: neglect the uu_x term (which is the cause of the instability) whenever $u < 0$. For brevity this will be referred to as the FLARE approximation. It does in fact stabilize the calculation and evidently one could integrate indefinitely in the x direction by this means. Although the results could not be expected to give the true structure, they were suggestive enough to lead to the correct form of expansion. As can be seen in Fig.2, the agreement with the accurate solution is very good, especially for the skin friction. However, as will be seen in section 3, it is not so good as regards the profile shape.

2. ASYMPTOTIC PROFILE

2.1 The expansions. The FLARE results suggested that p does in fact tend to a constant P_0 and that therefore A(x) is ultimately linear. The wall boundary layer breaks away from the wall to become a free shear layer centred on $y = A(x)$, and a weak back flow is generated by the entrainment into this shear layer. The downstream structure therefore assumed has an outer viscous layer centred on $y = A(x)$, an inner backward viscous layer on the wall, and an essentially inviscid region between them, as in Fig.3. With appropriate similarity variables as indicated the expansions for the downstream profile read

Outer variable: $$\xi = (y - A(x))/\, x^{1/3} \qquad (6)$$

$$u \sim x^{1/3}F_0' + x^{-1/3}F_2' + x^{-2/3}\log x\, F_3' + x^{-2/3}F_4' + \dots \qquad (7)$$

Inner variable: $$\eta = y/\, x^{2/3} \qquad (8)$$

$$u \sim x^{-1/3}f_0' + x^{-2/3}f_1' + x^{-1}f_2' + \dots \qquad (9)$$

These lead to an expansion for the pressure of the form

$$p \sim P_0 + P_2x^{-2/3} + P_3x^{-1} + P_4x^{-4/3} + P_5x^{-5/3}\log x + O(x^{-5/3}) \qquad (10)$$

and a corresponding expansion for A(x) beginning $A(x) \sim P_0x$. The F's are functions of ξ, and the f's functions of η, satisfying sequential systems of ordinary differential equations. As far as given the expansions are determined in terms of the single parameter P_0, apart from an eigenfunction in F_4 corresponding to an origin shift in x. Essentially the same asymptotic structure had already been obtained by Neiland (1971), but only to a first approximation. It turns out that at least the next term in the velocity profile should be taken into account because of the slow convergence of the expansions, even for fairly large values of x of the order of 50.

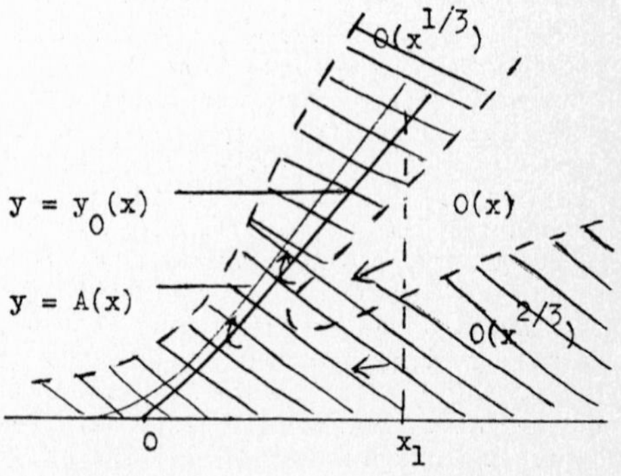

Figure 3. Asymptotic structure.

2.2 Application of the expansions. Although the f_2 term in expansion (9) is determined by P_0, it actually matches with later terms in the outer expansion (7), which have not in fact been found; it is therefore convenient to impose a downstream profile based on (7) and (9) without the last term in (9), then use the three-term inner expansion later as a check. This imposed profile is referred to as ASPRO; each term was represented as a Chebyshev series. In applying this downstream condition at each cycle of the iteration described in section 3, we need to match ASPRO to a given computed solution at some large value of x, say x_1. A convenient way of doing this is to simultaneously determine P_0 and an origin shift in x so as to match both $p(x_1)$ and $y_0(x_1)$, where $y = y_0(x_1)$ is the boundary between the forward and reverse flow regions (this lies in the outer layer corresponding to (7)).

3. ITERATIVE PROCEDURE DUIT

3.1 DUIT. In order to incorporate the above asymptotic structure into the computation of the solution downstream of separation, an iterative procedure was devised which involves downstream and upstream integrations (hence the abbreviation DUIT). It can be regarded as an extension of the FLARE method in that the neglect of the uu_x term in the reverse flow region is corrected for successively. Denoting the $u \doteq 0$ boundary by $y = y_0(x)$ and the uu_x term for $u < 0$ by $(uu_x)_-$, we define the following procedure:

1. With $(uu_x)_- = 0$, integrate downstream from a fixed separation profile at x = 0 (obtained from a previous calculation) to a prescribed large value of x, say x_1. This yields a first approximation for p(x) and $y_0(x)$ (the FLARE approximation).

2. From these values, compute P_0 and an asymptotic profile at $x = x_1$.

3. Starting from this profile, integrate back upstream in the reverse flow region only, keeping p(x) and $y_0(x)$ fixed.

4. Repeat from 1. with $(uu_x)_-$ now computed from the u values just obtained coming upstream, and so on, until the computed profile at $x = x_1$ agrees with the previous asymptotic profile.

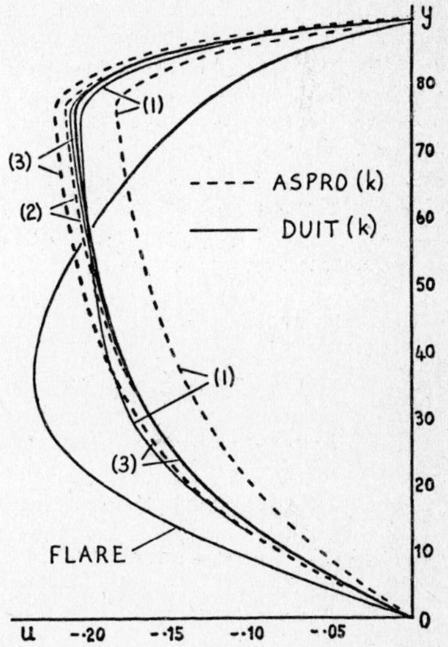

Figure 4. Convergence of reverse flow profiles.

3.2 Convergence. Table 1 shows that this procedure converges quite rapidly as far as P_0 is concerned. However, a difficulty arises with the reverse flow profiles, as can be seen in Fig.4. The quite large DUIT-ASPRO difference after the first cycle is dramatically reduced in the second cycle, DUIT(2) and ASPRO(2) being almost coincident. However, the next cycle increases the difference again, but now the ASPRO curve is further to the left. Subsequent cycles increase this difference still further, but the two sequences DUIT(k), ASPRO(k), k = 1,2,... do in fact converge (to different limits), DUIT(6) and DUIT(7) certainly being indistinguishable to graphical accuracy. The DUIT-ASPRO difference is ultimately as shown in Fig.5, and one might ask whether this difference is telling us something. Perhaps the DUIT sequence is trying to converge to the 'exact' solution and two-term ASPRO is not quite good

Table 1

k	P_0
0	1.8298
1	1.8007
2	1.7955
3	1.7963
4	1.7970
5	1.7970
6	1.7970
7	1.7971

enough for the chosen x_1. Fig.5 shows the three-term inner approximation as lying almost exactly halfway between the ASPRO and DUIT curves. Bearing in mind the slow convergence of the series, DUIT(7) could very well be quite close to the exact asymptotic profile evaluated at $x = x_1$. At any rate the DUIT-ASPRO difference is of the same order of magnitude as the third term. (Fig.5 also shows the one-term asymptotic curve for comparison.) Presumably the converged DUIT-ASPRO difference should be smaller for larger x_1, when ASPRO should be more accurate. However, when x_1 is increased to 97.25, it seems to remain of very much the same magnitude, as can be seen in Fig.5. Several explanations could be advanced for this. For example, it could be that there is fortuitous cancellation in the error term for the smaller x_1, so that the two-term expansion is more accurate than one might expect at $x = x_1$. Or it could be due to the longer range over which truncation error has been accumulating in the case $x_1 = 97.25$. A comparison for $x_1 = 58.12$ did in fact show that the DUIT-ASPRO difference is a little smaller for $\Delta y = 0.4$ than for $\Delta y = 0.8$, but the corresponding comparison for $x_1 = 97.25$ was not possible because of storage limits.

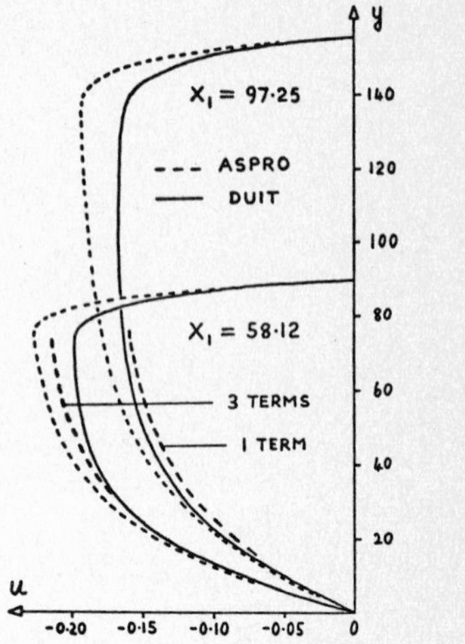

Figure 5. DUIT-ASPRO comparisons

4. BASIC STEP-BY-STEP PROCEDURE

4.1 Difference approximations. The method used in the earlier upstream computation was based on a Chebyshev polynomial representation of the velocity profile. However, for the implementation of the FLARE approximation, which involves a discontinuity in u_{yyy}, and its extension to the DUIT procedure, a difference method seemed more suitable. The algorithm used in Catherall, Stewartson and Williams, 1965, was therefore adapted to the present situation. The modification needed for the upstream integration was fairly trivial using linear extrapolation to impose the $u = 0$ condition at $y = y_0(x)$ (which is not necessarily a mesh point). We therefore restrict attention to the downstream integration. Here we have to deal with an unknown pressure and the boundary condition (3), which we use in the form of the pair of conditions

$$u_y \to 1, \quad u_x \to -p \quad \text{as} \quad y \to \infty . \tag{11}$$

For a general x station we denote the previous x station by $\overline{x} = x - \Delta x$, where Δx is the x step, and use a corresponding notation for other quantities. We denote by u_j ($j=0,1,\ldots,n+1$) the u values at uniform steps $\Delta y = h$ in y across the boundary layer, $y = (n+1)h$ being taken as the outer edge where the outer boundary conditions are imposed. This outer edge is allowed to move out as thesolution progresses, but for the moment we may regard it as fixed. Standard second-order difference replacements, as used by Leigh (1955), for example, then lead to a set of non-linear algebraic equations for p and the u_j. With the customary scaling these may be written

$$f_j(p,u_i) \equiv \delta^2 u_j + \delta^2 \overline{u}_j + d_j S_j - \alpha[(u_j^2 - \overline{u}_j^2)/2 + (p - \overline{p})] = 0, \tag{12}$$

which is to hold for $j = 1,\ldots,n$. Here δ is the central difference operator applied in the y direction, and

$$\alpha = 2h^2/\Delta x, \qquad S_j = 2 \sum_{i=0}^{j}{}'' (u_i - \overline{u}_i), \tag{13}$$

$$\beta = \alpha/8, \qquad d_j = \beta(u_{j+1} - u_{j-1} + \overline{u}_{j+1} - \overline{u}_{j-1}), \tag{14}$$

Σ'' denoting a trapezoidal sum with the first and last terms halved. To these must be added the condition $u_0 = 0$ and the two outer boundary conditions imposed at $y = (n+1)h$. We approximate $u_y = 1$ to first order in h, obtaining after scaling by h

$$u_{n+1} - u_n = h, \qquad (15)$$

which should be adequate if the outer boundary is far enough out. We replace $u_x = -p$ by $u_{n+1} - \overline{u}_{n+1} = -\frac{1}{2}\Delta x\,(p + \overline{p})$, which is equivalent to

$$f_{n+1} \equiv u_n - \overline{u}_n + \tfrac{1}{2}\Delta x\,(p + \overline{p}) = 0 \qquad (16)$$

if we use the fact that both u_j and $\overline{u}_j$ satisfy (15). With $u_0 = 0$ and $u_{n+1} = \overline{u}_{n+1} - \frac{1}{2}\Delta x\,(p + \overline{p})$ substituted in (12) we can regard (12),(16) as a system of n+1 equations for the n+1 unknowns $p, u_1, \ldots, u_n$.

4.2 <u>Newton iteration</u>. In solving these non-linear equations by Newton iteration we obtain the following linear system for the iterative corrections Δp, Δu_j :

$$f_j + (\partial f_j/\partial p)\Delta p + \sum_{i=1}^{n} (\partial f_i/\partial u_j)\Delta u_j = 0 \qquad (j=1,\ldots,n) \qquad (17)$$

$$f_{n+1} + \tfrac{1}{2}\Delta x\,\Delta p + \Delta u_n = 0, \qquad (18)$$

where $\partial f_j/\partial u_i = 0$ if $i > j+1$. It is convenient to write equations (17) in the form

$$c_j(\Delta u_1 + \ldots + \Delta u_{j-2}) + b_j\Delta u_{j-1} + a_j\Delta u_j + z_j\Delta u_{j+1} = -f_j - g_j\Delta p \qquad (19)$$

where

$$\left.\begin{aligned} z_j &= \partial f_j/\partial u_{j+1} && = 1 + \beta S_j \\ a_j &= \partial f_j/\partial u_j && = -2 - \alpha u_j + d_j \\ b_j &= \partial f_j/\partial u_{j-1} && = 1 - \beta S_j + 2 d_j \\ c_j &= \partial f_j/\partial u_i \quad (i < j-1) && = 2 d_j \\ g_j &= \partial f_j/\partial p && = -\alpha \end{aligned}\right\} \qquad (20)$$

for $j = 1,\ldots,n$, except for 'end effects', which may be accounted for by taking $c_1 = c_2 = b_1 = z_n = 0$ and $g_n = -\alpha - \frac{1}{2}\Delta x\,(1 + \beta S_n)$.

4.3 <u>Elimination algorithm</u>. We regard (19) as a system of equations for the Δu_j in terms of an unknown Δp, solve by elimination to obtain Δu_n in terms of Δp, then substitute in (17), which becomes an equation for Δp. In performing the elimination we start at the bottom, i.e. with the n-th equation; this allows us to take advantage of the special structure of (19). With Δu_{j+1} eliminated from each equation we obtain the system

$$C_j(\Delta u_1 + \ldots + \Delta u_{j-2}) + B_j\Delta u_{j-1} + A_j\Delta u_j = -F_j - G_j\Delta p, \qquad (21)$$

where the new coefficients (in capitals) are computed as follows:

$$\begin{aligned} m_j &= z_j/A_{j+1}, & C_j &= c_j - m_j C_{j+1}, & B_j &= 1 - \beta S_j + C_j, \\ A_j &= a_j - m_j B_{j+1}, & F_j &= f_j - m_j F_{j+1}, & G_j &= g_j - m_j G_{j+1} \end{aligned} \qquad (22)$$

for $j = n, n-1, \ldots, 1$ except that we should take $m_n = 0$. Back-substitution for each of the right-hand-side terms separately in (21) yields

$$\Delta u_j = \emptyset_j + \gamma_j \Delta p, \qquad (23)$$

where $\emptyset_j$ and γ_j are calculated for $j = 1,\ldots,n$ from

$$\phi_j = -[F_j + B_j\phi_{j-1} + C_j(\phi_{j-2} + \dots + \phi_1)]/A_j,$$
$$\gamma_j = -[G_j + B_j\gamma_{j-1} + C_j(\gamma_{j-2} + \dots + \gamma_1)]/A_j, \qquad (24)$$

except that the terms involving C_j do not occur for $j = 1,2$ and the terms involving B_j do not occur for $j = 1$. Substituting for Δu_n in (18) we obtain

$$\Delta p = -(f_{n+1} + \phi_n)/(\tfrac{1}{2}\Delta x + \gamma_n); \qquad (25)$$

thus Δp is now known and can be substituted back in (23) to find the Δu_j; finally we update p and the u_j.

4.4 Outer boundary. A check was made at the completion of each x step that the outer boundary was sufficiently far out. This consisted of testing that $u_y = 1$ to within a consistent tolerance for the last four points $j = n,\dots,n-3$. If not, n was increased by 1, new outer values inserted on the assumption that $u_y = 1$, then the complete x step recomputed.

4.5 Variable x step. Because of the large x values aimed at, it was desirable to use as large an x step as possible. However for uniform accuracy the shear layer should be covered in a roughly constant number of steps. Since the shear layer expands like $O(x^{1/3})$, the following ad hoc formula was used to achieve this:

$$\Delta x = (\Delta x)_0(1 + k_1 x)(1 + k_2 x)^{-2/3}$$

where k_1 and k_2 were chosen by experiment. The effect on P_0 of halving $(\Delta x)_0$ from the value actually used (≈ 0.14) was at a level lower than the third decimal.

A difficulty which occurs in starting the integration back upstream is that the solution is slightly singular since the imposed asymptotic starting profile does not quite match the prescribed pressure gradient (though presumably it would eventually if we had perfect convergence). This causes oscillations in the x direction, which die out fairly quickly upstream. If this was ignored, however, the oscillations would gradually increase in successive DUIT cycles. Therefore the u values obtained in integrating back upstream were smoothed before being used in computing (uu_x) for coming downstream. A comparison of pressure results with and without smoothing over one DUIT cycle showed very little effect.

5. RESULTS

5.1 Velocity profiles. Fig.6 shows the development of the reverse flow profiles into the asymptotic profile going downstream. For comparison the forward part of each profile is shown on a different scale (factor of 200). On this scale the reverse flow profiles would almost coincide with the axis, showing how the back flow actually is. These curves were drawn from a solution with $h = 0.4$, $(\Delta x)_0 \approx 0.14$ and $x_1 = 58.12$. This meant about 150 x steps and ultimately about 170 y steps, and took about 88 secs per DUIT on an IBM 360/65.

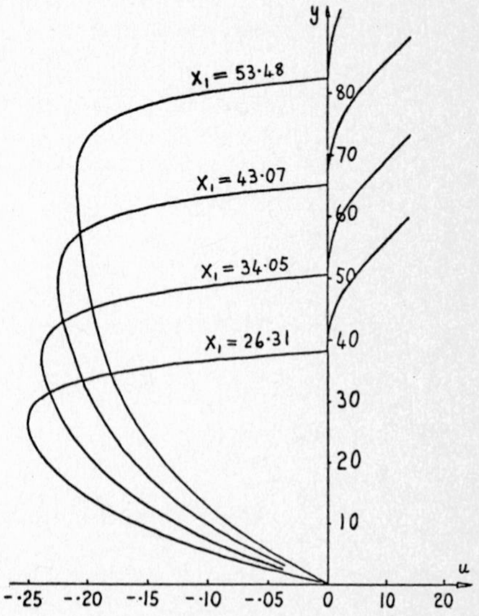

Figure 6. Reverse flow profiles.

5.2 Value of P_0. Table 2 gives values of P_0 obtained with various choices of h and x_1. The values for $h = 0.8$ show that $x_1 = 58.1$ is quite adequate. Note that the FLARE results compare much less favourably. Presumably this is because ASPRO (used to calculate P_0) is not the correct asymptotic expansion for the FLARE equations. The extrapolated result for DUIT(7) shows that it is difficult to decide whether P_0 is really different from the original guess of 1.8. Since other errors, due to finite x step and the initial separation profile, are estimated to be at the fourth decimal level, we take

$$P_0 = 1.800$$

Table 2.

	y	x_1=58.12	x_1=97.25
FLARE	0.8	1.8219	1.8187
	0.4	1.8298	
h^2-extrapolation		1.8324	
DUIT(7)	0.8	1.7891	1.7890
	0.4	1.7971	
h^2-extrapolation		1.7998	

5. REFERENCES

1. Catherall,D., Stewartson,K. and Williams,P.G. Proc. Roy. Soc. London A284, 370-396 (1965)

2. Flugge-Lotz,I. and Reyhner,T.A. Int. J. Non-linear Mech. 3, 173-199 (1968)

3. Leigh,D.C.F. Proc. Camb. Phil. Soc. 51, 320-332 (1955)

4. Neiland,V.Y. Izv. Akad. Nauk SSSR No.3, 19-24 (1971)

5. Stewartson,K. and Williams,P.G. Proc. Roy. Soc. London A312, 181-206 (1969)

6. Stewartson,K. and Williams,P.G. Mathematika 20, 98-108 (1973)

A FLOWFIELD SEGMENTATION METHOD FOR THE NUMERICAL SOLUTION OF VISCOUS FLOW PROBLEMS*

J. C. Wu, A. H. Spring, and N. L. Sankar

Georgia Institute of Technology, Atlanta, Ga., U.S.A.

INTRODUCTION

It is well known that, although progresses made during the past few years in developing finite-difference methods for the Navier-Stokes equations have been extensive, many important viscous flow problems are still beyond the scope of current methods. The limitation of the finite-difference method is due primarily to the often prohibitive computing effort required to obtain accurate solutions and is particularly acute for high Reynolds number viscous flows about finite solids of complex shapes. The application of finite-element techniques, which received emphasis in the recent literature, promises greater flexibility and easier accomodation of the boundary geometries, but has not relieved the excessive computing needs of the finite-difference methods.

In sharp contrast with the limited scope of current methods for the Navier-Stokes equations, a wide range of boundary layer problems are being treated routinely by various numerical methods. Boundary layers, of course, are specialized viscous flows at high Reynolds numbers. It is therefore paradoxical that the prospect of numerically simulating the general viscous flow problem appeared particularly discouraging at high Reynolds numbers. To be sure, the boundary layer equations are simpler than the Navier-Stokes equations. Nevertheless, the vastly different computing needs for the two types of equations that involve, after all, the same number of field variables could not be fully explained by this fact.

A different perspective becomes evident if one examines the principal attribute of the boundary layer concept - its ability to have the solution of the attached viscous region, i.e., the boundary layer, separated from that of the vastly larger region of essentially inviscid potential flow. This attribute, which led to the remarkable success of the boundary layer theory in both analytical and numerical studies, was not made available to the Navier-Stokes equations. In high Reynolds number flows, the viscous region, even with a separated flow present, is much smaller in size and involves much greater field-variable gradients than the potential region. A concurrent numerical solution of the two regions unavoidably involves the use of a huge number of data points, and hence also an immense amount of computing need, for reasonable solution resolution and accuracy. An inevitable question is then whether the solution field can still be confined to the viscous region when the full Navier-Stokes equations are to be solved numerically. In several articles by the first author of this paper and his co-workers [1 to 5], a new method that does possess the ability of confining the solution field for general incompressible viscous flows was described. This new method indeed produced very drastic improvements in solution speed as compared to previous methods.

The foundation of this new method is an integral representation of the kinematics of the flow. In the earlier work utilizing this integral representation simple and hence unsophisticated numerical procedures were employed by intention, the primary purpose being to verify the application of the method to various types of viscous flow problems. The present paper describes an important further development - a method of segmenting the solution field. This development is shown to produce a further drastic improvement in solution speed. In addition to discussing the segmentation method, this paper outlines a "kinematic method" for treating certain boundary conditions that are known to cause difficulties in prevailing methods [6]. The segmentation method and the kinematic method are both made

* This work is partly supported by the Langley Research Center of the National Aeronautics and Space Administration, Grant No. NSG-1004.

possible by the integral representation.

INTEGRAL REPRESENTATION

Employing the velocity $\vec{v}$, the vorticity $\vec{\omega}$, and the dilatation b of the flow as the prime field variables, the viscous flow problem can be conveniently partitioned into a kinematic aspect and a kinetic aspect [7]. The kinematic aspect permits each of the prime variables to be determined uniquely in terms of the other variables and the specified boundary conditions,

$$\vec{v} = \vec{v}_b \text{ on } B, \tag{1}$$

where B is the closed boundary of a fluid region R and $\vec{v}_b$ is a known function of the spatial and time coordinates. The boundary condition together with the familiar definitions of $\vec{\omega}$ and b, i.e.

$$\vec{\nabla} \times \vec{v} = \vec{\omega} \text{ and } \vec{\nabla} \cdot \vec{v} = b \tag{2}$$

constitute the differential formulation of the kinematics of the problem.

An integral representation for the kinematics of the compressible flow has been presented in [5]. This integral-representation contains Eqs. (2) as well as the velocity boundary condition (1), and therefore constitutes the entirety of the kinematics. For incompressible flow [2], b = 0 and the integral representation is

$$\vec{v} = -\frac{1}{A}\left[\int_R \frac{\vec{\omega}_o \times (\vec{r}_o - \vec{r})}{|\vec{r}_o - \vec{r}|^d} dR_o - \oint_B \frac{(\vec{v}_b \cdot \vec{n}_o)(\vec{r}_o - \vec{r}) - (\vec{v}_b \times \vec{n}_o) \times (\vec{r}_o - \vec{r})}{|\vec{r}_o - \vec{r}|^d} dB_o\right] \tag{3}$$

where the subscript "o" indicates that the variables and integrations are in the $\vec{r}_o$ space, e.g., $\vec{\omega}_o = \vec{\omega}(\vec{r}_o, t)$; $A = 4\pi$ and $d = 3$ in three-dimensional problems; $A = 2\pi$ and $d = 2$ in two-dimensional problems; and $\vec{n}$ is the unit outward normal vector on B.

For an exterior incompressible flow past a finite solid body, if R in Eq. (3) is taken to be the entire region occupied by the fluid, then B consists of two parts: the fluid-solid surface S on which the no-slip velocity condition, $\vec{v} = 0$, applies and a surface infinitely away from and enclosing S on which the freestream velocity boundary condition, $\vec{v} = \vec{v}_\infty$, applies. The second integral in Eq. (3) can then be evaluated, yielding the specialized integral representation

$$\vec{v} = -\frac{1}{A}\int_R \frac{\vec{\omega}_o \times (\vec{r}_o - \vec{r})}{|\vec{r}_o - \vec{r}|^d} dR_o + \vec{v}_\infty \tag{4}$$

Equation (4) can also be derived from the well known Biot-Savart Law [2] and is a specialized integral representation of the kinematics of the incompressible flow. This specialized representation was used by the present authors as well as in some other investigations. There are, however, essential differences between these efforts, as discussed in [4]. Eq. (3) is valid for any arbitrary velocity boundary condition and has an analogue in magnetostatics. This general representation has not been utilized by previous investigators to treat viscous flows.

The kinetic aspect of the incompressible viscous flow problem in the differential form is the familiar vorticity transport equation

$$\frac{\partial \omega}{\partial t} = \vec{\nabla} \times (\vec{v} \times \vec{\omega}) + \nu \nabla^2 \vec{\omega} \tag{5}$$

where ν is the kinematic viscosity, taken to be a constant in the present study. This equation describes the kinetic processes of vorticity diffusion and convection in the fluid. The generation of vorticity on the solid surface S, however, are governed by the kinematics of the problem. If the vorticity associated with a flow is separated into free vorticities interior of the fluid domain and boundary vorticities on the fluid-solid contact surface, S, then the kinematics of the problem requires a unique distribution of the boundary vorticity corresponding to any given free vorticity distribution and the velocity boundary conditions. Because of the limitation of space, a proof of this fact, given in [8], is not presented here. The basic procedure of the kinematic treatment, however, is outlined in the Numerical Procedure Section of this paper. It should be noted that this kinematic treatment, made possible by the integral representation, eliminates the need to employ a one-sided difference formula for estimating the boundary vorticity values. This kinematic treatment, unlike the one-sided formulae, accurately describes the local generation of vorticity and yields stable results at low and high Reynolds numbers alike.

For steady flows, Eq. (5) becomes elliptic and an integral representation for the kinetics of the problem has been found [5]. The result is that the entire problem is formulated as integral equations ideally suited for the application of the finite-element technique. The segmentation method, described in this paper for the time-dependent problem, is also applicable to the steady flow problem.

NUMERICAL PROCEDURES

Equations (3) and (5) constitute a complete description of the incompressible time-dependent viscous flow and are referred to as the integro-differential formulation. They do not contain simplifications or assumptions not in the differential formulation of the problem. Many options are available for the numerical solution of Eqs. (3) and (5). A solution procedure is outlined below for the two-dimensional exterior flow problem, using a finite-element technique for Eq. (3) and a finite-difference method for Eq. (5). Procedures for the interior and three-dimensional problems are similar to that outlined here.

The solution field is mapped into finite elements for the numerical quadrature of Eq. (3). Element-interpolation functions are introduced. For each element, the values of each of the field variables v_1, v_2, and ω (v_1 and v_2 are the velocity components in a Cartesian coordinate system with the flow in the x_1-x_2 plane, ω is the magnitude of the vorticity) are expressed in terms of nodal coordinates of the element and the nodal values of the field variable in question. The integrals in Eq. (3) are replaced by sums of integrals over individual elements. The latter integrals are evaluated, yielding

$$v_{\ell,i}^{n+1} = \sum_{j=1}^{Q} F_{\ell,ij} \; \omega_j^{n+1} + \sum_{m=1}^{2} \sum_{k=1}^{P} G_{m,ik} v_{m,k} \tag{6}$$

where "ℓ" is either 1 or 2; "i" refers to the node for which the value of v_ℓ is to be computed; "j" refers to any non-zero vorticity node; "k" boundary nodes; $F_{\ell,ij}$ and $G_{\ell,ik}$ are geometric functions depending on the relative positions of the nodes i and j or i and k; Q is the number of non-zero vorticity nodes in R; P is the number of boundary nodes on B; the superscript "n+1" refers to the (n+1)th time step in the computation.

A grid system is employed for a finite-difference representation of Eq. (5). For convenience, the grid points are made to coincide with the nodes of the finite element. The solution field is segmented into compartments each containing a suitable number of elements and nodes. In the kinetic part of the computation, the compartment boundaries that are adjacent to S are arranged to be one grid point away from S. That is, the compartments do not include any portion of S. For the kinematic part of the computation, however, the compartments adjacent to S are extended so that the surfaces S become parts of the compartment boundaries. The basic steps of the numerical procedure are given below.

1. Vorticity values on compartment boundaries. Using known values of v^n and ω^n throughout the solution field, including those on S, the values of ω^{n+1} at grid points on the compartment boundaries are computed using an explicit finite-difference representation of Eq. (5).

2. Vorticity values interior of each compartment. Either an explicit or an implicit finite-difference representation of Eq. (5) can be employed for this step. The values of ω^{n+1} at the compartment boundary points, determined in Step 1, can be

used to "upgrade" the solution for the interior points.

3. Vorticity values on solid surfaces S. Placing the values of ω^{n+1} computed in Steps 1 and 2 into Eq. (6) and applying the equation at the P number of nodes on S, where the velocities are specified (zero for the present problem), one obtains a set of 2P linear algebraic equations containing the P values of ω^{n+1} on S as unknowns. (The number of equations is 2P since there are two velocity components at each node.) Of these 2P equations, P and only P equations are independent, as can be shown by the use of the principle of maximums. The P unknown values of ω^{n+1} on S are obtained from the set of P independent algebraic equations.

4. Velocity values on compartment boundaries. v^{n+1} values on S are known as specified velocity boundary conditions. v^{n+1} values on other nodes of the compartment boundaries are computed using Eq. (6) and the ω^{n+1} values obtained in Steps 1, 2, and 3. The sum with respect to j is taken over all nodes in the flowfield with non-negligible vorticity. (That is, R is taken to be the entire region occupied by the fluid in Eq. 3.) The sum with respect to k is set to be $v_{\ell,\infty}$, in accordance with Eq. (4).

5. Velocity values interior of each compartment. Equation (6) is again used. The sum with respect to j now is taken over the finite elements contained in the compartment only, i.e., R is now the region occupied by the compartment in question. The sum with respect to k is taken over the boundary of the compartment, values of v^{n+1} in the sum have already been obtained in Step 4.

The advantage of employing the segmentation method can be easily seen from the different numbers of algebraic operations needed for computing each velocity values in Step 4 and in Step 5. For illustration, consider a flow in which there are 4000 nodes on which the values of vorticity and its derivatives are non-negligible. Velocity component values need to be computed only for these 4000 nodes. In Step 4, the computation of each $v_{\ell,i}^{n+1}$ value on a compartment boundary requires the multiplication of 4000 geometric functions $F_{\ell,ij}$ with 4000 nodal values of ω_j and the additions of the products. If the solution field is segmented into compartments each containing, say, 400 nodes, then, in Step 5, the computation of each interior $v_{\ell,i}^{n+1}$ value requires only 400 multiplications and 400 additions. If the solution field is not segmented, then each value of $v_{\ell,i}^{n+1}$ at every node requires 4000 multiplication and 4000 additions. Obviously, the segmentation of the solution field offers a very drastic improvement in solution speed. This improvement can, in fact, be made even greater by the use of successive segmentation. That is, one may first compute the velocity values on boundaries of larger compartments each containing, say, 1600 nodes. Once this is accomplished, the computation of the velocity values on the boundaries of the 400-node compartments (but are interior of the 1600-node compartments) can be accomplished much more rapidly. An additional important feature of the segmentation method is that, except for regions immediately adjacent to solid surfaces of irregular shapes, the compartments can be chosen to be of a standard shape (such as a rectangle). Therefore, various methods that are known to possess great solution speed and accuracy but are difficult to use in conjunction with complex boundaries can now be adopted for the interior nodes computations. This flexibility introduced by segmenting the solution field in fact assures the superiority of the integro-differential method, for optimum methods, including the ones that may appear in the future, can be incorporated into the numerical procedure. The advantages of segmenting the solution field are clearly applicable to the kinetic part of the computation (Steps 1 and 2) as they are to the kinematic part (Steps 4 and 5). The method also offers a possibility of utilizing the parallel programming capability of the most modern computers, for the computation of field variable values interior of each compartment is performed independently of that of the other compartments.

Steps 1 to 5 complete a computation loop. With known initial velocity and vorticity distributions, (e.g., the potential flow solution of an impulsively started finite body with a concentrated layer of vorticity on its surface) the numerical solution progresses with time by repeated applications of the computation loop.

RESULTS AND DISCUSSIONS

Several problems were treated in the earlier development of the integro-differential method without segmenting the solution field. They include the two-dimensional problem of a flow past a circular cylinder with and without a splitter plate, the three-dimensional problems of circular and elliptical jets exhausting from a flat plate and interacting with a cross flow and of flows about rectangular slabs at angles of attack. These earlier results [1 to 5] established conclusively the versatility of the integro-differential method as well as its drastic advantages in solution speed. It is noted that in the earlier work, the computer time needed, as quoted in [4], is quite large because of an inefficient method of handling the geometric functions. Although [4] appeared in the open literature only recently, its conclusions regarding the computing needs are no longer correct, as evidenced by the results given here without segmenting the solution field.

To verify the advantages of segmenting the solution field, the two-dimensional problem of a flow past a finite flat plate has been studied. This problem was selected because of its simple geometry and because previous finite-difference solutions [9] and Oseen's solution for low Reynolds number steady flow cases [10] are available for comparison. In the present work and in [9], the plate is set into impulsive motion from rest in its own plane. Steady state solutions were obtained asymptotically at large time.

In [9], a steady state solution was obtained for a Reynolds number of 4 using an alternating direction implicit (ADI) method for the vorticity computation. The computer time used was 90 minutes on a UNIVAC 1108 computer. An explicit finite-difference method was also used in [9] for the vorticity compution, but the solution was not carried to a steady state. The authors of [9] estimated that their ADI method was two to three times faster than the explicit method. Therefore, the explicit method of [9] would require between 180 and 270 minutes for the steady state solution. Steady state solutions were obtained in the present study using an explicit method for vorticity computation. The computer time used, also on a UNIVAC 1108, was 40 minutes without segmentation and 14 minutes with segmentation. The estimated computer time using an ADI method for vorticity computation is less than 7 minutes with segmentation.

The drastic reduction in computer time without segmenting the solution field is due to the ability of the integro-differential method to confine the solution field to the viscous region and thus to require a much smaller number of grid points in the computation (about 2000 points compared with 6000 in [9], the grid spacings being identical). The segmentation method further reduces the computer time by almost a factor of 3. For problems involving a greater number of data points, the advantage of the segmentation method obviously will be even greater. Numerical results obtained with and without segmentation are practically identical. The tangential velocity profile, u, at mid-plate obtained in this study are compared with the "best run" results of [9] and Oseen's results [10] in Fig. 1. The three sets of results are in good agreement near the plate. Far from the plate, where no Oseen's results are available from [10], the present results show a more pronounced overshoot ($u > u_\infty$) than that of [9]. The overshoot is expected and is attributable to the favorable pressure gradient along the plate caused by the boundary layer displacement-thickness. The deviation of the results of [9] from the present results is also expected because of the differences in specifying the "far field" velocity boundary condition. In [9], u is set to be u_∞, the freestream velocity, on a boundary one plate length upstream of the plate's leading edge. The value of u at this upstream boundary was computed using the integral representation, Eq. (4), at steady state. The results, shown in Fig. 2, indicates that the upstream boundary values of u assumed in [9] is in significant error.

Additional results have been obtained for the case of a Reynolds number of 1000. The steady state mid-plate velocity profile obtained, as expected, agrees well with the Blasius semi-infinite plate profile at a distance half the finite plate length from the leading edge. The velocity overshoot is still present, but is smaller in magnitude than that for the Reynolds number 4 case. This behavior is consistent with the decrease in displacement thickness with increasing Reynolds number. The mid-plate velocity profile obtained in [9] for the case of a Reynolds number of 993 deviated greatly from the corresponding Blasius' profile. It is noted that in [9] a one-sided difference formula was used to estimate the boundary vorticity values whereas the present, apparently more accurate, results were obtained using the kinematic treatment.

The present study and the earlier work using the integro-differential formulation indicate that the new formulation and the segmentation method have led to a novel approach for the numerical solution of Navier-Stokes equations that transcends the limitations of the prevailing finite-difference and finite-element methods. Further studies to develop the full potential of this novel approach is warranted.

REFERENCES

1. Wu, J. C. and Thompson, J. F., Fluid Dynamics of Unsteady Three-Dimensional and Separated Flows, Marshall, F. J., Editor, Project SQUID, Purdue Univ., Indiana, pp. 253-284 (1971).
2. Wu, J. C. and Thompson, J. F., Computers and Fluids, Vol. 1, pp. 197-215 (1973).
3. Wu, J. C., Finite Element Methods in Flow Problems, Oden, J. T. et al, Editors, Univ. of Alabama, Huntsville, Alabama, pp. 769-770 (1974).
4. Thompson, J. F., Shanks, S. P., and Wu, J. C., AIAA Jour., Vol. 12, pp. 787-794 (1974).
5. Wu, J. C., Finite Element Methods in Engineering, Univ. of New South Wales, Australia, 14 pages, (1974, in print).
6. Roach, P. J., Computational Fluid Dynamics, Hermosa Publishers, Alberquerque, N. M. (1972).
7. Lagerstrom, P. A., Theory of Laminar Flows, Moore, F. K., Ed., Princeton Univ., Princeton, New Jersey, Part B, Chap. 1, (1964).
8. Wu, J. C., Extraneous Boundary Conditions in Numerical Solution of Incompressible Viscous Flows, Georgia Inst. of Tech., Report - AE, Atlanta, Georgia (1974).
9. Pao, Y. H. and Daugherty, R. J., Time-Dependent Viscous Incompressible Flow Past a Finite Flat Plate, Boeing Scientific Research Lab. Rept. D1-82-0822(1969).
10. Tomokita, A. and Aoi, A., Quar. J. Mech. & Applied Math., Vol. 6, pp. 290-312 (1953).

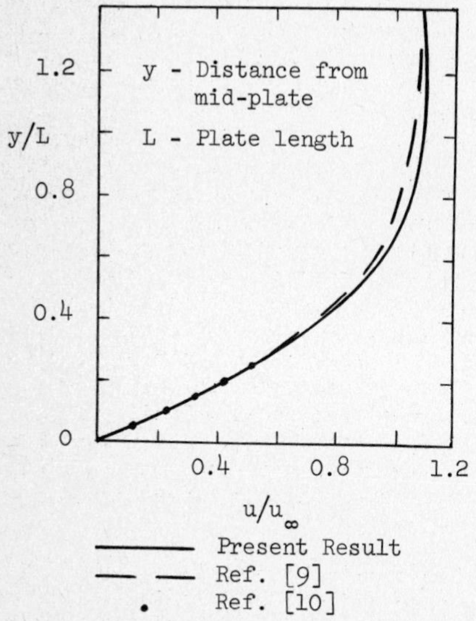

FIG. 1 MID-PLATE VELOCITY PROFILE

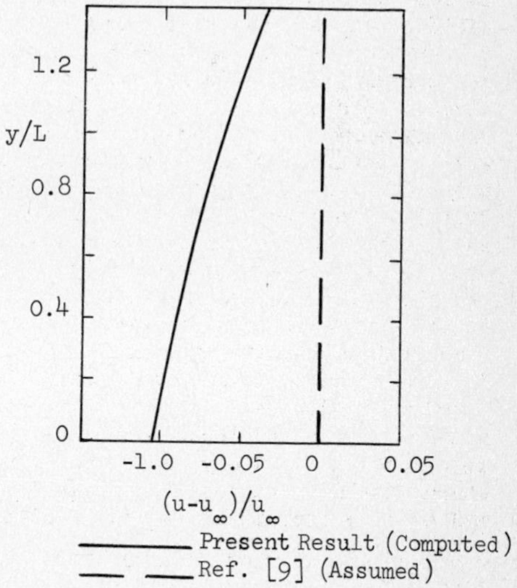

FIG. 2 FAR FIELD VELOCITY

Lecture Notes in Physics

Bisher erschienen / Already published

Vol. 1: J. C. Erdmann, Wärmeleitung in Kristallen, theoretische Grundlagen und fortgeschrittene experimentelle Methoden. 1969. DM 20,–

Vol. 2: K. Hepp, Théorie de la renormalisation. 1969. DM 18,–

Vol. 3: A. Martin, Scattering Theory: Unitarity, Analyticity and Crossing. 1969. DM 16,–

Vol. 4: G. Ludwig, Deutung des Begriffs physikalische Theorie und axiomatische Grundlegung der Hilbertraumstruktur der Quantenmechanik durch Hauptsätze des Messens. 1970. DM 28,–

Vol. 5: M. Schaaf, The Reduction of the Product of Two Irreducible Unitary Representations of the Proper Orthochronous Quantummechanical Poincaré Group. 1970. DM 16,–

Vol. 6: Group Representations in Mathematics and Physics. Edited by V. Bargmann. 1970. DM 24,–

Vol. 7: R. Balescu, J. L. Lebowitz, I. Prigogine, P. Résibois, Z. W. Salsburg, Lectures in Statistical Physics. 1971. DM 18,–

Vol. 8: Proceedings of the Second International Conference on Numerical Methods in Fluid Dynamics. Edited by M. Holt. 1971. DM 28,–

Vol. 9: D. W. Robinson, The Thermodynamic Pressure in Quantum Statistical Mechanics. 1971. DM 16,–

Vol. 10: J. M. Stewart, Non-Equilibrium Relativistic Kinetic Theory. 1971. DM 16,–

Vol. 11: O. Steinmann, Perturbation Expansions in Axiomatic Field Theory. 1971. DM 16,–

Vol. 12: Statistical Models and Turbulence. Edited by M. Rosenblatt and C. Van Atta. 1972. DM 28,–

Vol. 13: M. Ryan, Hamiltonian Cosmology. 1972. DM 18,–

Vol. 14: Methods of Local and Global Differential Geometry in General Relativity. Edited by D. Farnsworth, J. Fink, J. Porter and A. Thompson. 1972. DM 18,–

Vol. 15: M. Fierz, Vorlesungen zur Entwicklungsgeschichte der Mechanik. 1972. DM 16,–

Vol. 16: H.-O. Georgii, Phasenübergang 1. Art bei Gittergasmodellen. 1972. DM 18,–

Vol. 17: Strong Interaction Physics. Edited by W. Rühl and A. Vancura. 1973. DM 28,–

Vol. 18: Proceedings of the Third International Conference on Numerical Methods in Fluid Mechanics, Vol. I. Edited by H. Cabannes and R. Temam. 1973. DM 18,–

Vol. 19: Proceedings of the Third International Conference on Numerical Methods in Fluid Mechanics, Vol. II. Edited by H. Cabannes and R. Temam. 1973. DM 26,–

Vol. 20: Statistical Mechanics and Mathematical Problems. Edited by A. Lenard. 1973. DM 22,–

Vol. 21: Optimization and Stability Problems in Continuum Mechanics. Edited by P. K. C. Wang. 1973. DM 16,–

Vol. 22: Proceedings of the Europhysics Study Conference on Intermediate Processes in Nuclear Reactions. Edited by N. Cindro, P. Kulišić and Th. Mayer-Kuckuk. 1973. DM 26,–

Vol. 23: Nuclear Structure Physics. Proceedings of the Minerva Symposium on Physics. Edited by U. Smilansky, I. Talmi, and H. A. Weidenmüller. 1973. DM 26,–

Vol. 24: R. F. Snipes, Statistical Mechanical Theory of the Electrolytic Transport of Non-electrolytes. 1973. DM 20,–

Vol. 25: Constructive Quantum Field Theory. The 1973 "Ettore Majorana" International School of Mathematical Physics. Edited by G. Velo and A. Wightman. 1973. DM 26,–

Vol. 26: A. Hubert, Theorie der Domänenwände in geordneten Medien. 1974. DM 28,–

Vol. 27: R. Kh. Zeytounian, Notes sur les Ecoulements Rotationnels de Fluides Parfaits. 1974. DM 28,–

Vol. 28: Lectures in Statistical Physics. Edited by W. C. Schieve and J. S. Turner. 1974. DM 24,–

Vol. 29: Foundations of Quantum Mechanics and Ordered Linear Spaces. Advanced Study Institute Held in Marburg 1973. Edited by A. Hartkämper and H. Neumann. 1974. DM 26,–

Vol. 30: Polarization Nuclear Physics. Proceedings of a Meeting held at Ebermannstadt October 1–5, 1973. Edited by D. Fick. 1974. DM 24,–

Vol. 31: Transport Phenomena. Sitges International School of Statistical Mechanics, June 1974. Edited by G. Kirczenow and J. Marro. 1974. DM 39,–

Vol. 32: Particles, Quantum Fields and Statistical Mechanics. Proceedings of the 1973 Summer Institute in Theoretical Physics held at the Centro de Investigacion y de Estudios Avanzados del IPN – Mexico City. Edited by M. Alexanian and A. Zepeda. 1975. DM 18,–

Vol. 34: One Dimensional Conductors, GPS Summer School, Proceedings 1974. Edited by H. G. Schuster. 1975. DM 32,–

Vol. 35: Proceedings of the Fourth International Conference on Numerical Methods in Fluid Dynamics, June 24–28, 1974, University of Colorado. Edited by R. D. Richtmyer. 1975. DM 37,–